Engineering Hydrology: Processes and Modeling

Engineering Hydrology: Processes and Modeling

Editor: Marcus Gardner

RCALLISTO
REFERENCE

www.callistoreference.com

Callisto Reference,
118-35 Queens Blvd., Suite 400,
Forest Hills, NY 11375, USA

Visit us on the World Wide Web at:
www.callistoreference.com

ISBN: 978-1-64116-297-5 (Hardback)

Cataloging-in-Publication Data

Engineering hydrology : processes and modeling / edited by Marcus Gardner.
    p. cm.
Includes bibliographical references and index.
ISBN 978-1-64116-297-5
1. Hydrology. 2. Hydraulic engineering. 3. Water-supply engineering. 4. Environmental engineering. I. Gardner, Marcus.
GB665 .E54 2020
551.48--dc23

Table of Contents

Permissions

List of Contributors

Index

Preface

The scientific study of movement, distribution and quality of water on Earth and other planets is known as hydrology. It includes the study of the water cycle and water resources as well as environmental watershed sustainability. Hydrology draws on various fields such as environmental science, physical geography, environmental engineering and geology. Various scientific techniques and analytical methods are used in engineering hydrology. It works to solve water-related problems such as natural disasters, water management and environmental preservation. Remote sensing of hydrological processes is used to measure various constituents of terrestrial water balance, such as soil moisture, precipitation, surface water storage, evapotranspiration and snow and ice. The sources of remote sensing are land-based, remote-based and satellite sensors that are capable of capturing microwave, thermal and near-infrared data. The hydrological models are conceptual representations of a part of the hydrological cycle. It is classified into two types such as models based on data and models based on process descriptions. This book contains some path-breaking studies in the field of engineering hydrology. The various studies that are constantly contributing towards advancing technologies and evolution of this field are examined in detail. For someone with an interest and eye for detail, this book covers the most significant topics in the field of engineering hydrology.

The information shared in this book is based on empirical researches made by veterans in this field of study. The elaborative information provided in this book will help the readers further their scope of knowledge leading to advancements in this field.

Finally, I would like to thank my fellow researchers who gave constructive feedback and my family members who supported me at every step of my research.

Editor

Generalized combination equations for canopy evaporation under dry and wet conditions

J. P. Lhomme and C. Montes

IRD (UMR LISAH), 2 Place Viala, 34060 Montpellier, France

Correspondence to: J. P. Lhomme (jean-paul.lhomme@ird.fr)

Abstract. The formulation of canopy evaporation is investigated on the basis of the combination equation derived from the Penman equation. All the elementary resistances (surface and boundary layer) within the canopy are taken into account, and the exchange surfaces are assumed to be subject to the same vapour pressure deficit at canopy source height. This development leads to generalized combination equations: one for completely dry canopies and the other for partially wet canopies. These equations are rather complex because they involve the partitioning of available energy within the canopy and between the wet and dry surfaces. By making some assumptions and approximations, they can provide simpler equations similar to the common Penman–Monteith model. One of the basic assumptions of this down-grading process is to consider that the available energy intercepted by the different elements making up the canopy is uniformly distributed and proportional to their respective area. Despite the somewhat unrealistic character of this hypothesis, it allows one to retrieve the simple formulations commonly and successfully used up to now. Numerical simulations are carried out by means of a simple one-dimensional model of the vegetation–atmosphere interaction with two different leaf area profiles. In dry conditions and when the soil surface is moist (low surface resistance), there is a large discrepancy between the generalized formulation and its simpler Penman–Monteith form, but much less when the soil surface is dry. In partially wet conditions, the Penman–Monteith-type equation substantially underestimates the generalized formulation when leaves are evenly distributed, but provides better estimates when leaves are concentrated in the upper half of the canopy.

1 Introduction

The combination equation, which expresses the evaporation from natural surfaces, has certainly been one of the most successful breakthroughs in our understanding of evaporation. It is obtained by combining the energy balance equation with expressions of the convective fluxes of sensible and latent heat. The first equation of this type is the original Penman formula, initially derived to estimate the evaporation from a completely wet surface such as open water (Penman, 1948). It was extended by Monteith (1963) to describe the rate of evaporation from a dry surface characterized by a surface resistance (r_s) to vapour transfer added to the resistance of the air (r_a). The surface resistance is opposed to the transfer of water vapour between the level where evaporation takes place and the interface with the open air (source or sink of sensible heat). Provided both levels are at the same temperature, the Penman–Monteith equation is written as

$$\lambda E = \frac{\Delta A + \rho c_p D_a / r_a}{\Delta + \gamma (1 + r_s / r_a)}, \qquad (1)$$

where A is the available energy of the surface and D_a the vapour pressure deficit of the air. A familiar example is a thin dry layer covering a wet soil or a single leaf with its epidermis exchanging sensible heat and its stomatal cavities acting as a source of water vapour. Equation (1) simplifies into Penman equation when $r_s = 0$.

Monteith (1963, 1965) extended Eq. (1) to a stand of vegetation assuming the canopy to exchange sensible and latent heat with the atmosphere from a theoretical surface located at the same level as the effective sink of momentum: $z_m = d + z_0$ (d: displacement height; z_0: roughness length). The aerodynamic resistance r_a (assumed to be the same for sensible and latent heat) is calculated between this level and the reference height, where D_a is measured. The original idea of Monteith to place the source surface at level z_m is a priori questionable, because no real theoretical basis supports it. Thom (1972) showed that the transfer of heat and mass encounters greater aerodynamic resistance than momentum; therefore, the effective source of sensible and latent heat should be located at a lower level: $d + z_{0h}$ with $z_{0h} < z_0$ (e.g. Garrat and Hicks, 1973). The excess resistance $(r_{a,ex})$, associated with the boundary-layer resistances for the transfer of water vapour and sensible heat, is commonly expressed as $B^{-1}/u*$, where B^{-1} is a dimensionless bulk parameter and $u*$ is the friction velocity: B^{-1} is linked to z_{0h} by $B^{-1} = \ln(z_0/z_{0h})/k$. According to Monteith (1965), the surface resistance (r_s) is expected to be a plant factor depending on the stomatal resistance of individual leaves and on foliage area (soil evaporation being neglected). It is interpreted as the effective stomatal resistance of all the leaves acting as resistances in parallel (Shuttleworth, 1976b):

$$\frac{1}{r_s} = \sum_i \frac{1}{r_{s,i}}, \qquad (2)$$

$r_{s,i}$ being the stomatal resistance of an individual leaf i. The Penman–Monteith equation is often called "big-leaf model" because the whole canopy is assimilated to a big leaf located at level $d + z_0$ and with stomatal resistance r_s. The transfer processes through the air surrounding the leaves, supposedly negligible, are not taken into account or indirectly through the excess resistance. The lack of theoretical foundation of Eq. (1) applied to a canopy of leaves was apparent in a controversy which occurred in the 1970s about the formulation of evaporation from partially wet canopies (Shuttleworth, 1976a, 1977; Monteith, 1977). The Penman–Monteith equation was considered not to be able to represent the transition between dry and wet canopies, because the definition of canopy resistance according to Penman–Monteith (Eq. 2) implies that, if only a small part of the canopy is wet $(r_{s,i} = 0)$, the canopy resistance r_s should be equal to zero, which is unrealistic.

In this context, the main objectives of the paper are to investigate, under dry and wet conditions, the theoretical foundations of the combination equation applied to a canopy of leaves and concurrently to examine the different ways of aggregating the in-canopy resistances (surface and air) in a general single-source formulation of canopy evaporation. The basic principles used in the study are similar to those established by Shuttleworth (1978) in his simplified description of the vegetation–atmosphere interaction: the whole canopy (soil surface included) is assumed to be subject to the same vapour pressure deficit D_m at the mean source height $z_m (d + z_0)$, as in the original Penman–Monteith model and in two-source models (Shuttleworth and Wallace, 1985). Our investigation follows up previous works made on the formulation of evaporation from heterogeneous and sparse canopies (Lhomme et al., 2012, 2013). We show that the generalized formulation derived by Lhomme et al. (2013, Eq. 12)

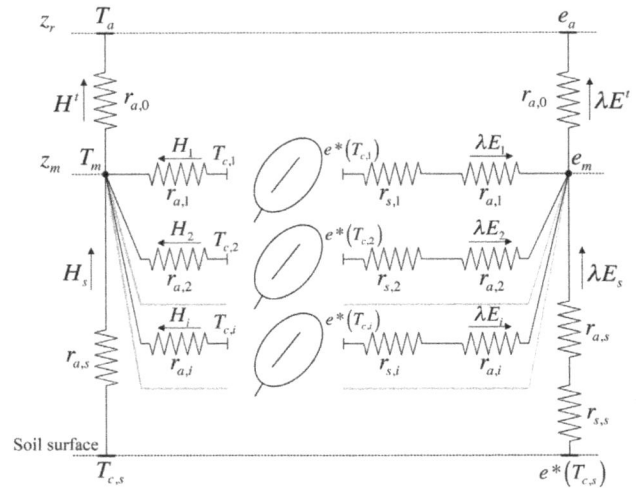

Fig. 1. Resistance network and potentials for a canopy represented by its elementary exchange surfaces (see list of symbols). All the component fluxes (sensible heat H_i and latent heat λE_i) converge at canopy source height (z_m). $T_{c,s}$ is soil surface temperature.

for multi-component canopies can be applied to a simple canopy, where the individual leaves and soil surface constitute the different components, and can be rewritten in a form similar to a combination equation. Different levels of approximation are identified to transform the general formulation of evaporation into the common Penman–Monteith equation. In this way, a bridge is established between a complex multi-source representation and the common practice based on the Penman–Monteith model with two bulk resistances (air and surface). The question of the "excess" resistance, linked to the exact location of the canopy source height, is also indirectly dealt with. Finally, the errors made when applying simple equations of the Penman–Monteith type instead of the more general ones are numerically assessed.

2 Evaporation from a dry canopy

2.1 General formulation

The canopy exchanges sensible and latent heat with the atmosphere through its leaf area and its soil surface. The modelling framework describing this interaction is similar to the one used by Lhomme et al. (2013) to derive the formulation of evaporation from a canopy made of n different components; however, the individual components or elements are represented here by the different leaves of the canopy and the soil surface, as shown in Fig. 1. The elementary evaporation (λE_i) per unit area of exchange surface (each side of a leaf being considered separately) is calculated from an equation of the Penman–Monteith type. It involves the saturation deficit of the air at canopy source height (D_m) and the available energy (A_i) for element i within the canopy (Lhomme et al., 2012):

$$\lambda E_i = \frac{\Delta A_i + \rho c_p D_m / r_{a,i}}{\Delta + \gamma (1 + r_{s,i}/r_{a,i})}. \tag{3}$$

In Eq. (3), for the canopy leaves, A_i is the net radiation per unit area of leaf, $r_{s,i}$ the leaf stomatal resistance (one side) per unit area of leaf and $r_{a,i}$ the corresponding leaf boundary-layer resistance for sensible and latent heat. For the soil surface symbolized by subscript $i = s$, A_s is the net radiation minus the soil heat flux per unit area of soil, $r_{s,s}$ being the soil surface resistance to evaporation and $r_{a,s}$ the air resistance between the soil surface and the canopy source height (z_m), defined by integrating the reciprocal of the appropriate eddy diffusivity (Choudhury and Monteith, 1988). Canopy leaf area index (LAI) being noted L_t, the total exchange surface area per unit area of soil is $S_t = 2 L_t + 1$ and total evaporation is obtained by summing the contributions of each individual exchange surface (soil and leaves):

$$\lambda E^t = \sum_{i \in S_t} \lambda E_i. \tag{4}$$

The vapour pressure deficit (D_m) in Eq. (3) is calculated from the vapour pressure deficit at reference height (D_a) (Shuttleworth and Wallace, 1985; Lhomme et al., 2013):

$$D_m = D_a + \left[\Delta A - (\Delta + \gamma) \lambda E^t \right] r_{a,0} / (\rho c_p), \tag{5}$$

where A is the available energy of the whole canopy and $r_{a,0}$ the aerodynamic resistance between the mean source height (z_m) and the reference height (z_r). Defining

$$R_i = r_{s,i} + \left(1 + \frac{\Delta}{\gamma} \right) r_{a,i}, \tag{6}$$

and introducing Eq. (5) into Eq. (3) and Eq. (3) into Eq. (4) leads to

$$\lambda E^t = \frac{\Delta \left[A + (R_c/r_{a,0}) \sum_{i \in S_t} \left(A_i r_{a,i}/R_i \right) \right] + \rho c_p D_a / r_{a,0}}{\Delta + \gamma \left(1 + R_c/r_{a,0} \right)}, \tag{7}$$

where R_c is expressed as

$$\frac{1}{R_c} = \sum_{i \in S_t} \frac{1}{R_i}. \tag{8}$$

Equations (7) and (8) represent a kind of generalized combination equation, where all the within-canopy resistances (air and surface) are taken into account. R_c defines a bulk canopy resistance which includes the surface resistances (leaves and soil) and the air resistances within the canopy. The temperature of each exchange surface can be determined from the above equations, as detailed in Appendix B.

If the boundary-layer resistances ($r_{a,i}$) within the canopy are neglected, assuming they are small compared to their stomatal counterpart (which is the assumption made in the Penman–Monteith equation), Eqs. (7) and (8) can be easily simplified. Excluding the soil component and putting $r_{a,i} = 0$, the summation in the right-hand term of Eq. (8) defines the canopy stomatal resistance in the sense of Monteith denoted by $r_{s,c}$:

$$\frac{1}{R_c} = \frac{1}{r_{s,c}} = \sum_{i \in 2L_t} 1/r_{s,i} \approx \frac{2L_t}{\langle r_{s,l} \rangle}, \tag{9}$$

where $\langle r_{s,l} \rangle$ is the harmonic mean of leaf stomatal resistances (per unit one-sided leaf area). The different leaves of the canopy acting as parallel resistors for the transfer of sensible heat and water vapour, harmonic means should be chosen when combining the elementary resistances (whereas arithmetic means would be used if conductances were considered). For a hypostomatous canopy, $2L_t$ should be replaced by L_t. Hence Eq. (7) becomes

$$\lambda E^t = \frac{\Delta A + \rho c_p D_a / r_{a,0}}{\Delta + \gamma (1 + r_{s,c}/r_{a,0})}. \tag{10}$$

Equation (10) is the well-known Penman–Monteith equation, which appears now as a particular case of a more general equation (Eq. 7), when all the air resistances within the canopy are set to zero and soil surface is neglected.

The case of a completely wet canopy can also be inferred from Eq. (7). When all the exchange surfaces (leaves and soil surface) are wet, the surface resistances ($r_{s,i}$) are nil and $R_i = (1 + \Delta/\gamma)r_{a,i}$. Noting that $\sum_{i \in S_t} A_i = A$ and after some manipulations, Eq. (7) transforms into a Penman-type equation:

$$\lambda E^t = \frac{\Delta A + \rho c_p D_a / (r_{a,0} + r_{a,c})}{\Delta + \gamma}, \tag{11}$$

where

$$\frac{1}{r_{a,c}} = \sum_{i \in S_t} \frac{1}{r_{a,i}} \approx \frac{2L_t}{\langle r_{a,l} \rangle} + \frac{1}{r_{a,s}}, \tag{12}$$

$\langle r_{a,l} \rangle$ being the harmonic mean of leaf boundary-layer resistances and $r_{a,s}$ the air resistance between the soil surface and the canopy source height. There is no surface resistance in the denominator of Eq. (11), as in the original Penman equation, but an additional air resistance ($r_{a,c}$) is added to the common aerodynamic resistance above the canopy ($r_{a,0}$). This additional resistance is the parallel sum of individual air resistances and encapsulates the bulk canopy resistance to heat and water vapour transfer from the wet exchange surfaces (leaves and soil) to the canopy source height.

2.2 Penman–Monteith-type formulation

The general combination equation derived above (Eq. 7) does not follow the exact form of the Penman–Monteith equation since an additional term mixing resistances with available energy partitioning is added to the total available energy (A). This section investigates under which conditions and approximations this general formula of canopy evaporation can be put in the simplified form of a Penman–Monteith equation, without neglecting the air resistances within the canopy. The approximations made below are essentially dictated by the result we aim at: i.e. the common form of the Penman–Monteith equation with bulk resistances expressed in a simple way.

The variable A_i giving the partition of available energy within the canopy is assumed to be in the form $A_i = A\Phi(i)$, where A is the total available energy for the whole canopy and $\Phi(i)$ is a function resulting from the radiative transfers within the canopy and depending on canopy structure and leaf area distribution. Beer's law, which is commonly used to express the attenuation of net radiation within the canopy, is typically a function of this kind. This assumption on the repartition of available energy is certainly a crude approximation. It is required, however, to mathematically derive a Penman–Monteith-type equation from the generalized form of Eq. (7), which means that the former implicitly includes this assumption. Consequently, after some manipulations, it can be shown that canopy evaporation (Eq. 7) can be written as

$$\lambda E^t = \frac{\Delta A + \rho c_p D_a/(r_{a,0} + r_{a,c})}{\Delta + \gamma\left[1 + r_{s,c}/(r_{a,0} + r_{a,c})\right]}, \tag{13}$$

where the bulk resistances $r_{a,c}$ and $r_{s,c}$ are defined as

$$r_{a,c} = R_c \sum_{i \in S_t} \Phi_i \frac{r_{a,i}}{R_i}, \tag{14}$$

$$r_{s,c} = R_c \left[1 - \left(1 + \frac{\Delta}{\gamma}\right) \sum_{i \in S_t} \Phi_i \frac{r_{a,i}}{R_i}\right]. \tag{15}$$

The resistances defined above involve air and surface resistances and the distribution function of available energy within the canopy. In order to get simpler formulations, some approximations are made substituting average values to summations. Introducing the harmonic mean of surface resistances per unit area of exchange surface $\langle r_{s,i}\rangle$ and the corresponding harmonic mean of leaf boundary-layer resistances noted $\langle r_{a,i}\rangle$, Eq. (8) can be written as

$$\frac{1}{R_c} \approx \frac{S_t}{\langle R_i\rangle}. \tag{16}$$

Summation in Eqs. (14) and (15) can be approximated using means denoted by angle brackets:

$$\sum_{i \in S_t} \Phi_i \frac{r_{a,i}}{R_i} \approx \langle\frac{\Phi_i r_{a,i}}{R_i}\rangle S_t \approx \frac{S_t}{\langle R_i\rangle}\langle\Phi_i r_{a,i}\rangle. \tag{17}$$

Substituting Eqs. (16) and (17) into Eqs. (14) and (15) leads to the following approximate expressions for bulk canopy resistances:

$$r_{a,c} \approx \langle\Phi_i r_{a,i}\rangle, \tag{18}$$

$$r_{s,c} \approx \frac{\langle R_i\rangle}{S_t} - \left(1 + \frac{\Delta}{\gamma}\right)\langle\Phi_i r_{a,i}\rangle. \tag{19}$$

These expressions still depend upon available energy partitioning, but it is interesting to note that if available energy is equally distributed within the canopy (soil included), i.e. $\Phi_i = 1/S_t$, the bulk air and surface resistances reduce to simple expressions independent of available energy. Although this assumption is not really realistic and constitutes a priori a strong approximation, it has been used by Shuttleworth (1978) in his "simplified general model". Using this assumption and separating the soil and leaves components ($S_t = 2L_t + 1$), the bulk canopy resistances can be rewritten in a way similar to Eqs. (12) and (9):

$$\frac{1}{r_{a,c}} \approx \frac{S_t}{\langle r_{a,i}\rangle} = \frac{2L_t}{\langle r_{a,l}\rangle} + \frac{1}{r_{a,s}}, \tag{20}$$

$$\frac{1}{r_{s,c}} \approx \frac{S_t}{\langle R_i\rangle - \left(1 + \frac{\Delta}{\gamma}\right)\langle r_{a,i}\rangle} = \frac{S_t}{\langle r_{s,i}\rangle} = \frac{2L_t}{\langle r_{s,l}\rangle} + \frac{1}{r_{s,s}}, \tag{21}$$

where $\langle r_{a,l}\rangle$ and $\langle r_{s,l}\rangle$ are the harmonic means of leaf boundary-layer resistances and stomatal resistances respectively. If the canopy is hypostomatous and if the average stomatal resistance $\langle r_{s,l}\rangle$ applies to the lower side of the leaves, $2L_t$ should be replaced by L_t in Eq. (21). Equation (13) appears now as a typical Penman–Monteith equation with its bulk resistances defined in the conventional way. The canopy surface resistance ($r_{s,c}$) accounts for all surface resistances, including leaves and soil. The "extra" resistance ($r_{a,c}$), added to the common aerodynamic resistance above the canopy ($r_{a,0}$), accounts for the air resistances opposed to heat and water vapour transfer within the canopy and can be perceived as similar to the excess resistance ($B^{-1}/u*$) introduced by Thom (1972) in the formulation of canopy evaporation.

3 Evaporation from a partially wet canopy

The partially wet canopy is taken here in the sense of "double canopy limit" described by Shuttleworth (1976b, 1978), all the individual elements being considered either totally dry or totally wet. It is opposed to the "single canopy limit", where the distribution of surface water resembles that of stomata, as when droplets of fog and mist impact the leaves. The "double canopy" is the most realistic case applicable to canopies which are drying out or in the process of wetting up by rainfall.

3.1 General formulation

The whole canopy is divided into two parts assumed to be independent: one is dry (with exchange surface S_d) and the other wet (with exchange surface S_w) and $S_t = S_d + S_w$. The assumption of independence is certainly questionable, but, as expressed by Shuttleworth (1978, p. 8): "such an assumption is certainly essential if theoretical progress is to be made in this field". Consequently, Eq. (4) can be rewritten in the following way:

$$\lambda E^t = \lambda E^d + \lambda E^w = \sum_{i \in S_d} \lambda E_i + \sum_{i \in S_w} \lambda E_i. \tag{22}$$

After substituting the expression of D_m (Eq. 5) into Eq. (3), elementary evaporation can be rewritten as

$$\lambda E_i = \left[\frac{\Delta}{\gamma} \left(r_{a,i} A_i + r_{a,0} A \right) + \left(\frac{\rho c_p}{\gamma} \right) D_a - \left(1 + \frac{\Delta}{\gamma} \right) r_{a,0} \lambda E^t \right] / R_i, \tag{23}$$

where R_i is given by Eq. (6). Bulk canopy resistances for the dry and wet parts of the canopy will be respectively defined as

$$\frac{1}{R_{c,d}} = \sum_{i \in S_d} \frac{1}{R_i} \text{ and } \frac{1}{R_{c,w}} = \sum_{i \in S_w} \frac{1}{r_{a,i}}. \tag{24}$$

With these definitions, the evaporation from the dry part of the canopy can be written as

$$\lambda E^d = \frac{\Delta}{\gamma} \left(\frac{r_{a,0}}{R_{c,d}} A + \sum_{i \in S_d} \frac{r_{a,i} A_i}{R_i} \right)$$
$$+ \frac{\rho c_p}{\gamma} \cdot \frac{D_a}{R_{c,d}} - \left(1 + \frac{\Delta}{\gamma} \right) \frac{r_{a,0}}{R_{c,d}} \lambda E^t, \tag{25}$$

and the contribution of the wet part is

$$\lambda E^w = \frac{\Delta}{\Delta + \gamma} \left(\frac{r_{a,0}}{R_{c,w}} A + \sum_{i \in S_w} A_i \right) + \frac{\rho c_p}{\Delta + \gamma} \cdot \frac{D_a}{R_{c,w}} - \frac{r_{a,0}}{R_{c,w}} \lambda E^t. \tag{26}$$

After some rearrangement, putting $A_w = \sum_{i \in S_w} A_i$ and defining a bulk canopy resistance for a partially wet canopy as

$$\frac{1}{R_{c,pw}} = \frac{1}{R_{c,d}} + \frac{\gamma}{\Delta + \gamma} \frac{1}{R_{c,w}}, \tag{27}$$

Eq. (22) becomes

$$\lambda E^t = \frac{\Delta \left\{ A + \frac{R_{c,pw}}{r_{a,0}} \left[\left(\frac{\gamma}{\Delta + \gamma} \right) A_w + \sum_{i \in S_d} \frac{r_{a,i} A_i}{R_i} \right] \right\} + \rho c_p \frac{D_a}{r_{a,0}}}{\Delta + \gamma (1 + R_{c,pw}/r_{a,0})}. \tag{28}$$

The contribution of each part of the canopy (wet and dry) to total evaporation is obtained by replacing λE^t by its expression in Eqs. (25) and (26). As could be expected, the limit of Eq. (28) when the canopy becomes completely dry is Eq. (7), and it is Eq. (11) when it becomes entirely wet. Consequently, Eq. (28) constitutes a kind of generalized combination equation applicable in all conditions (dry, wet or partially wet canopy). It is also a different and simpler writing of the single-source limit of the general model developed by

Shuttleworth (1976b, 1978). It is worthwhile noting that neglecting the air resistances within the canopy (i.e. $r_{a,i} = 0$) would lead to an inconsistency, as is the case for the Penman–Monteith equation applied in partially wet conditions. The bulk resistance $R_{c,pw}$ would become zero and Eq. (28) would turn into a simple Penman equation, which is not realistic.

3.2 Penman–Monteith-type formulation

This section examines under which conditions and approximations the general evaporation formula for partially wet canopies (Eq. 28) can be put in a form similar to the Penman–Monteith equation with simply defined resistances. Considering an amphistomatous canopy, the same assumptions as those made by Shuttleworth (1978) to derive the "Double Canopy Limit of the Simplified General model" are used here: (i) soil surface is neglected; (ii) a proportion W of the canopy is taken as wet, which means that $S_w = WS_t$ and $S_d = (1 - W)S_t$ with $S_t = 2L_t$; (iii) as discussed above, available energy is assumed to be equally distributed amongst the exchange surfaces, which implies that the available energy of each part (wet and dry) is proportional to its area: $A_w = AW$ and $A_d = A(1-W)$. Substituting average values to summations, Eq. (24) can be approximated by

$$\frac{1}{R_{c,w}} \approx \frac{2W L_t}{\langle r_{a,l,w} \rangle}, \tag{29}$$

where $\langle r_{a,l,w} \rangle$ represents the average value of leaf boundary-layer resistances for the wet part of the canopy, and

$$\frac{1}{R_{c,d}} \approx \frac{2(1 - W) L_t}{\langle r_{s,l,d} \rangle + \left(1 + \frac{\Delta}{\gamma} \right) \langle r_{a,l,d} \rangle}, \tag{30}$$

where $\langle r_{a,l,d} \rangle$ and $\langle r_{s,l,d} \rangle$ represent the average values of leaf resistances (air and surface) for the dry part of the canopy. Two bulk air resistances ($r_{a,c,w}$ and $r_{a,c,d}$), respectively for the wet and dry parts of the canopy, and a bulk surface resistance ($r_{s,c,d}$) for the dry part, are defined in the following way:

$$\frac{1}{r_{a,c,w}} = \frac{2W L_t}{\langle r_{a,l,w} \rangle}, \tag{31}$$

$$\frac{1}{r_{a,c,d}} = \frac{2(1 - W) L_t}{\langle r_{a,l,d} \rangle}, \tag{32}$$

$$\frac{1}{r_{s,c,d}} = \frac{2(1 - W) L_t}{\langle r_{s,l,d} \rangle}. \tag{33}$$

Consequently, Eq. (27) can be rewritten as a function of the bulk resistances defined above:

$$\frac{1}{R_{c,pw}} = \frac{r_{s,c,d} + (1 + \frac{\Delta}{\gamma})(r_{a,c,w} + r_{a,c,d})}{(1 + \frac{\Delta}{\gamma}) r_{a,c,w} \left[r_{s,c,d} + (1 + \frac{\Delta}{\gamma}) r_{a,c,d} \right]}. \tag{34}$$

The assumption on the equal distribution of available energy ($A_i = A/(2L_t)$) leads to

$$\sum_{i \in S_d} \frac{r_{a,i} A_i}{R_i} \approx \frac{A}{2L_t} \cdot \frac{\langle r_{a,l,d} \rangle}{\langle r_{s,l,d} \rangle + (1 + \frac{\Delta}{\gamma}) \langle r_{a,l,d} \rangle} . 2(1 - W)L_t$$

$$= \frac{A(1 - W)r_{a,c,d}}{r_{s,c,d} + (1 + \frac{\Delta}{\gamma})r_{a,c,d}}. \tag{35}$$

The generalized equation in partially wet conditions (Eq. 28) can be rewritten in a form similar to the Penman–Monteith equation as

$$\lambda E^t = \frac{\Delta A + \rho c_p D_a / (r_{a,0} + r_{a,pw})}{\Delta + \gamma \left[1 + \frac{r_{s,pw}}{r_{a,0} + r_{a,pw}}\right]}, \tag{36}$$

where the parameters $r_{a,pw}$ and $r_{s,pw}$ have the dimension of resistance and are defined as

$$r_{a,pw} = \frac{r_{a,c,w} \left[W r_{s,c,d} + (1 + \frac{\Delta}{\gamma})r_{a,c,d}\right]}{r_{s,c,d} + \left(1 + \frac{\Delta}{\gamma}\right)(r_{a,c,d} + r_{a,c,w})}, \tag{37}$$

$$r_{s,pw} = \frac{\left(1 + \frac{\Delta}{\gamma}\right)(1 - W)r_{a,c,w} r_{s,c,d}}{r_{s,c,d} + \left(1 + \frac{\Delta}{\gamma}\right)(r_{a,c,d} + r_{a,c,w})}. \tag{38}$$

Equation (36), however, has not the strict form of the Penman–Monteith equation, where an air resistance divides D_a in the numerator and where the ratio between a surface resistance and an air resistance appears in the denominator, because $r_{a,pw}$ includes a surface resistance ($r_{s,c,d}$) and consequently is not a "pure" air resistance. Additional assumptions should be made if we want to derive a strict Penman–Monteith equation. First, the mean (harmonic) leaf boundary-layer resistance should be assumed to be the same for the dry and wet parts of the canopy and equal to that of the whole canopy, which means that the bulk resistances can be rewritten as $r_{a,c,w} = r_{a,c}/W$ and $r_{a,c,d} = r_{a,c}/(1-W)$ with $r_{a,c}$ the bulk air resistance of the whole canopy (defined as in Eq. (20) without the soil component). Second, the mean (harmonic) leaf surface resistance for the dry part of the canopy should be assumed to be equal to that of the whole canopy, which leads to $r_{s,c,d} = r_{s,c}/(1 - W)$ with $r_{s,c}$ the bulk surface resistance of the whole canopy (defined as in Eq. (21) without the soil component). Under these conditions, $r_{a,pw}$ simplifies into $r_{a,c}$, and $r_{s,pw}$ can be rewritten in a simpler way as

$$r_{s,pw} = \frac{(1 - W)r_{a,c} r_{s,c}}{r_{a,c} + \frac{\gamma}{\Delta + \gamma} W r_{s,c}}. \tag{39}$$

Equation (36) becomes

$$\lambda E^t = \frac{\Delta A + \rho c_p D_a / (r_{a,0} + r_{a,c})}{\Delta + \gamma \left[1 + \frac{r_{s,pw}}{r_{a,0} + r_{a,c}}\right]}. \tag{40}$$

Equation (40) has now the typical form of a Penman–Monteith equation, because $r_{a,0} + r_{a,c}$ is a "pure" air resistance. Equation (39), which represents the canopy surface resistance in partially wet conditions, was initially derived by Shuttleworth (1978, Eq. 32; 2007, Eq. 16) using a different procedure, but with similar assumptions (those specified above and "intrinsic" resistances being disregarded). When $W = 1$ (totally wet canopy), $r_{s,pw} = 0$ and Eq. (40) reduces to a Penman-type equation (Eq. 11), as could be expected. When $W = 0$ (totally dry canopy), $r_{s,pw} = r_{s,c}$ and Eq. (40) reduces to the Penman–Monteith equation defined by Eq. (13).

4   Numerical simulations

In order to illustrate the different equations developed above and to assess the errors made when using simplified equations instead of the more comprehensive ones, numerical simulations were undertaken. Table 1 summarizes the different formulations or methods and specifies the corresponding equations and their notations. The simulations are based upon a simple one-dimensional model describing the vegetation–atmosphere interaction.

4.1   The simulation process

In the modelling approach, the crop canopy is considered as horizontally homogeneous with a mean height z_h. It is divided into several parallel layers (width Δz_i) counted from 1 to n from the top of the canopy to the soil surface. The different components or unit exchange surfaces (i) of the system are represented here by the different layers of vegetation making up the canopy plus the soil surface. This modelling approach is different from the traditional multi-layer approach (Waggoner and Reifsnyder, 1968) in the sense that each layer is subject to the same saturation deficit (D_m) without the inclusion of aerodynamic resistances in relation to the vertical transfer of sensible heat and water vapour. The parameterizations used for the microclimatic profiles, leaf area distribution, air and surface resistances are given in Appendix C. The available energy for each layer (see Eq. C2) is expressed as

$$A_i = c R_{n,a} \exp(-c L_i) \Delta L_i, \tag{41}$$

where L_i is the cumulative leaf area above layer i, $R_{n,a}$ the net radiation of the whole canopy and $\Delta L_i = l(z_i)\Delta z_i$ the leaf area of the corresponding layer, $l(z_i)$ being the leaf area density at height z_i. Component air and stomatal resistances (amphistomatous case) are expressed as

$$r_{a,i} = \frac{r_{a,l}(z_i)}{2\Delta L_i} \quad \text{and} \quad r_{s,i} = \frac{r_{s,l}(z_i)}{2\Delta L_i}, \tag{42}$$

where $r_{a,l}(z_i)$ and $r_{s,l}(z_i)$ are respectively the leaf boundary layer resistance and the leaf stomatal resistance (per

Table 1. Methods used in the numerical simulations to calculate canopy total evaporation with their corresponding equations and their symbol.

Method	Equations	Symbol
Dry canopy		
General equation without any assumption	Eq. (7) with R_c given by Eq. (8)	GE_d
Simplified equation: available energy is equally distributed (soil surface included).	Eq. (13) with resistances given by Eqs. (20) and (21)	SE_d
Common Penman–Monteith equation without soil surface contribution	Eq. (10) with surface resistance given by Eq. (9)	PM_d
Partially wet canopy		
General equation without any assumption	Eq. (28) with $R_{c,pw}$ given by Eq. (27)	GE_w
Penman–Monteith-type equation without soil surface contribution	Eq. (40) with surface resistance given by Eq. (39)	PM_w

unit area of leaf) at each height within the canopy, given by Eqs. (C5) and (C8) respectively. The soil surface resistance $r_{s,s}$ has a fixed value depending on soil surface moisture, and the corresponding air resistance $r_{a,s}$ (between the soil surface and the canopy source height) is given by Eq. (C7). Calculations are made for an amphistomatous canopy with $z_h = 1.2$ m and $L_t = 4$ under the following weather conditions at a reference height $z_r = 3$ m: incoming solar radiation $R_{s,a} = 700$ W m^{-2}, air temperature $T_a = 25\,°$C, vapour pressure deficit $D_a = 10$ hPa, and wind speed $u_a = 2$ m s^{-1}. Two types of leaf area profile are considered, as detailed in Appendix C and shown in Fig. 2: a profile constant with height (noted A) and another with a high leaf area density in the top layers and a lower density in the bottom layers (B). The canopy is divided into 20 layers plus the soil surface.

4.2 Numerical results

The differences among the predictions in relation to different formulations are assessed. In Fig. 3, the generalized combination equation giving canopy evaporation in dry conditions (GE_d) is compared with two simplified formulations (see Table 1): SE_d derived assuming available energy to be equally distributed amongst the exchange surfaces and the common Penman–Monteith equation (PM_d). The comparison is made as a function of canopy water stress for two leaf area profiles (A and B). In parallel, the figure shows the variation of canopy surface resistance $r_{s,c}$ calculated with methods SE_d and PM_d. When the soil surface is dry ($r_{ss} = 2000$ s m^{-1}), the simplified equations for canopy evaporation (SE_d and PM_d) approximate fairly well the complete formulation: for LAI profile A, the three estimates are practically mingled and the two surface resistances are very close to each other; for profile B, there is a slight underestimation of the simplified formulations SE_d and PM_d (Fig. 3a, b). This clearly

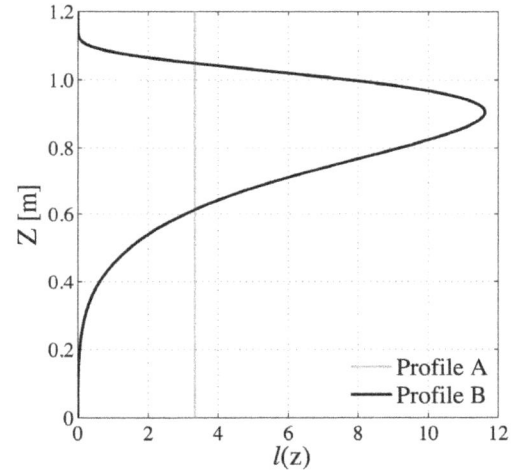

Fig. 2. Profiles of leaf area density considered in the simulation process: (**A**) constant profile and (**B**) profile adapted from a gamma function with a higher leaf area density in the top layers.

justifies the use of the Penman–Monteith equation in such conditions. However, when the soil surface becomes wetter ($r_{s,s} = 100$ s m^{-1}) (Fig. 3c, d), there is a large discrepancy between the formulations: the common Penman–Monteith equation (PM_d) clearly underestimates canopy evaporation (GE_d), as could be anticipated, and SE_d tends to overestimate it. In parallel, the surface resistances depart from each other with a departure greater for leaf area profile B.

In Fig. 4, the generalized combination equation established in partially wet conditions (GE_w) is compared with its simpler form (PM_w) based upon a series of simplifying assumptions. This comparison is made as a function of the fractional surface wetness W, assuming the wetting process begins by the top layers, as generally occurs during

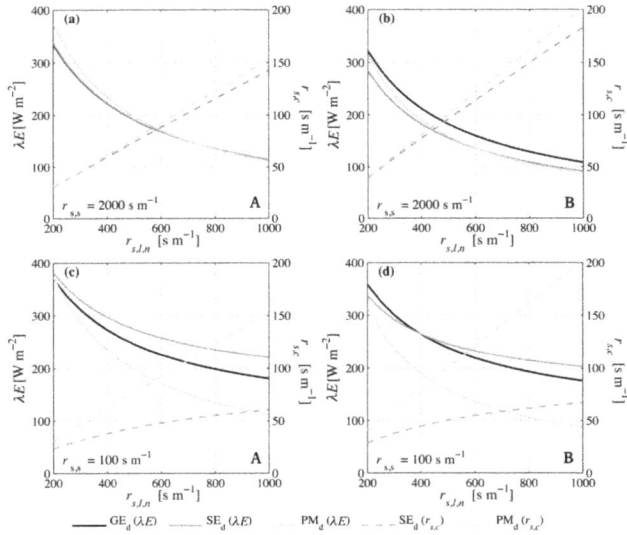

Fig. 3. For a dry canopy, latent heat flux (λE) and canopy surface resistance ($r_{s,c}$) as a function of minimal stomatal resistance ($r_{s,l,n}$) (Eq. C8) representing the canopy water stress. Comparison of three methods (GE_d, SE_d, PM_d) for two profiles of leaf area density (**A** and **B**) and two different values of soil surface resistance: (**a**) and (**b**) $r_{s,s} = 2000\,\mathrm{s\,m}^{-1}$ (dry soil); (**c**) and (**d**) $r_{s,s} = 100\,\mathrm{s\,m}^{-1}$ (moist soil).

Fig. 4. Latent heat flux (λE) from a partially wet canopy as a function of its fractional surface wetness (W). Comparison of two formulations (GE_w and PM_w) for two profiles of leaf area density (**A** and **B**) and two different values of minimal stomatal resistance $r_{s,l,n}$ representing canopy water stress: (**a**) and (**b**) $r_{s,l,n} = 100\,\mathrm{s\,m}^{-1}$ (unstressed canopy); (**c**) and (**d**) $r_{s,l,n} = 1000\,\mathrm{s\,m}^{-1}$ (stressed canopy). Soil surface resistance r_{ss} is set to $500\,\mathrm{s\,m}^{-1}$.

rainy events. With leaf area profile A, the Penman–Monteith-type equation (PM_w) underestimates the true evaporation rate (GE_w) by up to $200\,\mathrm{W\,m}^{-2}$ for a water stressed canopy (Fig. 4c), and the discrepancy decreases when the canopy becomes wetter (W close to 1). With leaf profile B, where most of the leaves are concentrated in the upper half of the canopy, the agreement is better: PM_w overestimates or underestimates GE_w depending on surface wetness (W) and canopy water status ($r_{s,l,n}$). A reason for this relative agreement could be that leaf profile B is closer to the "big-leaf" model represented by the Penman–Monteith equation. Canopy surface resistance rapidly decreases with the wetting process: when $W = 0.5$, $r_{s,c}$ is already close to zero for both profiles.

As previously noticed, the "extra" resistance ($r_{a,c}$) (Eq. 20), added to the aerodynamic resistance ($r_{a,0}$) in the Penman–Monteith form of the combination equations (Eqs. 13 and 40), plays the same role as the excess resistance ($r_{a,ex} = B^{-1}/u*$) introduced by Thom (1972) and mentioned in the introduction. The dimensionless parameter B^{-1} can be estimated by equating $r_{a,c}$ to $r_{a,ex}$: $kB^{-1} = \ln(z_0/z_{0h}) = ku*.r_{a,c}$. In Fig. 5a the extra resistance $r_{a,c}$ is plotted vs. wind speed at reference height for different LAI using leaf area profile A; $r_{a,c}$ is also compared with the rough approximation based on $B^{-1} = 4$, which is a typical value for permeable vegetation (Thom, 1972). The extra resistance $r_{a,c}$ is a decreasing function of wind speed (as could be anticipated) and also of LAI, with values close to the ones predicted by Thom's approximation ($r_{a,ex} = 4/u*$). In Fig. 5b,

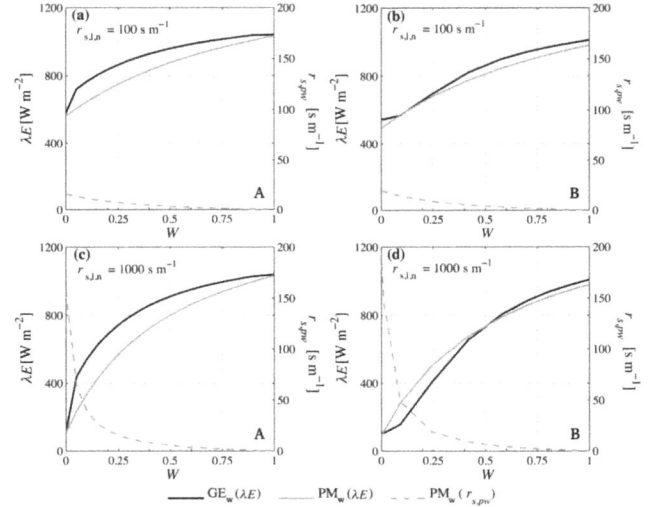

kB^{-1} appears as an increasing function of wind speed and a decreasing function of LAI with values ranging approximately from 0.5 to 2. Compared with the values given by Garrat (1992, Fig. 4.4) for different surface types, our results exhibit a slight underestimation. The same figures drawn using leaf area profile B provide almost identical results (results not shown): this should mean that different leaf area profiles do not lead to substantial change in bulk canopy air resistance. In interpreting these results, it is necessary to keep in mind that (i) the fact of representing canopy elements by layers necessarily restricts the theoretical space; (ii) the model used for simulating the vegetation–atmosphere interaction is itself relatively crude; (iii) the evaluation was done without addressing sensitivity to assumed canopy conditions; (iv) the equation defining $r_{a,c}$ (Eq. 20) is a simplified version of a more complex one (Eq. 18); and (v) the kB^{-1} concept itself is questionable and not really physically based (Verhoef et al., 1997).

5 Conclusions

The present paper sets a theoretical framework for canopy evaporation through the development of two generalized combination equations – one for completely dry canopies (Eq. 7) and the other for partially wet canopies (Eq. 28) – the former being included in the latter. These general equations are derived assuming that all the exchange surfaces are subject to the same vapour pressure deficit at canopy source

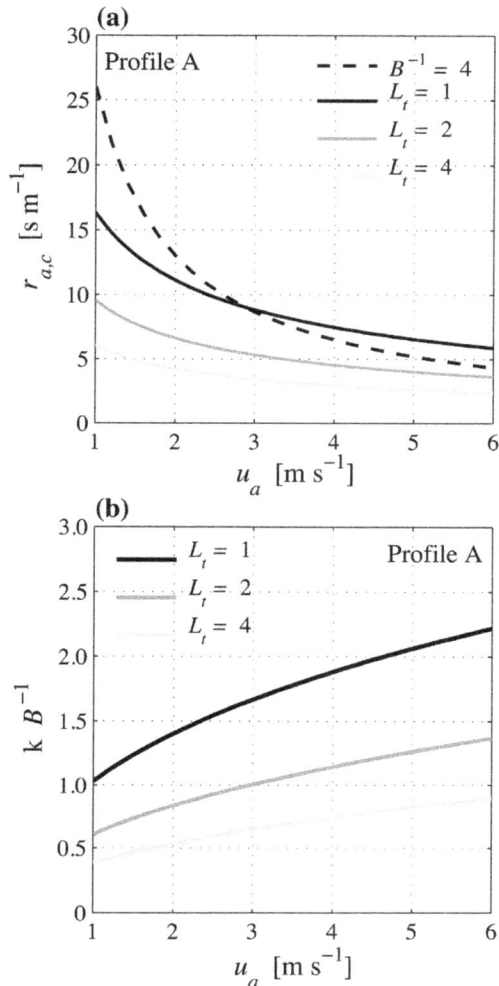

Fig. 5. (a) The additional aerodynamic resistance ($r_{a,c}$) given by Eq. (20) is plotted as a function of wind speed at reference height (u_a) for different LAI (L_t) and compared with the rough estimate based on $B^{-1} = 4$ (leaf area profile A); **(b)** the bulk parameter kB^{-1} (inferred from the value of $r_{a,c}$) is plotted as a function of wind speed (u_a) for different LAI (L_t) and the same profile of LAI.

height. In this sense, as already said, the modelling approach is different from the common multi-layer approach, where the whole canopy is divided into parallel layers, each one subject to a different air saturation deficit, with an additional aerodynamic resistance in relation to the vertical transfer of heat and mass. Comprehensive combination equations have been derived using this approach (Lhomme, 1988a, b), but they are more complex than the equations derived here. Despite their relative simplicity, the present generalized combination equations cannot be easily applied in an operational way, since the available energy partition (within the canopy and between wet and dry surfaces) is required as input. To provide equations easier to handle, assumptions and approximations can be made. In this down-grading process, one of

the basic assumptions is to consider that the available energy is equally distributed amongst the exchange surfaces. This hypothesis appears to be rather unrealistic, both in dry and wet conditions, but it leads to simple formulations of the Penman–Monteith type (Eqs. 13 and 40, respectively), which have been successfully used up to now. The numerical simulations, based on a simple one-dimensional model with two types of leaf area profile, confirm that the Penman–Monteith equation performs well in dry conditions, when the soil surface does not evaporate. In partially wet conditions, a discrepancy with the comprehensive formulation exists, but it tends to be less when the leaves are concentrated in the upper part of the canopy.

Acknowledgements. The autors are very grateful to two anonymous reviewers for their useful comments and constructive suggestions.

Edited by: M. Weiler

References

Choudhury, B. J. and Monteith, J. L.: A four-layer model for the heat budget of homogeneous land surfaces, Q. J. Roy. Meteorol. Soc., 114, 373–398, 1988.

Garrat, J. R.: The Atmospheric Boundary Layer, CUP, Cambridge, UK, 1992.

Garrat, J. R. and Hicks, B. B.: Momentum, heat and water vapour transfer to and from natural and artificial surfaces, Q. J. Roy. Meteorol. Soc., 99, 680–687, 1973.

Inoue, E.: On the turbulent structure of air flow within crop canopies, J. Meteorol. Soc. Japan, 41, 317–326, 1963.

Lhomme, J. P.: Extension of Penman's formulae to multi-layer models, Bound.-Lay. Meteorol., 42, 281–291, 1988a.

Lhomme, J. P.: A generalized combination equation derived from a multi-layer micrometeorological model, Bound.-Lay. Meteorol., 45, 103–115, 1988b.

Lhomme, J. P., Rocheteau, A., Ourcival, J. M., and Rambal, S.: Non-steady state modeling of water transfer in a Mediterranean evergreen canopy, Agr. Forest Meteorol., 108, 67–83, 2001.

Lhomme, J. P., Montes, C., Jacob, F., and Prévot, L.: Evaporation from heterogeneous and sparse canopies: on the formulations related to multi-source representations, Bound.-Lay. Meteorol., 144, 243–262, 2012.

Lhomme, J. P., Montes, C., Jacob, F., and Prévot, L.: Evaporation from multi-component canopies: generalized formulations, J. Hydrol., 486, 315–320, 2013.

Monteith, J. L.: Gas exchange in plant communities, in: Environmental Control of Plant Growth, edited by: Evans, L. T., Academic Press, New York, 95–112, 1963.

Monteith, J. L.: Evaporation and the environment, Symp. Soc. Experiment. Biol., 19, 205–234, 1965.

Monteith, J. L.: Resistance of a partially wet canopy: whose equation fails?, Bound.-Lay. Meteorol., 12, 379–383, 1977.

Penman, H. L.: Natural evaporation from open water, bare soil and grass, Proc. R. Soc. London Ser. A, 193, 120–145, 1948.

Shuttleworth, W. J.: Experimental evidence for the failure of the Penman-Monteith equation in partially wet conditions, Bound.-Lay. Meteorol., 10, 91–94, 1976a.

Shuttleworth, W. J.: A one-dimensional theoretical description of the vegetation-atmosphere interaction, Bound.-Lay. Meteorol., 10, 273–302, 1976b.

Shuttleworth, W. J.: Comments on "Resistance of a partially wet canopy: whose equation fails?", Bound.-Lay. Meteorol., 12, 385–386, 1977.

Shuttleworth, W. J.: A simplified one-dimensional theoretical description of the vegetation-atmosphere interaction, Bound.-Lay. Meteorol., 14, 3–27, 1978.

Shuttleworth, W. J.: Putting the "vap" into evaporation, Hydrol. Earth Syst. Sci., 11, 210–244, doi:10.5194/hess-11-210-2007, 2007.

Shuttleworth, W. J. and Wallace, J. S.: Evaporation from sparse crops – an energy combination theorym, Q. J. Roy. Meteorol. Soc.m 111, 839–855, 1985.

Thom, A. S.: Momentum, mass and heat exchange of vegetation, Q. J. Roy. Meteorol. Soc., 98, 124–134, 1972.

Verhoef, A., de Bruin, H. A. R., and van den Hurk, B. J. J. M.: Some practical notes on the parameter kB^{-1} for sparse vegetation, J. Appl. Meteorol., 36, 560–572, 1997.

Waggoner, P. E. and Reifsnyder, W. E.: Simulation of the temperature, humidity and evaporation profiles in a leaf canopy, J. Appl. Meteorol., 7, 400–409, 1968.

Appendix A

Table A1. List of symbols.

A	Available energy for the whole canopy (W m^{-2})
A_i	Available energy for the unit area of exchange surface i (W m^{-2})
$R_{s,a}$	Incoming solar radiation (W m^{-2})
$R_{n,a}$	Net radiation of the whole canopy (W m^{-2})
G	Soil heat flux (W m^{-2})
H^t	Sensible heat flux from the whole canopy (W m^{-2})
H_i	Sensible heat flux from the unit area of exchange surface i (W m^{-2})
λE^t	Latent heat flux from the whole canopy (W m^{-2})
λE^d	Latent heat flux from the dry part of the canopy (W m^{-2})
λE^w	Latent heat flux from the wet part of the canopy (W m^{-2})
λE_i	Latent heat flux from the unit area of exchange surface i (W m^{-2})
D_a	Vapour pressure deficit at reference height ($e*(T_a) - e_a$) (Pa)
D_m	Vapour pressure deficit at canopy source height ($e*(T_m) - e_m$) (Pa)
e_a	Vapour pressure at reference height (Pa)
e_m	Vapour pressure at canopy source height (Pa)
$e*(T)$	Saturated vapour pressure at temperature T (Pa)
T_a	Air temperature at reference height (°C)
T_m	Air temperature at canopy source height (°C)
$T_{c,i}$	Surface temperature of the unit area of exchange surface i (°C)
u_a	Wind speed at reference height (m s^{-1})
$u*$	Friction velocity (m s^{-1})
k	von Karman's constant (0.41)
c_p	Specific heat of air at constant pressure (J kg^{-1} K^{-1})
ρ	Air density (kg m^{-3})
γ	Psychrometric constant (Pa K^{-1})
Δ	Slope of the saturated vapour pressure curve (Pa K^{-1})

Canopy physical characteristics		:
d	Zero plane displacement height (m)	
L_t	Leaf area index of the whole canopy (m^2 m^{-2})	
S_t	Canopy exchange surface area per unit area of soil (m^2 m^{-2})	
ΔL_i	Leaf area of the vegetation layer i with width Δz_i (m^2 m^{-2})	
$r_{a,0}$	Aerodynamic resistance between the source height and the reference height (s m^{-1})	
$r_{a,i}$	Boundary-layer resistance for sensible heat and water vapour of the unit area of exchange surface i (s m^{-1})	
$r_{s,i}$	Surface resistance per unit area of exchange surface (s m^{-1})	
$r_{a,l}$	Boundary-layer resistance for sensible heat and water vapour of the unit area of leaf (one side) (s m^{-1})	
$r_{s,l}$	Leaf stomatal resistance per unit area of leaf (one side) (s m^{-1})	
$r_{s,l,n}$	Minimal leaf stomatal resistance (Eq. C8) (s m^{-1})	
$r_{a,s}$	Air resistance between the soil surface and the canopy source height (s m^{-1})	
$r_{s,s}$	Soil surface resistance to evaporation per unit area of soil (s m^{-1})	
$r_{a,c}$	Bulk air resistance of the canopy (s m^{-1})	
$r_{a,c,d}$	Bulk air resistance for the dry part of the canopy (s m^{-1})	
$r_{a,c,w}$	Bulk air resistance for the wet part of the canopy (s m^{-1})	
$r_{s,c}$	Bulk surface resistance of the canopy (s m^{-1})	
$r_{s,c,d}$	Bulk surface resistance for the dry part of the canopy (s m^{-1})	
$r_{s,pw}$	Canopy surface resistance in partially wet conditions defined by Eq. (39) (s m^{-1})	
W	Wet proportion of the canopy expressed as a fraction of 1	
z_r	Reference height (m)	
z_h	Mean canopy height (m)	
z_m	Mean canopy source height ($= d + z_0$) (m)	
z_0	Canopy roughness length for momentum (m)	
z_{0h}	Canopy roughness length for sensible and latent heat (m)	

Appendix B

Expressing the temperature of exchange surfaces ($T_{c,i}$)

The basic equations for the transfer of sensible heat are

$$H_i = \frac{\rho c_p \left(T_{c,i} - T_m\right)}{r_{a,i}} \quad \text{with} \quad H_i = A_i - \lambda E_i, \tag{B1}$$

$$H^t = \frac{\rho c_p \left(T_m - T_a\right)}{r_{a,0}} \quad \text{with} \quad H^t = A - \lambda E^t. \tag{B2}$$

Surface temperature is inferred from Eqs. (B1) and (B2):

$$T_{c,i} = T_a + + \frac{(A_i - \lambda E_i)\, r_{a,i}}{\rho c_p} + \frac{\left(A - \lambda E^t\right) r_{a,0}}{\rho c_p}. \tag{B3}$$

Elementary flux λE_i is given by Eq. (3) with D_m expressed by Eq. (5). Substituting and rearranging gives the following expression of $T_{c,i}$ as a function of λE^t (Eq. 7):

$$T_{c,i} - T_a = \frac{1}{\rho c_p} \left\{ \left[\left(A - \lambda E^t \right) r_{a,0} + A_i r_{a,i} \right] \left(1 - \frac{\Delta}{\gamma} \frac{r_{a,i}}{R_i} \right) \right\} \\ + \frac{r_{a,i}}{R_i} \left(\frac{\lambda E^t r_{a,0}}{\rho c_p} - \frac{D_a}{\gamma} \right). \tag{B4}$$

Appendix C

Parameterizations used in the simulation process

Solar radiation R_s and net radiation R_n are assumed to decrease within the canopy as exponential functions of the cumulative leaf area index $L(z)$ (Beer's law) counted from the top of the canopy

$$R_s(z) = R_{s,a}\, \exp\ \left[-c\, L(z) \right], \tag{C1}$$

$$R_n(z) = R_{n,a}\, \exp\ \left[-c\, L(z) \right]. \tag{C2}$$

The attenuation coefficient is assumed to be the same for both profiles: $c = 0.60$. Net radiation above the canopy $R_{n,a}$ is calculated as 60 % of global radiation $R_{s,a}$ and soil heat flux G, as half the net radiation reaching the soil surface. The profile of wind speed within the canopy is given by

$$u(z) = u(z_h)\, \exp\ \left[-\beta L(z) \right], \tag{C3}$$

where $u(z_h)$ is the wind speed at canopy height z_h (inferred from wind speed u_a at reference height z_r using a simple logarithmic profile) and $\beta = 0.5$ (Inoue, 1963). Two profiles of leaf area density $l(z)$ are considered: one is constant with height $l(z) = L_t/z_h$ (profile A) and the other (profile B) uses a gamma-type function to represent a canopy with a higher leaf area density in the top layers, as frequently occurs,

$$l(z) = \Lambda_0 u^{\gamma-1} \exp\left(-u\right) \quad \text{with} \quad u = \frac{z_h}{z} - 1. \tag{C4}$$

The shape parameter γ is taken equal to 4 and Λ_0 is determined as $L_t / \int_0^{z_h} l(z)\,dz$ to obtain a canopy LAI equal to L_t. Leaf boundary-layer resistance (per unit one-sided leaf area) is calculated as a function of wind speed and leaf width w (0.01 m) as (Choudhury and Monteith, 1988)

$$r_{a,l}(z) = \alpha \left[w/u(z) \right]^{1/2}, \tag{C5}$$

with $\alpha = 200$ in SI units. For the sake of convenience, the aerodynamic resistance above the canopy is expressed as a simple function of wind speed without stability correction:

$$r_{a,0} = \frac{1}{ku*} \ln\!\left[\frac{z_r - d}{z_0} \right], \tag{C6}$$

where $u* = k u_a / \ln[(z_r - d)/z_0]$ with $d = 0.63 z_h$ and $z_0 = 0.13 z_h$. The air resistance between the soil surface and the canopy source height is given by Choudhury and Monteith (1988):

$$r_{a,s} = \frac{z_h \exp\ (\omega)}{\omega K(z_h)} \left\{ \exp\left[-\omega z_{0,s}/z_h \right] \right. \\ \left. - \exp\ \left[-\omega(d + z_0)/z_h \right] \right\}, \tag{C7}$$

where $K(z_h) = k^2 u_a (z_h - d)/\ln[(z_r - d)/z_0]$ is the value of eddy diffusivity at canopy height, $\omega = 2.5$ (dimensionless) and $z_{0,s} = 0.01$ m. The profile of leaf stomatal resistance (per unit one-sided leaf area) is made a function of solar radiation within the canopy following a Jarvis-type formulation:

$$r_{s,l}(z) = \frac{r_{s,l,n}}{1 - \exp\left[-\nu R_s(z) \right]}, \tag{C8}$$

where $r_{s,l,n}$ is a minimal stomatal resistance, which depends on available soil water, and $\nu = 0.009$ with R_s expressed in $\mathrm{W\,m^{-2}}$ (Lhomme et al., 2001).

Regional frequency analysis of extreme rainfall in Belgium based on radar estimates

Edouard Goudenhoofdt[1], Laurent Delobbe[1], and Patrick Willems[2]

[1]The Royal Meteorological Institute of Belgium, Brussels, Belgium
[2]Department of Civil Engineering – Hydraulics Division, University of Leuven, Leuven, Belgium

Correspondence to: Edouard Goudenhoofdt (edouard.goudenhoofdt@meteo.be)

Abstract. In Belgium, only rain gauge time series have been used so far to study extreme rainfall at a given location. In this paper, the potential of a 12-year quantitative precipitation estimation (QPE) from a single weather radar is evaluated. For the period 2005–2016, 1 and 24 h rainfall extremes from automatic rain gauges and collocated radar estimates are compared. The peak intensities are fitted to the exponential distribution using regression in Q-Q plots with a threshold rank which minimises the mean squared error. A basic radar product used as reference exhibits unrealistic high extremes and is not suitable for extreme value analysis. For 24 h rainfall extremes, which occur partly in winter, the radar-based QPE needs a bias correction. A few missing events are caused by the wind drift associated with convective cells and strong radar signal attenuation. Differences between radar and gauge rainfall values are caused by spatial and temporal sampling, gauge underestimations and radar errors. Nonetheless the fit to the QPE data is within the confidence interval of the gauge fit, which remains large due to the short study period. A regional frequency analysis for 1 h duration is performed at the locations of four gauges with 1965–2008 records using the spatially independent QPE data in a circle of 20 km. The confidence interval of the radar fit, which is small due to the sample size, contains the gauge fit for the two closest stations from the radar. In Brussels, the radar extremes are significantly higher than the gauge rainfall extremes, but similar to those observed by an automatic gauge during the same period. The extreme statistics exhibit slight variations related to topography. The radar-based extreme value analysis can be extended to other durations.

1 Introduction

Localised rainfall extremes can have a strong impact on human activities especially in urban areas (Ootegem et al., 2016). For flood management applications (e.g. sewer system and dam design) it is needed to know the probability that rainfall exceeds a given amount. This probability is often expressed as the rainfall level which, on average, will be exceeded once over a given period of T years, which is defined as the return period. For infrastructure design applications, one is interested in return periods from 50 to 100 years. Such long return periods often exceed the available observation period and a model is needed.

Extreme values are often extracted from a time series using block maxima, typically over one year (AM) for meteorological data. The performance of the statistical modelling applied to AM data is limited by the number of years available. The peak-over-threshold (POT) method, where values exceeding a given threshold are kept, allows to increase the number of samples. The extreme value theory showed that under some hypotheses – including independence – of the random variables, the AM and POT series can asymptotically converge only to the 3-parameter distributions known as GEV and GPD, respectively.

Different fitting methods to the extreme value distributions have been developed in the literature. The maximum likelihood estimator (MLE) is the most widely used fitting method but for small samples it can lead to unrealistic parameter estimates. This problem is partially addressed with the generalized MLE proposed by Martins and Stedinger (2000) or the L-moments method (Overeem et al., 2009). The above methods do not focus on the tail of the distribution, which is the most relevant for risk analysis. For this goal, Willems et al. (2007) proposed a method based on regression in Q-Q plots.

To reduce the uncertainty associated with the limited number of data at a single site, regional frequency analysis (RFA) methods have been proposed (Svensson and Jones, 2010). The RFA is characterized by the selection of the regions and the parameter estimation approach applied to each region (Buishand, 1991). There are numerous studies of RFA for rainfall extremes based on rain gauge datasets. The index flood approach, which considers that only the location parameter varies in the region, is very popular (Gellens, 2000; Sveinsson et al., 2001; Rulfova et al., 2014). Uboldi et al. (2014) used a bootstrap technique to randomly select data from neighbouring locations with a probability depending on the distance and altitude difference with the target location. The combined use of POT and RFA methods is recommended by Roth et al. (2015).

One of the challenges in RFA is the intersite dependence (e.g. Hosking and Wallis, 1988). Even for a 1 h duration, rainfall maxima exhibit a spatial correlation (Vannitsem and Naveau, 2007). Using the sum of the length of all sites is common but causes an underestimation of the extremes (e.g. Bardet et al., 2011). Several approaches have been proposed to deal with this problem (e.g. Castellarin, 2007; Weiss et al., 2014).

To obtain the rainfall statistics at any given point, spatial models have been developed using geographical and climatological covariates (e.g. Cooley et al., 2007). In Belgium, Van de Vyver (2012) derived a spatial GEV model depending linearly on the altitude. Rulfova et al. (2014) found for 6 h rainfall in the Czech Republic that the assumption of a linear model might be too restrictive, especially for convective precipitation.

The rain gauge network can capture rainfall extremes for widespread situations. However, they can only catch a small part of rainfall extremes caused by convective storms, which exhibit strong spatial variations over short distances. The use of high-resolution gridded rainfall datasets to study rainfall extremes is still in its infancy. This can be explained by their unavailability, their processing requirements and their limited quality. Precipitation estimations from satellite offer global and relatively long records suitable for extreme value analysis (Marra et al., 2017) but still suffer from large uncertainties (Sapiano and Arkin, 2009). The best potential is currently provided by radar-based quantitative precipitation estimation (QPE) products. It should be noted that the radar estimates represent the averaged precipitation over a given area (typically a square of 1 km). While this area is much bigger than the gauge area, we will consider it as representative for small-scale precipitation. It has been shown that the sub-pixel variability of rainfall extremes is significant, especially for short durations (Peleg et al., 2016). The relatively short record of radar datasets is an issue if the extreme statistics depend only on time (i.e. are completely dependent spatially). While this is a reasonable assumption for larger durations (e.g. 1 day), it is difficult to prove for short durations (e.g. 1 h). In case of significant climatic variations, a short record will be more representative of the extreme statistics.

In a pioneering work, Overeem et al. (2009) showed that a 11-year radar dataset is suitable to derive depth-duration-frequency (DDF) curves for the Netherlands. But some differences with rain gauge results were found for short durations. Based on a unique 23-year radar dataset in Israel, Marra and Morin (2015) found that the DDF curves were generally overestimated but 60 % of them lay within the rain gauge DDF confidence intervals. In Ontario (Canada), Paixao et al. (2015) demonstrate the potential to integrate radar (Digital Precipitation Array product) to rain gauge analysis, especially to identify homogeneous regions of extreme rainfall. Saito and Matsuyama (2015) used a 26-year radar-gauge dataset (without RFA) to study the spatial variation of hourly rainfall extremes in Japan. They found significant spatial patterns but also large uncertainties in the radar datasets. Different index flood approaches were tested by Eldardiry et al. (2015) in Louisiana, who defined a region as a square window of 44 km size. They found for Louisiana (USA) that the relatively short period (13 years) explains the high uncertainty of the analysis, that the index flood method is recommended and that a systematic underestimation is associated with the radar products (its spatial resolution is 4×4 km). Haberlandt and Berndt (2016) found that the operational DWD product is only suitable for studies on longer durations after bias correction. Using a 10-year high-resolution radar rainfall dataset, Wright et al. (2014b) performed a RFA using stochastic storm transposition. They found that the radar-based intensity-duration-frequency (IDF) estimates generally reproduce conventional gauge-based IDF estimates but overestimate these for longer return periods and shorter durations.

The potential of the radar data can be fully exploited by studying the extremes of the mean (or maximum) rainfall over areas. With the goal of deriving alert thresholds for 159 regions in Switzerland, Panziera et al. (2016) studied the areal rainfall maxima (with sizes from the pixel to the region). Using RFA on squares, Overeem et al. (2010) derived areal rainfall DDF curves for the Netherlands. Wright et al. (2014b) applied a similar methodology but on different catchments in Louisiana.

In this study, we want to demonstrate the potential of high-resolution radar-based QPE to derive rainfall extreme statistics at a given location. To our knowledge none of the previous studies combine high-quality radar-based QPE with a high-quality reference rain gauge measurements. At the Royal Meteorological Institute of Belgium (RMIB), a QPE has been derived from the reprocessing of raw volumetric radar measurements. This dataset has been used for various applications such as case studies and model verification. The methodology to derive this dataset has been verified for the period 2005–2014 against an independent rain gauge network (Goudenhoofdt and Delobbe, 2016). RMIB also has a unique 40 year dataset of 10 min rain gauge measurements, which has been used in extreme value studies (Vannitsem and Naveau, 2007; Van de Vyver, 2012).

Unlike existing radar studies, we select our data using the POT approach and use a regression in Q-Q plot (QQR) fitting method. Radar-based extreme statistics for 1 and 24 h durations are compared with the ones derived from rain gauge data covering the same period. We propose a new RFA which makes use of independent radar data in a predefined neighbourhood. The results are compared with those obtained using the long-term rain gauge network. Finally, the regional approach is applied at each radar pixel on the whole of Belgium to study the spatial variations of the rainfall extremes.

2 Rainfall data

2.1 Rain gauge measurements

Over the years, Belgium (Fig. 1) has been covered by several rain gauge networks for different purposes.

Since the end of the 19th century, RMIB maintains a network (CLIM) of non-recording rain gauges from which rainfall measurements are taken at 08:00 LT. The data are carefully controlled and used for climate applications (Journée et al., 2015).

A Hellmann–Fuess pluviograph has been in operation in Uccle (RMIB) from 1898 to 2008 and has enabled the compilation of a continuous time series of 10 min rainfall (Demarée, 2003). The 10 min rainfall values had to be manually extracted from line graphs on papers. Starting from the fifties, additional rain gauges were installed to constitute a network (BUL) for hydrological research. Since the rain gauges underestimate the rainfall by 5–10 % due to its mechanism, its records have been calibrated using a collocated gauge from the CLIM network.

For weather forecast purposes, the RMIB maintains a network of automatic weather stations (AWS) in Belgium. These stations provide rainfall measurements at 10 min temporal resolution. The tipping-bucket gauges are progressively replaced by weighted gauges (the first one was installed in Uccle on 10 February 2009). The data are available since 2002–2004 and have been quality controlled.

The hydrological service of the Walloon region (SPW) maintains a dense network of hourly (every 5 min since 2012) rainfall measurements. The tipping-bucket gauges are progressively replaced by weighted gauges since 2015. The data have been quality controlled by RMIB since April 2004.

It is important to know the limitations of the respective rain gauges in case of extreme rainfall. It is known (Nystuen, 1999; Duchon and Biddle, 2010) that tipping buckets underestimate high rainfall rates. The use of weighted gauges for extreme rainfall is discussed in Colli et al. (2012). Every 10 mm, the pluviograph has to be emptied which results in an underestimation in case of extreme rainfall. The calibration

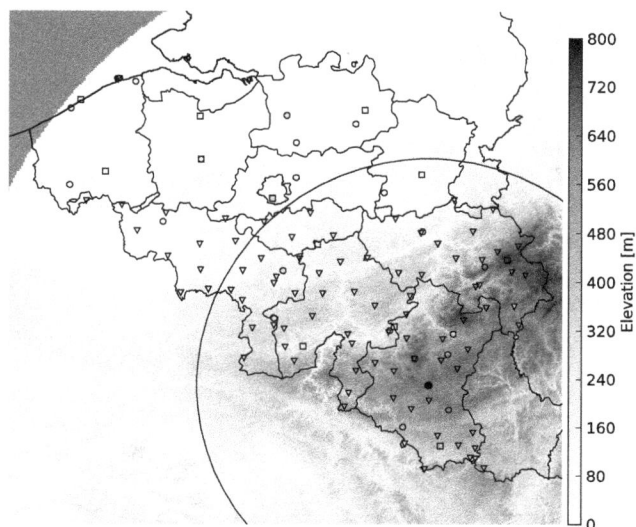

Figure 1. Elevation map centred on Belgium with the Wideumont radar (black dot) covering a 240 km range (the black circle denotes the 120 km range) with AWS (square), SPW (triangle) and BUL (circle) rain gauge networks. The gauge locations selected in this paper are in cyan. Country borders with France, Luxembourg, Germany and the Netherlands are also displayed.

of the pluviograph is probably not sufficient for sub-daily extremes. Finally, the quality controls, albeit conscientious, can never be considered as perfect.

2.2 Radar estimation

The QPE available on a 1 km grid every 5 min is made using an elaborated processing chain from the radar volumetric reflectivity measurements. The quality of the radar volume is controlled using several algorithms:

- a static clutter map: pixels with unrealistically high probability of rainfall are identified as clutter

- a beam blockage map: the percentage of the beam blocked by topography is computed using a simple propagation model

- a first clutter identification based on reflectivity differences between radar beam elevations

- a second clutter identification based on strong deviations of a pixel from its neighbourhood and unrealistic lines

- a third clutter identification for radar echoes in cloud-free areas determined by satellite observations

A maximum threshold for reflectivity is set to 55 dBZ to mitigate higher reflectivity values due to hail. The rainfall rate estimates are obtained using stratiform-convective classification, a 40 min averaged vertical profile of reflectivity

(VPR), a bright band identification and a specific transformation to rain rates for the two precipitation regimes. The detailed procedure is described in Goudenhoofdt and Delobbe (2016). As a reference for the QPE product, the CAP product is defined as the interpolation at 800 m above the radar level. It makes use of a standard $Z–R$ relationship, which comes from the hypothesis that the drop size distribution follows the distribution of Marshall–Palmer, as discussed in Uijlenhoet and Pomeroy (2001).

Consecutive rain rate estimates are integrated to obtain 10 min accumulations (5 min gaps are tolerated) to match the lowest resolution of the rain gauge data. Hourly accumulations are combined with the SPW gauges using a mean field bias correction. This method applied to the QPE product is referred to as the MFB product from now on. A more complex merging method (i.e. external drift Kriging) was tested but found to be unstable for some time moments.

It is important to mention the limitations of the radar products in case of extreme precipitation. The most important impact of the QPE processing on extreme values is the 55 dBZ reflectivity threshold used to mitigate hail. Using the convective $Z–R$ relationship, this corresponds to a maximum rainfall rate of 80 mm h^{-1}. Higher values of about 100 mm h^{-1} are possible when the standard $Z–R$ relationship is used for stratiform areas. This can only happen close to the radar where convective precipitation can not be identified. Slightly higher thresholds have been used by Overeem et al. (2009) (100 mm h^{-1}) and Wright et al. (2014b) (105 mm h^{-1}). A higher threshold is used by Marra and Morin (2015) (150 mm h^{-1}) but for a Mediterranean climate. Only half of the AWS gauges recorded up to 3 times more than 100 mm h^{-1} in 10 min. Given the sub-pixel spatial variability, one can assume that this threshold will never be exceeded for the pixel average. This threshold can only partly correct for the overestimation due to hail. The second most important error is related to signal attenuation especially in the case of well organised convective systems. This is why extremes might be underestimated the further the distance from the radar. In addition, the increasing radar sample volume will produce an underestimation of small-scale extremes. The uncertainty in the $Z–R$ relation is another important source of error.

2.3 Comparison framework

In this study, we will only consider validated rain gauge data. Given that the SPW network is used for merging, the radar dataset for 2005–2016 is used. To perform a direct comparison, the gauge data of AWS and SPW for the same period are used. For comparison against the reference BUL network, the gauge data for the period 1965–2010 are used. The time series of the BUL and CLIM networks have been tested for homogeneity by Van de Vyver (2012) and a selection of useful stations has been made. Gellens (2000) and Vannitsem and Naveau (2007) found that the vast majority of the CLIM

and BUL time series are stationary for summer rainfall. However, the existence of a multi-decadal oscillation in rainfall extremes has been found in the Uccle time series (Ntegeka and Willems, 2008; Willems, 2013).

The 10 min rainfall accumulation from the gauge networks (AWS, BUL) and radar products (CAP, QPE) are summed to obtain sliding 1 h rainfall accumulations. Such a duration is associated with convective storms, which can only be properly seen on radar images. The hourly bias obtained by the MFB method could be applied to the 10 min accumulations. However, it will not be used due to the possible risk of representativity errors related to convective storms and the small benefits expected.

The hourly rainfall from the SPW network and the radar products (CAP, QPE, MFB) are summed to obtain sliding 24 h rainfall accumulations. The SPW network is preferred to the AWS network because it is denser and more homogeneous. Such a duration is mainly associated with widespread precipitation for which the benefit of merging methods is clear. The risk of instability with MFB (e.g. in case of strong spatial variation of the bias) is tolerated given the significant expected benefit for widespread precipitation events.

It should be noted that using the lowest available duration for each network would result in an underestimation of the extremes due to the discrete time sampling (Marra and Morin, 2015). Additionally, random errors and time sampling difference can be compensated by performing the sum. For both the radar and the gauge, no missing data is tolerated in the sum to avoid underestimation. Furthermore, only timestamps with both radar and gauge data are kept.

Due to the amount of stations, it is not possible to analyse in detail the results at each station. Therefore, a few stations are picked at different distances from the radar (see Table 1 and Fig. 1). The Uccle station is chosen because it is included in the three networks, which makes inter-comparison possible. The availability of the 1 h accumulation data is about 95 % for the radar products and close to 100 % for the AWS gauges. The radar availability of the 24 h accumulation is lower than the 1 h accumulation because a significant part of the intervals without data are short. The availability of the SPW gauges is around 90 % but this is mainly due to the removal of snow events, when no extreme precipitation is expected. The availability of the BUL stations for the period 1965–2010 is highest at Uccle with 96.3 %, then about 86 % at Deurne and Gosselies. The station of Nadrin has only 60 % of availability (for the period 1965–2010) because it was started in 1978.

3 On-site frequency analysis

3.1 Methodology

It has been shown by Pickands III (1975) that the extreme values converge asymptotically to a generalized Pareto dis-

Table 1. Rain gauge stations used for comparison and availability of the extreme rainfall datasets. The last column is the percentage of time when both radar and gauge data are available. DNG is the *Deuxième nivellement général*.

Station	Altitude (DNG)	Distance to radar (km)	Duration	Avail. gauge (%)	Avail. radar (%)	Avail. both (%)
Humain (AWS)	296	36	1 h	98.5	94.8	93.5
Uccle (AWS)	100	128	1 h	99.9	94.8	94.7
Uccle (SPW)	100	128	24 h	90.6	86.0	78.2
St-Vith (SPW)	456	61	24 h	89.2	86.0	76.7
Deurne (BUL)	12	161	1 h	86.0	–	–
Uccle (BUL)	100	128	1 h	96.3	–	–
Gosselies (BUL)	187	97	1 h	85.7	–	–
Nadrin (BUL)	403	30	1 h	59.3	–	–

tribution (GPD):

$$F_{(\xi,\mu,\sigma)}(x) = \begin{cases} 1 - \left(1 + \frac{\xi(x-\mu)}{\sigma}\right)^{-1/\xi} & \text{for } \xi \neq 0, \\ 1 - \exp\left(-\frac{x-\mu}{\sigma}\right) & \text{for } \xi = 0, \end{cases} \tag{1}$$

with ξ, μ and σ commonly defined as the shape, location and scale parameters, respectively. The special case when the shape parameter is equal to zero is defined as the exponential distribution (EXP).

The choice of the threshold has an important impact on the estimation of the distribution parameters. When the number of selected values increases, the variance naturally decreases but the bias increases (due to the deviation from the theoretical distribution). It is more practical to use a threshold rank instead of a threshold value to control the sample size.

To apply the theory, the extreme values have to be independent but successive peaks within the same time window can be observed due to the nature of precipitation. For the 1 h duration, two peaks are considered dependent if the time interval is less than 12 h as proposed by Ntegeka and Willems (2008). This choice is consistent with the characteristics of convective storms analysed in Goudenhoofdt and Delobbe (2013). Jakob et al. (2011) used a separation time of 24 h but found little sensitivity when taking lower or higher values. We also found that using 3 days hardly affects the selection of the 1 h extremes. For the 24 h duration, we use a time interval of 3 days, which is the typical scale of synoptic regimes. These choices are consistent with Roth et al. (2014) who empirically found a temporal dependence of 1 and 2 days for winter and summer precipitation, respectively. In practice, a peak is kept if it is the maximum compared to its dependent peaks (if any).

The type of the distribution can be derived by looking for the Q-Q plot where the extremes behave in an asymptotic linear way. Willems (2000) found for the Uccle series that the tail of the distribution has an exponential behaviour for all durations. In the gauge datasets used in this study, we also found a tendency for the EXP distribution. The EXP distribution is preferred for short periods since estimating the shape parameter is very uncertain. Blanchet et al. (2015) found that GPD fails to robustly estimate the tail of the distribution be-

cause of a lack of data and unrealistic return levels for very long return periods (when the shape parameter is positive). An additional argument for the EXP model is that it is less affected by observational errors, which plays an important role here.

In this study we use the QQR method based on regression in Q-Q plots proposed by Willems et al. (2007). The exponential Q-Q plot is the extremes x versus minus the logarithmically transformed exceedance probability $1 - G(x)$. The EXP distribution appears as a line in this plot, with slope equal to the scale parameter σ:

$$x = x_t - \sigma \ln(1 - G(x)), \tag{2}$$

where x_t is the threshold level. The same properties hold for the plot of the return level x_T against the return period T when the latter is plotted on a logarithmic scale:

$$x_T = x_t + \sigma \ln(T \cdot M/n), \tag{3}$$

where M is the number of extremes and n the length of the time series.

The estimators for the slope are based on linear regression in the Q-Q plot above the specific threshold level x_t. Amongst the available estimators for σ we used an unconstrained and unweighted linear regression.

The optimal threshold rank t is found by minimization of the mean squared error (MSE) of the calibration. With our datasets, this rank is chosen between 18 and 30 considering the uncertainties and the relatively short period, respectively. Confidence intervals for the scale parameter are computed using a parametric bootstrap technique. The fitted distribution is used to generate 1000 extreme value series with a size corresponding to the optimal rank. The fitting procedure is applied to each of the 1000 series to obtain 1000 simulated scale parameters. The 10 and 90 percentiles of the simulated parameters are taken as the 10 and 90 % confidence interval bounds for the true scale parameter.

3.2 Comparison of 1 h extremes

The extreme events as seen by both the radar and the gauge are compared in Table 2. Since the focus is on the tail of

Table 2. Comparison of the 10 highest 1 h rainfall extremes from the gauge (AWS) and radar (QPE) at Humain and Uccle stations. The events with a high probability of hail have their number in bold. The events are ordered by the maximum of the gauge and radar values.

Humain				
Event	Date yyyy-mm-dd	Time	Gauge (mm h^{-1})	Radar (mm h^{-1})
1	2016-06-07	18:50	57.65	45.25
2	2005-07-30	00:40	28.60	11.62
3	2014-04-24	15:40	27.00	20.35
4	2014-06-10	21:40	15.60	26.40
5	2007-06-14	01:20	25.80	16.32
6	2009-05-25	13:10	24.10	25.17
7	2008-05-14	17:40	13.10	24.35
8	2015-07-19	01:00	22.87	15.47
9	2009-06-27	14:30	20.40	19.83
10	2009-07-22	21:20	19.80	12.08
11	2010-07-14	15:40	19.80	–
12	2012-06-12	22:20	18.30	15.61
13	2013-03-23	07:40	–	17.30
14	2005-06-28	22:20	–	16.74

Uccle				
Event	Date yyyy-mm-dd	Time	Gauge (mm h^{-1})	Radar (mm h^{-1})
1	2016-06-07	15:20	18.08	38.21
2	2011-08-23	08:40	35.50	23.22
3	2009-10-07	18:40	30.79	33.32
4	2012-05-20	16:30	12.37	29.79
5	2005-09-10	19:40	29.10	17.54
6	2011-08-18	15:50	28.98	14.77
7	2007-06-14	14:50	21.90	25.88
8	2011-09-03	22:40	25.34	18.46
9	2016-06-11	18:50	–	24.88
10	2005-07-29	19:10	24.29	–
11	2010-07-14	15:20	24.15	–
12	2014-07-29	16:10	20.10	18.17
13	2013-07-27	22:20	20.07	–
14	2008-07-26	10:40	16.60	18.30

the distribution, only the 10 highest values from either the gauge or the radar data are selected. The events for which the probability of hail is high (i.e. when the threshold was applied) are highlighted. An event is considered as problematic if the corresponding radar or gauge extreme rank is below 30. For these events, the underlying precipitation patterns are analysed using the radar images. This comparison allows for identifying the weaknesses of the gauge and radar datasets.

The maximum at Humain has been observed by both the radar and the gauge on 7 June 2016. This relatively high value can be due to randomness and the short period of records. But it is also possible that the other quantiles are underestimated (the maximum was recorded by the new weighted gauge). There is generally a good match between

the radar and the gauge quantiles except for the following events:

– event 2: the radar underestimates globally

– event 7: the gauge is located at the boundary of the convective cell

– event 11: the radar signal is strongly attenuated by a mesoscale convective system

– event 13: there was probably snow in the gauge

– event 14: the gauge is located at the boundary of a convective cell.

The second highest quantile at Uccle has been observed by both the radar and the gauge on the 7 October 2009. There is generally a good match between the two datasets. A few events are problematic:

– event 1, 4: the gauge is at the boundary of a cell

– event 9: there is a stationary storm underestimated by the gauge

– event 10: the gauge is at the boundary of a cell and the radar is attenuated (same as event 2 in Humain)

– event 11: the radar signal is strongly attenuated (same as event 11 in Humain)

– event 13: the radar is attenuated.

The problems with cell boundaries are easily explained: the radar estimation is taken at a given height above ground and the rain is subject to wind drift before reaching the ground. This effect increases with the distance to the radar. Due to its randomness, it should not affect the statistics. The other problematic events can be considered as missing data. Since the level of missingness is limited, the impact on the statistics is expected to be small.

Figure 2 shows the results of the extreme value analysis for the 1 h rainfall accumulation. The return levels are obtained using formulas from Willems et al. (2007), which are based on the Weibull plotting position. Numerical values of the temporal independence, the optimal rank, the location parameter and the scale parameter can be found in Table 3. The percentage of independent peaks (among peaks exceeding the threshold) is around 20 % for both the radar and the gauges at the two locations. This low value is mainly due to the fact that five consecutive values at 10 min resolution are correlated.

The empirical quantiles of the QPE product are systematically slightly lower than those for the AWS gauges. This may be expected as we compare point rainfall observations with rainfall averaged on a 1 km square. However, the underestimation of very high rainfall rates by tipping-bucket gauges can compensate for this effect. One also notes small groups

Table 3. Results of the extreme value distribution fitting at two locations of the AWS network. The tables successively show the temporal independence, optimal rank, the location parameter and the scale parameter. A value is indicated as missing when its extreme rank is below 30.

temporal independence (%)				
Station	Gauge	CAP	QPE	MFB
Humain	25.6	20.7	22.6	–
Uccle	20.8	19.4	21.0	–

optimal rank				
Station	Gauge	CAP	QPE	MFB
Humain	30	30	28	–
Uccle	29	23	30	–

location parameter ($mm\,h^{-1}$)				
Station	Gauge	CAP	QPE	MFB
Humain	12.2	11.0	10.7	–
Uccle	12.3	13.9	12.3	–

Station	Gauge	CAP	QPE	MFB
Humain	7.5	8.9	6.6	–
Uccle	6.8	10.8	6.4	–

(a)

(b)

Figure 2. Return levels for 1 h duration at location Humain (**a**) and Uccle (**b**) of the AWS gauge (red stars) compared to CAP (blue triangles) and QPE (magenta squares) radar products. The extreme value distribution (solid line) fitted to the extremes and its confidence intervals (dashed line) are also displayed.

of similar values for both the radar and the gauge, which are mainly associated with hail events. This can be explained by the effect of hail threshold and the rainfall rate limit, respectively. The extremes tend to be heavy tailed but this can be at least partially explained by the observation biases described above.

The fit of the EXP distribution is relatively good for the two locations with a relatively low MSE (not shown). The scale parameter tends to be higher for the gauge data than the radar data. In general, the uncertainty for the scale parameter remains high and this results in wide confidence intervals for higher return periods.

When using the CAP product, the higher quantiles are overestimated especially for Uccle. This can be mainly attributed to the effect of hail. This results in an overestimation of the scale parameter.

3.3 Comparison of 24 h extremes

The comparison of the 10 highest extremes from either the radar (MFB) or the gauge (SPW) can be seen in Table 4. For Uccle, most extreme values occurred during summer and are therefore associated with convective storms. There is a good match between the gauge and the radar except for a few events:

– event 8, 11: the gauge is at the boundary of a convective cell

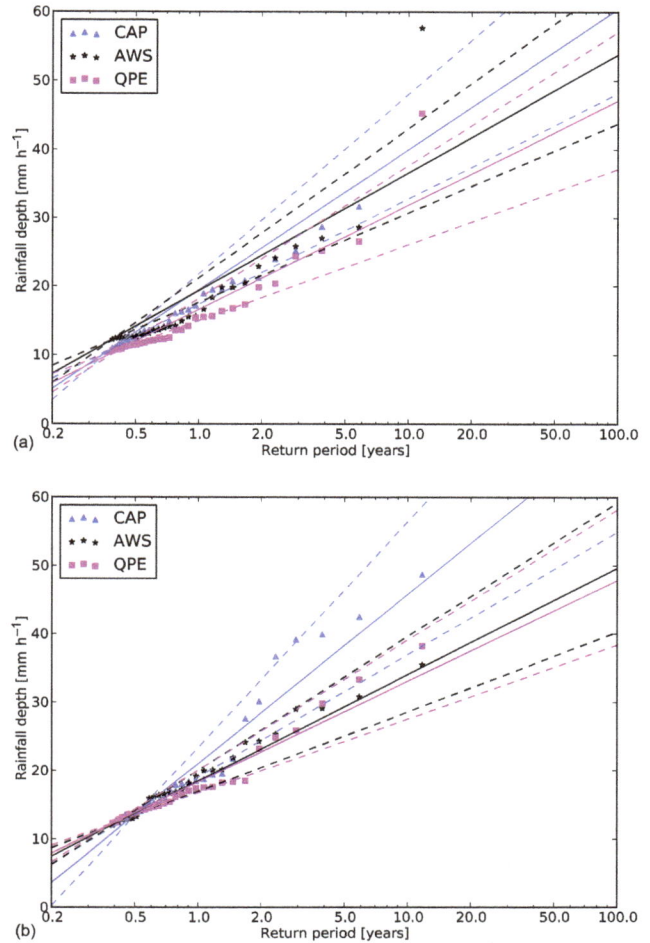

– event 13: strong radar attenuation by a mesoscale convective system

– event 14: snow episode probably underestimated by the radar.

For Saint-Vith, the extreme values occurred either in summer or in winter with therefore a mix of convective and widespread precipitation episodes. The match is very good except for the following events:

– event 2: at the boundary of a cell (probably with hail)

– event 3: slight overestimation due to snow melting (QPE); overestimation due to non-uniform bias (MFB)

– event 13: at the boundary of a cell.

Table 4. Comparison of the 10 highest 24 h rainfall extremes from the gauge (SPW) and radar (MFB) at Uccle and Saint-Vith stations. A value is indicated as missing when its extreme rank is below 30. The events are ordered by the maximum of the gauge and radar values.

		Uccle		
Event	Date	End time	Gauge mm day^{-1}	Radar mm day^{-1}
1	2010-08-16	23:00	63.30	48.99
2	2009-10-07	23:00	52.50	61.83
3	2011-08-23	15:00	59.31	61.00
4	2006-08-03	23:00	43.00	58.44
5	2016-05-30	23:00	35.30	53.34
6	2014-08-26	15:00	45.30	48.51
7	2012-10-04	08:00	34.60	45.63
8	2012-06-12	11:00	–	44.87
9	2016-06-12	17:00	31.30	39.45
10	2011-09-04	21:00	38.70	26.10
11	2015-08-16	03:00	–	37.75
12	2007-06-15	11:00	36.99	33.91
13	2014-07-10	04:00	36.90	–
14	2016-01-16	02:00	36.30	–

		Saint-Vith		
Event	Date	End time	Gauge mm day^{-1}	Radar mm day^{-1}
1	2007-01-18	16:00	74.60	56.88
2	2009-07-03	16:00	37.90	61.68
3	2011-12-16	23:00	–	56.62
4	2012-07-28	21:00	53.60	46.72
5	2012-10-04	12:00	49.70	39.86
6	2007-08-22	19:00	47.50	48.73
7	2010-08-16	03:00	45.80	55.50
8	2006-08-05	06:00	43.70	41.10
9	2007-12-03	08:00	43.40	46.09
10	2007-09-28	08:00	42.40	38.87
11	2014-09-21	14:00	34.00	40.71
12	2016-05-31	02:00	40.01	33.44
12	2016-07-23	21:00	40.00	–

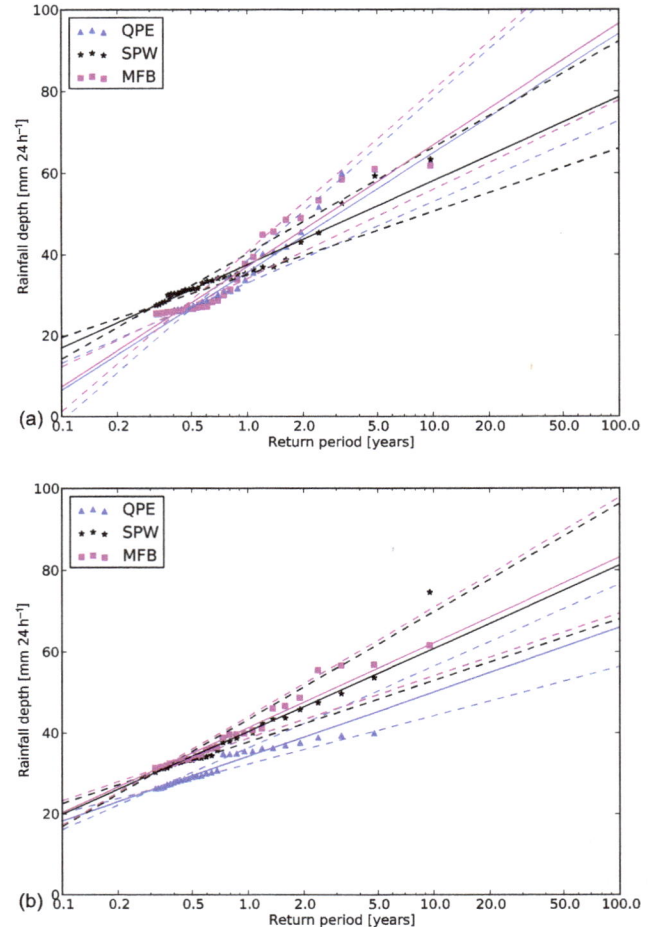

Figure 3. Return levels for 24 h duration at location Uccle (**a**) and Saint-Vith (**b**) of the SPW gauge (red stars) compared to QPE (blue triangles) and MFB (magenta squares) radar products. The extreme value distribution (solid line) fitted to the extremes and its confidence intervals (dashed line) are also displayed.

The problematic events not related to boundary effects can be considered as missing data. Since they are limited it is expected that they only slightly affect the statistics.

Figure 3 shows the results of the extreme value analysis for the 24 h rainfall accumulation. Numerical values can be found in Table 5. The percentage of independent peaks (amongst peaks exceeding the threshold) is between 6 and 9 % for the two locations and for all datasets. This is what we expect from a 24 h accumulation available every hour.

For Uccle there are not many differences between QPE and MFB because most events are associated with convective storms. Compared to the gauge quantiles, the radar quantiles are lower below 1-year and higher between 1- and 5-year return periods. This can be attributed mainly to hail overestimation by the radar and gauge losses. It results in a higher

scale for the radar, which is close to the upper bound of the gauge confidence interval.

For Saint-Vith, there is a clear effect of the bias correction (MFB) to remove the underestimation of the QPE product. As for Uccle, the radar quantiles are higher for return periods higher than 2 years but the effect is limited because less convective storms are involved. The final result is a good match of the two distributions for this station.

For the two stations, no significant instability in the MFB values have been found.

For Uccle, the CAP product overestimates the scale parameter and underestimates the location parameter due to hail and VPR errors, respectively. For Saint-Vith, the quantiles (not shown) are similar to QPE except for a very high unrealistic maximum.

Table 5. Results of the extreme value distribution fitting at two locations of the SPW network. The tables shows successively the temporal independence, optimal rank, the location parameter and the scale parameter.

	temporal independence (%)			
Station	Gauge	CAP	QPE	MFB
Uccle	7.1	6.0	6.6	6.7
St-Vith	7.4	8.4	9.0	8.4
Station	Gauge	CAP	QPE	MFB
Uccle	30	26	19	23
St-Vith	30	30	30	28
Station	Gauge	CAP	QPE	MFB
Uccle	27.2	25.0	27.2	27.5
St-Vith	30.2	25.8	26.3	31.5
	scale parameter (mm day^{-1})			
Station	Gauge	CAP	QPE	MFB
Uccle	9.0	13.5	12.7	12.9
St-Vith	8.9	8.2	6.9	9.1

4 RFA

4.1 Methodology

As in Overeem et al. (2009) and Wright et al. (2014b) we consider that the extreme statistics are the same within the region. The region should be sufficiently large to have a large sample size (many extremes) and small enough to neglect extreme statistic variability. No strong variability is expected in Belgium because it is a relatively flat country. Therefore we define our region as the radius of 20 km around the target location. A similar size has been used in other radar studies (e.g. Overeem et al., 2009; Wright et al., 2014b; Eldardiry et al., 2015).

We also consider that the extremes observed within the 20 km radius during a time window of 12 h are dependent. As in Wright et al. (2014b), we keep only the maximum amongst dependent values. We therefore implicitly assume that the regional maximum follows the same distribution as the local extremes. The possible benefit of taking one extreme value at random is an open question. It is important to remember that we are interested in the extreme statistics of any given pixel in the region. This is different from studying the extreme statistics of the maximum rainfall over the region as in Panziera et al. (2016). We also tested the hypothesis that 1 h extremes are independent after a certain distance which is set to 10 km. This distance corresponds to the maximum expected size of a convective cell (Goudenhoofdt and Delobbe, 2013). If this is true it allows to reduce the uncertainty of the

analysis. In the text, we will refer to these datasets by the names REG and R10, respectively.

Due to the spatial dependence, the effective length n_{eff} of the pooled time series is smaller than the total length of the records. The total length is obtained by multiplying the number of years n by the number of pixels N:

$$n_{max} = n \times N. \tag{4}$$

In this study n_{eff} is computed by multiplying n_{max} by the fraction of spatially independent peaks, amongst peaks exceeding the threshold. The latter is obtained by dividing the number of independent peaks by the total number of peaks. It can be shown that this is the same as the method based on the averaged exceedance rate found in Wright et al. (2014b) and explained in detail by Weiss et al. (2014). The large number of peaks available from the radar data allows us to choose a higher threshold rank. This increase in sample size leads to a more reliable extreme value analysis, which is the final goal of this radar-based RFA. Accordingly the QQR method is applied for threshold ranks between 30 and 100 and the optimal rank is found.

4.2 Comparison with rain gauges

Figures 4 and 5 show the results of the RFA for 1 h rainfall accumulation at the four locations selected from the BUL network. The results of the on-site frequency analysis for the gauge and collocated radar pixels are shown as reference. Numerical values can be found in Table 6. The percentage of temporally independent extremes for the gauge is close to 30 % for Deurne and Uccle, while it is slightly above 20 % for the two other stations. This suggests that there are larger clusters which might be related to altitude. Above the threshold, the percentage of spatially independent extremes (REG) ranges from 1.1 % (Uccle) to 2.6 % (Nadrin). The effective period length of the pooled dataset is then between 200 and 500 years. Using a decorrelation distance of 10 km results in twice as much data, which is more than one expects from randomness. It suggests that convection can be organized on large spatial scales.

The radar images associated with each maximum of the radar-based RFA is analysed:

- Deurne and Uccle (28 July 2006): several supercells in the whole of Belgium

- Gosselies (22 August 2011): a squall line moving parallel to the flow

- Nadrin (26 July 2008): a stationary convective cell.

The highest extremes exhibit abrupt variations in the form of steps for both the gauge and radar. This could be explained by the siphonage of the gauge and hail threshold, respectively. Since Nadrin is close to the radar, the standard $Z-R$ relationship is used instead of the convective $Z-R$ relationship. This permits higher rain rates (i.e. 100 mm h^{-1}).

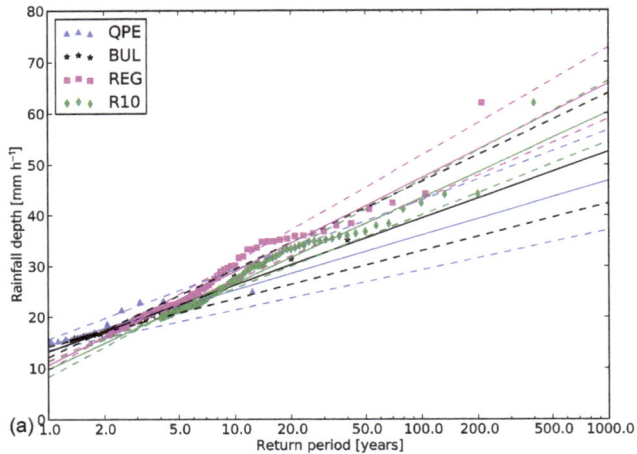

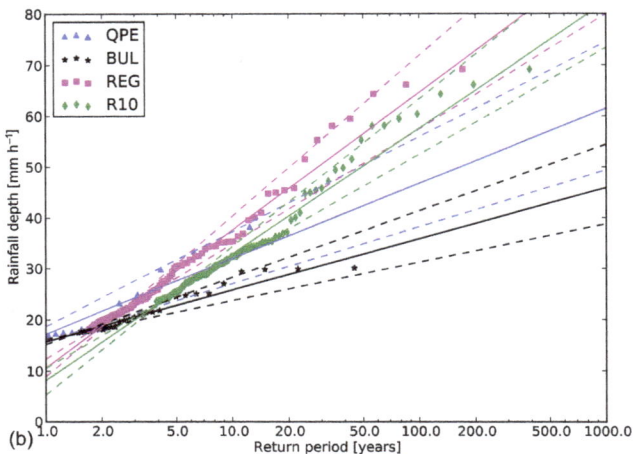

Figure 4. Return levels for 1 h duration at location Deurne (**a**) and Uccle (**b**) from the BUL gauge data (red stars) compared to the on-site QPE (blue triangle), REG (purple square) and R10 (green diamond) radar data. The extreme value distribution (solid line) fitted to the extremes and its confidence intervals (dashed line) are also displayed.

Figure 5. Return levels for 1 h duration at location Gosselies (**a**) and Nadrin (**b**) from the BUL gauge data (red stars) compared to the QPE (blue triangle), REG (purple square) and R10 (green diamond) radar data. The extreme value distribution (solid line) fitted to the extremes and its confidence intervals (dashed line) are also displayed.

The gauge extremes are relatively low at Deurne and Uccle compared to Gosselies and Nadrin. The radar extremes are lower at Deurne compared to the other stations. This can be at least partially attributed to the large sample volume at this range. The match between the gauge and the radar (REG and R10) is good except at Uccle with much higher radar extremes. The REG exhibits higher extremes than R10 suggesting some dependence beyond 10 km. Indeed the results should be similar if the hypothesis of independence after 10 km was valid.

This can be partially attributed to hail but the similar four highest extremes suggest a gauge limitation. It is also striking that half of the 20 highest gauge extremes occurred during the period 1999–2008 (not shown). This positive trend for Uccle is possibly related to the urban heat island effect (Hamdi and Van de Vyver, 2011). The uncertainty of the

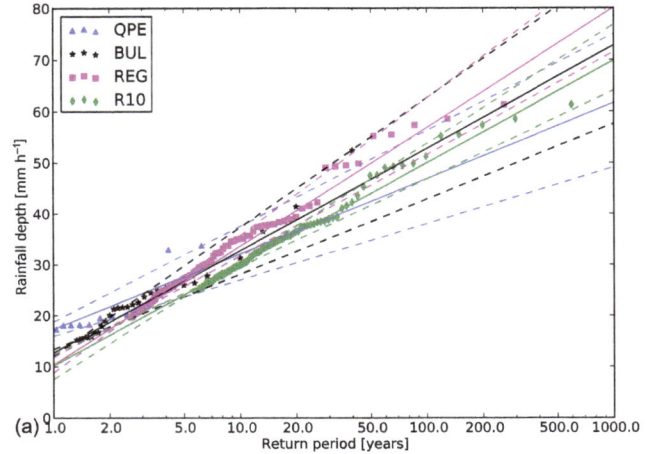

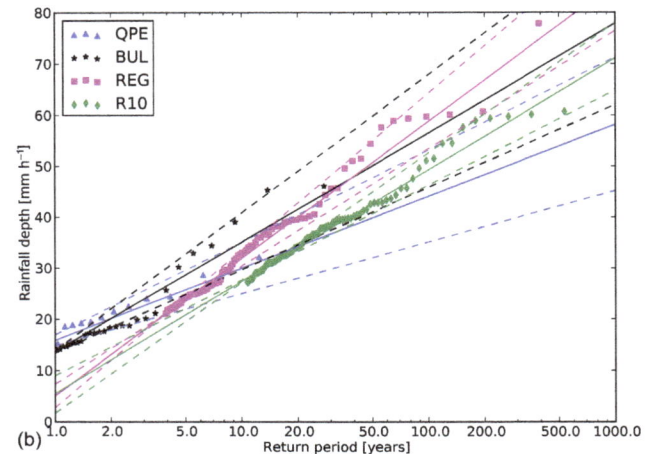

radar fit is low because of the larger sample size, due to which a higher rank can be chosen. Furthermore, the fit is less impacted by the potentially large errors of the highest extremes. The location parameter (corresponding to the threshold) increases with the sample size of the products.

Except for the Uccle station, the scale parameter is the lowest for the QPE dataset due to the bias as a result of the small sample size. The scale parameter of the pooled radar datasets is slightly higher at Deurne and significantly higher in Uccle. For Gosselies and Nadrin, the R10 and BUL data have similar scales while it is slightly higher for the REG data. The fit to the REG and R10 data is within the uncertainty bound of the fit to the BUL data. For those two stations, the fit to the BUL data is even in the small uncertainty bound of the fit to the REG data.

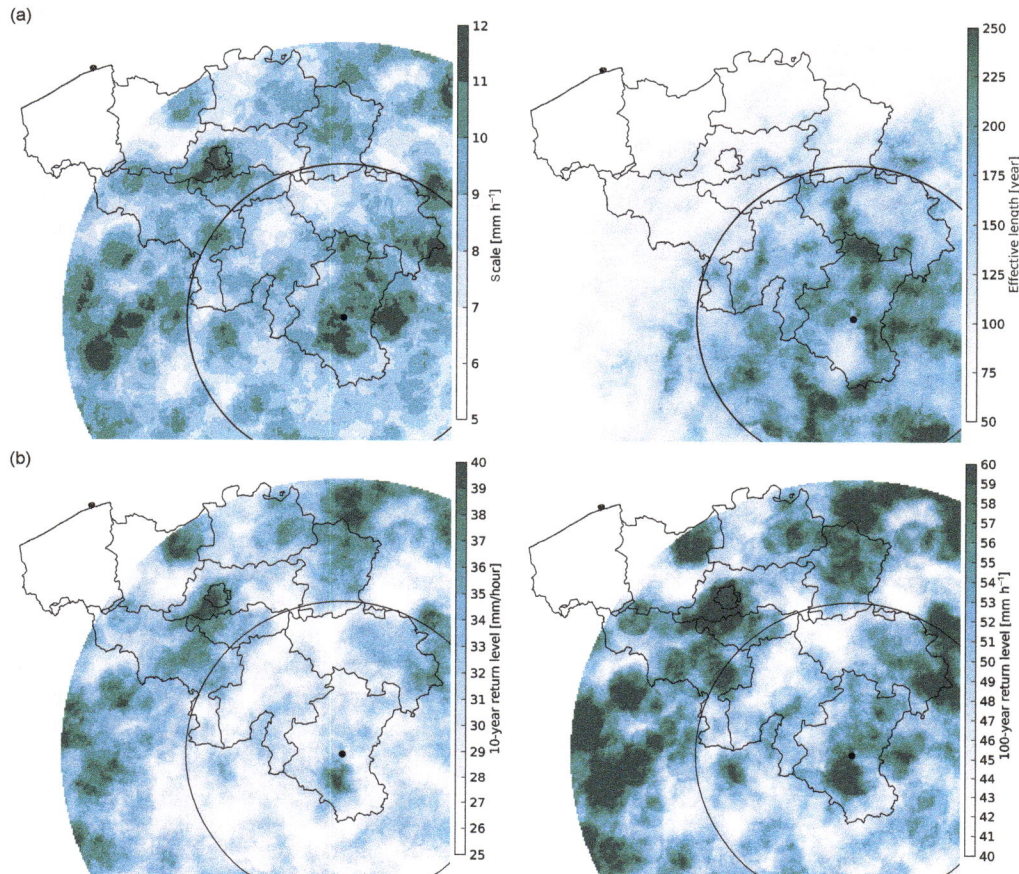

Figure 6. Results of the regional frequency analysis (RFA) for 1 h duration applied over Belgium up to 180 km from the radar. The scale parameter and the effective length are shown in **(a)**. The levels corresponding to a 10- and 100-year return periods are shown in **(b)**. A circle with a radius of 100 km centred at the radar is also drawn.

4.3   Spatial maps

We apply the RFA described above for 1 h duration to all pixel locations in Belgium with some modifications. We use a smaller radius of 10 km to reduce the computation cost and consider that all pixels are spatially dependent. This smaller radius improves the resolution of the maps at the expense of a higher uncertainty. Several pixels in the radar dataset are affected by permanent non-meteorological echoes. They can be identified by an unrealistic high frequency of extremes. In practice one looks at the distribution of the number of values exceeding 12 mm h^{-1}. The pixels with more than 50 exceedances have been found as outliers and removed. To make the comparison easier, we choose a fixed threshold rank of 60. No larger ranks have been considered due to computational limitations.

Figure 6 shows the results of the RFA applied to Belgium. The provinces of Belgium are also displayed to help comparison between the maps. No values are shown beyond the 180 km range because the quality of the radar QPE is significantly reduced. The return periods are computed using Eq. (3) and therefore depend on the scale parameter and the effective length. The higher the scale the higher the difference between the 10- and 100-year return levels.

Some artifacts due to the radar and the regional approach can be seen on the maps. The effective length decreases significantly beyond 100 km meaning that the spatial dependence increases. This is due to the fact that the actual radar sample is larger than the 1 km pixel at those ranges. Circular patterns appear on the maps due to the influence of the pixels located at their centres. The high values are caused by pixels contaminated by non-meteorological echoes (e.g. at the German border) and hail. A stronger filter for non-meteorological echoes is not used because it could remove actual precipitation information. The circular effect might be reduced by using a larger radius or a higher threshold rank but this is computationally expensive. Areas with a 10-year return level exceeding 30 mm are mainly located beyond 100 km. This is probably due to an increased contamination by hail with the distance to the radar (and the height of the measurements). The small-scale variability in the study area can be explained by uncertainties due to the sample size.

Table 6. Results of the extreme value distribution fitting for the RFA. The tables shows successively the independence (temporal or spatial), the optimal rank, the location parameter and the scale parameter.

| independence (%) | | | |
Station	QPE	BUL	REG	R10
Deurne	–	27.5	1.4	2.6
Uccle	–	28.0	1.1	2.6
Gosselies	–	22.2	1.7	3.9
Nadrin	–	19.9	2.6	7.0

| optimal rank (%) | | | |
Station	QPE	BUL	REG	R10
Deurne	28	22	100	99
Uccle	30	30	70	88
Gosselies	29	30	96	90
Nadrin	23	30	100	91

| location parameter ($mm\,h^{-1}$) | | | |
Station	QPE	BUL	REG	R10
Deurne	10.8	16.7	16.5	20.0
Uccle	11.5	17.5	21.1	24.2
Gosselies	11.9	15.2	20.4	26.5
Nadrin	12.2	12.9	21.0	29.0

| scale parameter ($mm\,h^{-1}$) | | | |
Station	QPE	BUL	REG	R10
Deurne	4.7	5.7	8.0	7.3
Uccle	6.4	4.4	11.7	10.7
Gosselies	6.4	8.7	10.1	8.6
Nadrin	6.1	9.3	11.7	9.5

There is some correlation between the 10-year return level and the scale parameter. Therefore the spatial patterns between the two return periods are similar. Within the 100 km radius, the maps are only slightly influenced by the topography and the mean annual rainfall (Journée et al., 2015). This suggests that applying our regional approach is valid, at least for 1 h duration. Van de Vyver (2012) obtained slightly lower values for the 10-year return level but a slightly higher 100-year return level due to the positive shape parameter. One notes that the scale is very high around the Brussels region where the Uccle station is located.

5 Conclusions

5.1 Results

The potential of a radar-based precipitation dataset to study extreme rainfall at a given location is evaluated. The QPE is obtained by a careful processing of the volumetric reflectivity measurements from a single weather radar in Belgium. The radar dataset covers the period 2005–2016, has a resolution of 1 km and is available every 5 min.

The first evaluation is based on a comparison of the extreme statistics between the radar dataset and two automatic rain gauge networks with 10 min and 1 h resolution, respectively. For each network, two locations are chosen to study sliding 1 and 24 h extremes using the collocated radar estimation. A regression method in Q-Q plots is used to fit an exponential distribution to independent peaks. This method has the property to focus on the tail of the extreme value distribution, which is of interest when studying extremes. An optimal threshold rank is selected by minimizing the MSE of the regression.

The 10 highest 1 h extremes occurred in summer and are well captured by both the radar and the gauge. A few problematic events are caused by wind drift or severe radar signal attenuation and should be considered as missing data. Differences up to 30 % between the gauge and radar values are observed and can be explained by spatial sampling and estimation errors. The radar extremes tend to be lower than the gauge extremes especially for short return periods. This is consistent with the results of Peleg et al. (2016) on the small-scale spatial variability of extreme rainfall. In particular, tipping-bucket gauges underestimate the heavy rainfall rate and can be blocked by accumulated snow. The radar underestimates due to signal attenuation and overestimates in the case of hail. Additional radar uncertainties come from time sampling and the $Z-R$ relationship. Despite the uncertainties in the datasets, the fitting of the exponential distribution to the QPE product is within the large uncertainty bound of the AWS one. This result is in accordance with the fact that the temporal variability (related to the sample size) is higher than the spatial variability (Peleg et al., 2017).

For the 24 h accumulation there is a mix of summer and winter events, with more of the latter for stations with higher altitude. There is a clear benefit of bias correction for the highest station, making the distribution fits similar for both stations. For both 1 and 24 h accumulations, the basic radar product exhibits unrealistic high extremes, which results in an overestimated scale parameter. Such a product is therefore not suitable for an extreme value analysis.

In the second evaluation, a RFA is applied to 1 h radar data at the location of four pluviographs with recordings from 1965 to 2010. Spatially independent extremes within a circle of 20 km are selected using a novel approach. They are fitted with a maximum threshold rank extended from 30 to 100 thanks to the increased sample size. There is a good agreement between the radar and the gauge for the two closest stations. The most important result is that the uncertainty is significantly lower using the available radar data. The extremes are lower when a decorrelation distance of 10 km is assumed suggesting that this hypothesis is not valid. In Uccle, the radar extremes and therefore the scale parameter are significantly higher. This can be attributed partially to radar overestimation due to hail and gauge underestimations, but the increasing urban heat island effect should not be ruled out. The decreasing tail of the radar extremes is at least partially caused by the hail threshold but a physical limit for the Belgium climate could play a role.

For each of the rain gauge networks, only a few stations have been selected and presented in this paper. The results from these stations are representative of the variability of the results obtained from the other stations.

The regional approach has been applied all over the study area using a 10 km radius and a fixed threshold rank of 60. The extreme statistics for the 1 h duration are slightly influenced by the topography. The reliability of the radar results beyond the 100 km range is questionable.

5.2 Prospects

There is still some room to improve the quality of the radar and gauge datasets. The recently installed weighted gauges are able to cope with intense rainfall and snowfall. One will have to wait a few decades before it can produce reliable statistics. Radar calibration errors can be mitigated by computing a monthly bias using rain gauges. The attenuation can be solved easily by using other radars when available. To avoid overestimation of the extremes, an advection correction can be used for the time sampling error. Dual-polarization radars can potentially provide a better estimation for high rainfall rates (Figueras i Ventura and Tabary, 2013). However uncertainties related to the relation between the radar measurements and the rainfall rate remain, especially in the case of hail. In this study we considered all data as the amount of liquid water at the ground. For some applications it could be necessary to take the melting of snow and hail into account. Identification of hail at ground level is a challenging problem using radar and ground station networks (Lukach et al., 2017).

Since the paper focuses on the comparison between radar and rain gauges, the extreme value analysis has been kept simple. While the EXP distribution was found to fit generally well with the empirical data, the generalized Pareto distribution should be considered as well for the RFA. The analysis of longer durations can be refined by taking into consideration the effect of the type of rainfall (e.g. Rulfova et al., 2014; Panziera et al., 2016). A bias correction should also be considered for a proper handling of the asymptotic behaviour of the distribution (Willems et al., 2007).

The extreme value theory was applied to the radar datasets by removing the spatially dependent extremes in the region of analysis. This is performed using a simple technique based on a decorrelation distance. Evin et al. (2016) decided not to use such a method because it reduces the sample size. Better performance is expected using recently proposed statistical models (Buishand et al., 2008; Davison et al., 2012).

The radar-based RFA can be extended to other durations to derive IDF curves. Note that the hypothesis of constant parameter over the region might not be valid for longer du-rations. In many applications in hydrology, it is the averaged rainfall over a given area which is relevant. A popular technique is to apply areal reduction factors to point-based statistics. The radar dataset can be used directly to derive areal rainfall statistics (e.g. Durrans et al., 2002; Overeem et al., 2010; Wright et al., 2014a).

Code and data availability. The code used in this study is part of the RMIB radar library. The rain gauge rainfall measurements and radar-based precipitation estimates are archived at the RMIB.

Competing interests. The authors declare that they have no conflict of interest.

Acknowledgements. The authors would like to thank Francesco Marra, Luca Panziera and referee no. 3 for their very constructive comments which allowed us to significantly improve the quality of the paper. We thank the hydrological service of the Walloon region (SPW) for providing their rain gauge data. The comments of Michel Journee and Hans Van de Vyver are highly appreciated.

Edited by: Uwe Ehret

References

Bardet, L., Duluc, C.-M., Rebour, V., and L'Her, J.: Regional frequency analysis of extreme storm surges along the French coast, Nat. Hazards Earth Syst. Sci., 11, 1627–1639, https://doi.org/10.5194/nhess-11-1627-2011, 2011.

Blanchet, J., Touati, J., Lawrence, D., Garavaglia, F., and Paquet, E.: Evaluation of a compound distribution based on weather pattern subsampling for extreme rainfall in Norway, Nat. Hazards Earth Syst. Sci., 15, 2653–2667, https://doi.org/10.5194/nhess-15-2653-2015, 2015.

Buishand, T.: Extreme rainfall estimation by combining data from several sites, Hydrolog. Sci. J., 36, 345–365, 1991.

Buishand, T. A., de Haan, L., and Zhou, C.: On spatial extremes: With application to a rainfall problem, Ann. Appl. Stat., 2, 624–642, https://doi.org/10.1214/08-AOAS159, 2008.

Castellarin, A.: Probabilistic envelope curves for design flood estimation at ungauged sites, Water Resour. Res., 43, W04406, https://doi.org/10.1029/2005WR004384, 2007.

Colli, M., Lanza, L. G., and La Barbera, P.: An evaluation of the uncertainty of extreme events statistics at the WMO/CIMO Lead Centre on precipitation intensity, AGU Fall Meeting Abstracts, 2012.

Cooley, D., Nychka, D., and Naveau, P.: Bayesian Spatial Modeling of Extreme Precipitation Return Levels, J. Am. Stat. Assoc., 102, 824–840, https://doi.org/10.1198/016214506000000780, 2007.

Davison, A. C., Padoan, S. A., and Ribatet, M.: Statistical Modeling of Spatial Extremes, Stat. Sci., 27, 161–186, https://doi.org/10.1214/11-sts376, 2012.

Demarée, G.: Le pluviographe centenaire du plateau d'Uccle: son histoire, ses données et ses applications, La Houille Blanche, 4, 95–102, https://doi.org/10.1051/lhb/2003082, 2003.

Duchon, C. E. and Biddle, C. J.: Undercatch of tipping-bucket gauges in high rain rate events, Adv. Geosci., 25, 11–15, https://doi.org/10.5194/adgeo-25-11-2010, 2010.

Durrans, S. R., Julian, L. T., and Yekta, M.: Estimation of depth-area relationships using radar-rainfall data, J. Hydrol. Eng., 7, 356–367, 2002.

Eldardiry, H., Habib, E., and Zhang, Y.: On the use of radar-based quantitative precipitation estimates for precipitation frequency analysis, J. Hydrol., 531, 441–453, https://doi.org/10.1016/j.jhydrol.2015.05.016, 2015.

Evin, G., Blanchet, J., Paquet, E., Garavaglia, F., and Penot, D.: A regional model for extreme rainfall based on weather patterns subsampling, J. Hydrol., 541, 1185–1198, https://doi.org/10.1016/j.jhydrol.2016.08.024, 2016.

Figueras i Ventura, J. and Tabary, P.: The new French operational polarimetric radar rainfall rate product, J. Appl. Meteorol. Clim., 52, 1817–1835, 2013.

Gellens, D.: Trend and Correlation Analysis of k-Day Extreme Precipitationover Belgium, Theor. Appl. Climatol., 66, 117–129, https://doi.org/10.1007/s007040070037, 2000.

Goudenhoofdt, E. and Delobbe, L.: Statistical Characteristics of Convective Storms in Belgium Derived from Volumetric Weather Radar Observations, J. Appl. Meteorol. Clim., 52, 918–934, https://doi.org/10.1175/JAMC-D-12-079.1, 2013.

Goudenhoofdt, E. and Delobbe, L.: Generation and Verification of Rainfall Estimates from 10-Yr Volumetric Weather Radar Measurements, J. Hydrometeorol., 17, 1223–1242, https://doi.org/10.1175/JHM-D-15-0166.1, 2016.

Haberlandt, U. and Berndt, C.: The value of weather radar data for the estimation of design storms – an analysis for the Hannover region, Proceedings of the International Association of Hydrological Sciences, 373, 81–85, 2016.

Hamdi, R. and Van de Vyver, H.: Estimating urban heat island effects on near-surface air temperature records of Uccle (Brussels, Belgium): an observational and modeling study, Adv. Sci. Res., 6, 27–34, https://doi.org/10.5194/asr-6-27-2011, 2011.

Hosking, J. R. M. and Wallis, J. R.: The effect of intersite dependence on regional flood frequency analysis, Water Resour. Res., 24, 588–600, https://doi.org/10.1029/WR024i004p00588, 1988.

Jakob, D., Karoly, D. J., and Seed, A.: Non-stationarity in daily and sub-daily intense rainfall – Part 1: Sydney, Australia, Nat. Hazards Earth Syst. Sci., 11, 2263–2271, https://doi.org/10.5194/nhess-11-2263-2011, 2011.

Journée, M., Delvaux, C., and Bertrand, C.: Precipitation climate maps of Belgium, Adv. Sci. Res., 12, 73–78, https://doi.org/10.5194/asr-12-73-2015, 2015.

Lukach, M., Foresti, L., Giot, O., and Delobbe, L.: Estimating the occurrence and severity of hail based on 10 years of observations from weather radar in Belgium, Meteorol. Appl., 24, 250–259, 2017.

Marra, F. and Morin, E.: Use of radar {QPE} for the derivation of Intensity-Duration-Frequency curves in a range of climatic regimes, J. Hydrol., 531, 427–440, https://doi.org/10.1016/j.jhydrol.2015.08.064, 2015.

Marra, F., Morin, E., Peleg, N., Mei, Y., and Anagnostou, E. N.: Intensity-duration-frequency curves from remote sensing rainfall estimates: comparing satellite and weather radar over the eastern Mediterranean, Hydrol. Earth Syst. Sci., 21, 2389–2404, https://doi.org/10.5194/hess-21-2389-2017, 2017.

Martins, E. S. and Stedinger, J. R.: Generalized maximum-likelihood generalized extreme-value quantile estimators for hydrologic data, Water Resour. Res., 36, 737–744, https://doi.org/10.1029/1999WR900330, 2000.

Ntegeka, V. and Willems, P.: Trends and multidecadal oscillations in rainfall extremes, based on a more than 100-year time series of 10 m in rainfall intensities at Uccle, Belgium, Water Resour. Res., 44, W07402, https://doi.org/10.1029/2007WR006471, 2008.

Nystuen, J. A.: Relative Performance of Automatic Rain Gauges under Different Rainfall Conditions, J. Atmos. Ocean. Tech., 16, 1025–1043, https://doi.org/10.1175/1520-0426(1999)016<1025:RPOARG>2.0.CO;2, 1999.

Ootegem, L. V., Herck, K. V., Creten, T., Verhofstadt, E., Foresti, L., Goudenhoofdt, E., Reyniers, M., Delobbe, L., Tuyls, D. M., and Willems, P.: Exploring the potential of multivariate depth-damage and rainfall-damage models, J. Flood Risk Manage., https://doi.org/10.1111/jfr3.12284, online first, 2016.

Overeem, A., Buishand, T. A., and Holleman, I.: Extreme rainfall analysis and estimation of depth-duration-frequency curves using weather radar, Water Resour. Res., 45, W10424, https://doi.org/10.1029/2009WR007869, 2009.

Overeem, A., Buishand, T. A., Holleman, I., and Uijlenhoet, R.: Extreme value modeling of areal rainfall from weather radar, Water Resour. Res., 46, W09514, https://doi.org/10.1029/2009WR008517, 2010.

Paixao, E., Mirza, M. M. Q., Shephard, M. W., Auld, H., Klaassen, J., and Smith, G.: An integrated approach for identifying homogeneous regions of extreme rainfall events and estimating {IDF} curves in Southern Ontario, Canada: Incorporating radar observations, J. Hydrol., 528, 734–750, https://doi.org/10.1016/j.jhydrol.2015.06.015, 2015.

Panziera, L., Gabella, M., Zanini, S., Hering, A., Germann, U., and Berne, A.: A radar-based regional extreme rainfall analysis to derive the thresholds for a novel automatic alert system in Switzerland, Hydrol. Earth Syst. Sci., 20, 2317–2332, https://doi.org/10.5194/hess-20-2317-2016, 2016.

Peleg, N., Marra, F., Fatichi, S., Paschalis, A., Molnar, P., and Burlando, P.: Spatial variability of extreme rainfall at radar subpixel scale, J. Hydrol., https://doi.org/10.1016/j.jhydrol.2016.05.033, online first, 2016.

Peleg, N., Blumensaat, F., Molnar, P., Fatichi, S., and Burlando, P.: Partitioning the impacts of spatial and climatological rainfall variability in urban drainage modeling, Hydrol. Earth Syst. Sci., 21, 1559–1572, https://doi.org/10.5194/hess-21-1559-2017, 2017.

Pickands III, J.: Statistical inference using extreme order statistics, Ann. Stat., 3, 119–131, 1975.

Roth, M., Buishand, T., Jongbloed, G., Tank, A. K., and van Zanten, J.: Projections of precipitation extremes based on a regional, non-stationary peaks-over-threshold approach: A case study for the Netherlands and north-western Germany, Weather and Climate

Extremes, 4, 1–10, https://doi.org/10.1016/j.wace.2014.01.001, 2014.

Roth, M., Jongbloed, G., and Buishand, T.: Threshold selection for regional peaks-over-threshold data, J. Appl. Stat., 43, 1291–1309, https://doi.org/10.1080/02664763.2015.1100589, 2015.

Rulfova, Z., Buishand, A., Kysely, J., and Roth, M.: Two-Component Extreme Value Distributions for Convective and Stratiform Precipitation, AGU Fall Meeting Abstracts, 2014.

Saito, H. and Matsuyama, H.: Probable Hourly Precipitation and Soil Water Index for 50-yr Recurrence Interval over the Japanese Archipelago, SOLA, 11, 118–123, https://doi.org/10.2151/sola.2015-028, 2015.

Sapiano, M. R. P. and Arkin, P. A.: An Intercomparison and Validation of High-Resolution Satellite Precipitation Estimates with 3-Hourly Gauge Data, J. Hydrometeorol., 10, 149–166, https://doi.org/10.1175/2008JHM1052.1, 2009.

Sveinsson, O. G. B., Boes, D. C., and Salas, J. D.: Population index flood method for regional frequency analysis, Water Resour. Res., 37, 2733–2748, https://doi.org/10.1029/2001wr000321, 2001.

Svensson, C. and Jones, D. A.: Review of rainfall frequency estimation methods, J. Flood Risk Manage., 3, 296–313, https://doi.org/10.1111/j.1753-318X.2010.01079.x, 2010.

Uboldi, F., Sulis, A. N., Lussana, C., Cislaghi, M., and Russo, M.: A spatial bootstrap technique for parameter estimation of rainfall annual maxima distribution, Hydrol. Earth Syst. Sci., 18, 981–995, https://doi.org/10.5194/hess-18-981-2014, 2014.

Uijlenhoet, R.: Raindrop size distributions and radar reflectivity-rain rate relationships for radar hydrology, Hydrol. Earth Syst.

Sci., 5, 615–628, https://doi.org/10.5194/hess-5-615-2001, 2001.

Van de Vyver, H.: Spatial regression models for extreme precipitation in Belgium, Water Resour. Res., 48, W09549, https://doi.org/10.1029/2011WR011707, 2012.

Vannitsem, S. and Naveau, P.: Spatial dependences among precipitation maxima over Belgium, Nonlin. Processes Geophys., 14, 621–630, https://doi.org/10.5194/npg-14-621-2007, 2007.

Weiss, J., Bernardara, P., and Benoit, M.: Modeling intersite dependence for regional frequency analysis of extreme marine events, Water Resour. Res., 50, 5926–5940, https://doi.org/10.1002/2014WR015391, 2014.

Willems, P.: Compound intensity/duration/frequency-relationships of extreme precipitation for two seasons and two storm types, J. Hydrol., 233, 189–205, https://doi.org/10.1016/S0022-1694(00)00233-X, 2000.

Willems, P.: Multidecadal oscillatory behaviour of rainfall extremes in Europe, Climatic Change, 120, 931–944, https://doi.org/10.1007/s10584-013-0837-x, 2013.

Willems, P., Guillou, A., and Beirlant, J.: Bias correction in hydrologic GPD based extreme value analysis by means of a slowly varying function, J. Hydrol., 338, 221–236, https://doi.org/10.1016/j.jhydrol.2007.02.035, 2007.

Wright, D. B., Smith, J. A., and Baeck, M. L.: Critical Examination of Area Reduction Factors, J. Hydrol. Eng., 19, 769, https://doi.org/10.1061/(ASCE)HE.1943-5584.0000855, 2014a.

Wright, D. B., Smith, J. A., and Baeck, M. L.: Flood frequency analysis using radar rainfall fields and stochastic storm transposition, Water Resour. Res., 50, 1592, https://doi.org/10.1002/2013WR014224, 2014b.

3

Development of streamflow drought severity–duration–frequency curves using the threshold level method

J. H. Sung[1] and E.-S. Chung[2]

[1]Ministry of Land, Infrastructure and Transport, Yeongsan River Flood Control Office, Gwangju, Republic of Korea
[2]Department of Civil Engineering, Seoul National University of Science & Technology, Seoul, 139-743, Republic of Korea

Correspondence to: E.-S. Chung (eschung@seoultech.ac.kr)

Abstract. This study developed a streamflow drought severity–duration–frequency (SDF) curve that is analogous to the well-known depth–duration–frequency (DDF) curve used for rainfall. Severity was defined as the total water deficit volume to target threshold for a given drought duration. Furthermore, this study compared the SDF curves of four threshold level methods: fixed, monthly, daily, and desired yield for water use. The fixed threshold level in this study is the 70th percentile value (Q_{70}) of the flow duration curve (FDC), which is compiled using all available daily streamflows. The monthly threshold level is the monthly varying Q_{70} values of the monthly FDC. The daily variable threshold is Q_{70} of the FDC that was obtained from the antecedent 365 daily streamflows. The desired-yield threshold that was determined by the central government consists of domestic, industrial, and agricultural water uses and environmental in-stream flow. As a result, the durations and severities from the desired-yield threshold level were completely different from those for the fixed, monthly and daily levels. In other words, the desired-yield threshold can identify streamflow droughts using the total water deficit to the hydrological and socioeconomic targets, whereas the fixed, monthly, and daily streamflow thresholds derive the deficiencies or anomalies from the average of the historical streamflow. Based on individual frequency analyses, the SDF curves for four thresholds were developed to quantify the relation among the severities, durations, and frequencies. The SDF curves from the fixed, daily, and monthly thresholds have comparatively short durations because the annual maximum durations vary from 30 to 96 days, whereas those from the desired-yield threshold have much longer durations of up to 270 days. For the additional analysis, the return-period–duration curve was

also derived to quantify the extent of the drought duration. These curves can be an effective tool to identify streamflow droughts using severities, durations, and frequencies.

1 Introduction

The rainfall deficiencies of sufficient magnitude over prolonged durations and extended areas and the subsequent reductions in the streamflow interfere with the normal agricultural and economic activities of a region, which decreases agriculture production and affects everyday life. Dracup et al. (1980) defined a drought using the following properties: (1) nature of water deficit (e.g., precipitation, soil moisture, or streamflow); (2) basic time unit of data (e.g., month, season, or year); (3) threshold to distinguish low flows from high flows while considering the mean, median, mode, or any other derived thresholds; and (4) regionalization and/or standardization. Based on these definitions, various indices were proposed over the years to identify drought. Recent studies have focused on such multi-faceted drought characteristics using various indices (Palmer, 1965; Rossi et al., 1992; McKee et al., 1993; Byun and Wilhite, 1999; Tsakiris et al., 2007; Pandey et al., 2008a, b, 2010; World Meteorological Organization, 2008; Nalbantis and Tsakiris, 2009; Wang et al., 2011; Tabari et al., 2013; Tsakiris et al., 2013).

The American Meteorological Society (1997) groups the drought definitions and types into four categories: meteorological or climatological, agricultural, hydrological, and socioeconomic droughts. The meteorological drought is a result of the absence or reduction of precipitation and short-term dryness results in an agricultural drought that severely reduces crop yields. Precipitation deficits over a prolonged period that reduce streamflow, groundwater, reservoir, and lake levels result in a hydrological drought. If hydrological droughts continue until the supply and demand of numerous economic goods are damaged, a socioeconomic drought occurs (Heim Jr., 2002).

Hydrological and socioeconomic droughts are notably difficult to approach. Nalbantis and Tsakiris (2009) defined a hydrological drought as "a significant decrease in the availability of water in all its forms, appearing in the land phase of the hydrological cycle". These forms are reflected in various hydrological variables such as streamflows, which include snowmelt and spring flow, lake and reservoir storage, recharge of aquifers, discharge from aquifers, and baseflow (Nalbantis and Tsakiris, 2009). Therefore, Tsakiris et al. (2013) described that streamflow is the key variable in describing hydrological droughts because it considers the outputs of surface runoff from the surface water subsystem, subsurface runoff from the upper and lower unsaturated zones, and baseflow from the groundwater subsystem. Furthermore, streamflow crucially affects the socioeconomic drought for several water supply activities such as hydropower generation, recreation, and irrigated agriculture, where crop growth and yield largely depends on the water availability in the stream (Heim Jr., 2002). Hence, hydrological and socioeconomic droughts are related to streamflow deficits with respect to hydrologically normal conditions or target water supplies for economic growth and social welfare.

For additional specification, Tallaksen and van Lanen (2004) defined a streamflow drought as a "sustained and regionally extensive occurrence of below average water availability". Thus, threshold level approaches, which define the duration and severity of a drought event while considering the daily, monthly, seasonal, and annual natural runoff variations, are widely applied in drought analyses (Yevjevich, 1967; Sen, 1980; Dracup et al., 1980; Dalezios et al., 2000; Kjeldsen et al., 2000; American Meteorological Society, 1997; Hisdal and Tallaksen, 2003; Wu et al., 2007; Pandey et al., 2008a; Yoo et al., 2008; Tigkas et al., 2012; van Huijgevoort et al., 2012). These approaches provide an analytical interpretation of the expected availability of river flow; a drought occurs when the streamflow falls below the threshold level. This level is frequently considered a certain percentile flow for a specific duration and assumed to be steady during the considered month, season, or year. Therefore, Kjeldsen et al. (2000) applied three variable threshold level methods using seasonal, monthly, and daily streamflows.

There has been a growing need for new planning and design of natural resources and environment based on the aforementioned scientific trends. For design purposes, rainfall intensity-duration-frequency (IDF) curves have been used for a long time to synthesize the design storm. Therefore, many studies have integrated drought severity and duration based on the multivariate theory (Bonaccorso et al., 2003; González

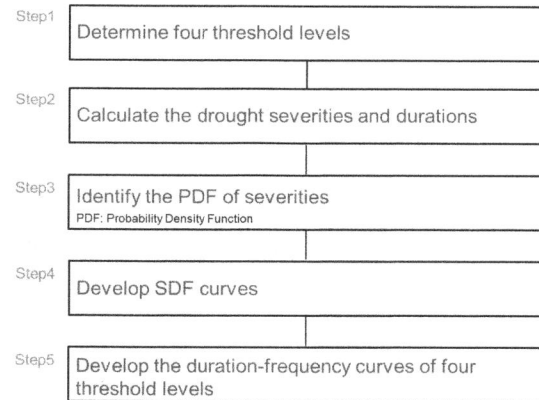

Figure 1. Procedure in this study.

and Valdés, 2003; Mishra and Singh, 2009; Song and Singh, 2010a, b; De Michele et al., 2013). However, these studies cannot fully explain droughts without considering the frequency, which resulted in the development of drought iso-severity curves for certain return periods and durations for design purposes.

Thus, based on the typical drought characteristics (water deficit and duration) and threshold levels, this study developed quantitative relations among drought parameters, namely, severity, duration, and frequency. This study quantified the streamflow drought severity, which is closely related to hydrological and socioeconomic droughts, using fixed, monthly, daily, and desired-yield threshold levels. Furthermore, this study proposed a streamflow SDF curve using the traditional frequency analyses. In addition, this study also developed duration frequency curves of four threshold levels from the occurrence probabilities of various duration events using a general frequency analysis because the deficit volume is not sufficient to explain the extreme droughts. This framework was applied to the Seomjin River basin in South Korea.

2 Methodology

2.1 Procedure

This study consists of five steps as shown in Fig. 1. Step 1 determines the threshold levels for the fixed, monthly, daily, and desired-yield levels for water use. The threshold selection description is shown in Sect. 2.3. Step 2 calculates the severities (total water deficits) and durations for all drought events at the four threshold levels. The method to derive the severity and duration is shown in Sect. 2.2. Step 3 derives the annual maxima of severity and duration and identifies the best-fit probability distribution functions using the L-moment ratio diagrams (Hosking and Wallis, 1997). The calculation procedure is shown in Sect. 2.4 using related equations and

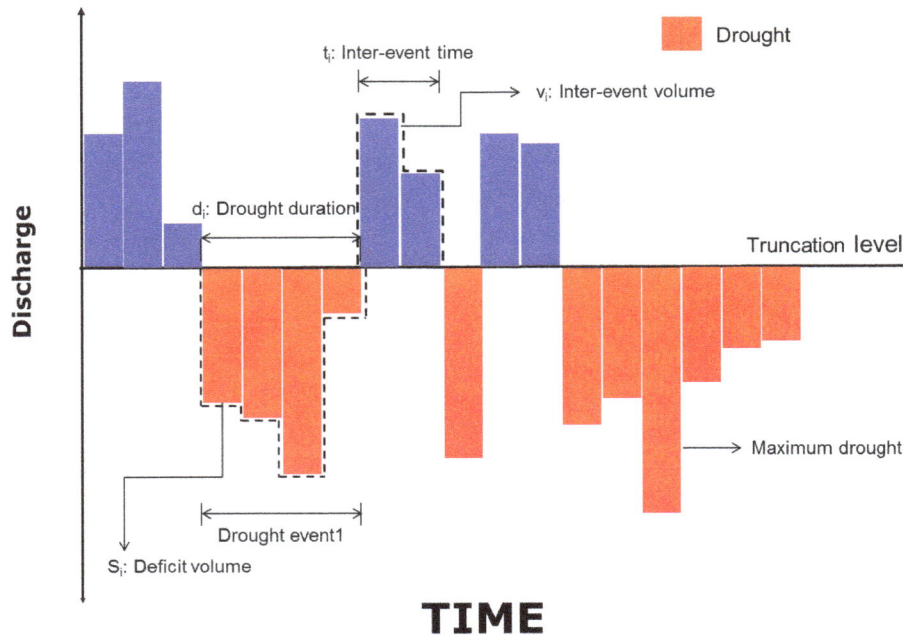

Figure 2. Definition sketch of a general drought event.

descriptions. Step 4 calculates the streamflow drought severities using the selected probability distribution with the best-fit parameters and develops the SDF curves. This step is described in Sect. 2.5. Step 5 develops the duration–frequency curves of the four threshold levels using an appropriate probability distribution.

2.2 Streamflow drought severity

In temperate regions where the runoff values are typically larger than zero, the most widely used method to estimate a hydrological drought is the threshold level approach (Yevjevich, 1967; Fleig et al., 2006; Tallaksen et al., 2009; Van Loon and Van Lanen, 2012). The streamflow drought severity with the threshold level method has the following advantages over the standardized precipitation index (SPI) in meteorology (Yoo et al., 2008) and the Palmer drought severity index (PDSI) in meteorology and agriculture (Dalezios et al., 2000): (1) no a priori knowledge of probability distributions is required, and (2) the drought characteristics such as frequency, duration, and severity are directly determined if the threshold is set using drought-affected sectors.

A sequence of drought events can be obtained using the streamflow and threshold levels. Each drought event is characterized by its duration D_i, deficit volume (or severity) S_i, and time of occurrence T_i as shown by the definition sketch in Fig. 2. With a prolonged dry period, the long drought spell is divided into several minor drought events. Because these droughts are mutually dependent, Tallaksen et al. (1997) proposed that an independent sequence of drought events must be described using some type of pooling as described below.

If the "inter-event" time t_i between two droughts of duration d_i and d_{i+1} and severity s_i and s_{i+1}, respectively, are less than the predefined critical duration t_c and the pre-allowed inter-event excess volume z_c, then the mutually dependent drought events are pooled to form a drought event as (Zelenhasic and Salvai, 1987; Tallaksen et al., 1997)

$$d_{\text{pool}} = d_i + d_{i+1} + t_c$$
$$s_{\text{pool}} = s_i + s_{i+1} - z_c. \tag{1}$$

This study assumed $t_c = 3$ days and $z_c = 10\%$ of d_i or d_{i+1} for simplicity.

2.3 Threshold selection

The threshold may be fixed or vary over the course of a year. A threshold is considered fixed if a constant value is used for the entire series and variable if it varies over the year based on the monthly and daily variable levels (Hisdal and Tallaksen, 2003). If the threshold is derived from the flow duration curve (FDC), the entire streamflow record is used in its derivation. As shown in Fig. 3, which is obtained from the study area, fixed and monthly thresholds can be obtained from an FDC and twelve monthly FDCs based on the entire record period. The daily varying threshold can be derived using the antecedent 365-day streamflow.

The threshold choice is influenced by the study objective, region, and available data. In general, a percentile of the data can be used as the threshold. Relatively low thresholds in the range of Q_{70}–Q_{95} are often used for perennial rivers (Kjeldsen et al., 2000). The fixed threshold level in this study

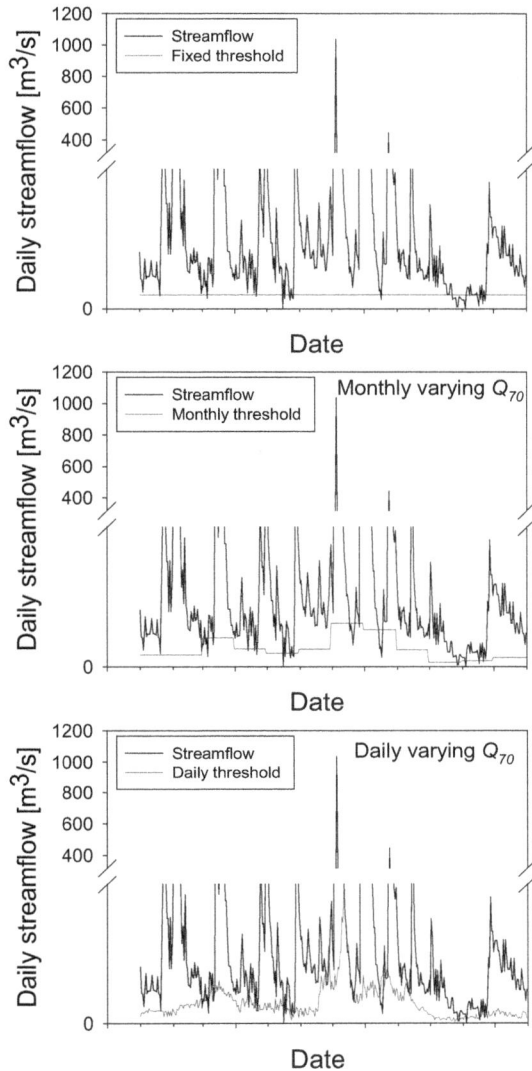

Figure 3. Examples of threshold levels: fixed (top panel), monthly varying (middle panel), and daily varying (bottom panel).

is the 70th percentile value (Q_{70}) of FDC, which is compiled using all available daily streamflows, and the monthly threshold level is the monthly varying Q_{70}s of each month's FDC. The daily variable threshold is the Q_{70} value of the FDC, which is obtained from the antecedent 365 daily streamflows. However, the threshold selection should be further analyzed because it is not clear that Q_{70} should be used as a representative threshold for rivers in a monsoon climate.

The time resolution, i.e., whether to apply a series of annual, monthly, or daily streamflows, depends on the hydrologic regime in the region of interest. In a temperate zone, a given year may include both severe droughts (seasonal droughts) and months with abundant streamflow, which indicates that the annual data do not often reveal severe droughts. Dry regions are more likely to experience droughts that last

for several years, i.e., multi-year droughts, which supports the use of a monthly or annual time step. Hence, different time resolutions may lead to different results regarding the drought event selection. This study used the daily streamflow data, and various time resolutions (30, 60, 90, 120, 150, 180, 210, 240, and 270 days) were selected to identify the temporal characteristics.

The variable threshold approach is adapted to detect streamflow deviations for both high- and low-flow seasons. Lower than average flows during high-flow seasons may be important for later drought development. However, periods with relatively low flow either during the high-flow season, which can be caused by a delayed onset of a snowmelt flood, are not commonly considered a drought. Therefore, the events that are defined with the varying threshold should be called streamflow deficiencies or streamflow anomalies instead of streamflow droughts (Hisdal et al., 2004). In contrast, the desired yield for sufficient water supply and environmental in-stream flow can be an effective method to identify a streamflow drought by considering hydrological and socioeconomic demands because environmental in-stream flow has become important in recent years.

2.4 Probability distribution function

An L-moment diagram for various goodness-of-fit techniques was used to evaluate the best probability distribution function for data sets in several recent studies (Hosking, 1990; Chowdhury et al., 1991; Vogel and Fennessey, 1993; Hosking and Wallis, 1997). The L-moment ratio diagram is a graph where the sample L-moment ratios, L-skewness (τ_3), and L-kurtosis (τ_4) are plotted as a scatterplot and compared with the theoretical L-moment ratio curves of the candidate distributions. The L-moment ratio diagrams were suggested as a useful graphical tool to discriminate amongst candidate distributions for a data set (Hosking and Wallis, 1997). The sample average and line of best fit were used to select statistical distributions, and they can be plotted on the same graph to select the best-fit distribution.

When plotting an L-moment ratio diagram, the relation among the parameters and the L-moment ratios τ_3 and τ_4 for several distributions are required. For a generalized extreme value (GEV) distribution, the three-parameter GEV distribution described by Stedinger et al. (1993) has the following probability density function (PDF, $f(x)$) and cumulative distribution function (CDF, $F(x)$):

$$f(x) = \frac{1}{\alpha} \left\{ 1 - \frac{\kappa}{\alpha}(x - \xi) \right\}^{1/\kappa - 1}$$
$$\cdot \exp\left[-\left\{ 1 - \frac{\kappa}{\alpha}(x - \xi) \right\}^{1/k} \right] \quad \kappa \neq 0, \tag{2a}$$

$$f(x) = \frac{1}{\alpha} \exp\left\{ -\frac{x-\xi}{\alpha} - \exp\left(-\frac{x-\xi}{\alpha} \right) \right\} \quad \kappa = 0, \tag{2b}$$

$$F(x) = \exp\left[-\left\{ 1 - \frac{\kappa}{\alpha}(x - \xi) \right\}^{1/\kappa} \right] \quad \kappa \neq 0, \tag{3a}$$

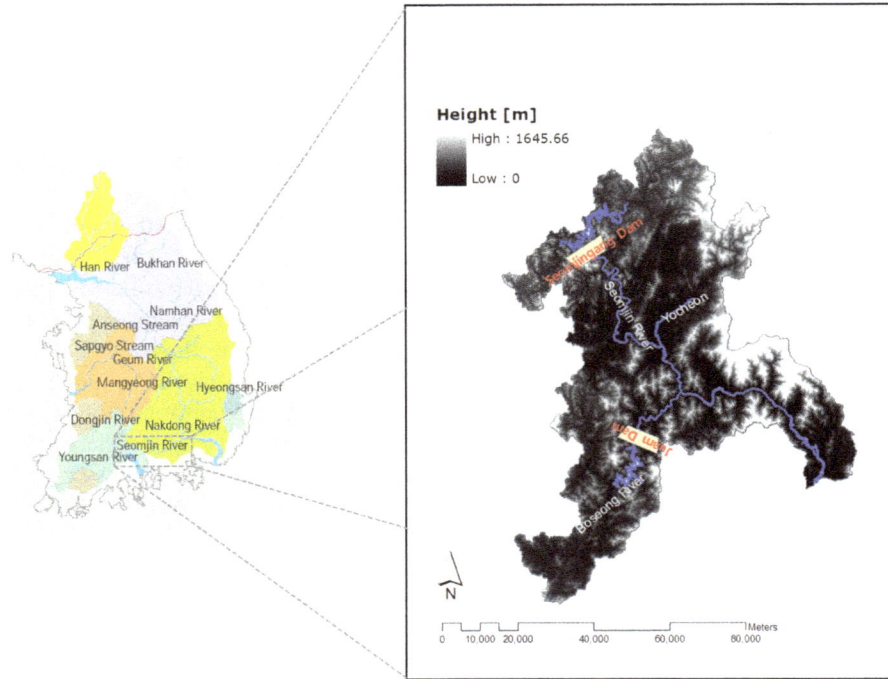

Figure 4. Location of the selected river basin, including elevation and rivers.

$$F(x) = \exp\left\{-\exp\left(-\frac{x-\xi}{\alpha}\right)\right\} \quad \kappa = 0, \tag{3b}$$

where $\xi + \alpha/\kappa \leq x \leq \infty$ for $\kappa < 0$, $-\infty \leq x \leq \infty$ for $\kappa = 0$, and $-\infty \leq x \leq \xi + \alpha/\kappa$ for $\kappa > 0$. Here, ξ is a location, α is a scale, and κ is a shape parameter. For $\kappa = 0$, the GEV distribution reduces to the classic Gumbel (EV1) distribution with $\tau_3 = 0.17$. Hosking and Wallis (1997) provided more detailed information regarding the GEV distribution. The relation among the parameters and τ_3 and τ_4 for the GEV distribution of the shape parameters can be obtained as follows (Hosking and Wallis, 1997):

$$\tau_3 = \frac{2\left(1 - 3^{-\kappa}\right)}{\left(1 - 2^{-\kappa}\right)} - 3 \tag{4a}$$

$$\tau_4 = \frac{5\left(1 - 4^{-\kappa}\right) - 10\left(1 - 3^{-\kappa}\right) + 6\left(1 - 2^{-\kappa}\right)}{\left(1 - 2^{-\kappa}\right)}. \tag{4b}$$

2.5 Development of the SDF relationships

The IDF or depth–duration–frequency (DDF) curves can be defined to "allow calculation of the average design rainfall intensity (or depth) for a given exceedance probability over a range of durations" (Stedinger et al., 1993). Statistical frequency analyses such as rainfall analyses are frequently used for drought events. However, this method cannot fully explain droughts without considering the severity and duration, which resulted in the development of the SDF curve. Thus,

extreme drought events can be specified using the frequency, duration, and either depth or mean intensity (i.e., severity). The frequency is usually described by the return period of the drought. Because its magnitude is given by the total depth that occurs in a particular duration, the SDF relation can be derived. To estimate the return periods of drought events of a particular depth and duration, the frequency distributions can be used (Dalezios et al., 2000).

3 Study region

The Seomjin River basin is located in southwestern Korea (Fig. 4). The area and total length of Seomjin River are approximately 4911.9 km^2 and 212.3 km, respectively. The altitude range is notably large, spanning from approximately 0 to 1646 m (Fig. 4). The climate of South Korea is characterized by extreme seasonal variations. Winter is cold and dry under the dominant influence of the Siberian air mass, whereas the summer is hot and humid with frequent heavy rainfalls, which are associated with the East Asian monsoon. In the Seomjin River basin, the measured precipitation is mainly concentrated in summer, and the measured mean annual precipitation varied from $< 1350\,\text{mm}\,\text{yr}^{-1}$ (in the northern region) to $> 1600\,\text{mm}\,\text{yr}^{-1}$ (in the southeastern region) during the 1975–2012 observation period. In general, approximately 60 % of the annual precipitation occurs during the wet season (July through September) in South Korea. This extreme seasonality in the precipitation causes periodic shortages of

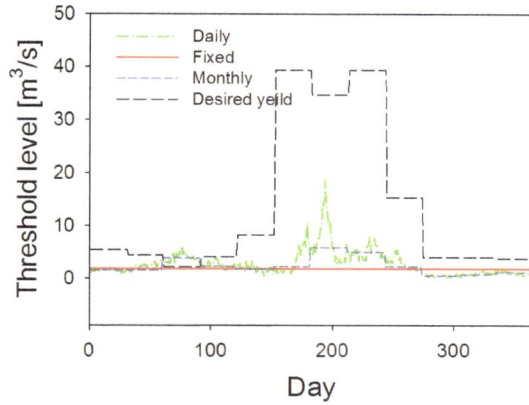

Figure 5. Comparison of the four threshold levels in this study.

Table 1. Monthly average of the four threshold levels.

	Threshold level [$m^3 s^{-1}$]			
	Fixed	Monthly	Daily	Desired yield
Jan	1.9	1.6	1.5	5.4
Feb	1.9	1.6	2.4	4.5
Mar	1.9	3.9	3.9	2.2
Apr	1.9	2.4	2.5	4.1
May	1.9	1.8	1.9	8.2
Jun	1.9	2.4	3.4	39.4
Jul	1.9	5.9	7.1	34.7
Aug	1.9	5.0	5.1	39.4
Sep	1.9	2.3	2.9	15.4
Oct	1.9	0.6	0.7	4.0
Nov	1.9	0.8	0.9	4.0
Dec	1.9	1.2	1.2	3.8

water during the dry season (October through March) and flood damage during the wet season.

The administrative districts where the basin is located cover three provinces, four cities, and 11 counties (Namwon City, Jinan County, Imsil County, and Sunchang County in the northern Jeolla Province; Suncheon City, Gwangyang City, Damyang County, Gokseong County, Gurye County, Hwasun County, Boseong County, and Jangheung County in the southern Jeolla Province; and Hadong County in the southern Gyeongsang Province). The influx rates into the basin from these province are 47 % (southern Jeolla Province), 44 % (northern Jeolla Province), and 9 % (southern Gyeongsang Province), and a total of 321 104 residents, who occupy 129 322 households, live in these areas.

The land use consists of arable land (876.29 km^2), forest land (3400.61 km^2), urban area (67.12 km^2), and other land uses (567.86 km^2). Major droughts occurred in the southern Jeolla Province from 1967 to 1968 and from 1994 to 1995. The Seomjin River basin had <1,000 mm of precipitation on average in 1977, 1988, 1994, and 2008. Among these years, the annual precipitation in 1988 was only 782.7 mm (56.5 %) of the annual average of 1385.5 mm from 1967 to 2008, which represents a severe drought.

4 Results

4.1 Determination of the threshold levels

This study used four threshold levels. The fixed threshold level is Q_{70} of the FDC, which resulted from 37 year daily streamflows. The monthly thresholds are twelve Q_{70} values of monthly FDCs, which incorporated the data of all daily streamflows from January to December for the past 37 years. The daily threshold is Q_{70} of the FDCs, which resulted from the antecedent 365 daily streamflows. Thus, the daily threshold level smoothly varies everyday. The desired-yield threshold for a sufficient water supply and environmental in-stream flow was determined by the Korean central government. This

threshold is related to social and economic droughts because it associates the supply and demand of a number of economic goods and environmental safety. The desired-yield threshold is considerably different from the other levels and represents more realistic conditions because the desired yield is equivalent to the planned water supply.

The four calculated thresholds are presented in Fig. 5, and the specific monthly averaged values are listed in Table 1. The average levels were 1.9, 2.5, 2.8, and 13.8 m^3 s^{-1} for the fixed, monthly, daily and desired-yield levels, respectively. The daily threshold levels, which significantly fluctuated because of the natural streamflow variations during the antecedent 365 days, were the largest among the four threshold levels because a summer period (June, July, and August) was considered. The desired-yield level was larger than the fixed, monthly, and daily thresholds. This phenomenon occurred during the winter in Korea, which significantly decreased both the water demand and natural runoff during the winter (December, January, and February). However, the thresholds for the daily, monthly, and desired-yield levels during the summer were much higher than those during the other seasons. The desired yield during May and June had much higher threshold levels than the other thresholds because this season had the highest agricultural water demand.

4.2 Calculations of the streamflow drought severity and duration

The durations and severities for all streamflow drought events were calculated based on the streamflow drought concept and threshold levels. The annual maxima values of duration and severity are shown in Fig. 6, and the summarized values are listed in Table 2. The maximum durations from the desired-yield threshold approach were considerably higher than those from the other thresholds because the desired yields were highest during June and July for agricultural

(a) Duration.

(b) Severity.

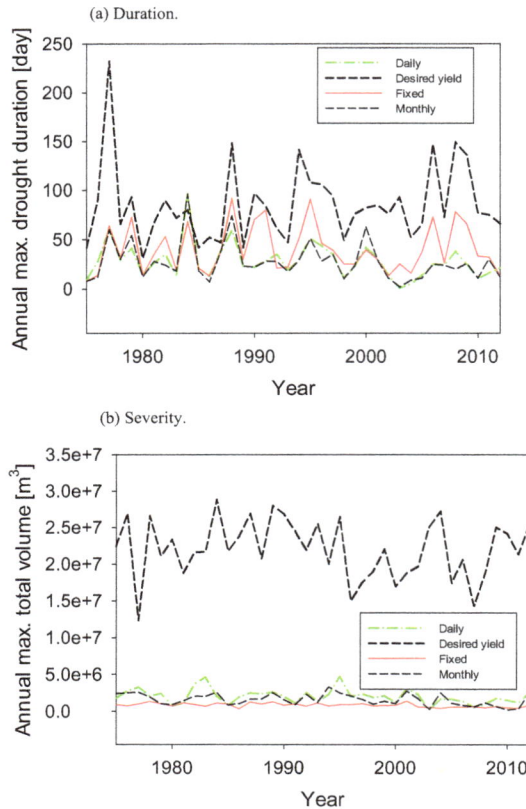

Figure 6. Time series of the annual maxima values of duration and severity.

Table 2. Summary of the four threshold approaches.

Threshold level method	Maximum duration (days)	Maximum severity (m^3)
Fixed	92	9 304 762
Monthly	96	10 774 642
Daily	96	18 457 943
Desired yield	232	285 854 400

Table 3. Correlations between the durations and the severities of the four threshold levels.

	Fixed	Monthly	Daily	Desired yield
		Duration		
Fixed	1			
Monthly	0.632	1		
Daily	0.632	0.923	1	
Desired yield	0.677	0.420	0.475	1
		Severity		
Fixed	1			
Monthly	0.441	1		
Daily	0.414	0.853	1	
Desired yield	0.281	0.551	0.599	1

water use. Similar to the results for the drought duration, the severities showed much higher values.

To compare the differences among the four threshold levels, the correlation coefficients among the water deficits from four different threshold levels were calculated as shown in Table 3. Similar trends were observed for the monthly and daily threshold levels. However, the durations and severities from the desired-yield threshold level were completely different from those for the fixed, monthly, and daily levels. In other words, the drought identification techniques based on general threshold levels cannot reflect the socioeconomic drought in terms of the water supply and demand. Therefore, two-way approaches that are categorized using the time periods (fixed, monthly, and daily) for hydrological drought and the desired-yield threshold for socioeconomic droughts should be separately included to identify specific drought characteristics.

4.3 Determination of the probability distribution function

The L-moment diagrams of various goodness-of-fit techniques were used to evaluate the best probability distribution function for the data sets. To develop a streamflow

drought SDF curve, the proper probability distribution function should be determined based on the statistical results as described in Sect. 2.4.

The L-moment ratio diagrams were derived for the four threshold approaches and are shown in Fig. 7. Among the examined distribution models, three parameter distributions (the Pearson Type 3 (PT3), Generalized Normal (GNO), and GEV distributions) appeared consistent with their data sets. In the frequency analysis that addressed extreme values, the distributions that use three parameters were required to express the upper tail. The PT3, GNO, and GEV distributions can be applied in this study. As shown in Fig. 7, this study selected the GEV distribution for a representative probability distribution because most observations are appropriate for the GEV.

4.4 Development of SDF curves

Streamflow drought SDF curves were developed using the derived probability distribution functions as shown in Fig. 8. The SDF curves described the streamflow drought severities with respect to durations and frequencies. The severity increases with increasing frequency and duration. For these plots, 10-, 20-, 50-, 80-, and 100-year-frequency severities were calculated at 30-, 60-, 90-, 120-, 150-, 180-, 210-, and 270-day durations. Because the amount of available data only corresponds to 37 years, we calculated up to a

(a) Fixed.

(b) Daily.

(c) Monthly.

(c) Desired yield.

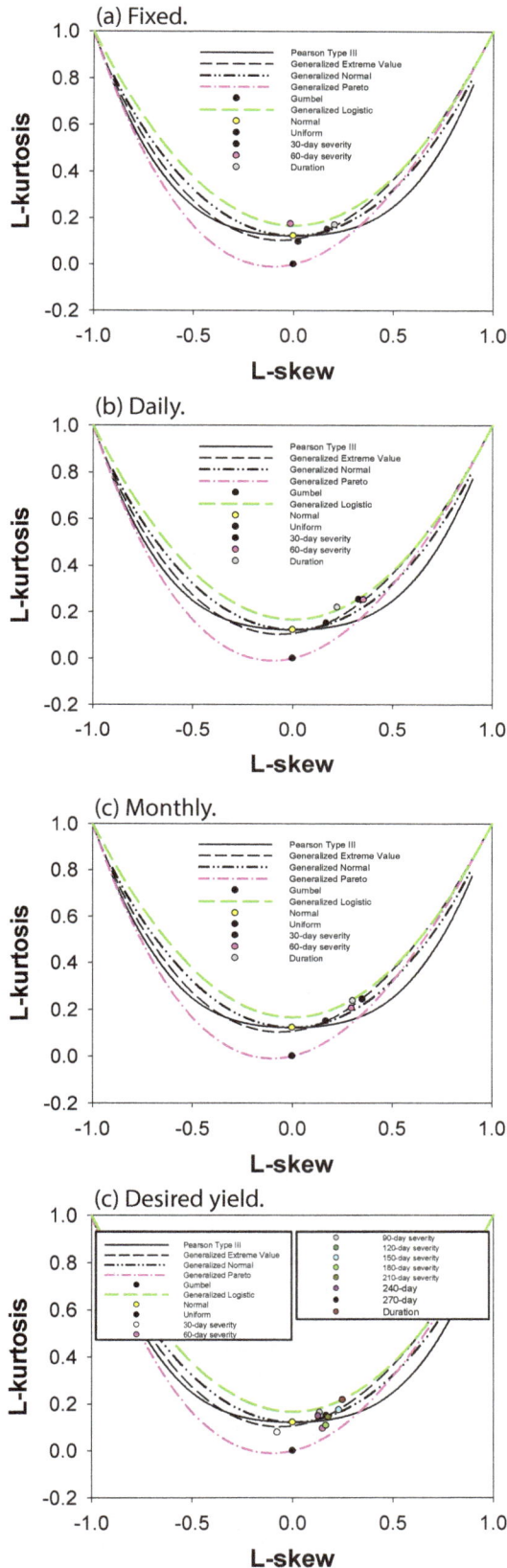

Figure 7. L-moment diagram to identify the probability distribution.

Table 4. Severity–duration–frequency of the desired yield in the Seomjin River basin.

Duration	Return period [yr]				
[day]	10	20	50	80	100
30	60.7	66.4	73.1	75.9	77.2
60	82.4	95.9	112.5	120.8	124.9
90	95.6	112.8	133.7	144.6	149.3
120	106.8	132.7	170.0	189.7	200.1
150	116.6	145.2	186.6	208.7	220.3
180	126.0	155.5	197.5	220.8	231.7
210	134.3	168.7	217.7	243.1	257.6
240	141.0	174.2	223.9	248.8	261.3
270	144.6	182.0	233.3	258.9	272.9

100-year frequency. However, the SDF curves from the fixed, daily, and monthly thresholds were calculated using comparatively short durations because the annual maximum durations vary from 30 to 96 days. Nonetheless, the SDF curve from the desired-yield levels showed the water deficits for much longer durations of 30–270 days. In addition, the water deficits from the desired-yield levels are much higher than those from other levels even for the same duration.

For a specific description, Table 4 compares all severities to specific frequencies and durations for the desired-yield threshold. When the duration increases, the severity differences among the return periods significantly increase. Therefore, because the streamflow drought severity should be more crucial when the drought continues for a longer period, the frequency of long droughts should be approached with caution.

4.5 Development of duration–frequency curve

Using the same traditional frequency analysis, the duration–frequency curves for four threshold levels were developed as shown in Fig. 9. In other words, the annual maxima durations are derived based on the four threshold level methods. As shown in the SDF relationship, the GEV distribution was selected from the L-moment ratio diagram. For these plots, 2-, 3-, 5-, 10-, 20-, 30-, 50-, 70-, 80-, and 100-year-frequency severities were calculated. Similar to the SDF curves, the durations for the desired-yield threshold were much higher than those for the other three thresholds.

5 Summary and conclusions

This study developed a useful concept to describe the characteristics of streamflow droughts using threshold level methods. The SDF curves for streamflow droughts were developed to quantify a specific volume based on a specific duration and frequency. This study compared the SDF curves of four threshold level methods: fixed, monthly, daily, and

Engineering Hydrology: Processes and Modeling

(a) Fixed.

(b) Daily.

(c) Monthly.

(d) Desired yield.

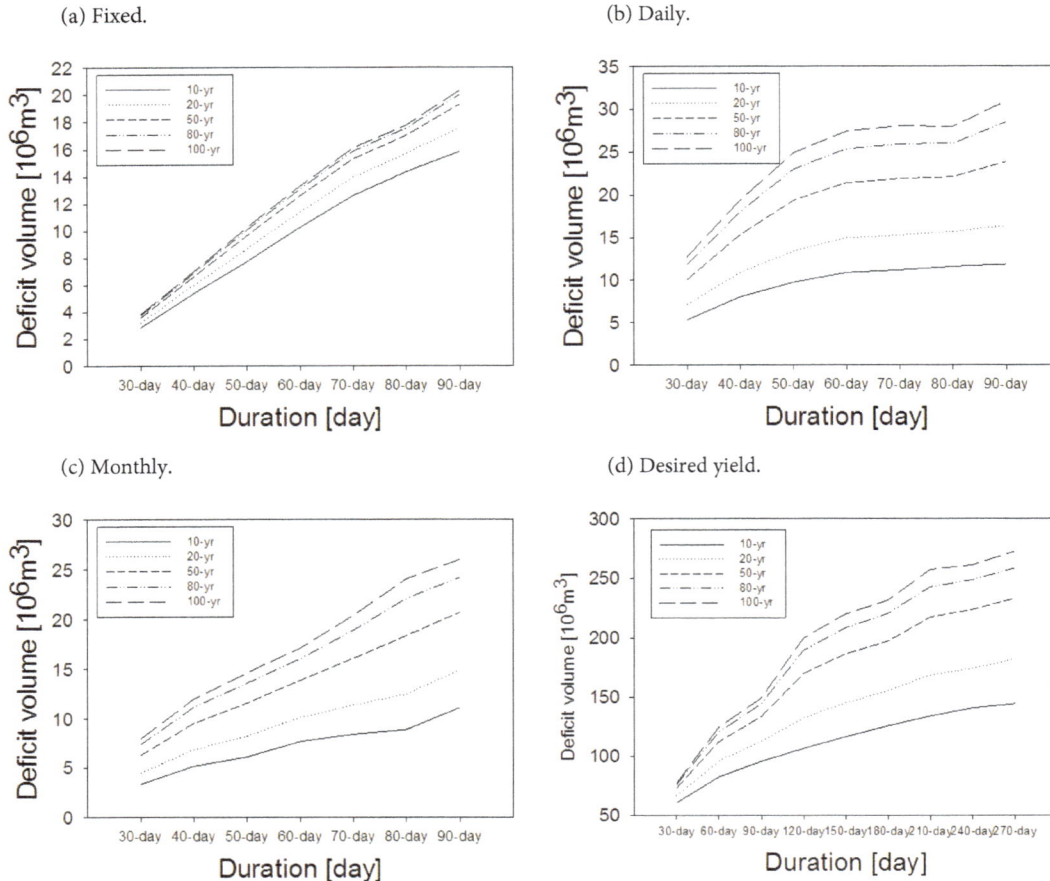

Figure 8. SDF curves of the four threshold approaches in the Seomjin River basin.

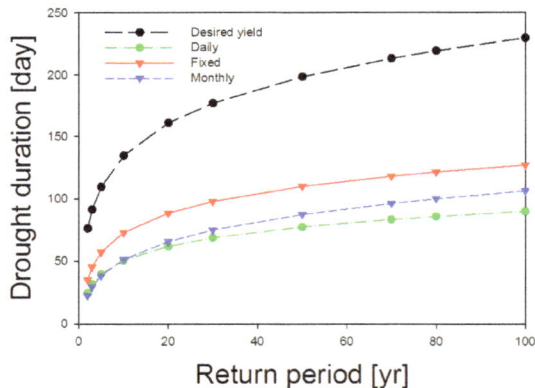

Figure 9. Duration–frequency curves of the four threshold level approaches in the Seomjin River basin.

desired-yield levels for water use. In addition, the duration–frequency curves for four thresholds were used to derive the relationship between the drought duration and the drought frequency. This study used the severity, which represents the total water deficit for specific durations. From this study, we can make the following conclusions:

1. The daily threshold levels significantly fluctuated because of the natural streamflow variations for the antecedent 365 days and were the largest threshold level because a summer period (June, July, and August) was considered. The desired-yield level was larger than the fixed, monthly, and daily thresholds. This phenomenon occurred during the winter in Korea; thus, both the water demand and natural runoff during the winter (December, January, and February) were notably small.

2. The durations and severities from the desired-yield threshold level were completely different from those for the fixed, monthly, and daily levels. In other words, the desired-yield threshold can identify streamflow droughts using the total water deficit to the hydrological and socioeconomic targets, whereas the fixed, monthly, and daily streamflow thresholds derive the deficiencies or anomalies from the average of historical streamflow.

3. The GEV distribution for a representative probability distribution was selected for the streamflow drought severities because most observations are appropriate for the GEV.

4. The severities increased with increasing duration and frequency. However, these values were notably different because the four threshold level approaches defined the streamflow drought differently. The SDF curves from the fixed, daily, and monthly thresholds were calculated using comparatively short durations because the annual maximum durations vary from 30 to 96 days. However, the SDF curve from the desired-yield levels shows the water deficits for longer durations of 30–270 days. In addition, the water deficits from the desired-yield levels are significantly higher than those from the others even in the same duration.

5. For the SDF curve of the desired-yield threshold, when the duration increases, the severity differences among return periods significantly increase. Therefore, because the streamflow drought severity should be more crucial when the drought continues for a longer period, the frequency of long droughts should be approached with caution.

6. Duration–frequency curves for four threshold levels were also developed to quantify the streamflow drought duration. Similar to the SDF curves, the desired-yield level had much longer durations for the other three thresholds.

7. In the end, the drought identification techniques based on the general threshold levels cannot reflect the socio-economic drought in terms of water supply and demand. Therefore, the two-way approaches that are categorized by the time periods (fixed, monthly, and daily) for hydrological drought and the desired-yield threshold for socioeconomic drought should be separately included to identify specific drought characteristics.

The streamflow drought SDF curves that were developed in this study can be used to quantify the water deficit for natural streams and reservoirs. In addition, these curves will be extended to allow for regional frequency analyses, which can estimate the streamflow drought severity at ungauged sites. Therefore, they can be an effective tool to identify any streamflow droughts using the severity, duration, and frequency.

Acknowledgements. This study was supported by funding from the Basic Science Research Program of the National Research Foundation of Korea (2010-0010609).

Edited by: C. De Michele

References

American Meteorological Society: Meteorological drought – Policy statement, B. Am. Meteorol. Soc., 78, 847–849, 1997.

Bonaccorso, B., Cancelliere, A., and Rossi, G.: An analytical formulation of return period of drought severity, Stoch. Environ. Res. Risk A., 17, 157–174, 2003.

Byun, H.-R. and Wilhite, D. A.: Objective quantification of drought severity and duration, J. Climate, 12, 747–756, 1999.

Chowdhury, J. U., Stedinger, J. R., and Lu, L.-H.: Goodness-of-fit tests for regional generalized extreme value flood distributions, Water Resour. Res., 27, 1765–1776, 1991.

Dalezios, N., Loukas, A., Vasiliades, L., and Liakopolos, E.: Severity-duration-frequency analysis of droughts and wet periods in Greece, Hydrolog. Sci. J., 45, 751–769, 2000.

De Michele, C., Salvadori, G., Vezzoli, R., and Pecora, S.: Multivariate assessment of droughts: Frequency analysis and dynamic return period, Water Resour. Res., 49, 6985–6994, 2013.

Dracup, J. A., Lee, K. S., and Paulson Jr., E. G.: On the statistical characteristics of drought events, Water Resour. Res., 16, 289–296, 1980.

Fleig, A. K., Tallaksen, L. M., Hisdal, H., and Demuth, S.: A global evaluation of streamflow drought characteristics, Hydrol. Earth Syst. Sci., 10, 535–552, doi:10.5194/hess-10-535-2006, 2006.

González, J. and Valdés, J. B.: Bivariate drought recurrence analysis using tree ring reconstructions, J. Hydrol. Eng., 8, 247–258, 2003.

Heim Jr., R. R.: A review of twentieth-century drought indices used in the United States, B. Am. Meteorol. Soc., 83, 1149–1165, 2002.

Hisdal, H. and Tallaksen, L. M.: Estimation of regional meteorological and hydrological drought characteristics: A case study for Denmark, J. Hydrol., 281, 230–247, 2003.

Hisdal, H., Tallaksen, L. M., Clausen, B., and Alan, E. P.: Ch. 5 Hydrological drought characteristics, in: Hydrological Droughts: Processes and Estimation Methods for Streamflow and Groundwater, Developments in Water Science, Elsevier, Amsterdam, 139–198, 2004.

Hosking, J. R. M.: L-moments: Analysis and estimation of distributions using linear combinations of order statistics, J. Roy. Stat. Soc. Ser. B, 52, 105–124, 1990.

Hosking, J. R. M. and Wallis, J. R.: Regional Frequency Analysis: An Approach Based on L-Moments, Cambridge Univ. Press, New York, 1997.

Kjeldsen, T. R., Lundorf, A., and Dan, R.: Use of two component exponential distribution in partial duration modeling of hydrological droughts in Zimbabwean rivers, Hydrolog. Sci. J., 45, 285–298, 2000.

McKee, T. B., Doesken, N. J., and Kleist, J.: The relationship of drought frequency and duration to time scales, Proc. 8th Conf. Appl. Climatol., American Meteor. Soc., Boston, 179–184, 1993.

Mishra, A. K. and Singh, V. P.: Analysis of drought severity.area.frequency curves using a general circulation model and scenario uncertainty, J. Geophys. Res., 114, D06120, doi:10.1029/2008JD010986, 2009.

Nalbantis, I. and Tsakiris, G.: Assessment of hydrological drought revisited, Water Resour. Manage., 23, 881–897, 2009.

Palmer, W.C.: Meteorological drought, Research Paper No. 45, US Department of Commerce Weather Bureau, Washington, D.C., 1965.

Pandey, R. P., Mishra, S. K., Singh, R., and Ramasastri, K. S.: Streamflow drought severity analysis of Betwa river system (INDIA), Water Resour. Manage., 22, 1127–1141, 2008a.

Pandey, R. P., Sharma, K. D., Mishra, S. K., Singh, R., and Galkate, R. V.: Assessing streamflow drought severity using ephemeral streamflow data, Int. J. Ecol. Econ. Stat., 11, 77–89, 2008b.

Pandey, R. P., Pandey, A., Galkate, R. V., Byun, H.-R., and Mal, B. C.: Integrating hydro-meteorological and physiographic factors for assessment of vulnerability to drought, Water Resour. Manage., 24, 4199–4217, 2010.

Rossi, G., Benedini, M., Tsakins, G., and Giakoumakis, S.: On regional drought estimation and analysis, Water Resour. Manage., 6, 249–277, 1992.

Sen, Z.: Statistical analysis of hydrologic critical droughts, J. Hydraul. Div.-ASCE, 106, 99–115, 1980.

Song, S. B. and Singh, V. P.: Frequency analysis of droughts using the Plackett copula and parameter estimation by genetic algorithm, Stoch. Environ. Res. Risk A., 24, 783–805, 2010a.

Song, S. B. and Singh, V. P.: Meta-elliptical copulas for drought frequency analysis of periodic hydrologic data, Stoch. Environ. Res. Risk A., 24, 425–444, 2010b.

Stedinger, J. R., Vogel, R. M., and Foufoula-Georgiou, E.: Frequency Analysis of Extreme Events, in: Chapter 18, Handbook of Hydrology, edited by: Maidment, D., McGraw-Hill, Inc., New York, 1993.

Tabari, H., Nikbakht, J., and Talaee, P. H.: Hydrological drought assessment in Northwestern Iran based on streamflow drought index (SDI), Water Resour. Manage., 27, 137–151, 2013.

Tallaksen, L. M. and van Lanen, H. A. J.: Hydrological drought: processes and estimation methods for streamflow and groundwater, Developments in Water Science, Elsevier Science B. V., Amsterdam, 2004.

Tallaksen, L. M., Madsen, H., and Clusen, B.: On the definition and modeling of streamflow drought duration and deficit volume, Hydrolog. Sci. J., 42, 15–33, 1997.

Tallaksen, L. M., Hisdal, H., and van Lanen, H. A. J.: Space-time modelling of catchment scale drought characteristics, J. Hydrol., 375, 363–372, 2009.

Tigkas, D., Vangelis, H., and Tsakiris, G.: Drought and climatic change impact on streamflow in small watersheds, Sci. Total Environ., 440, 33–41, 2012.

Tsakiris, G., Pangalou, D., and Vangelis, H.: Regional drought assessment based on the Reconnaissance Drought Index (RDI), Water Resour. Manage., 21, 821–833, 2007.

Tsakiris, G., Nalbantis, I., Vangelis, H., Verbeiren, B., Huysmans, M., Tychon, B., Jacquemin, I., Canters, F., Vanderhaegen, S., Engelen, G., Poelmans, L., De Becker, P., and Batelaan, O.: A System-based Paradigm of Drought Analysis for Operational Management, Water Resour. Manage., 27, 5281–5297, 2013.

van Huijgevoort, M. H. J., Hazenberg, P., van Lanen, H. A. J., and Uijlenhoet, R.: A generic method for hydrological drought identification across different climate regions, Hydrol. Earth Syst. Sci., 16, 2437–2451, doi:10.5194/hess-16-2437-2012, 2012.

Van Loon, A. F. and Van Lanen, H. A. J.: A process-based typology of hydrological drought, Hydrol. Earth Syst. Sci., 16, 1915–1946, doi:10.5194/hess-16-1915-2012, 2012.

Vogel, R. M. and Fennessey, N. M.: L-moment diagrams should replace product moment diagrams, Water Resour. Res., 29, 1745–1752, 1993.

Wang, A., Lettenmaier, D. P., and Sheffield, J.: Soil moisture drought in China, 1950–2006, J. Climate, 24, 3257–3271, 2011.

World Meteorological Organization: Manual on Low-flow Estimation and Prediction, Operational Hydrology Report No. 50, Geneva, 2008.

Wu, J., Soh, L. K., Samal, A., and Chen, X. H.: Trend analysis of streamflow drought events in Nebraska, Water Resour. Manage., 22, 145–164, 2007.

Yevjevich, V.: An objective approach to definition and investigation of continental hydrological droughts, Hydrology Paper No. 23, Colorado State University, Fort Collins, Colorado, USA, 1967.

Yoo, C., Kim, D., Kim, T. W., and Hwang, K. N.: Quantification of drought using a rectangular pulses Poisson process model, J. Hydrol., 355, 34–48, 2008.

Zelenhasic, E. and Salvai, A.: A method of streamflow drought analysis, Water Resour. Res., 23, 156–168, 1987.

4

A comprehensive one-dimensional numerical model for solute transport in rivers

Maryam Barati Moghaddam, Mehdi Mazaheri, and Jamal Mohammad Vali Samani

Department of Water Structures, Tarbiat Modares University, Tehran, Iran

Correspondence to: Mehdi Mazaheri (m.mazaheri@modares.ac.ir)

Abstract. One of the mechanisms that greatly affect the pollutant transport in rivers, especially in mountain streams, is the effect of transient storage zones. The main effect of these zones is to retain pollutants temporarily and then release them gradually. Transient storage zones indirectly influence all phenomena related to mass transport in rivers. This paper presents the TOASTS (third-order accuracy simulation of transient storage) model to simulate 1-D pollutant transport in rivers with irregular cross-sections under unsteady flow and transient storage zones. The proposed model was verified versus some analytical solutions and a 2-D hydrodynamic model. In addition, in order to demonstrate the model applicability, two hypothetical examples were designed and four sets of well-established frequently cited tracer study data were used. These cases cover different processes governing transport, cross-section types and flow regimes. The results of the TOASTS model, in comparison with two common contaminant transport models, shows better accuracy and numerical stability.

1 Introduction

First efforts to understand the solute transport subject led to a longitudinal dispersion theory which is often referred to as the classical advection–dispersion equation (ADE; Taylor, 1954). This equation is a parabolic partial differential equation derived from a combination of a continuity equation and Fick's first law. The one-dimensional ADE equation is as follows:

$$\frac{\partial (AC)}{\partial t} = -\frac{\partial (QC)}{\partial x} + \frac{\partial}{\partial x}\left(AD\frac{\partial C}{\partial x}\right) - \lambda AC + AS, \quad (1)$$

where A is the flow area, C the solute concentration, Q the volumetric flow rate, D the dispersion coefficient, λ the first-order decay coefficient, S the source term, t the time and x the distance. When this equation is used to simulate the transport in prismatic channels and rivers with relatively uniform cross-sections, accurate results can be expected; but field studies, particularly in mountain pool-and-riffle streams, indicate that observed concentration–time curves have a lower peak concentration and longer tails than the ADE equation predictions (Godfrey and Frederick, 1970; Nordin and Sabol, 1974; Nordin and Troutman, 1980; Day, 1975). Thus a group of researchers, based on field studies, stated that to accomplish more accurate simulations of solute transport in natural rivers and streams, the ADE equation should be modified. They added some extra terms to it for consideration of the impact of stagnant areas that were so-called storage zones (Bencala et al., 1990; Bencala and Walters, 1983; Jackman et al., 1984; Runkel, 1998; Czernuszenko and Rowinski, 1997; Singh, 2003). Transient storage zones mainly include eddies, stream poolside areas, stream gravel bed, streambed sediments, porous media of river bed and banks and stagnant areas behind flow obstructions such as big boulders, stream side vegetation, woody debris and so on (Jackson et al., 2013).

In general, these areas affect pollutant transport in two ways: on the one hand, temporary retention and gradual release of solute cause an asymmetric shape in the observed concentration–time curves, which could not be explained by the classical advection–dispersion theory; on the other hand, it is also affected by the opportunity for reactive pollutants to be frequently contacted with streambed sediments that indirectly affect solute sorption, especially in low-flow condi-

Table 1. Comparison of the features of the three models used in this study.

Model	Model features				
	Limitations on input parameters	Irregular cross-sections	Unsteady flow	Transient storage	Kinetic sorption
OTIS	Yes	No	No	Yes	Yes
MIKE 11	No	Yes	Yes	No	No
TOASTS	No	Yes	Yes	Yes	Yes

Table 2. Comparison of numerical methods used in the three models.

Model	Numerical methods			
	Discretization scheme	Order of accuracy	Stability	Numerical dispersion
TOASTS	Centered Time–QUICK Space (CTQS)	2nd-order in time 3rd-order in space	$Pe < \frac{8}{3}$	0
OTIS	Centered Time–Centered Space (CTCS)	2nd-order in time 2nd-order in space	$Pe < 2$	0
MIKE 11	Backward Time–Centered Space (BTCS)	1st-order in time 2nd-order in space	$Pe < 2$	$\frac{U^2 \Delta t}{2}$

* $Pe = \frac{U \Delta x}{D}$.

Table 3. Error indices of verification by the analytical solution for continuous boundary condition.

Index	With storage			Without storage
	50 m	75 m	100 m	100 m
R^2 (%)	99.97	99.96	99.96	99.99
RMSE (mg m^{-3})	0.021	0.026	0.033	0.009
MAE (mg m^{-3})	0.017	0.023	0.029	0.006
MRE (%)	0.450	0.780	1.20	0.640

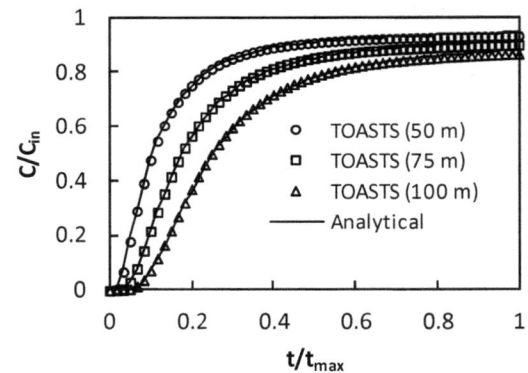

Figure 1. Results of the TOASTS model verification by the analytical solution for continuous boundary condition ($\alpha \neq 0$).

tions (Bencala, 1983, 1984; Bencala et al., 1990; Bencala and Walters, 1983).

In the literature, several approaches have been proposed to simulate solute transport in the rivers with storage areas, that one of the most commonly used is the transient storage model (TSM). TSM has been developed to consider solute movement from the main channel to stagnant zones and vice versa. The simplest form of the TSM is the one-dimensional advection–dispersion equation with an additional term to consider transient storage (Bencala and Walters, 1983). After the introduction of the TSM, transient storage processes have been studied in a variety of small mountain streams, as well as large rivers, and it was shown that simulation results of tracer study data considering the transient storage impact have good agreement with observed data. Also, it was shown that the interaction between the main channel and storage

zones, especially in mountain streams, has a great effect on solute transport behavior (D'Angelo et al., 1993; DeAngelis et al., 1995; Morrice et al., 1997; Czernuszenko et al., 1998; Chapra and Runkel, 1999; Chapra and Wilcock, 2000; Laenen and Bencala, 2001; Fernald et al., 2001; Keefe et al., 2004; Ensign and Doyle, 2005; Van Mazijk and Veling, 2005; Gooseff et al., 2007; Jin et al., 2009).

In this study, a comprehensive model, called TOASTS (third-order accuracy simulation of transient storage), able to obviate shortcomings of current models of contaminant transport, is presented. The TOASTS model uses high-order accuracy numerical schemes and considers transient storage

Table 4. Error indices of verification by the analytical solution for Heaviside boundary condition.

Index	With storage			Without storage
	50 m	75 m	100 m	100 m
R^2 (%)	99.98	99.97	99.96	99.99
RMSE (mg m^{-3})	0.034	0.045	0.058	0.0094
MAE (mg m^{-3})	0.031	0.044	0.056	0.007
MRE (%)	3.5	4.2	5	1.49

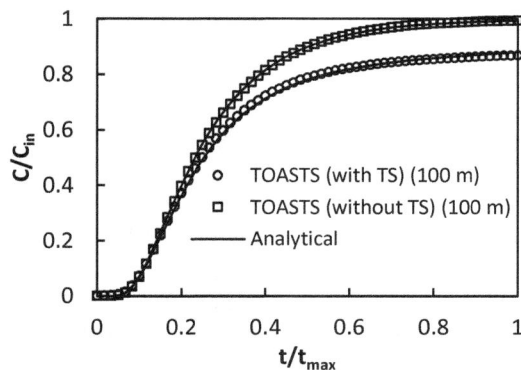

Figure 2. Results of the TOASTS model verification by the analytical solution for continuous boundary condition ($\alpha = 0$).

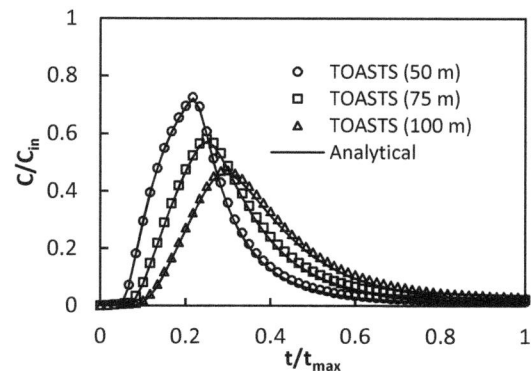

Figure 3. Results of the TOASTS model verification by the analytical solution for Heaviside boundary condition ($\alpha \neq 0$).

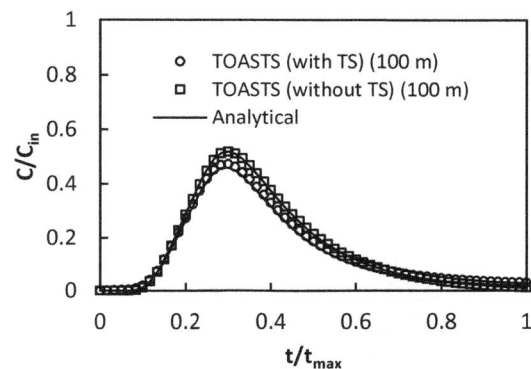

Figure 4. Results of the TOASTS model verification by the analytical solution for Heaviside boundary condition ($\alpha = 0$).

in rivers with irregular cross-sections under non-uniform and unsteady flow regimes. This model presents a comprehensive modeling framework that links three sub-models for calculating geometric properties of irregular cross-sections, solving unsteady flow equations and solving transport equations with transient storage and kinetic sorption.

To demonstrate the applicability and accuracy of the TOASTS model, results of two hypothetical examples (designed by the authors) and four sets of well-established tracer study data, are compared with the results of two existing frequently used solute transport models: the MIKE 11 model, developed by the Danish Hydraulic Institute (DHI), and the OTIS (one-dimensional transport with inflow and storage) model that today is the only existing model for solute transport with transient storage (Runkel, 1998). The TOASTS model and the two other model features are listed in Table 1. It should be noted that the OTIS model, in simulating solute transport in irregular cross-sections under unsteady flow regimes, has to rely on external stream routing and geometric programs. By contrast, in the TOASTS and MIKE 11 models, geometric properties and unsteady flow data are directly evaluated from river topography, bed roughness, flow initial and boundary conditions data. Another important point is in the numerical scheme which has been used in the TOASTS model solution. The key and basic difference of the TOASTS model is spatial discretization of

the transport equation. This model uses the control-volume approach and QUICK (quadratic upstream interpolation for convective kinematics) scheme in spatial discretization of the advection–dispersion equation considering transient storage and kinetic sorption; whereas the two other models employ central spatial differencing. More detailed comparison of numerical schemes used in the structure of three subjected models is given in Table 2.

As many researchers claim, central spatial differencing is incapable of simulation of pure advection problems and does not introduce good performance in this regard (it leads to non-convergent results with numerical oscillations; Zhang and Aral, 2004; Szymkiewicz, 2010; Versteeg and Malalasekera, 2007). It should be mentioned that, in recent years, the QUICK scheme has been widely used in numerical solutions of partial differential equations due to its high-order accuracy, very small numerical dispersion and higher stability range (Neumann et al., 2011; Lin and Medina, 2003). Hence, usage of the QUICK scheme in numerical discretization of the transport equation leads to significantly better results, especially in advection-dominant problems.

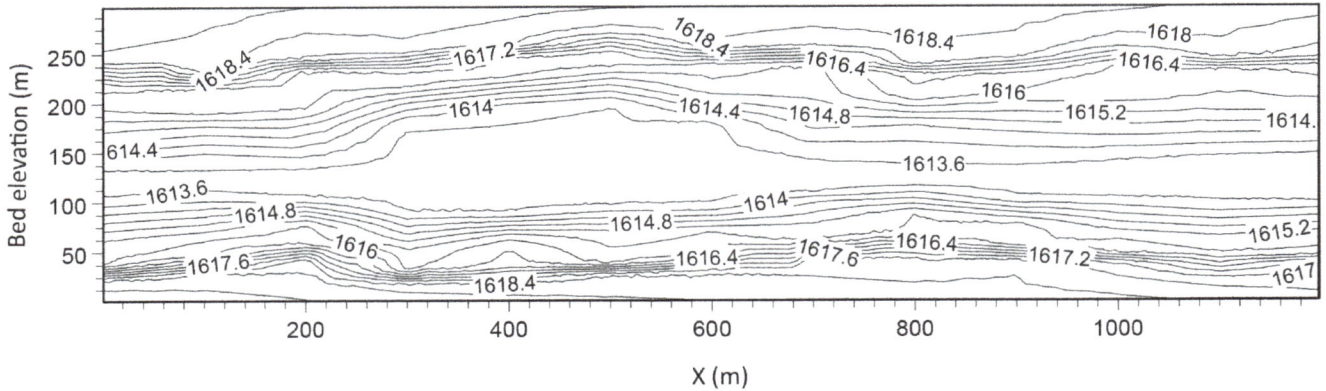

Figure 5. Bed elevation contours of the 2-D hypothetical example.

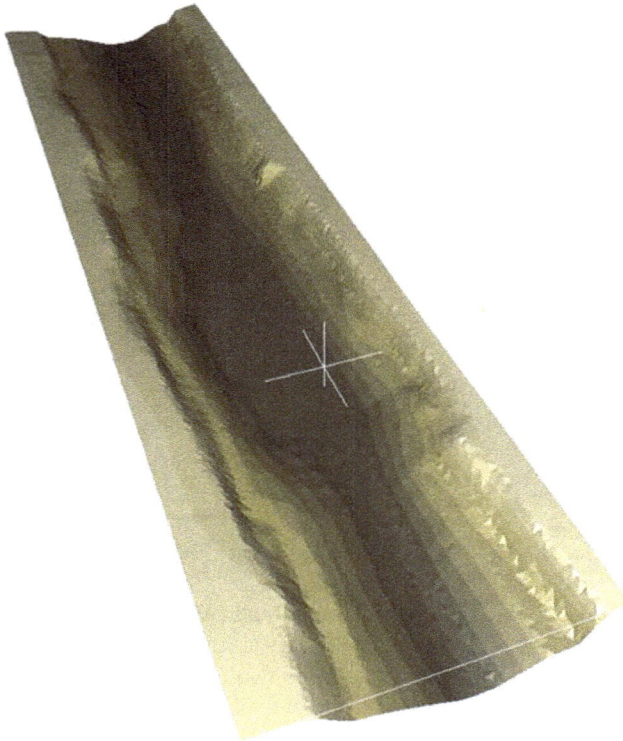

Figure 6. Bed elevation three-dimensional view of the 2-D hypothetical example.

Table 5. Error indices of verification by the 2-D model.

Index	With storage	Without storage
R^2 (%)	99.97	99.91
RMSE (mg m^{-3})	0.095	1.88
MAE (mg m^{-3})	0.066	0.77
MRE (%)	3.1	36.5

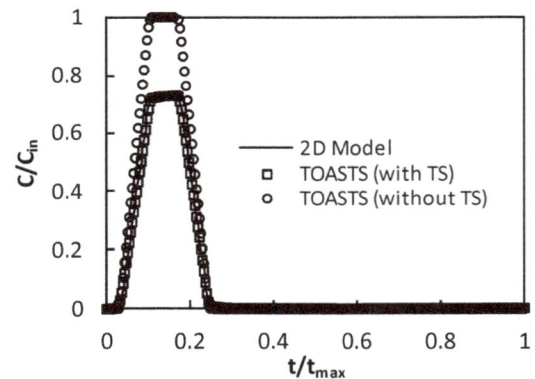

Figure 7. Results of the TOASTS model verification using the 2-D model.

2　Methodology

2.1　Governing equations

There are several equations for solute transport with transient storage, the most well known being the TSM presented by Bencala and Walters (1983). By writing conservation of mass principle for solute in the main channel and storage zone and doing some algebraic manipulation, a coupled set of differential equations is derived:

$$\frac{\partial C}{\partial t} = -\frac{Q}{A}\frac{\partial C}{\partial x} + \frac{1}{A}\frac{\partial}{\partial x}\left(AD\frac{\partial C}{\partial x}\right)$$
$$+ \frac{q_{\mathrm{LIN}}}{A}(C_{\mathrm{L}} - C) + \alpha(C_{\mathrm{S}} - C) \tag{2}$$

$$\frac{dC_{\mathrm{S}}}{dt} = \alpha\frac{A}{A_{\mathrm{S}}}(C - C_{\mathrm{S}}), \tag{3}$$

where A and A_{S} are the main channel and storage zone cross-sectional area respectively, C, C_{L} and C_{S} are the main chan-

Table 6. Properties of the test cases used for the TOASTS model application.

			Solute transport processes					
			Physical				Chemical	
					Transient storage			
Example no.	Section type	Flow regime	Advection	Dispersion	Surface	Hyporheic exchange	First-order decay	Kinetic sorption
1	Regular	Steady Uniform	Yes	No	No	No	No	No
2	Regular	Steady Uniform	Yes	Yes	No	No	Yes	No
3	Irregular	Steady Non-uniform	Yes	Yes	Yes	No	No	No
4	Irregular	Steady Non-uniform	Yes	Yes	Yes	No	No	Yes
5	Irregular	Steady Non-uniform	Yes	Yes	Yes	No	No	No
6	Irregular	Unsteady Non-uniform	Yes	Yes	No	Yes	No	No

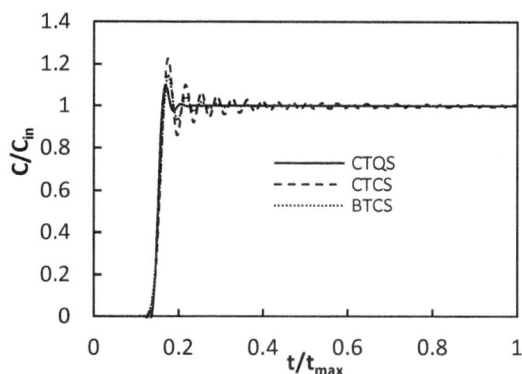

Figure 8. Comparison of the CTQS, CTCS and BTCS schemes for the pure advection test case.

Table 7. Simulation parameters related to test case 2.

L (m)	D (m^2 s^{-1})	λ (s^{-1})	Case	Space step (m)	Peclet number
2200	5	0.00002	1	10	0.24
			2	100	2.4
			3	100	10

Table 8. Error indices of concentration time series in test case 2.

| | | Model | | |
	Index	TOASTS	OTIS	MIKE 11
Pe = 0.24	R^2 (%)	99.93	99.93	99.98
	RMSE (mg m^{-3})	0.460	0.460	0.850
	MAE (mg m^{-3})	0.236	0.238	0.480
	MRE (%)	0.9	1.0	1.7
Pe = 2.4	R^2 (%)	98.26	97.82	97.75
	RMSE (mg m^{-3})	2.66	2.98	3.24
	MAE (mg m^{-3})	1.42	1.55	1.73
	MRE (%)	3.77	4.11	4.93
Pe = 10	R^2 (%)	98.8	98.2	98.24
	RMSE (mg m^{-3})	3.60	4.41	4.46
	MAE (mg m^{-3})	0.80	1.12	1.17
	MRE (%)	1.25	1.95	2.15

nel, lateral inflow and storage zone solute concentration, respectively, q_{LIN} is the lateral inflow rate and α is the storage zone exchange coefficient. For reactive solute, considering two types of chemical reactions (kinetic sorption and first-order decay) Eqs. (2) and (3) are rewritten as:

$$\frac{\partial C}{\partial t} = L(C) + \rho\hat{\lambda}(C_{\text{sed}} - K_{\text{d}}C) - \lambda C \tag{4}$$

$$\frac{dC_{\text{S}}}{dt} = S(C_{\text{S}}) + \hat{\lambda}_{\text{S}}\left(\hat{C}_{\text{S}} - C_{\text{S}}\right) - \lambda_{\text{S}}C_{\text{S}} \tag{5}$$

$$\frac{dC_{\text{sed}}}{dt} = \hat{\lambda}(K_{\text{d}}C - C_{\text{sed}}), \tag{6}$$

where $\hat{C}_{\text{S}}$ is the background storage zone solute concentration, C_{sed} is the sorbate concentration on the streambed sediment, K_{d} is the distribution coefficient, λ and λ_{S} are the main channel and storage zone first-order decay coefficients respectively, $\hat{\lambda}$ and $\hat{\lambda}_{\text{S}}$ are the main channel and storage zone sorption rate coefficients respectively, ρ is the mass of acces-

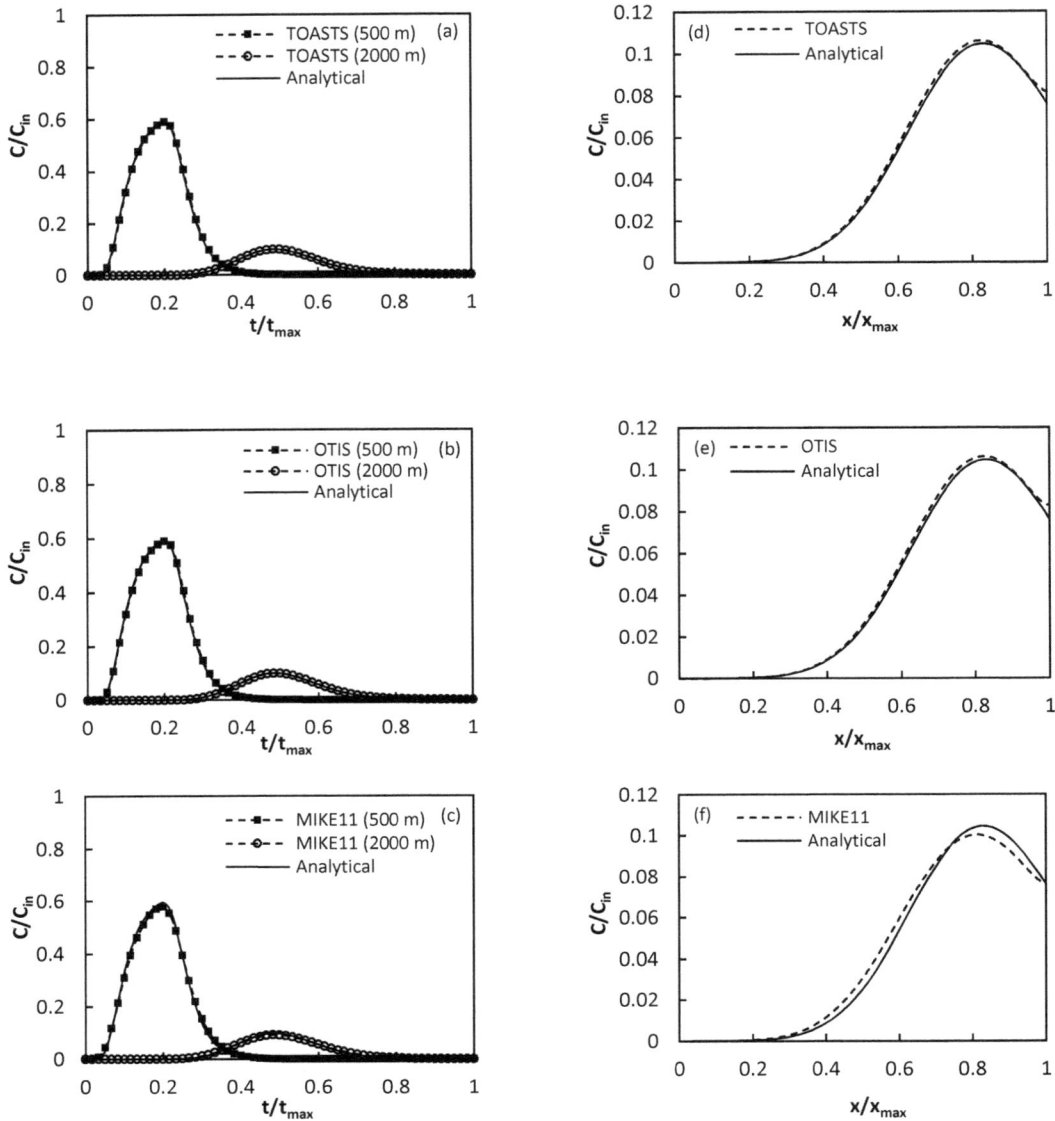

Figure 9. Comparison of the TOASTS, OTIS and MIKE 11 models in test case 2 for Pe $= 0.24$.

sible sediment/volume water and L and S are the right-hand side differential operator of Eqs. (2) and (3) respectively.

2.2   Numerical solution scheme

Numerical solution of Eqs. (4) to (6) in this study are based on the control-volume method and centered time–QUICK space (CTQS) scheme. The spatial derivatives are discretized by the QUICK scheme, which is based on quadratic upstream interpolation of discretization of the advection–dispersion equation (Leonard, 1979). In this scheme, face values are computed using quadratic function passing through two upstream nodes and a downstream node. For an equally spaced grid, the values of a desired quantity, φ, on the cell faces are given by the following equations:

$$\phi_{\text{face}} = \frac{6}{8}\phi_{i-1} + \frac{3}{8}\phi_i - \frac{1}{8}\phi_{i-2} \tag{7}$$

$$\phi_w = \frac{6}{8}\phi_W + \frac{3}{8}\phi_P - \frac{1}{8}\phi_{WW} \tag{8}$$

$$\phi_e = \frac{6}{8}\phi_P + \frac{3}{8}\phi_E - \frac{1}{8}\phi_W, \tag{9}$$

where P denotes an unknown node with neighbor nodes W (at left) and E (at right). It should be noted that the corresponding cell faces are denoted by the lowercase letters, w and e. Gradient at cell faces can be estimated using the following relationships:

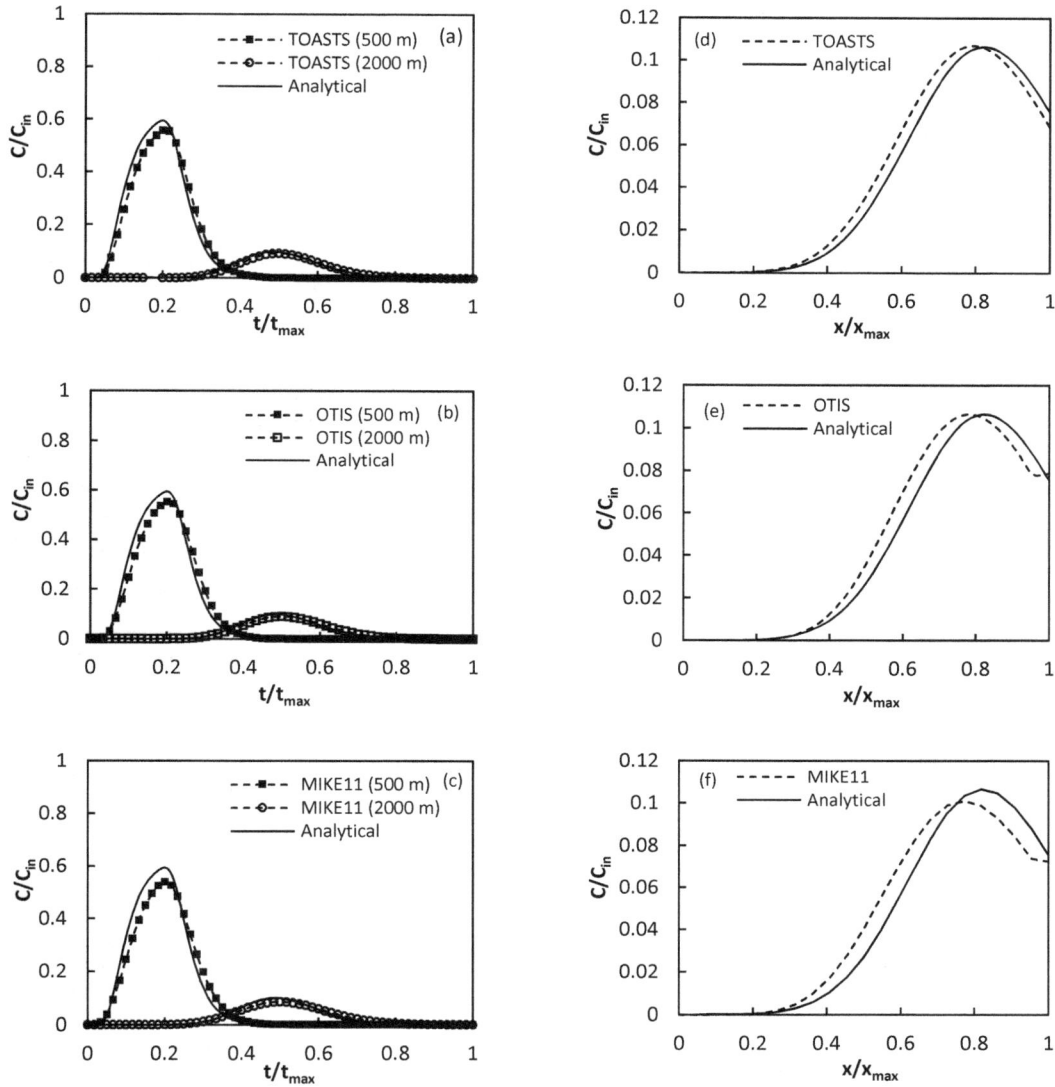

Figure 10. Comparison of the TOASTS, OTIS and MIKE 11 models in test case 2 for Pe = 2.4.

$$\left(\frac{\partial \phi}{\partial x}\right)_w = \frac{\phi_P - \phi_W}{\Delta x} \qquad (10)$$

$$\left(\frac{\partial \phi}{\partial x}\right)_e = \frac{\phi_E - \phi_P}{\Delta x}. \qquad (11)$$

Finally, the difference equations related to the Eqs. (4) to (6) can be derived as follows:

$$\frac{C_P^{n+1} - C_P^n}{\Delta t} = \frac{1}{2}\left[\left(\frac{-Q_P}{A_P \Delta x}(C_e - C_w)\right)^{n+1}\right.$$
$$\left. + \left(\frac{-Q_P}{A_P \Delta x}(C_e - C_w)\right)^n\right]$$

$$+ \frac{1}{2}\left\{\frac{1}{A_P^{n+1}\Delta x}\left[\left(AD\frac{\partial C}{\partial x}\right)_e - \left(AD\frac{\partial C}{\partial x}\right)_w\right]^{n+1}\right.$$
$$+ \frac{1}{A_P^n \Delta x}\left[\left(AD\frac{\partial C}{\partial x}\right)_e - \left(AD\frac{\partial C}{\partial x}\right)_w\right]^n\right\}$$
$$+ \frac{1}{2}\left[\frac{q_{\text{LIN}}^{n+1}}{A_P^{n+1}}(C_L - C_P)^{n+1} + \frac{q_{\text{LIN}}^n}{A_P^n}(C_L - C_P)^n\right]$$
$$+ \frac{\alpha}{2}\left[(C_S - C_P)^{n+1} + (C_S - C_P)^n\right]$$
$$+ \frac{\rho\hat{\lambda}}{2}\left[(C_{\text{sed}} - K_d C_P)^{n+1} + (C_{\text{sed}} - K_d C_P)^n\right]$$
$$- \frac{\lambda}{2}\left(C_P^{n+1} + C_P^n\right) \qquad (12)$$

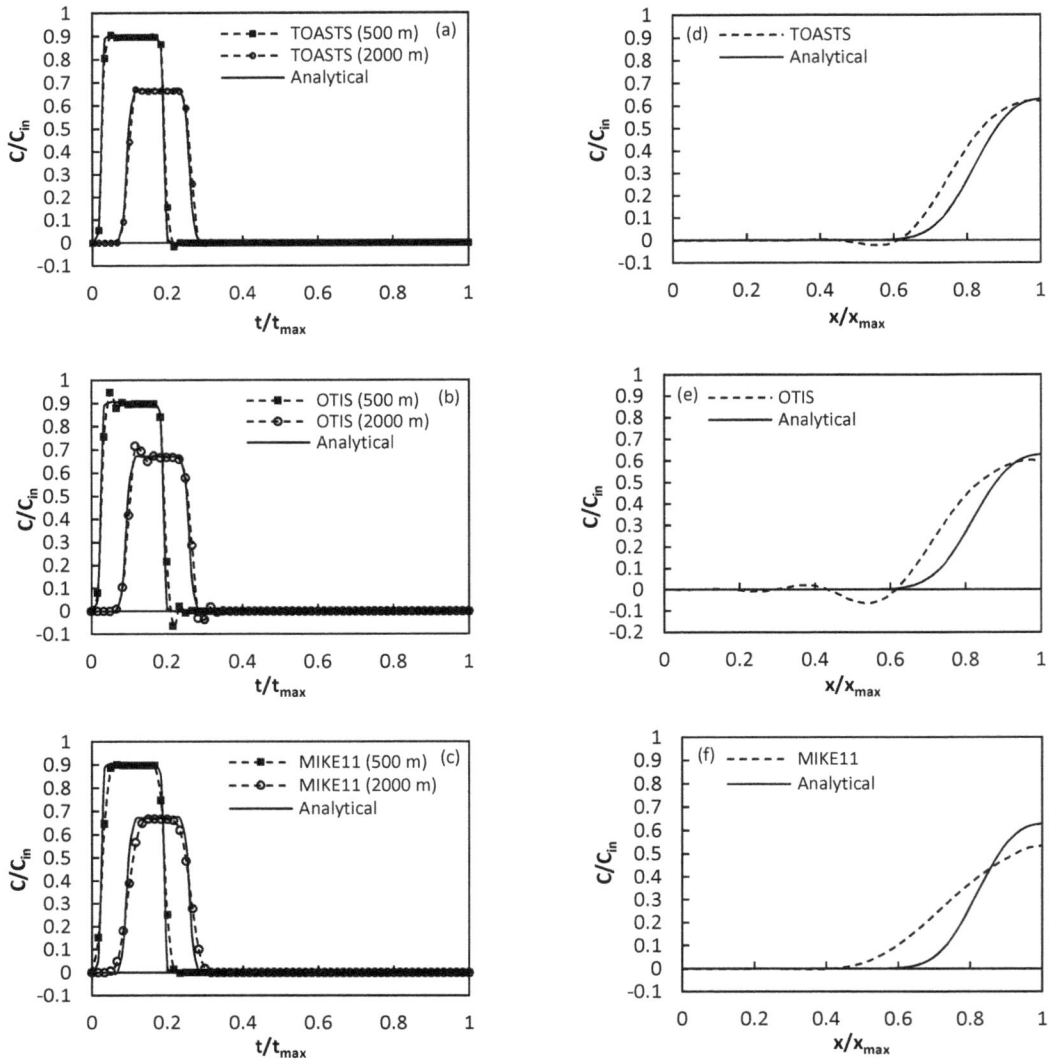

Figure 11. Comparison of the TOASTS, OTIS and MIKE 11 models in test case 2 for Pe = 10.

$$
\frac{C_S^{n+1} - C_S^n}{\Delta t} = \frac{1}{2}\left[\left(\alpha\frac{A_P}{A_S}(C_P - C_S)\right.\right.
$$
$$
\left. + \hat{\lambda}_S\left(\hat{C}_S - C_S\right) - \lambda_S C_S\right)^{n+1}
$$
$$
+ \left(\alpha\frac{A_P}{A_S}(C_P - C_S)\right.
$$
$$
\left.\left. + \hat{\lambda}_S\left(\hat{C}_S - C_S\right) - \lambda_S C_S\right)^n\right] \tag{13}
$$

$$
\frac{C_{sed}^{n+1} - C_{sed}^n}{\Delta t} = \frac{1}{2}\left[\left(\hat{\lambda}\left(K_d C_P - C_{sed}\right)\right)^{n+1}\right.
$$
$$
\left. + \left(\hat{\lambda}\left(K_d C_P - C_{sed}\right)\right)^n\right]. \tag{14}
$$

Writing Eqs. (12) to (14) for all control-volumes in the solution domain and applying the boundary conditions, a system of linear algebraic equations will be introduced:

$$
a_{WW}C_{WW}^{n+1} + a_W C_W^{n+1} + a_P C_P^{n+1} + a_E C_E^{n+1} = R_P, \tag{15}
$$

where a_{WW}, a_W, a_P, a_E and R_P are the corresponding coefficients and the right-hand side term. Solving this system, main channel concentrations in $n+1$ time level will be computed. Having main channel concentration values, the storage zone and streambed sediment concentrations could be calculated.

Table 9. Error indices of concentration longitudinal profiles in test case 2.

		Model		
	Index	TOASTS	OTIS	MIKE 11
Pe = 0.24	R^2 (%)	99.9	99.9	99.9
	RMSE (mg m^{-3})	0.146	0.154	0.360
	MAE (mg m^{-3})	0.105	0.108	0.280
	MRE (%)	1.91	1.97	3.20
Pe = 2.4	R^2 (%)	98.6	98	96
	RMSE (mg m^{-3})	0.53	0.65	0.86
	MAE (mg m^{-3})	0.40	0.47	0.64
	MRE (%)	5.40	6.56	11.20
Pe = 10	R^2 (%)	95.7	92	88.4
	RMSE (mg m^{-3})	5.46	7.24	7.88
	MAE (mg m^{-3})	3.02	4.47	5.05
	MRE (%)	6.27	12.44	13.50

2.3 Damköhler Index

The Damköhler number is a dimensionless number that reflects the exchange rate between the main channel and storage zones (Jin et al., 2009; Harvey and Wagner, 2000; Wagner and Harvey, 1997; Scott et al., 2003). For a stream or channel this number is defined as:

$$DaI = \alpha \left(1 + \frac{A}{A_S}\right) \frac{L}{u}, \qquad (16)$$

where L is the main channel length, u the average flow velocity and DaI the Damköhler number. When DaI is much greater than unity (e.g., 100), the exchange rate between the main channel and storage zone is too fast and it could be assumed that these two segments are in balance. Accordingly, when DaI is much lower than unity (e.g., 0.01) the exchange rate between main channel and storage zone is very low and negligible. In other words, in such a stream where DaI is very low, there is practically no significant exchange between the main channel and storage zone, and transient storage zones do not affect downstream solute transport. Therefore, for reasonable estimation of transient storage model parameters, the DaI value must be within 0.1 to 10 range (Fernald et al., 2001; Wagner and Harvey, 1997; Ramaswami et al., 2005).

3 Model verification

In this section the TOASTS model is verified using several test cases. These test cases include analytical solutions of constant-coefficient governing equations for two types of upstream boundary condition (continuous and Heaviside) and also by comparing the model results with the 2-D model. Complementary explanations for each case are given below.

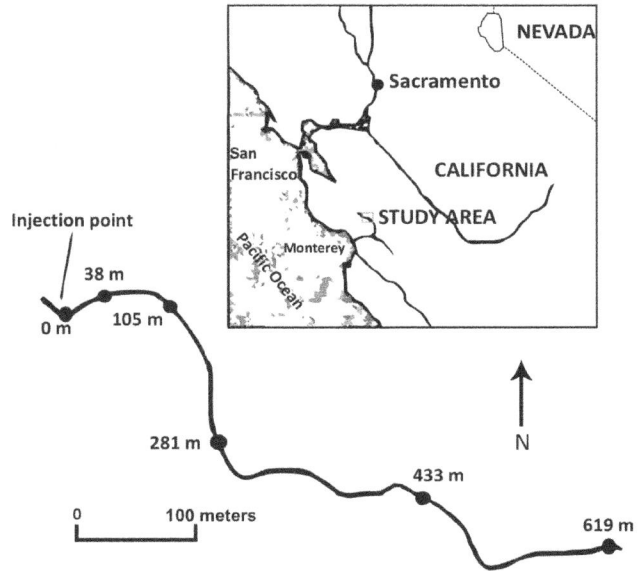

Figure 12. Uvas Creek (Santa Clara County, California) tracer study site map (Bencala and Walters, 1983).

3.1 Verification by analytical solutions

In this section, model verification is carried out using analytical solutions presented by Kazezyılmaz-Alhan (2008). The designed example is a 200 m length channel with constant cross-sectional area equal to 1 m^2. The flow discharge, dispersion coefficient, storage zone area and exchange coefficient are 0.01 m^3 s^{-1}, 0.2 m^2 s^{-1}, 1 m^2 and 0.00002 s^{-1}, respectively. The DaI number can be calculated from the Eq. (19) equal to 0.8. This example is implemented for two different types of upstream boundary conditions: (a) continuous and (b) Heaviside.

3.1.1 (a) Continuous boundary condition

In this case, a solute concentration of 5 mg m^{-3} is injected continuously for 10 h at the inlet. The time and space steps are considered equal to 30 s and 1 m, respectively. Figure 1 shows the TOASTS model results compared to the analytical solution at 50, 75 and 100 m from the inlet. Note that both axes have been nondimensionalized with respect to the maximum values. Also, square of correlation coefficient (R^2), root mean square error (RMSE), mean absolute error (MAE) and mean relative error (MRE) are given in Table 3. According to Fig. 1 and the error indices given in Table 3, it is clear that the trends of numerical and analytical solutions are similar, and the TOASTS model shows a good accuracy in this example.

In order to show the model capability and assess the model accuracy in a case without transient storage, the model is executed for $\alpha = 0$ for this example, and the result at the distance of 100 m from the inlet is compared to the analytical solution

Table 10. Simulation parameters for the Uvas Creek experiment (test case 3).

Reach (m)	Flow discharge ($m^3\,s^{-1}$)	Dispersion coefficient ($m^2\,s^{-1}$)	Cross-sectional areas		Exchange coefficient (s^{-1})	
			Main channel	Storage zone		
0–38	0.0125	0.12	0.30		0	0
38–105	0.0125	0.15	0.42		0	0
105–281	0.0133	0.24	0.36	0.36	3×10^{-5}	
281–433	0.0136	0.31	0.41	0.41	1×10^{-5}	
433–619	0.0140	0.40	0.52	1.56	4.5×10^{-5}	

Table 11. Error indices of simulation of the Uvas Creek experiment (test case 3).

Index	38 m			281 m			433 m		
	TOASTS	OTIS	MIKE 11	TOASTS	OTIS	MIKE 11	TOASTS	OTIS	MIKE 11
R^2 (%)	94.30	94.20	94.10	99.40	99.31	99.10	98.84	98.8	97.82
RMSE ($mg\,m^{-3}$)	0.727	0.728	0.730	0.180	0.183	0.340	0.203	0.205	0.440
MAE ($mg\,m^{-3}$)	0.202	0.203	0.212	0.108	0.109	0.205	0.121	0.125	0.280
MRE (%)	3.50	3.55	3.68	2.07	2.08	3.60	2.27	2.40	5.30

of the classical advection–dispersion equation. The results are shown in Fig. 2 and Table 3. Figure 2 also illustrates that in the case of transient storage, the concentration–time curve has a lower peak than the one without storage ($\alpha = 0$), which matches the previously mentioned transient storage concept.

3.1.2 (b) Heaviside boundary condition

In this case a solute concentration of $5\,mg\,m^{-3}$ is injected at the inlet for a limited time of 100 min. The time and space steps are considered equal to 30 s and 1 m, respectively. Comparison of the model results and the analytical solution at the distance of 50, 75 and 100 m from the inlet is presented in Fig. 3 and Table 4. Also, corresponding results at the distance of 100 m for the case without storage ($\alpha = 0$) are given in Fig. 4 and Table 4. It is obvious that the TOASTS model results in both cases (with and without storage) have a reasonable agreement with the analytical solution.

3.2 Verification by 2-D model

The main cause of transient storage phenomena is velocity difference between the main channel and storage zones. 2-D depth-averaged models consider velocity variations in two dimensions and give more accurate predictions of solute transport behavior in reality. Hence, they could be used for verification of the presented 1-D model as a benchmark. For this purpose, a hypothetical example was designed. To do so, a 1200 m long river, with irregular cross-sections, is considered. Figures 5 and 6 show bed topography of the hypothetical river. In order to take into account a hypothetical storage zone, the distance between 300 and 600 m of the river has

been widened. The flow conditions in the river are considered to be non-uniform and unsteady. The solute concentration in the main channel and storage zone, at the beginning of the simulation (initial conditions), is assumed to be zero. In calculations of both flow and transport models, space and time steps are considered equal to 100 m and 1 min respectively. The dispersion coefficient, storage zone area and exchange coefficient are $10\,m^2\,s^{-1}$, $22\,m^2$ and $1.8 \times 10^{-4}\,s^{-1}$, respectively. For this example the *DaI* number is calculated equal to 0.4. The upstream boundary condition for transport submodel is a 3 h lasting step loading pulse with $20\,mg\,m^{-3}$ pick concentration. The results of the TOASTS model for simulating with and without transient storage were compared to the 2-D model at the distance of 800 m from the inlet. Figure 7 and Table 5 illustrates these results. This figure shows that with appropriate choice of A_S and α, concentration–time curves given by the TOASTS model are close to those given by the 2-D model. These results also imply the necessity of considering transient storage term in the advection–dispersion equation for more accurate simulation of solute transport, especially in natural rivers and streams.

4 Application

In this section, the applications of the TOASTS model using a variety of hypothetical examples and several sets of observed data are presented. Some properties of these test cases are given in Table 6. As shown in this table, the test cases include a wide variety of solute transport simulation applications at different conditions.

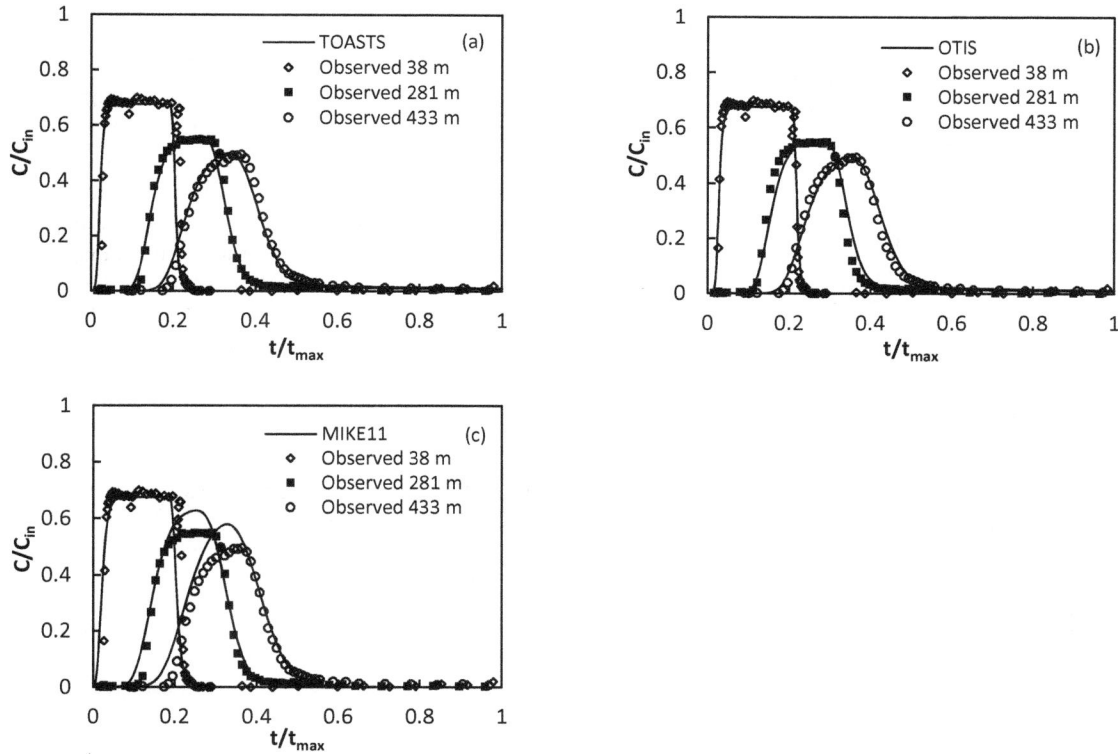

Figure 13. Observed and simulated chloride concentrations in the main channel (test case 3).

Table 12. Simulation parameters related to test case 4.

Distribution coefficient ($m^2 s^{-1}$)	Sorption rate coefficient (s^{-1})		Background concentration ($mg\,L^{-1}$)			Input concentration ($mg\,L^{-1}$)
	Main channel	Storage zone	Main channel	Storage zone	Bed sediments	
70	56×10^{-6}	1	0.13	0.13	9.1×10^{-3}	1.73

4.1 Test case 1: pure advection

In order to show the advantage of the numerical scheme used in the TOASTS model, for advection-dominant problems, a hypothetical example was designed and three numerical schemes were applied: CTQS (centered time–quick space), CTCS (centered time–centered space) and BTCS(backward time–centered space). To do so, steady flow by velocity of $1\,m\,s^{-1}$ was assumed. Total simulation time was 5 h and space and time steps were 100 m and 10 s respectively. Note that advection is the only transport mechanism. The results of this test case are depicted in Fig. 8. It is clear that, for the pure advection simulation, the CTQS scheme has less oscillation than the other two schemes. In particular, this figure indicates that the result of the CTCS scheme, which is used in the OTIS model, shows high oscillations. Therefore, it can be concluded that for advection-dominant sim-

ulation the TOASTS model has a better performance. It is interesting to note that in mountain rivers where the transient storage mechanism is more observed, due to relatively high slope, higher flow velocities occur which lead to advection-dominant solute transport.

4.2 Test case 2: transport with first-order decay

This example illustrates the application of the TOASTS model in solute transport simulation by first-order decay. A decaying substance enters the stream with steady and uniform flow during a 2 h period. The solute concentration at the upstream boundary is $100\,mg\,m^{-3}$. Also, in order to assess the TOASTS model capability in the case of high-flow velocity and advection-dominant transport, this example implemented for three cases with different Peclet numbers. The simulation parameters for different cases are given in Table 7.

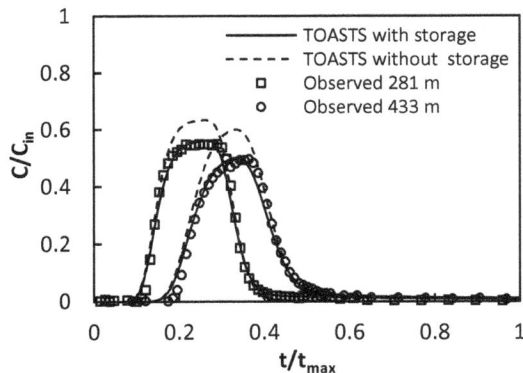

Figure 14. The TOASTS model results for simulation with and without transient storage (test case 3).

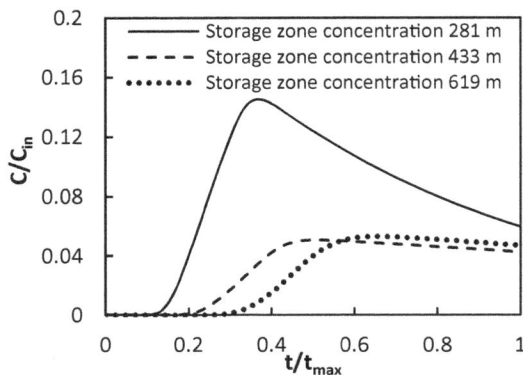

Figure 15. Observed and simulated storage zone concentrations computed by the TOASTS model (test case 3).

Figures 9–11 show simulation results of the three numerical models in comparison with analytical solution. Error indices are given in Tables 8 and 9. It is obvious from Fig. 9a–c that in the first case (Peclet number less than 2), all methods simulated concentration–time curves accurately. Also, Fig. 9d–f show that the MIKE 11 model cannot simulate a concentration longitudinal profile accurately, because it does not consider the transient storage effect on solute transport.

In the second case, by increasing the computational space step, all methods show a drop in the peak concentration, that its amount for the MIKE 11 model is more and for the TOASTS model is less than the others (Fig. 10a–c). Figure 10d–f and Table 9 show that the results of the models that use the central differencing scheme in spatial discretization of transport equations show more discrepancy in comparison with the analytical solution.

In the third case, flow velocity increased about four times. As illustrated in Fig. 11c, by increasing the Peclet number, the OTIS model results show more oscillations. This model also shows very intense oscillations in the longitudinal concentration profile in the form of negative concentrations

(Fig. 11e), while observed oscillations in the TOASTS model are very small compared to the OTIS model (Fig. 11d). However, the QUICK scheme oscillations in advection-dominant cases are less likely to corrupt the solution. Also the MIKE 11 model results, in comparison with the TOASTS model, have greater difference with the analytical solution.

The main reason for the difference between the obtained results in the three cases is actually related to how advection and dispersion affect the solute transport. The dispersion process affects the distribution of solute in all directions, whereas advection acts only in the flow direction. This fundamental difference manifests itself in the form of limitation in computational grid size.

4.3 Test case 3: conservative solute transport with transient storage

This example shows the TOASTS model application to field data, by using the conservative tracer (chloride) injection experiment results, which was conducted in Uvas Creek, a small mountain stream in California (Fig. 12). Details of the experiments can be found in Avanzino et al. (1984). Table 10 shows simulation parameters for the Uvas Creek experiment (Bencala and Walters, 1983). For assessing efficiency and accuracy of the three discussed models in simulation of the impact of physical processes on solute transport in a mountain stream, they are implemented for this set of observed data. Figure 13a–c illustrates simulated chloride concentration in the main channel. It can be seen from these figures and Table 11 that the TOASTS model simulated the experiment results slightly better than the two other models. Comparison of Fig. 13a and b shows that the TOASTS and OTIS models have good accuracy in modeling the peak concentration and the TOASTS model has a slightly better performance in simulation of a rising tail of concentration–time curve, particularly in the 281 m station. Figure 13c shows MIKE 11 model results. It shows significant discrepancies with the observed data, particularly in peak concentrations. However, at the 38 m station, where transient storage has not still affected solute transport, the results of the three models have little difference with the observed data (Table 11). Figure 14 depicts the TOASTS model results for the Uvas Creek experiment for simulations with and without transient storage at the 281 and 433 m stations. This figure shows that in simulation with transient storage, the results have more fitness with the observed data in general shape of the concentration–time curve, peak concentration and peak arrival time. Figure 15 shows the simulated chloride concentrations in the storage zone. The concentration–time curves in the storage zone have longer tails in comparison with the main channel. That means some portions of the solute mass remain in the storage zones and gradually return to the main channel.

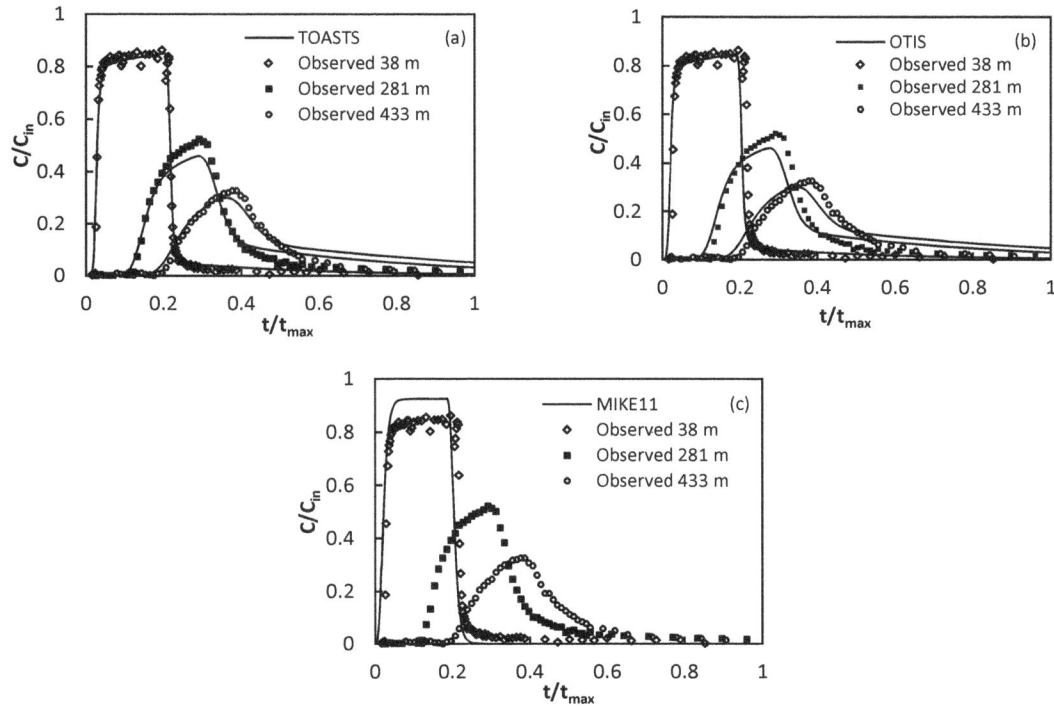

Figure 16. Observed and simulated strontium concentrations in the main channel (test case 4).

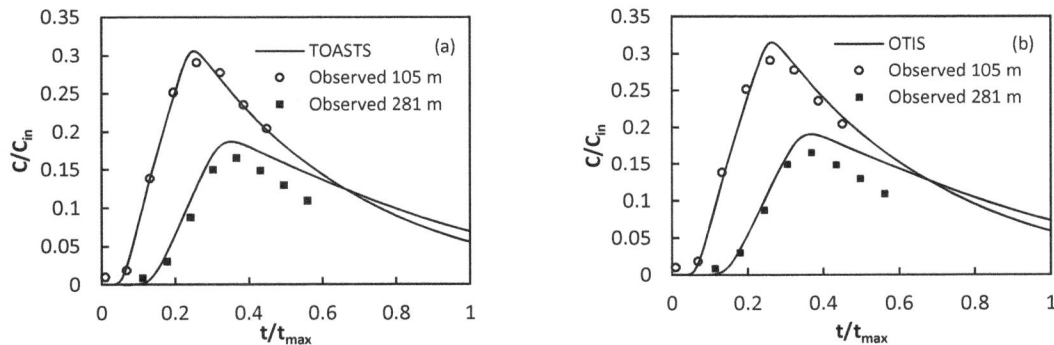

Figure 17. Observed and simulated sorbate strontium concentrations in the Uvas Creek experiment (test case 4).

4.4 Test case 4: non-conservative solute transport with transient storage

The objective of this test case is to demonstrate the capability of the TOASTS model in non-conservative solute transport modeling in natural rivers. For this purpose, the field experiment of the 3 h reactive tracer (strontium) injection into the Uvas Creek was used. The experiment was conducted at low-flow conditions and, due to the high opportunity of solute having frequent contact with relatively immobile streambed materials, solute and streambed interactions and solute sorption into bed sediments were more intense than during the high-flow conditions. Hence, the sorption process must be

considered in simulation of this experiment (Bencala, 1983). Some of the simulation parameters are given in Table 12 and the other parameters are the same as those given in Table 10. Figure 16a–c and Table 13 show solute transport simulation results of the three subjected models in comparison with the observed data. According to these figures it could be said that the TOASTS model shows better fitness with the observed data. Figure 16c shows that simulation without taking into account the transient storage and kinetic sorption in the MIKE 11 model leads to very poor results. The zero exchange coefficient at the 38 m station causes reasonable results by this model at this station. Figure 17 illustrates the TOASTS and OTIS model results for sorbate concentrations

Table 13. Error indices of simulation of the Uvas Creek experiment (test case 4).

| | Main channel concentration | | | | | | Sorbate concentration | | | |
| | 38 m | | | 281 m | | | 105 m | | 281 m | |
Index	TOASTS	OTIS	MIKE 11	TOASTS	OTIS	MIKE 11	TOASTS	OTIS	TOASTS	OTIS
R^2 (%)	99.30	93.17	93.00	99.00	96.00	90.80	99.40	99.30	99.16	98.6
RMSE (mg m^{-3})	0.05	0.12	0.17	0.055	0.070	0.200	1.05	1.64	2.67	2.86
MAE (mg m^{-3})	0.021	0.044	0.086	0.048	0.055	0.115	0.75	1.50	2.40	2.41
MRE (%)	6.40	11.80	24.60	13.60	18.00	27.40	3.04	5.66	10.50	10.80

Table 14. Error indices of the Athabasca River experiment (test case 5).

| | Distance from upstream, 1850 m | | |
Index	TOASTS	OTIS	MIKE 11
R^2 (%)	99.75	99.8	62.5
RMSE (mg m^{-3})	0.030	0.047	0.50
MAE (mg m^{-3})	0.020	0.025	0.260
MRE (%)	1.70	4.77	28.60

Table 15. Simulation parameters related to test case 6.

Reach (m)	Dispersion coefficient (m^2 s^{-1})	Storage zone area (m^2)	Exchange coefficient (s^{-1})
0–213	0.50	0.20	1.07×10^{-3}
213–457	0.50	0.25	5.43×10^{-4}
457–726	0.50	0.14	1.62×10^{-2}

on the streambed sediments versus the observed data at the 105 and 281 m stations. It is clear from this figure and Table 13 that the TOASTS model is slightly better fitted to the observed data.

4.5   Test case 5: solute transport with transient storage in a river with irregular cross-sections

This test case shows the TOASTS model application for a river with irregular cross-sections under non-uniform flow conditions. The real data set for this test case was collected in a tracer experiment which has been done in the Athabasca River near Hinton, Alberta, Canada. Details of the experiments can be found in Putz and Smith (2000). In this study, the simulation reach length is 8.3 km, between 4.725 to 13.025 km of the river. The main reason for selecting this reach is that it has common geometric properties of rivers with storage zones. Total simulation time is 10 h, space and time steps are considered equal to 25 m and 1 min, respectively. The exchange coefficient is assumed equal to 6×10^{-4} s^{-1} by calibration. According to the estimated parameters, DaI is calculated equal to 3.8 which is in the acceptable range and therefore transient storage zones affect

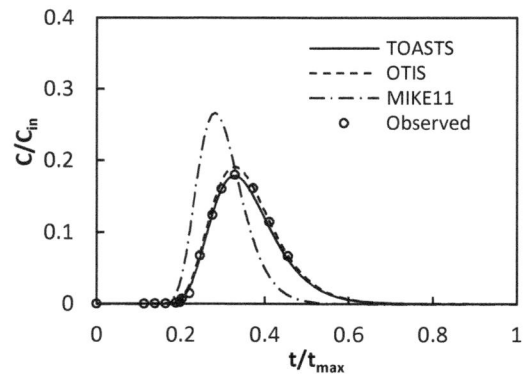

Figure 18. Simulation results for the Athabasca River experiment (test case 5).

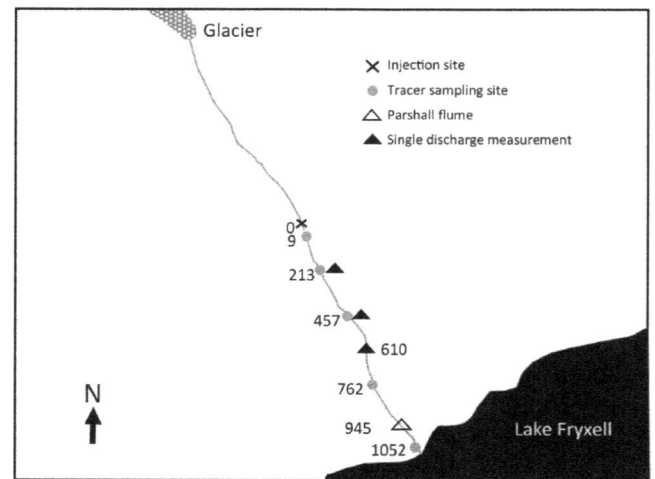

Figure 19. Huey Creek tracer study site map (Runkel et al., 1998).

downstream solute transport in the simulation reach. Since samples were collected only in four cross-sections downstream of the injection site, the observed concentration–time curve at 4.725 km was used as an upstream boundary condition of the transport model and the observed concentration–time curve at 11.85 km was used to compare the model results with the observed data. Figure 18 and Table 14 repre-

Table 16. Huey Creek experiment error indices (test case 6).

Index	213 m			457 m		
	TOASTS	OTIS	MIKE 11	TOASTS	OTIS	MIKE 11
R^2 (%)	68.6	67	84	78	63.5	94
RMSE (mg m^{-3})	0.673	0.674	0.740	0.48	0.63	0.62
MAE (mg m^{-3})	0.28	0.30	0.54	0.23	0.28	0.52
MRE (%)	7.14	7.32	20.40	6.46	7.60	15

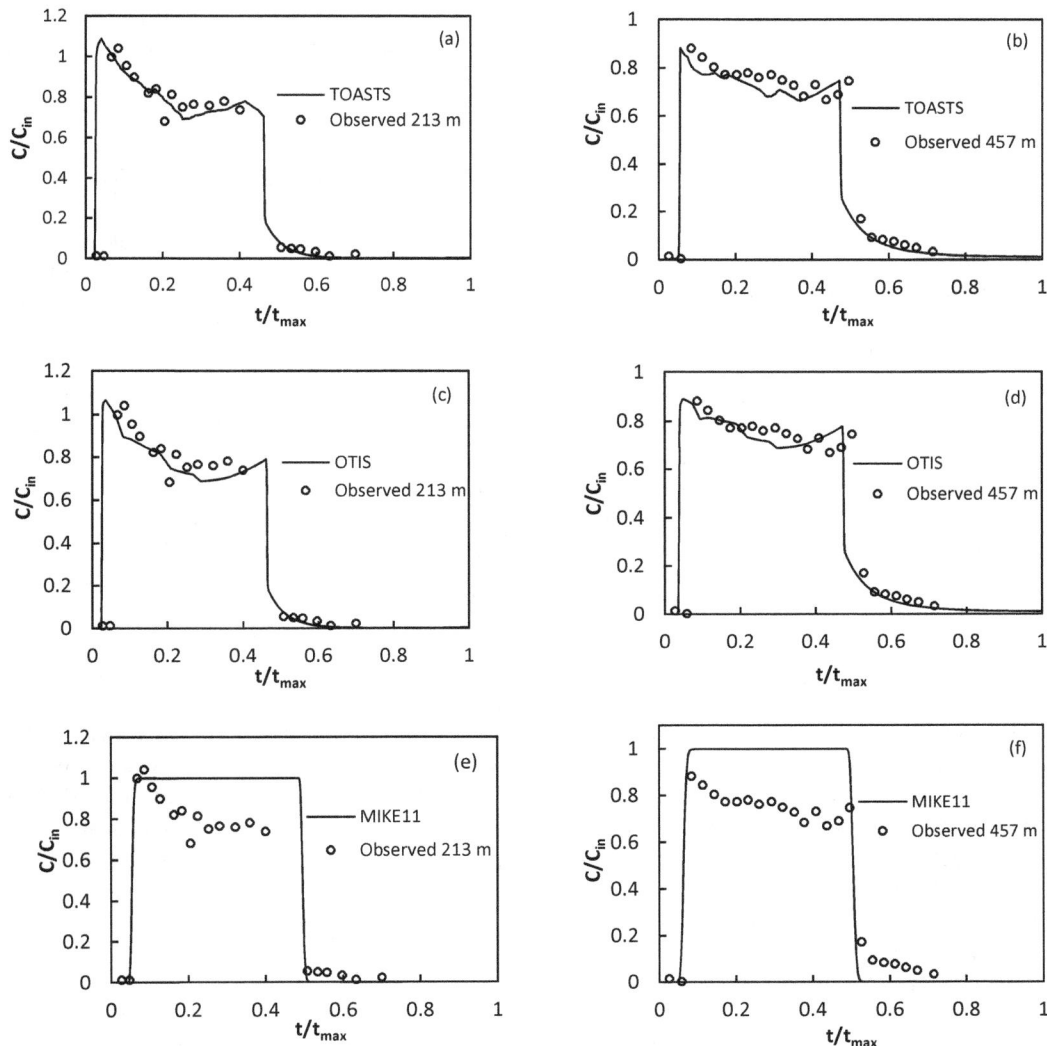

Figure 20. Observed and simulated main channel lithium concentrations (test case 6).

sent Athabasca experiment simulation results. It is clear that the concentration–time curves simulated by the TOASTS and OTIS models fit very well with the observed data; but again the MIKE 11 model failed to reproduce an accurate result, which means a poor performance of the classical advection–dispersion equation in simulation of solute transport in natural rivers.

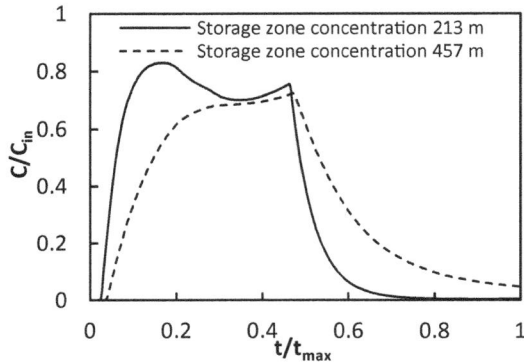

Figure 21. Simulated storage zone concentrations (test case 6).

4.6 Test case 6: solute transport with hyporheic exchange under unsteady flow conditions

This test case shows an application of the TOASTS model to simulate solute transport in a stream with irregular cross-sections, under an unsteady flow regime. In most of solute transport models, for simplification, flow is considered to be steady, while in most natural rivers unsteady flow condition is common, and neglecting temporal flow variations may lead to inaccurate results for solute transport simulation.

Tracer study that is used in this section, conducted in January 1992 at Huey Creek, located in McMurdo valleys, Antarctica (Fig. 19). The flow rate was variable from 1 to $4\,\mathrm{cfs^{-1}}$ (cubic feet per second) during the experiments. Since this stream does not have obvious surface storage zones, the cross-sectional area of storage zones and exchange rate of this area actually represent the rate of hyporheic exchange and interaction between surface and subsurface water (Runkel et al., 1998). Details of the experiments can be found in Runkel et al. (1998). Table 15 shows the simulation parameters. Figure 20a–c and Table 16 represent simulation results of lithium concentration at the 213 and 457 m stations, by the three subjected models. The results of the TOASTS model have a slightly better fitness to the observed data than the two other models. This figure also indicates that the general shape of the concentration–time curve for this example is a little different from the other examples. Figure 20c represents the results of the MIKE 11 model. As seen in this figure, results have large differences with the observed data in peak concentrations and general shape of the curve. Figure 21 shows the corresponding storage zone concentrations at 213 and 457 m stations. It can be seen that solute concentration–time curves in the storage zone have lower peaks and much longer tails, which imply longer residence time of solute in these areas compared to the main channel.

5 Conclusions

In this study a comprehensive model was developed that combines numerical schemes with high-order accuracy for solution of the advection–dispersion equation, considering transient storage zones term in rivers. In developing the subjected model (TOASTS), for achieving better accuracy and applicability, irregular-cross sections and an unsteady flow regime were considered. For this purpose the QUICK scheme, due to its high stability and low approximation error, has been used for spatial discretization.

The presented model was verified successfully using several analytical solutions and 2-D hydrodynamics and transport model as benchmarks. Also, its validation and applications were proved using several hypothetical examples and four sets of well-established tracer experiments data under different conditions. The main concluding remarks of this research are as the following:

The numerical scheme used in the TOASTS model (i.e., CTQS scheme), in cases where advection is the dominant transport process (higher Peclet numbers), has less numerical oscillations and higher stability compared to the CTCS and BTCS numerical schemes.

For a specified level of accuracy, TOASTS can provide larger grid size, while other models based on the central scheme face step limitation that leads to more computational cost.

As shown by other researchers, the inclusion of transient storage and sorption in a classical advection–dispersion equation, in many cases, leads to more accurate simulation results.

The TOASTS model is a comprehensive and practical model, that has the ability of solute transport simulation (reactive and non-reactive), with and without storage, under both steady and unsteady flow regimes, in rivers with irregular cross-sections that from this aspect is unique compared to the other existing models. Thus, it could be suggested as a reliable alternative to current popular models in solute transport studies in natural rivers and streams.

6 Data availability

In order to access the data, we kindly ask researchers to contact the corresponding author.

Edited by: Y. Chen

References

Avanzino, R. J., Zellweger, G., Kennedy, V., Zand, S., and Bencala, K.: Results of a solute transport experiment at Uvas Creek, September 1972, USGS Open-File Report, 84–236, 1984.

Bencala, K. E.: Simulation of solute transport in a mountain pool-and-riffle stream with a kinetic mass transfer model for sorption, Water Resour. Res., 19, 732–738, doi:10.1029/WR019i003p00732, 1983.

Bencala, K. E.: Interactions of solutes and streambed sediment: 2. A dynamic analysis of coupled hydrologic and chemical processes that determine solute transport, Water Resour. Res., 20, 1804–1814, doi:10.1029/WR020i012p01804, 1984.

Bencala, K. E., Mcknight, D. M., and Zellweger, G. W.: Characterization of transport in an acidic and metal-rich mountain stream based on a lithium tracer injection and simulations of transient storage, Water Resour. Res., 26, 989–1000, 1990.

Bencala, K. E. and Walters, R. A.: Simulation of Solute Transport in a Mountain Pool-and-Riffle Stream: A Transient Storage Model, Water Resour. Res., 19, 718–724, doi:10.1029/wr019i003p00718, 1983.

Chapra, S. C. and Runkel, R. L.: Modeling impact of storage zones on stream dissolved oxygen, J. Environ. Eng., 125, 415–419, doi:10.1061/(ASCE)0733-9372(1999)125:5(415), 1999.

Chapra, S. C. and Wilcock, R. J.: Transient storage and gas transfer in lowland stream, J. Environ. Eng., 126, 708–712, doi:10.1061/(ASCE)0733-9372(2000)126:8(708), 2000.

Czernuszenko, W., Rowinski, P.-M., and Sukhodolov, A.: Experimental and numerical validation of the dead-zone model for longitudinal dispersion in rivers, J. Hydraul. Res., 36, 269–280, doi:10.1080/00221689809498637, 1998.

Czernuszenko, W. and Rowinski, P.: Properties of the dead-zone model of longitudinal dispersion in rivers, J. Hydraul. Res., 35, 491–504, doi:10.1080/00221689709498407, 1997.

D'Angelo, D., Webster, J., Gregory, S., and Meyer, J.: Transient storage in Appalachian and Cascade mountain streams as related to hydraulic characteristics, J. N. Am. Benthol. Soc., 223–235, doi:10.2307/1467457, 1993.

Day, T. J.: Longitudinal dispersion in natural channels, Water Resour. Res., 11, 909–918, doi:10.1029/WR011i006p00909, 1975.

DeAngelis, D., Loreau, M., Neergaard, D., Mulholland, P., and Marzolf, E. Modelling nutrient-periphyton dynamics in streams: the importance of transient storage zones, Ecol. Modell., 80, 149–160, doi:10.1016/0304-3800(94)00066-Q, 1995.

Ensign, S. H. and Doyle, M. W.: In-channel transient storage and associated nutrient retention: Evidence from experimental manipulations, Limnol. Oceanogr., 50, 1740–1751, doi:10.4319/lo.2005.50.6.1740, 2005.

Fernald, A. G., Wigington, P., and Landers, D. H.: Transient storage and hyporheic flow along the Willamette River, Oregon: Field measurements and model estimates, Water Resour. Res., 37, 1681–1694, doi:10.1029/2000WR900338, 2001.

Godfrey, R. G. and Frederick, B. J.: Stream dispersion at selected sites, US Government Printing Office, 1970.

Gooseff, M. N., Hall, R. O., and Tank, J. L.: Relating transient storage to channel complexity in streams of varying land use in Jackson Hole, Wyoming, Water Resour. Res., 43, doi:10.1029/2005WR004626, 2007.

Harvey, J. W. and Wagner, B.: Quantifying hydrologic interactions between streams and their subsurface hyporheic zones, Streams and ground waters, 344 pp., 2000.

Jackman, A., Walters, R., and Kennedy, V.: Transport and concentration controls for Chloride, Strontium, potassium and lead in Uvas Creek, a small cobble-bed stream in Santa Clara County,

California, USA: 2, Mathematical modeling, J. Hydrol., 75, 111–141, doi:10.1016/0022-1694(84)90046-5, 1984.

Jackson, T. R., Haggerty, R., and Apte, S. V.: A fluid-mechanics based classification scheme for surface transient storage in riverine environments: quantitatively separating surface from hyporheic transient storage, Hydrol. Earth Syst. Sci., 17, 2747–2779, doi:10.5194/hess-17-2747-2013, 2013.

Jin, L., Siegel, D. I., Lautz, L. K., and Otz, M. H.: Transient storage and downstream solute transport in nested stream reaches affected by beaver dams, Hydrol. Process., 23, 2438–2449, doi:10.1002/hyp.7359, 2009.

Kazezyılmaz-Alhan, C. M.: Analytical solutions for contaminant transport in streams, J. Hydrol., 348, 524–534, doi:10.1016/j.jhydrol.2007.10.022, 2008.

Keefe, S. H., Barber, L. B., Runkel, R. L., Ryan, J. N., Mcknight, D. M., and Wass, R. D.: Conservative and reactive solute transport in constructed wetlands, Water Resour. Res., 40, doi:10.1029/2003WR002130, 2004.

Laenen, A. and Bencala, K. E.: transient storage assessments of dye-tracer injections in rivers of the Willamette basin, Oregon, JAWRA Journal of the American Water Resources Association, 37, 367–377, doi:10.1111/j.1752-1688.2001.tb00975.x, 2001.

Leonard, B. P.: A stable and accurate convective modelling procedure based on quadratic upstream interpolation, Comput. Method. Appl. M., 19, 59–98, doi:10.1016/0045-7825(79)90034-3, 1979.

Lin, Y.-C. and Medina Jr., M. A.: Incorporating transient storage in conjunctive stream–aquifer modeling, Adv. Water Resour., 26, 1001–1019, doi:10.1016/S0309-1708(03)00081-2, 2003.

Morrice, J. A., Valett, H., Dahm, C. N., and Campana, M. E.: Alluvial characteristics, groundwater–surface water exchange and hydrological retention in headwater streams, Hydrol. Process., 11, 253–267, 1997.

Neumann, L., Šimunek, J., and Cook, F.: Implementation of quadratic upstream interpolation schemes for solute transport into HYDRUS-1-D, Environ. Modell. Softw., 26, 1298–1308, doi:10.1016/j.envsoft.2011.05.010, 2011.

Nordin, C. F. and Sabol, G. V.: Empirical data on longitudinal dispersion in rivers, WRI, 74–20, 372 pp., 1974.

Nordin, C. F. and Troutman, B. M.: Longitudinal dispersion in rivers: The persistence of skewness in observed data, Water Resour. Res., 16, 123–128, doi:10.1029/WR016i001p00123, 1980.

Putz, G. and Smith, D. W.: Two-dimensional modelling of effluent mixing in the Athabasca River downstream of Weldwood of Canada Ltd., Hinton, Alberta, University of Alberta, 2000.

Ramaswami, A., Milford, J. B., and Small, M. J.: Integrated environmental modeling: pollutant transport, fate, and risk in the environment, J. Wiley, 2005.

Runkel, R. L.: One-dimensional transport with inflow andstorage (otis): a solute transport model for streams and rivers, Water-Resources Investigations Report, 1998.

Runkel, R. L., Mcknight, D. M., and Andrews, E. D.: Analysis of transient storage subject to unsteady flow: Diel flow variation in an Antarctic stream, J. N. Am. Benthol. Soc., 17, 143–154, doi:10.2307/1467958, 1998.

Scott, D. T., Gooseff, M. N., Bencala, K. E., and Runkel, R. L.: Automated calibration of a stream solute transport model: implications for interpretation of biogeochemical parameters, J. N. Am. Benthol. Soc., 22, 492–510, doi:10.2307/1468348, 2003.

Singh, S. K.: Treatment of stagnant zones in riverine advection-dispersion, J. Hydraul. Eng., 129, 470–473, doi:10.1061/(ASCE)0733-9429(2003)129:6(470), 2003.

Szymkiewicz, R.: Numerical modeling in open channel hydraulics, Springer, 2010.

Taylor, G.: The dispersion of matter in turbulent flow through a pipe, P. Roy. Soc. A, 223, 446–468, doi:10.1098/rspa.1954.0130, 1954.

Van Mazijk, A. and Veling, E.: Tracer experiments in the Rhine Basin: evaluation of the skewness of observed concentration distributions, J. Hydrol., 307, 60–78, doi:10.1016/j.jhydrol.2004.09.022, 2005.

Versteeg, H. K. and Malalasekera, W.: An introduction to computational fluid dynamics: the finite volume method, Pearson Education, 2007.

Wagner, B. J. and Harvey, J. W.: Experimental design for estimating parameters of rate-limited mass transfer: Analysis of stream tracer studies, Water Resour. Res., 33, 1731–1741, doi:10.1029/97WR01067, 1997.

Zhang, Y. and Aral, M. M.: Solute transport in open-channel networks in unsteady flow regime, Environ. Fluid Mech., 4, 225–247, doi:10.1023/B:EFMC.0000024237.17777.b1, 2004.

5

A comparison of the discrete cosine and wavelet transforms for hydrologic model input data reduction

Ashley Wright[1], Jeffrey P. Walker[1], David E. Robertson[2], and Valentijn R. N. Pauwels[1]

[1]Department of Civil Engineering, Monash University, Clayton, Victoria, Australia
[2]CSIRO, Land and Water, Clayton, Victoria, Australia

Correspondence to: Ashley Wright (ashley.wright@monash.edu)

Abstract. The treatment of input data uncertainty in hydrologic models is of crucial importance in the analysis, diagnosis and detection of model structural errors. Data reduction techniques decrease the dimensionality of input data, thus allowing modern parameter estimation algorithms to more efficiently estimate errors associated with input uncertainty and model structure. The discrete cosine transform (DCT) and discrete wavelet transform (DWT) are used to reduce the dimensionality of observed rainfall time series for the 438 catchments in the Model Parameter Estimation Experiment (MOPEX) data set. The rainfall time signals are then reconstructed and compared to the observed hyetographs using standard simulation performance summary metrics and descriptive statistics. The results convincingly demonstrate that the DWT is superior to the DCT in preserving and characterizing the observed rainfall data records. It is recommended that the DWT be used for model input data reduction in hydrology in preference over the DCT.

1 Introduction

Rainfall uncertainty is the biggest obstacle hydrologists face in their pursuit of accurate, precise and timely streamflow forecasts (McMillan et al., 2011). Unfortunately, errors in rainfall time series data may lead to hydrological model parameter estimates that produce adequate streamflow simulations only during the calibration period (Beven, 2006). This can lead to poor-quality streamflow predictions for independent periods and low confidence in the ability of streamflow forecasts. Consequently, a precise and accurate representation of rainfall uncertainty is paramount for robust hydrolog-

ical model parameter estimation, streamflow forecasting and quantitative precipitation forecasts (QPFs). Robertson et al. (2013) and Shrestha et al. (2015) have demonstrated that skill can be added to QPFs by postprocessing with past observations. As such, skill can be added to QPFs, and consequently flood forecasts, through developing a greater understanding of rainfall uncertainty.

The propagation of input errors in rainfall runoff modeling impedes the hydrologic community's ability to validate model structural error. Despite the vast amount of literature on rainfall measurement, estimation, statistical analysis (Testik and Gebremichael, 2010) and quality control procedures (World Meteorological Organization, 2014), a shroud of uncertainty still surrounds how rainfall and its associated uncertainty should be addressed in rainfall runoff modeling. The implementation of uncertainty analysis in many hydrological applications is also often limited by computational power.

Recent advancements in computational power as well as remote sensing have led to considerable improvements in availability and quality of hydrological observations (Cloke and Pappenberger, 2009). These improvements can be leveraged to increase the hydrological and flood forecasting knowledge base and consequently provide water policy decision makers and emergency management services with higher-quality information.

The advancement of computational power has also aided the search for hydrological model parameters that optimally simulate hydrological observations. These approaches initially focused on finding only the global optimum values of the parameters for a given objective function (Duan et al., 1994; Gan and Biftu, 1996; Thyer et al., 1999). However, in the past two decades, it has been recognized that the uncertainties in model parameters and predictions need to be estimated. Methods that seek to estimate parameter and prediction uncertainty include Bayesian recursive parameter estimation (Thiemann et al., 2001), the limits of acceptability approach (Beven, 2006; Blazkova and Beven, 2009), the Bayesian total error analysis (BATEA) framework (Kavetski et al., 2006a, b; Kuczera et al., 2006; Thyer et al., 2009; Renard et al., 2011), the simultaneous optimization and data assimilation (SODA) (Vrugt et al., 2005), the DREAM algorithm and its variations (Vrugt et al., 2005, 2008, 2009a, b; Vrugt and Ter Braak, 2011; Laloy and Vrugt, 2012; Sadegh and Vrugt, 2014), Bayesian model averaging (Butts et al., 2004; Ajami et al., 2007; Vrugt and Robinson, 2007), the hypothetico-inductive data-based mechanistic modeling framework of Young (2013) and Bayesian data assimilation (Bulygina and Gupta, 2011). It is through the development of these parameter estimation algorithms that hydrologists are able to explore input uncertainty.

Kavetski et al. (2006b) and Vrugt et al. (2008) identified the need to represent true catchment rainfall and its associated uncertainty using parameters, both applied a parametric approach to estimating true catchment rainfall and its associated uncertainty using a rainfall multiplier to storm events. The use of a parametric representation of rainfall with an effective sampling algorithm provides the ability to jointly estimate hydrologic model parameter distributions as well as input uncertainty. As in most hydrological problems, there is a lack of sufficient data to obtain a unique solution. However, Kavetski et al. (2006b) and Vrugt et al. (2008) found there were sufficient data to estimate both hydrological model parameters and rainfall input. Data reduction transformations offer the potential to reduce the dimensionality of the parameter estimation problem and thus enable a more robust inference. Signal transforms, such as Fourier and wavelet transforms, are examples of data reduction transformations that have been applied in hydrology; however, they have not previously been used to reduce the dimensionality of input data.

Fourier transforms use sinusoidal functions to represent the spectral component of an input signal; thus, a periodic signal could be represented using a smaller number of Fourier coefficients than the number of input data points. A pitfall of the Fourier transform is that it represents the spectral components of a signal, without any indication of the time localization of those specific spectral components. In order to account for this, the windowed Fourier transform (WFT), sometimes referred to as the short-time Fourier transform, segments the signal into discrete time windows before performing the Fourier analysis. A major drawback to this approach is that the uncertainty principle of signal processing imposes a limitation on the time and frequency resolutions that can be obtained for a given signal. As a response to this, Daubechies (1990) produced discrete basis functions with good time and frequency localization. In conjunction with the pyramid algorithm, as described by Mallat (1989), this work formed the basis for multi-resolution analysis with the discrete wavelet transform (DWT; Polikar, 1999). The DWT decomposes an input signal into high- and low-frequency components.

Wavelet analysis was first introduced to the geophysical sciences by Kumar and Foufoula-Georgiou (1997) and has been adopted for several different applications. Wavelet analysis has been used to assess the performance of hydrological models for parameter estimation (Schaefli and Zehe, 2009) to analyze changes over different time periods for both streamflow and precipitation data (Nalley et al., 2012). Various spectral methods have also been applied in hydrology, including the application of discrete Fourier transforms to calibrate water and energy balance models (Pauwels and De Lannoy, 2011) and for the calibration of the conceptual rainfall runoff model known as the probability distributed model (PDM) (De Vleeschouwer and Pauwels, 2013). While wavelet and spectral methods have been applied in the hydrological sciences, to date there have been no instances in which the suitability of different transforms has been compared for hydrological data reduction applications. Labat (2005) has pointed out that Fourier transforms and their derivatives are not well suited to reconstruct hydrologic data, which are generated by transient mechanisms. This is due to the Fourier transforms' poor capability to represent sporadic high-frequency events when dimensionally reduced. If model input data reduction techniques are to be accepted by the hydrologic community, it is of critical important that the transform used is able to reconstruct transient events. Through a comparative study, it will be shown that DWTs are a good multi-resolution alternative to the discrete cosine transform (DCT).

Traditionally, transform coefficients are the result of a convolution operation on an input signal. However, the aim of model input data reduction is to estimate these transform coefficients. Hence, they shall be referred to as transform parameters from herein. This paper provides novel theoretical and numerical comparisons of the DCT and DWT in a hydrological context. The ability of both transforms to reproduce key components of hydrological data sets is investigated. The extent to which each transform can reproduce hydrologic data using a decreasing number of parameters will serve as a metric upon which their ability to be used as a tool for model input data reduction for hydrological data will be evaluated. To address the requirements for hydrologic model input data reduction, this paper details (i) theoretical differences between the DCT and DWT, (ii) methodologies to reduce input rainfall to parameters and (iii) an evaluation of the proposed methodologies using several simulation performance summary metrics.

2 Model input data reduction theory

For this study, model input data reduction theory is introduced using a lumped conceptual watershed model. Consider a nonlinear model, $\mathcal{F}(\cdot)$, which simulates n discharge values, $\widehat{\boldsymbol{Y}} = \{\widehat{y}_1, \ldots, \widehat{y}_n\}$, in mm day^{-1} according to

$$\widehat{\boldsymbol{Y}} = \mathcal{F}\left(\boldsymbol{\theta}, \widetilde{\boldsymbol{x}}_0, \widehat{\boldsymbol{E}}, \widehat{\boldsymbol{R}}\right), \tag{1}$$

where the model input arguments are the $1 \times d$ vector $\boldsymbol{\theta}$, with arbitrary model parameter values, the $1 \times m$ vector $\widetilde{\boldsymbol{x}}_0$, with values of the initial states in millimeters and the $1 \times n$ vectors $\widehat{\boldsymbol{E}} = \{\widehat{e}_1, \ldots, \widehat{e}_n\}$ and $\widehat{\boldsymbol{R}} = \{\widehat{r}_1, \ldots, \widehat{r}_n\}$ which store the observed values of the potential evapotranspiration (PET) and rainfall in mm day^{-1}, respectively. Note that $\widehat{\boldsymbol{R}}$ is used to represent rainfall and not precipitation, as snow, hail and other forms of precipitation are not considered. The ⌢ (hat) symbol is used to denote measured quantities and the ˜ (tilde) symbol reflects variables that are either reconstructed or could, in theory, be observed in the field but due to their conceptual nature are difficult to determine accurately.

If the traditional hydrological perspective in which the inputs $\boldsymbol{E}$ and $\boldsymbol{R}$ are considered to be fixed and known quantities is relaxed, and rainfall is now considered unknown, then a new inference problem arises in which the input rainfall is estimated via the treatment of the input rainfall as a series of parameters. Inference problems in which the input is considered unknown can be dealt with using a Bayesian framework. Such inference problems have been considered by Kavetski et al. (2006a) and Vrugt et al. (2008) but are outside the scope of this paper. Consequently, for rainfall to be inferred, a suitable parametric representation of rainfall must be determined.

Given a daily rainfall data record with n observations in millimeters, n rainfall parameters could be used to represent the input hyetograph. This approach would be particularly elegant and parsimonious. Yet, for a 10-year record of daily discharge data, the inference problem would grow from d model parameters to roughly $10 \times 365 + d = 3650 + d$ parameters. These values would need to be estimated from the observed rainfall and discharge data record, respectively. As many hydrological models are already underdetermined, the introduction of additional parameters would make the model even less determinable. Additionally, an excessive amount of CPU time is required to solve for a 3600+ dimensional posterior parameter distribution. An alternative approach is therefore necessary.

Sparse transforms convey large amounts of data using fewer parameters than data points in the observed signal. An input rainfall signal can be reduced to sparse transform parameters. Doing so allows multiple rainfall observations to be modified using a single parameter. Some or all of these transform parameters can be altered before the transform is inverted to produce a new input signal for streamflow simulation and posterior analysis. The use of sparse transforms to represent input time series enables input uncertainty to be

explored in great detail. The ability of discrete wavelet and Fourier transformations to reduce hydrological input data to a set of parameters for uncertainty estimation is compared using theoretical and analytical methods.

2.1 Overview of the DCT and DWT

Wavelet and Fourier transforms are invertible transforms in which a forward convolution operation can be used to decompose a signal into various components. Similarly, a backwards deconvolution operation can be applied to retrieve the original signal. Fourier-based transforms decompose signals into frequency components and are best used for regular time-invariant signals that do not exhibit time-specific information. Alternatively, wavelet-based transforms decompose signals into frequency and time components. The advantage of using wavelet functions to transform data is that time-specific information about when higher frequency components occur can be preserved. To obtain time-specific information, Fourier-based transforms can be applied over pre-specified temporal windows. Yet, this approach is limited by the uncertainty principle of signal processing. The uncertainty principle of signal processing imposes a lower limit on obtainable resolutions in the time–frequency domain such that

$$\sigma_{\mathrm{t}} \sigma_{\omega} \geq \frac{1}{2}, \tag{2}$$

where σ_{t} (s) and σ_{ω} (s^{-1}) are the respective temporal and frequency widths used in the sparse transform.

Applying the uncertainty principle of signal processing (Eq. 2), it is clear that any attempt to narrow the temporal period analyzed to gain increased resolution in the time domain would be met by a widening of the frequency spectrum and consequently a loss of resolution in the frequency domain.

Considering that there is no time–frequency window that is able to obtain limitless resolution in both the time and frequency domains, it is clear that an alternative solution must be found. Wavelet transforms can be used to decompose a signal into different levels that consist of different time and frequency resolution windows. Thus, the wavelet transform is able to be configured to simultaneously obtain high levels of resolution in both the time and frequency domains. For a more detailed discussion on wavelets and sparse transforms, the reader is referred to Mallat (2009).

2.2 Discrete cosine transform

The DCT (Ahmed et al., 1974) is a version of the WFT that has advantageous properties for the field of data compression. Due to the boundary conditions of the cosine function, the DCT is well suited to represent an observed input signal with a minimal number of parameters: in this case, rainfall

$\widehat{\boldsymbol{R}}(t)$. The DCT parameters $\boldsymbol{p}(i)$ are calculated as

$$\boldsymbol{p}(i) = w(i) \sum_{t=1}^{n} \widehat{\boldsymbol{R}}(t) \cos\left[\frac{\pi}{2n}(2t-1)(i-1)\right], \tag{3}$$

where $i = 1, 2, \ldots, n$ and

$$w(i) = \begin{cases} \dfrac{1}{\sqrt{n}}, & i = 1 \\ \sqrt{\dfrac{2}{n}}, & 2 \leqslant i \leqslant n. \end{cases} \tag{4}$$

The convolution process can be reversed to reconstruct the observed signal using the inverse transform:

$$\widetilde{\boldsymbol{R}}(t) = \sum_{i=1}^{n} w(i)\,\boldsymbol{p}(i) \cos\left[\frac{\pi(2t-1)(i-1)}{2n}\right], \tag{5}$$

where $t = 1, 2, \ldots, \text{n}$.

2.3 Discrete wavelet transform

Using the pyramid algorithm, depicted in Fig. 1, Mallat (1989) first described the decomposition of an input signal into multi-resolution components using high- and low-pass filters. Each stage of decomposition is referred to as a level. An advantage of using wavelets is that decomposition can be performed using a variety of different wavelet families. This allows for signals with differing properties to be analyzed using the same methodology. The most commonly used wavelet family is the Daubechies wavelets (Daubechies, 1990). Each wavelet within each family consists of a scaling $h(m)$ and wavelet $w(m)$ function, where m denotes the length along the scaling and wavelet function. The scaling and wavelet functions are used in the low- and high-pass filtering sequences, respectively. Whilst there are numerous wavelet families that can be chosen for analysis, this study applies the most commonly used Daubechies wavelets. Depending on the choice of wavelet, stepwise convolutions of the input signal are performed over the filter length L. $j_{\max}$ imposes an upper limit on the level of decomposition j that a signal can be decomposed into, where

$$j_{\max} = \left\lfloor \log_2\left(\frac{n+L-1}{2}\right) \right\rfloor, \tag{6}$$

in which $\lfloor . \rfloor$ is the floor operator. The input signal is then convoluted by being passed through high- and low-pass filters, where

$$\boldsymbol{p}_j^{\text{L}}(i) = \begin{cases} \displaystyle\sum_{m=1}^{L} \widetilde{\boldsymbol{R}}(2i-m-1)w(m), & j = 1 \\ \displaystyle\sum_{m=1}^{L} \boldsymbol{p}_{j-1}^{L}(2i-m-1)w(m), & j > 1 \end{cases} \tag{7}$$

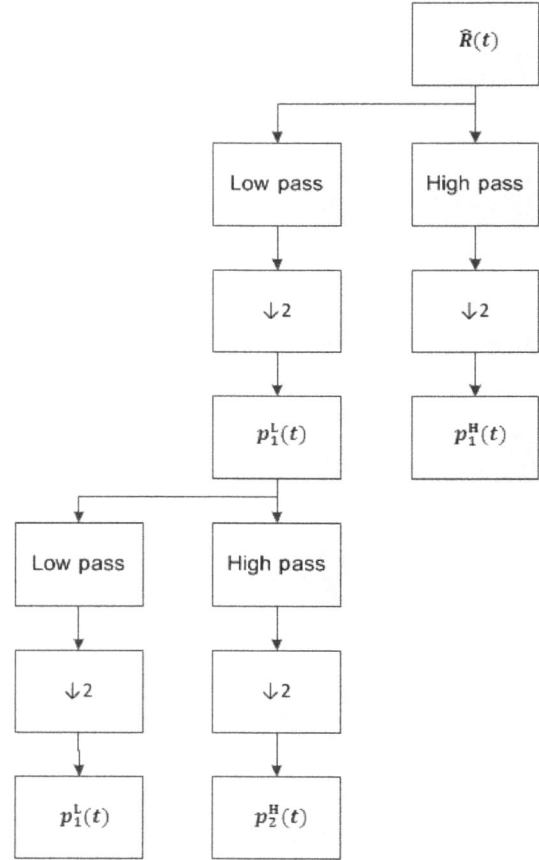

Figure 1. A schematic showing the pyramid algorithm used to decompose and downsample ($\downarrow 2$) an input signal ($\widehat{\boldsymbol{R}}$) into high- and low-frequency components. The input signal is filtered using the high- and low-pass filters described in Eqs. (7) and (8) before being downsampled to produce the level 1 high- and low-pass parameters. The low pass parameters are now used as input for the high- and low-pass filters. This process of filtering and downsampling is repeated until the desired level of decomposition is met.

is the low pass and

$$\boldsymbol{p}_j^{\text{H}}(i) = \begin{cases} \displaystyle\sum_{m=1}^{L} \widetilde{\boldsymbol{R}}(2i-m-1)h(m), & j = 1 \\ \displaystyle\sum_{m=1}^{L} \boldsymbol{p}_{j-1}^{L}(2i-m-1)h(m), & j > 1 \end{cases} \tag{8}$$

is the high pass, $i = 1, \ldots, n_{j-1} + L - 1$ and refers to the ith parameter, $j = 1, \ldots, j_{\max}$ and refers to the jth level, m refers to the mth filter coefficient. The resultant low-pass $\boldsymbol{p}_j^{\text{L}}(i)$ and high-pass $\boldsymbol{p}_j^{\text{H}}(i)$ parameters are commonly referred to as approximation and detail parameters, respectively. After the input signal is passed through the high- and low-pass filters there is an issue of redundancy that needs to be dealt with. The filters split the input signal into high- and low-frequency components that each contain roughly half the information of the input signal. As the length of each of the resultant approximation and detail parameter series is equivalent to the length of the input signal, each of the parameter series must be downsampled. The process of downsampling removes every other parameter. It is the process of high- and low-pass filtering followed by downsampling that enables the DWT to analyze multi-resolution components of a signal. After downsampling, the length of the resultant approximation and detail parameter series is

$$
n_j = \begin{cases} \left\lfloor \dfrac{n+L-1}{2} \right\rfloor, & j = 1 \\[2mm] \left\lfloor \dfrac{n_{j-1}+L-1}{2} \right\rfloor, & j > 1 \end{cases}, \tag{9}
$$

where n_j refers to the length of the series at the jth level. If further decomposition is required, the downsampled low pass may be fed back into the filters until the resultant parameters can no longer be split any further. An iteration of this process is shown in Fig. 1. To reverse the decomposition process and reconstruct a signal, upsampling is performed on the parameter series before the lower level parameters are obtained through

$$
p_{j-1}(i) = \sum_{m=\lceil i/2 \rceil}^{\lfloor (L-1+i)/2 \rfloor} \left(p_j^{\mathrm{H}}(i)h(2m-i) \right) \left(p_j(i)w(2m-i) \right),
$$
$$
j > 1, \tag{10}
$$

where $\lceil . \rceil$ is the ceiling operator and the input signal is reconstructed using

$$
\widetilde{\boldsymbol{R}}(i) = \sum_{m=\lceil i/2 \rceil}^{\lfloor (L-1+i)/2 \rfloor} \left(p_j^{\mathrm{H}}(i)h(2m-i) \right) \left(p_j(i)w(2m-i) \right),
$$
$$
j = 1. \tag{11}
$$

3 Data

This study utilizes data from the Model Parameter Estimation Experiment (MOPEX) data set. The 10 years of rainfall data spanning the 1990s for 438 catchments in the United States of America (USA) are used to compare the suitability of the DWT and DCT to represent rainfall time series. The catchments used in this study were chosen to ensure they had sufficient rain gauge density and represented a range of catchment sizes and climates. Rainfall for the Leaf River catchment (Collins, Mississippi), a catchment that is frequently used for hydrological studies (Sivakumar, 2001; Tang et al., 2006; Bulygina and Gupta, 2011), is used to compare the DWTs' and DCTs' ability to reconstruct high-magnitude rainfall events. A single rainfall product for each catchment is used for analysis at a daily time step. A complete description of the selection process and MOPEX data set is given by Schaake et al. (2006). No streamflow data are used in the experiment.

4 Experiment design

This experiment does not involve the use of any hydrological models. Due to this and the nature of the transforms, there are no calibration and evaluation periods. A major use of both the DWT and DCTs has been in image compression; consequently, the observed input signals were compressed and decompressed using a methodology similar to that used in image compression. In order to determine which transform's parameters are able to effectively store the most hydrological input data, both DWT and DCT parameters will be compressed to varying extents for the MOPEX rainfall time series.

The process undertaken involves a number of steps. Firstly, before any compression is applied, the original rainfall signal for a given catchment is transformed into DCT and DWT parameters using Eqs. (3) and (4) and Eqs. (7) to (9) for the DCT and DWT, respectively. Secondly, each transform is compressed by iteratively zeroing out parameters that provide a low degree of information; these parameters are those closest to zero. A threshold value T (mm) applies a lower limit for which transform parameters above the threshold are retained. This threshold is iteratively increased until the compressed transform is composed of the desired number of remaining parameters k and percent of original parameters (POP) is met.

$$
\mathrm{POP}(T) = 100 \times \left(\frac{k}{n} \right), \tag{12}
$$

where k becomes smaller as the threshold T increases and $\lim_{T \to \infty} \mathrm{POP} = 0$. The next step is to reconstruct the observed signal from the compressed transform parameters using Eqs. (5) and (11) for the DCT and DWT, respectively. After the reconstruction has been performed, a comparison between the reconstructed and observed rainfall can be made. Lastly, this process is iterated for different POPs as well as for each catchment within the data set.

To provide a meaningful comparison between the DCTs' and DWTs' ability to reproduce different rainfall time series with an increasing POP, a number of simulation performance summary metrics are used. Following Moriasi et al. (2007), a combination of graphical techniques and dimensionless and error index statistics that are widely accepted by the hydrological community were adopted for model evaluation. The Nash–Sutcliffe efficiency (NSE) and the root mean square error (RMSE) to standard deviation ratio (RSR) of the observed input signal ($\mathrm{RSR} = \mathrm{RMSE}/\sigma_{\mathrm{obs}}$) are used to compare the performance of the reconstructed rainfall signal with the observed rainfall signal. Once the reconstructed signals are obtained, further comparison with the observed rainfall will be made using the bias summary metric. The variance, kurtosis and skewness of the reconstructed signals will be compared with those of the observed signal. The bias is calculated as $\sum_{t=1}^{n} [\widehat{\boldsymbol{R}}(t) - \widetilde{\boldsymbol{R}}(t)]/n$, where $\widehat{\boldsymbol{R}}(t)$ and $\widetilde{\boldsymbol{R}}(t)$ are the observed and reconstructed rainfall signals, respectively. The reconstructed variance, kurtosis and skewness are all normalized by the observed input signals variance, kurtosis and skewness, respectively. The peak error (PE) is the peak rainfall error over the 10-year period. It is used to compare the reconstructed and observed signals for seasonal and flood forecasting situations. The PE is normalized by the peak height of the observed input signal. Further, the number of rain events missed is computed for each reconstruction by flagging original or reconstructed observations that exhibit no rainfall. Either the absolute difference between the reconstructed and original observation is less than 0.01 or the ratio of the reconstructed and original observation is equal to 0 or larger than 10. Lastly, reconstructed rainfall using the DCT and DWT will be presented for the Leaf River catchment to compare each transform's ability to reconstruct high-magnitude rainfall events.

5 Results

Figure 2 shows the relationships between RSR and the number of transform parameters using the DCT and DWT for three different catchments: Arroyo Chico, Skykomish River and Ohio Brush Creek. These catchments represent the smallest, largest and mean rainfall volumes for the MOPEX data set, respectively. It is clear that for all but the highest POP the DWT is able to reconstruct the observed signal with lower RSR than the DCT and that as the rainfall volume increases the RSR decreases. For intermediate POPs, the DWT is able to reconstruct the observed signal with significantly better RSR than the DCT. As the POP approaches both 100 and 0 %, there is little discernible difference between the DCT and DWT reconstructions.

By comparing the reconstructed DWT and DCT signals, using 20 POP and the observed rainfall signal as a reference, a histogram for the NSE is shown for all catchments in Fig. 3. Each frequency count in the histogram represents a catchment from the MOPEX data set. The reconstructed DWT signals are clearly able to better simulate the observed rainfall signal. All DWT reconstructed rainfall signals obtained a higher NSE than the DCT reconstructed rainfall signals. Table 1 shows that as the transforms are compressed and fewer parameters are used in the reconstruction, the mean NSE for the DWT stays much closer to the ideal value of 1 than the DCT. Further, the standard deviation of NSE becomes much larger for the DWT.

Figure 4 compares the RSR for the DCT and DWT using four different POPs. A 1 : 1 line is included in all subplots and each point represents a catchment from the data set. If the data points fall above the 1 : 1 line, then for that catchment and POP the DWT is able to reconstruct the input rainfall signal with lower RSR. Again, it is found that DWT is always able to reconstruct the original signal with lower RSR than the DCT reconstructions for all POPs. In a similar fashion to that discussed regarding Fig. 2, it is observed that as the POP

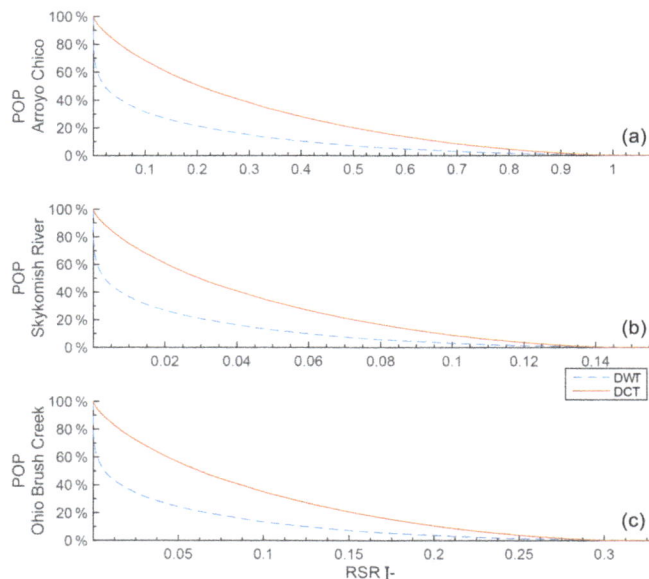

Figure 2. Empirical plots showing the relationship between RSR and the POP used for reconstructing an input rainfall signal using the DWT and DCT. The three catchments, from the top to the bottom of the figure, represent the smallest, largest and mean rainfall volumes throughout the 1990s for the MOPEX data set.

Table 1. The mean and standard deviation (SD) of NSE for the DWT and DCT using a different POP.

POP	NSE DWT		NSE DCT	
	Mean	SD	Mean	SD
40 %	0.988	0.007	0.918	0.010
30 %	0.965	0.017	0.844	0.016
20 %	0.905	0.036	0.729	0.025
10 %	0.746	0.070	0.522	0.037

approaches 0 % the difference between the DWT and DCT reconstructions becomes smaller.

The bias, variance and skewness observed in the reconstructed signals for each catchment are shown in Fig. 5 for different POPs. The DWT reconstructions are able to maintain a smaller bias than the DCT reconstructions at different POPs for all of the catchments. As the POP decreases, the bias becomes increasingly positive and negative for the DWT and DCT, respectively. The distribution of the bias becomes more dispersed for both the DCT and DWT as the POP decreases. The bias can be seen to be dependent on the transform and POP used as well as the catchment being analyzed. Both the DWT and DCT never reconstruct the observed signal with greater variance than that of the observed rainfall signal. As the POP decreases, the normalized variance for the DCT moves further away from unity than the normalized variance for the DWT. The reduction in normalized variance means that, as the POP decreases, both the DWT and espe-

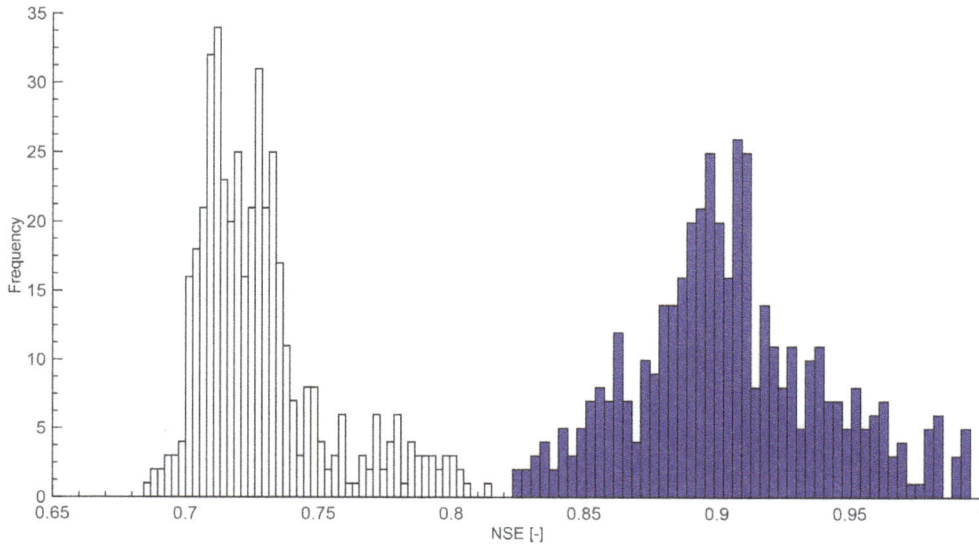

Figure 3. Histogram representing the reconstructed DWT (dark bins) and DCT (clear bins) NSE when compared to the observed rainfall signal. Rainfall is reconstructed after the input signal is compressed to 20 POP. Each frequency count represents a catchment from the MOPEX data set.

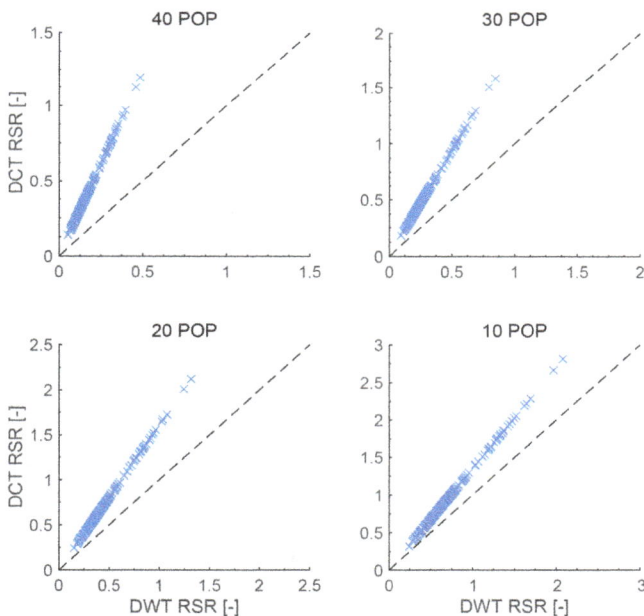

Figure 4. Comparative plots of RSR for the DCT and DWT using a different POP. Each data point represents a catchment.

cially the DCT reconstructions will have fewer extreme values when compared to the observed rainfall. The normalized skewness is a measure of symmetry that describes whether or not the reconstructed signal is more positively skewed (more than 1) or less positively skewed (less than 1) than the observed input signal. All of the reconstructed and observed signals had a positive normalized skewness. When compared to the observed signal, the DWT becomes increasingly skewed as the POP is reduced. The opposite of this is observed for the DCT. This indicates that, when compressed, the DWT and DCT will reconstruct the observed rainfall signal with a greater and lower number of values close to zero when compared to the observed signal, respectively. This does not mean that the total volume will be any lower than the total volume of rainfall observed. This is made evident by the low bias observed in Fig. 5.

The normalized kurtosis and PE for all catchments using different POPs are shown in Fig. 6. The measure of kurtosis describes how much the fraction of the distributions' variance is explained by extreme deviations. Consequently, a normalized kurtosis value larger than 1 indicates that the reconstructed signals variance is explained more by extreme deviations than the observed input signal. This is likely to be the result of more rainfall values being reconstructed at the extremities than those of the observed rainfall series. A value smaller than one indicates that the variance is described less by extreme deviations than the observed input signal. Similarly, this is likely to be the result of fewer rainfall values being reconstructed at the extremities than those of the observed rainfall series. It is worth noting that a reconstructed time series can have the same variance yet different kurtosis than the observed rainfall time series. As the POP decreases, the dispersion of normalized kurtosis and skewness increases, and the normalized kurtosis and skewness for the DWT and DCT reconstructions become larger and smaller than unity, respectively. With decreasing POP, the normalized PE for the reconstructed DWT signal remains small and relatively consistent when compared to the normalized PE for the reconstructed DCT signal.

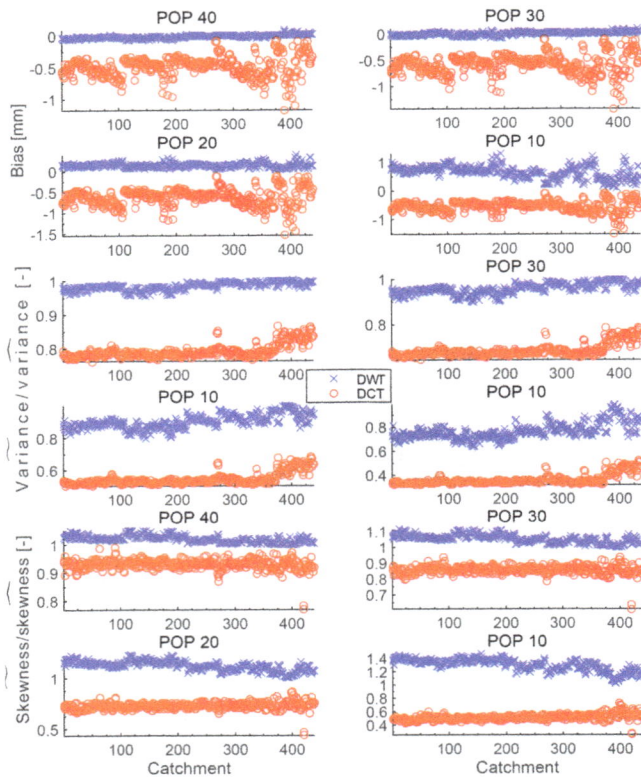

Figure 5. Bias and normalized variance and skewness of the reconstructed DWT and DCT signals for each catchment using a different POP.

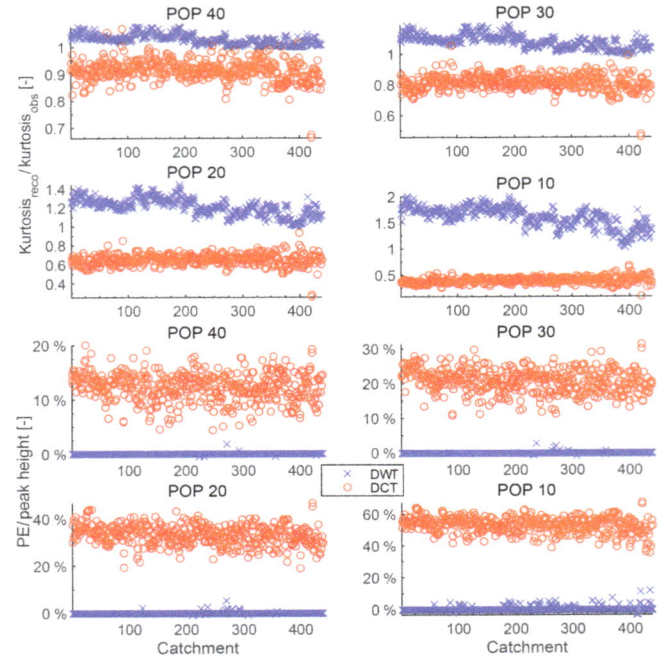

Figure 6. Normalized kurtosis of the reconstructed DWT and DCT signals and percentage PE for the reconstructed DWT and DCT signals for each catchment using a different POP.

6   Discussion

Figure 3 shows that the DWT and DCT are able to reconstruct the observed input signals with good efficiency using 20 POP. However, the DWT consistently outperforms the DCT. Fig. 2 shows that, as the POP decreases from 100 %, the DWT is able to reconstruct the input signal with increasingly lower RSR than the DCT; the gap in performance is largest for 40 POP. As the POP continues to decrease towards 0 %, the gap in RSR reduces to zero. It is interesting to note that the DWT perfectly reconstructs the observed rainfall signal with as many parameters as there are rainy days, whereas the DCT does not.

As the bias for the DWT is consistently close to zero, the use of the DWT for rainfall input data reduction is likely to be beneficial for hydrologic studies that have short time steps and involve rainfall as an input. Whilst modification of the DWT parameters may slightly overestimate input rainfall, it is not as significant as the consistent underestimation of input rainfall by the DCT. The diminishing ability of both the DWT and DCT to match the input rainfall signal variance indicates that both transforms smooth out input data towards the mean. This behavior is more significant for the DCT than the DWT. Consequently, when used as a technique for input data reduc-

tion, the DWT will reconstruct temporal variances better than the DCT. The increased skewness for the reconstructed DWT signals compared to the observed input signals indicates that there is an increased reconstruction of low-magnitude rainfall events. On the contrary, the decreased normalized skewness for the reconstructed DCT signals indicates that a number of the low-magnitude rainfall events are tending to be reconstructed towards the mean. The kurtosis results shown in Fig. 6 demonstrate that, when compared to the observed input signal, events of extreme deviation explain more of the variance for the reconstructed DWT and less of the variance for the reconstructed DCT. Consequently, as the nature of the extreme deviations is a critical piece of information, the use of the DCT for model input data reduction for hydrologic studies that have short time steps and involving rainfall as an input is not recommended. It is also seen in Fig. 6 that the DCT is more likely to miss peak rainfall height information. Consequently, care needs to be taken when choosing a transform when peak height is critical. Further, the DCT should not be used for studies involving flood forecasting situations where the accuracy of peak height is critical.

Whilst it is important that rain gauges measure high-magnitude rainfall events with accuracy and precision, it is also important that low-magnitude rainfall events are recorded. Consequently, when evaluating the merits of the DCT and DWT to reconstruct rainfall it would be prudent to analyze the frequency in which each transform is either unable to reconstruct a rainfall event or erroneously constructs

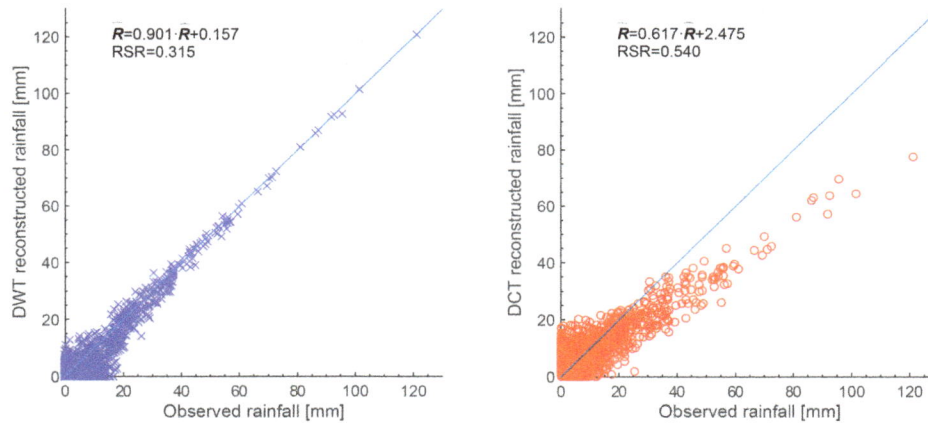

Figure 7. Comparison of the reconstructed DCT and DWT signal for the Leaf River (Collins) catchment using 20 POP.

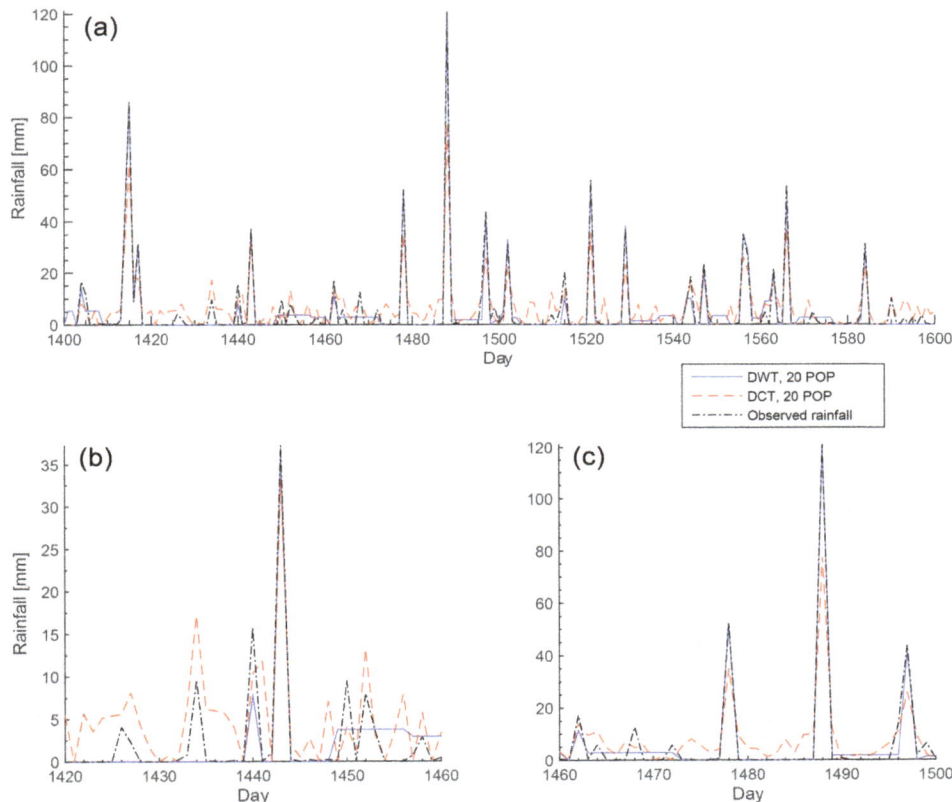

Figure 8. Panel **(a)** shows a time series comparison of the reconstructed DCT and DWT signals for the Leaf River (Collins) catchment using 20 POP for a period of 200 days. Panels **(b)** and **(c)** are smaller windows of the same time series during both low- and high-rainfall periods.

a rainfall event. Table 2 illustrates that, at times, both transforms will either fail to reconstruct a low-magnitude rainfall event or will erroneously construct a rainfall event when there was none observed in the original rainfall time series. In general, the DWT outperforms the DCT. The exception to this is at 10 POP. This is a result of the discrete nature of the DWT analysis function as opposed to the continuous analysis function used in the DCT. As the POP decreases towards zero, both transforms miss more rainfall events.

Due to rapid increases in rainfall intensity, high-magnitude rainfall events tend to have high-frequency components. In Fig. 7, the smoothing of high-frequency, high-magnitude rainfall events by the DCT is made evident by the lower slope of the linear least squares fit for the DCT reconstruc-

Table 2. The mean and standard deviation (SD) for the number of missed rainfall events for the DWT and DCT using a different number of parameters.

| | Number of missed rainfall events | | | |
| | DWT | | DCT | |
POP	Mean	SD	Mean	SD
40 %	239.004	117.317	587.934	155.375
30 %	398.005	138.793	645.495	159.378
20 %	581.591	145.769	696.288	168.524
10 %	852.340	168.590	748.075	184.910

tion of Leaf River observed rainfall data when compared to the DWT. This shows that the compressed DWT is able to retain more detail for high-magnitude rainfall events than the DCT. Using 20 POP, 730 DWT parameters are able to reconstruct observed rainfall with an RSR of 0.315, whereas 730 DCT parameters are able to reconstruct observed rainfall with an RSR of 0.540. Figures 7 and 8 show that the DWT often misses and sometimes smooths out low-magnitude rainfall events; the DCT, however, does reconstruct inaccurate rainfall at these times. Figure 8 also demonstrates that, at lower POPs, the DCT will smooth out and underestimate high-magnitude events whilst the DWT will maintain accuracy and precision.

7   Conclusions

Succinct descriptions of the DCT and DWT were provided to determine the suitability of each transform to be used as a tool for hydrologic model input data reduction. Due to their different construction, each transform provides different possibilities for use in model input data reduction. Since it is infeasible to estimate all transform parameters, the modeller could choose to estimate high- or low-frequency parameters of the DCT. This would result in minimal control of the temporal component being modified. Due to the multi-level decomposition of an input signal into high- and low-frequency parameters by the DWT, the modeller is able to specify the estimation of both time and frequency components. Hence, portions of the input data record can be targeted for estimation. The use of the DWT as a hydrologic model input data reduction technique allows the modeller more flexible options. A comparison of the DWTs' and DCTs' ability to reconstruct MOPEX rainfall data using standard simulation performance summary metrics, descriptive statistics and peak errors was then made, and it was found that the DWT is most efficient at preserving high-magnitude and transient rainfall events. Thus, it is recommended that the DWT be used as a model input data reduction technique for hydrologic studies that have short time steps and involve rainfall as an input. Considering that the bias for the reconstructed

DWT rainfall signal is consistently lower than that of the reconstructed DCT signal and that the skewness, kurtosis and variance are also closest to the input rainfall signal, it is recommended that the DWT also be used as a model input data reduction technique for hydrologic studies that have long time steps with rainfall as an input.

Author contributions. AW conducted the experimental work, contributed towards the theory and wrote the paper. JW and DR assisted in the writing process. VP contributed towards the theory and assisted in the writing process.

Acknowledgements. The authors would like to extend their gratitude to Jasper Vrugt, Hamid Bazargan and the anonymous reviewers for their comments and recommendations. This work was supported by the Multi-modal Australian Sciences Imaging and Visualisation Environment (MASSIVE) (http://www.massive.org.au), a Monash University Engineering Research Living Allowance stipend and a top-up scholarship from the Bushfire & Natural Hazards Cooperative Research Centre. Valentijn Pauwels is funded by ARC grant FT130100545.

Edited by: Insa Neuweiler

References

Ahmed, N., Natarajan, T., and Rao, K.: Discrete Cosine Transform, IEEE T. Comput., C-23, 90–93, https://doi.org/10.1109/T-C.1974.223784, 1974.

Ajami, N., Duan, Q., and Sorooshian, S.: An integrated hydrologic Bayesian multimodel combination framework: Confronting input, parameter, and model structural uncertainty in hydrologic prediction, Water Resour. Res., 43, W01403, https://doi.org/10.1029/2005WR004745, 2007.

Beven, K.: A manifesto for the equifinality thesis, J. Hydrol., 320, 18–36, https://doi.org/10.1016/j.jhydrol.2005.07.007, 2006.

Blazkova, S. and Beven, K.: A limits of acceptability approach to model evaluation and uncertainty estimation in flood frequency estimation by continuous simulation: Skalka catchment, Czech Republic, Water Resources Research, 45, W00B16, https://doi.org/10.1029/2007WR006726, 2009.

Bulygina, N. and Gupta, H.: Correcting the mathematical structure of a hydrological model via Bayesian data assimilation, Water Resour. Res., 47, WO5514, https://doi.org/10.1029/2010WR009614, 2011.

Butts, M., Payne, J., Kristensen, M., and Madsen, H.: An evaluation of the impact of model structure on hydrological modelling uncertainty for streamflow simulation, J. Hydrol., 298, 242–266, https://doi.org/10.1016/j.jhydrol.2004.03.042, 2004.

Cloke, H. L. and Pappenberger, F.: Ensemble flood forecasting: A review, J. Hydrol., 375, 613–626, 2009.

Daubechies, I.: The Wavelet Transform, Time-Frequency Localization and Signal Analysis, IEEE T. Inform. Theory, 36, 961–1005, https://doi.org/10.1109/18.57199, 1990.

De Vleeschouwer, N. and Pauwels, V. R. N.: Assessment of the indirect calibration of a rainfall-runoff model for ungauged catchments in Flanders, Hydrol. Earth Syst. Sci., 17, 2001–2016, https://doi.org/10.5194/hess-17-2001-2013, 2013.

Duan, Q., Sorooshian, S., and Gupta, V. K.: Optimal use of the SCE-UA global optimization method for calibrating watershed models, J. Hydrol., 158, 265–284, 1994.

Gan, T. Y. and Biftu, G. F.: Automatic calibration of conceptual rainfall-runoff models: Optimization algorithms, catchment conditions, and model structure, Water Resour. Res., 32, 3513–3524, 1996.

Kavetski, D., Kuczera, G., and Franks, S.: Bayesian analysis of input uncertainty in hydrological modeling: 2. Application, Water Resour. Res., 42, W03408, https://doi.org/10.1029/2005WR004376, 2006a.

Kavetski, D., Kuczera, G., and Franks, S.: Bayesian analysis of input uncertainty in hydrological modeling: 1. Theory, Water Resour. Res., 42, W03407, https://doi.org/10.1029/2005WR004368, 2006b.

Kuczera, G., Kavetski, D., Franks, S., and Thyer, M.: Towards a Bayesian total error analysis of conceptual rainfall-runoff models: Characterising model error using storm-dependent parameters, J. Hydrol., 331, 161–177, https://doi.org/10.1016/j.jhydrol.2006.05.010, 2006.

Kumar, P. and Foufoula-Georgiou, E.: Wavelet analysis for geophysical applications, Rev. Geophys., 35, 385–412, 1997.

Labat, D.: Recent advances in wavelet analyses: Part 1. A review of concepts, J. Hydrol., 314, 275–288, 2005.

Laloy, E. and Vrugt, J.: High-dimensional posterior exploration of hydrologic models using multiple-try DREAM (ZS) and high-performance computing, Water Resour. Res., 48, W01526, https://doi.org/10.1029/2011WR010608, 2012.

Mallat, S.: A Theory for Multiresolution Signal Decomposition: The Wavelet Representation, IEEE T. Pattern Anal., 11, 674–693, https://doi.org/10.1109/34.192463, 1989.

Mallat, S.: A Wavelet Tour of Signal Processing, third edition, Academic Press, Boston, USA, https://doi.org/10.1016/B978-0-12-374370-1.50001-9, 2009.

McMillan, H., Jackson, B., Clark, M., Kavetski, D., and Woods, R.: Rainfall uncertainty in hydrological modelling: An evaluation of multiplicative error models, J. Hydrol., 400, 83–94, 2011.

Moriasi, D. N., Arnold, J. G., Van Liew, M. W., Bingner, R. L., Harmel, R. D., and Veith, T. L.: Model evaluation guidelines for systematic quantification of accuracy in watershed simulations, T. ASABE, 50, 885–900, 2007.

Nalley, D., Adamowski, J., and Khalil, B.: Using discrete wavelet transforms to analyze trends in streamflow and precipitation in Quebec and Ontario (1954–2008), J. Hydrol., 475, 204–228, https://doi.org/10.1016/j.jhydrol.2012.09.049, 2012.

National Weather Service (NWS): Model Parameter Estimation Experiment (MOPEX), National Oceanic and Atmospheric Administration (NOAA), available at: ftp://hydrology.nws.noaa.gov/pub/gcip/mopex/US_Data/, last access: 4 January 2017.

Pauwels, V. and De Lannoy, G.: Multivariate calibration of a water and energy balance model in the spectral domain, Water Resour. Res., 47, W07523, https://doi.org/10.1029/2010WR010292, 2011.

Polikar, R.: The story of wavelets, Physics and modern topics in mechanical and electrical engineering, World Scientific and Engineering Society Press, Wisconsin, USA, 192–197, 1999.

Renard, B., Kavetski, D., Leblois, E., Thyer, M., Kuczera, G., and Franks, S.: Toward a reliable decomposition of predictive uncertainty in hydrological modeling: Characterizing rainfall errors using conditional simulation, Water Resour. Res., 47, W11516, https://doi.org/10.1029/2011WR010643, 2011.

Robertson, D. E., Shrestha, D. L., and Wang, Q. J.: Post-processing rainfall forecasts from numerical weather prediction models for short-term streamflow forecasting, Hydrol. Earth Syst. Sci., 17, 3587–3603, https://doi.org/10.5194/hess-17-3587-2013, 2013.

Sadegh, M. and Vrugt, J.: Approximate Bayesian Computation using Markov Chain Monte Carlo simulation: DREAM(ABC), Water Resour. Res., 50, 6767–6787, https://doi.org/10.1002/2014WR015386, 2014.

Schaake, J., Cong, S., and Duan, Q.: The US mopex data set, in: Large Sample Basin Experiments for Hydrological Model Parameterization: Results of the Model Parameter Experiment – MOPEX, edited by: Andréassian, V., Hall, A., Chahinian, N., and Schaake, J., 9–28, IAHS Publ. no. 307, IAHS Press, Wallingford, UK, 2006.

Schaefli, B. and Zehe, E.: Hydrological model performance and parameter estimation in the wavelet-domain, Hydrol. Earth Syst. Sci., 13, 1921–1936, https://doi.org/10.5194/hess-13-1921-2009, 2009.

Shrestha, D., Robertson, D., Bennett, J., and Wang, Q.: Improving precipitation forecasts by generating ensembles through postprocessing, Mon. Weather Rev., 143, 3642–3663, https://doi.org/10.1175/MWR-D-14-00329.1, 2015.

Sivakumar, B.: Rainfall dynamics at different temporal scales: A chaotic perspective, Hydrol. Earth Syst. Sci., 5, 645–652, https://doi.org/10.5194/hess-5-645-2001, 2001.

Tang, Y., Reed, P., and Wagener, T.: How effective and efficient are multiobjective evolutionary algorithms at hydrologic model calibration?, Hydrol. Earth Syst. Sci., 10, 289–307, https://doi.org/10.5194/hess-10-289-2006, 2006.

Testik, F. Y. and Gebremichael, M. (Eds.): Rainfall: State of the Science, Geoph. Monog. Series, 191, 1–287, https://doi.org/10.1029/gm191, 2010.

Thiemann, M., Trosset, M., Gupta, H., and Sorooshian, S.: Bayesian recursive parameter estimation for hydrologic models, Water Resour. Res., 37, 2521–2535, 2001.

Thyer, M., Kuczera, G., and Bates, B. C.: Probabilistic optimization for conceptual rainfall-runoff models: A comparison of the shuffled complex evolution and simulated annealing algorithms, Water Resour. Res., 35, 767–773, 1999.

Thyer, M., Renard, B., Kavetski, D., Kuczera, G., Franks, S., and Srikanthan, S.: Critical evaluation of parameter consistency and predictive uncertainty in hydrological modeling: A case study using Bayesian total error analysis, Water Resour. Res., 45, W00B14, https://doi.org/10.1029/2008WR006825, 2009.

Vrugt, J. A. and Robinson, B.: Treatment of uncertainty using ensemble methods: Comparison of sequential data assimilation and

Bayesian model averaging, Water Resour. Res., 43, W01411, https://doi.org/10.1029/2005WR004838, 2007.

Vrugt, J. A. and Ter Braak, C. J. F.: DREAM$_{(D)}$: an adaptive Markov Chain Monte Carlo simulation algorithm to solve discrete, noncontinuous, and combinatorial posterior parameter estimation problems, Hydrol. Earth Syst. Sci., 15, 3701–3713, https://doi.org/10.5194/hess-15-3701-2011, 2011.

Vrugt, J. A., Diks, C., Gupta, H., Bouten, W., and Verstraten, J.: Improved treatment of uncertainty in hydrologic modeling: Combining the strengths of global optimization and data assimilation, Water Resour. Res., 41, 1–17, https://doi.org/10.1029/2004WR003059, 2005.

Vrugt, J. A., Ter Braak, C., Clark, M., Hyman, J., and Robinson, B.: Treatment of input uncertainty in hydrologic modeling: Doing hydrology backward with Markov chain Monte Carlo simulation, Water Resour. Res., 44, W00B09, https://doi.org/10.1029/2007WR006720, 2008.

Vrugt, J. A., ter Braak, C., Gupta, H., and Robinson, B.: Equifinality of formal (DREAM) and informal (GLUE) Bayesian approaches in hydrologic modeling?, Stoch. Env. Res. Risk A., 23, 1011–1026, https://doi.org/10.1007/s00477-008-0274-y, 2009a.

Vrugt, J. A., ter Braak, C. J. F., Diks, C. G. H., Robinson, B. A., Hyman, J. M., and Higdon, D.: Accelerating Markov Chain Monte Carlo Simulation by Differential Evolution with Self-Adaptive Randomized Subspace Sampling, Int. J. Nonlin. Sci. Num., 10, 273–290, 2009b.

World Meteorological Organization: Guide to Meteorological Instruments and Methods of Observation, Geneva, Switzerland, 2014.

Young, P.: Hypothetico-inductive data-based mechanistic modeling of hydrological systems, Water Resour. Res., 49, 915–935, https://doi.org/10.1002/wrcr.20068, 2013.

6

Kalman filters for assimilating near-surface observations into the Richards equation – Part 1: Retrieving state profiles with linear and nonlinear numerical schemes

G. B. Chirico[1]**, H. Medina**[2]**, and N. Romano**[1]

[1]Department of Agricultural Engineering, University of Naples Federico II, Naples, Italy
[2]Department of Basic Sciences, Agrarian University of Havana, Havana, Cuba

Correspondence to: G. B. Chirico (gchirico@unina.it)

Abstract. This paper examines the potential of different algorithms, based on the Kalman filtering approach, for assimilating near-surface observations into a one-dimensional Richards equation governing soil water flow in soil. Our specific objectives are: (i) to compare the efficiency of different Kalman filter algorithms in retrieving matric pressure head profiles when they are implemented with different numerical schemes of the Richards equation; (ii) to evaluate the performance of these algorithms when nonlinearities arise from the nonlinearity of the observation equation, i.e. when surface soil water content observations are assimilated to retrieve matric pressure head values. The study is based on a synthetic simulation of an evaporation process from a homogeneous soil column. Our first objective is achieved by implementing a Standard Kalman Filter (SKF) algorithm with both an explicit finite difference scheme (EX) and a Crank-Nicolson (CN) linear finite difference scheme of the Richards equation. The Unscented (UKF) and Ensemble Kalman Filters (EnKF) are applied to handle the nonlinearity of a backward Euler finite difference scheme. To accomplish the second objective, an analogous framework is applied, with the exception of replacing SKF with the Extended Kalman Filter (EKF) in combination with a CN numerical scheme, so as to handle the nonlinearity of the observation equation. While the EX scheme is computationally too inefficient to be implemented in an operational assimilation scheme, the retrieval algorithm implemented with a CN scheme is found to be computationally more feasible and accurate than those implemented with the backward Euler scheme, at least for the examined one-dimensional problem. The UKF appears to be as feasible as the EnKF when one has to handle nonlinear numerical schemes or additional nonlinearities arising from the observation equation, at least for systems of small dimensionality as the one examined in this study.

1 Introduction

Soil water in the vadose zone exerts a large control on the water and energy balance of land-atmosphere systems over a wide range of space-time scales (e.g. Milly and Dunne, 1994; Entekhabi et al., 1996; Vrugt et al., 2001, 2003; Rodriguez-Iturbe and Porporato, 2005). With the increasing availability of near-surface data from remote and ground-based sensors, unique opportunities emerge to predict the soil water dynamics (McLaughlin, 2002; Vereecken et al., 2008). A key challenge is to identify the best approaches for efficiently integrating these data with the soil water dynamic models, in order to achieve more reliable and purposeful predictions. Hence, data assimilation has become a relatively important area of investigation, aiming at an efficient integration of remote-sensing techniques, ground-based sensors and soil water dynamic models (Hoeben and Troch, 2000; Heathman et al., 2003; de Lannoy et al., 2007; Matgen et al., 2010).

The physics of isothermal flow in unsaturated soils is commonly modelled with the Richards equation (Jury et al., 1991). Three standard forms of the unsaturated flow equation can be found in literature: (i) the "h-based form" and (ii) the "θ-based form", whether the dependent variable is matric pressure head, h [L], or soil water content θ [L^3 L^{-3}], respectively; (iii) the "mixed form" when both the dependent variables are employed. The water retention $\theta(h)$ and the hydraulic conductivity $K(\theta)$ [L T^{-1}] functions provide constitutive relationships between those two variables and the hydraulic conductivity K, allowing for conversion of one form of the equation to the other one.

A primary source of numerical difficulty when dealing with the Richards equation is its strongly nonlinear nature. The standard numerical approximations that are applied to the spatial domain are the finite difference method and the finite element method. For any Euler method other than the fully explicit forward method, nonlinear algebraic equations result and some linearization and/or iteration procedure must be implemented to solve the discrete equations (Celia et al., 1990).

The Kalman Filter (Kalman, 1960) is a sequential data assimilation technique, largely employed in hydrological applications to merge observations arising from different sources, such as remote-sensing instruments or ground-based sensors, with dynamic models (McLaughlin, 2002), and it can be considered as the most general estimator for linear dynamic systems (Vereecken et al., 2008). The Kalman Filter "optimally" weights the a priori model state predictions at a given time with the measurements available at the same time, according to a least squares approach, in order to provide a posteriori estimates of the state system evolution.

Although the standard Kalman Filter (SKF) was originally formulated for an optimal estimation of linear state space models with Gaussian uncertainties, more KF algorithms have been developed to handle nonlinear models. The Extended Kalman Filter (EKF), which relies on the linearization of model using first order approximation of Taylor series, was the first variant designed for dealing with nonlinear models. Katul et al. (1993) and Entekhabi et al. (1994) applied the EKF for the estimation of the vertical soil moisture profile with nonlinear numerical schemes of the Richards equation. Walker et al. (2001) compared SKF and direct insertion assimilation schemes within a synthetic study similar to that adopted by Entekhabi et al. (1994), but employing a linear explicit finite difference scheme of the Richards equation.

Further developments of the Kalman Filter have been suggested to overcome drawbacks of the EKF, which have been reported in case of strong nonlinearities and high dimensional applications (e.g. Reichle et al., 2002b; van der Merwe, 2004).

Evensen (1994) proposed the Ensemble Kalman Filter (EnKF), based upon Monte Carlo generations of an ensemble of states to approximate the propagation of the state prediction error statistics through the nonlinear model. The EnKF is relatively easy to implement and it can efficiently deal with high dimensional applications. It has become a popular choice for data assimilation in traditional hydrological applications like the estimation of streamflow (Moradkhani et al., 2005; Clark et al., 2008; Weerts and El Serafy, 2006; Xie and Zhang, 2010), land surface energy fluxes (Dunne and Entekhabi, 2006; Pipunic et al., 2008) and soil moisture (Re-

ichle and Koster, 2003; Reichle et al., 2007; De Lannoy et al., 2007). It has also been applied in subsurface models based on the numerical solver of the Richards equation (Das and Mohanty, 2006; Huang et al., 2007; Camporese et al., 2009).

A method less commonly applied in hydrological studies is the Unscented Kalman Filter (UKF) developed by Julier et al. (1995) and Julier and Uhlmann (1997, 2004), also based on propagating an ensemble of sample states, but chosen in a deterministic way. Compared with the EnKF, the UKF is also expected to be less computationally efficient for systems of large dimensionality and its implementation is less straightforward. The results of Luo and Moroz (2009, 2010) suggest that in small scale applications the UKF could perform slightly better than the EnKF in response to the relatively large biases and spurious modes of Monte Carlo approximations. To our knowledge, the UKF has not been implemented yet for retrieving soil water state profiles with a soil water transport model based on a numerical solution of the Richards equation.

The assimilation of near-surface information into the Richards equation can be treated by two alternative approaches: (i) using a Standard Kalman Filter (SKF), providing an optimal estimate of the mean and error variance of the state variable for a linear numerical solver of the Richards equation, or (ii) using a non-standard KF, such as the EKF, the UKF and the EnKF, which supplies an approximate solution of the first two moments of the state variable, but with a nonlinear numerical scheme. Examining the advantages and limitations of these two alternative approaches can be relevant to identify the best strategy for implementing assimilation algorithms in operational soil hydrological studies. Nonlinear numerical schemes outperform linear schemes, since they allow us to achieve numerical stability and accuracy with much larger time-steps than linear schemes (e.g. Haverkamp et al., 1977; Paniconi et al., 1991). However, the larger computational effort required for the application of linear schemes might be compensated by applying SKF analytic estimators for the propagation of the first two moments of the state profiles, with minor computational costs than the corresponding non-standard Kalman Filter algorithms.

We are not aware of previous studies that have explicitly examined the relative efficiency of these alternative approaches.

The general aim of this paper is to compare the efficiency of soil water state profile retrieval algorithms involving standard and non-standard Kalman Filter methods applied respectively to linear and nonlinear numerical schemes of the Richards equation. These analyses are conducted by repeating the same synthetic experiment implemented by Entekhabi et al. (1994) and Walker et al. (2001), simulating an evaporation process from a homogeneous soil column.

Two scenarios are examined in this paper. The first concerns the assimilation of surface matric head for retrieving matric head profiles with h-based numerical schemes of the Richards equation. Under this scenario, the following retrieval algorithms are compared: SKF applied to both the explicit and Crank–Nicolson linear numerical schemes; the UKF and EnKF applied to a nonlinear numerical scheme.

The second scenario differs from the first as the assimilated surface variable is the soil water content instead of the matric head. In this case, in order to handle the nonlinearity arising from the discrepancy between the observed and the retrieved variables, the SKF is replaced by the EKF in combination with the Crank-Nicolson linear numerical scheme, while the explicit numerical scheme is excluded from the comparison study.

The paper is structured as follows: Sect. 2 illustrates the Kalman Filter algorithms employed in this study; Sect. 3 presents the different numerical schemes of the Richards equation; Sect. 4 describes the results of the numerical experiments; Sect. 5 is devoted to the conclusions.

2   Kalman filtering

The Kalman Filter is a recursive filter that estimates the state of a dynamic system from a series of noise corrupted measurements. Its basic theory has originally been designed for linear systems (Kalman, 1960), but several variants have subsequently been proposed for studying the dynamics of nonlinear systems.

In the most general case, the dynamic system and the measurements are described by two sets of equations, discretised in the time domain (e.g. van der Merwe; 2004):

$$x_k = F_{k-1,k}(x_{k-1}, u_k) + v_{k-1} \qquad (1)$$
$$y_k = H_k(x_k) + \eta_k \qquad (2)$$

where $F_{k-1,k}$ is the dynamic system model that propagates the state vector x in time, assuming discrete time steps k; u_k represents the current exogenous input vector, which is assumed to be known; H_k is the measurement model, which describes how the current measurements vector y_k is related to the current state x_k. The dynamic system model is assumed to be corrupted by a zero mean additive Gaussian noise v_{k-1} with covariance Q_{k-1}. Similarly, the measurement model is assumed to be corrupted by a zero mean additive Gaussian noise vector η_k with covariance R_k.

With respect to the more general Bayesian theory, the system state x_k evolves over time according to a hidden Markov process, with a conditional probability density $p(x_k|x_{k-1})$ fully specified by $F_{k-1,k}$ and by the process noise distribution $p(v_{k-1})$. The observations y_k are conditionally independent given the state and are generated according to the conditional probability density $p(y_k|x_k)$, which is fully specified by H_k and the observation noise distribution $p(\eta_k)$.

The Kalman Filter provides a posteriori estimates of the first two moments of the state distribution:

– the mean state $\hat{x}_k = E[x_k]$, corresponding to the estimated state;

– the covariance of the state distribution $P_k = E\left[(x_k - \hat{x}_k)(x_k - \hat{x}_k)^T\right]$, which is equivalent to the error covariance matrix, i.e. a measure of the accuracy of the estimated state.

The two moments are computed according to two different phases: a prediction phase and an update phase. During the prediction phase, an a priori estimate of the state $\hat{x}_k^-$ and its covariance matrix P_k^- are provided based on the information available at time step t_{k-1}.

The update phase is activated as the measurements y_k become available. In this phase, an a posteriori state estimate $\hat{x}_k$ is provided by a linear combination of the a priori estimate $\hat{x}_k^-$ and the measurement innovation vector, equal to the difference between the actual measurements y_k and the a priori prediction of the measurements $\hat{y}_k^-$:

$$\hat{x}_k = \hat{x}_k^- + K_k\left(y_k - \hat{y}_k^-\right). \qquad (3)$$

In Eq. (3), the innovation vector is weighted through the matrix K_k, expressed as a function of the cross covariance matrix of the state prediction error and the observation prediction error $P_{xy,k}$, and the auto-covariance matrix of the predicted measurement $P_{yy,k}$:

$$K_k = P_{xy,k}\left[P_{yy,k} + R_k\right]^{-1}. \qquad (4)$$

The a posteriori error covariance P_k is estimated as follows:

$$P_k = P_k^- - K_k\left[P_{yy,k} + R_k\right]K_k^T. \qquad (5)$$

Equation (5) represents an expression of the system covariance alternative to the common one $P_k = P_k^- - K_k H_k P_k^-$ (Julier and Uhlmann, 2004; van der Merwe, 2004). This last relationship explicitly presumes that the measurement operator H_k is linear, hence represented by a matrix H_k. We prefer using the more general expression as in Eq. (5), since we also examine the case of nonlinear measurement operators in this study.

2.1   Standard Kalman Filter (SKF) and Extended Kalman Filter (EKF)

The SKF involves linear dynamic system and measurement operators and thus the operator $F_{k-1,k}$ and H_k are described by matrices, indicated as $F_{k-1,k}$ and H_k, respectively. The a posteriori state $\hat{x}_k$ is the optimal estimate in this case, with the minimum mean square error. The system covariance P_k^-, the cross-covariance matrix between the error in $\hat{x}_k^-$ and the error in $\hat{y}_k^-$, $P_{xy,k}$, and the covariance of the predicted measurements, $P_{yy,k}$, can be computed by the following closed linear relations:

$$\mathbf{P}_k^- = \mathbf{F}_{k-1,k}\,\mathbf{P}_{k-1}\,\mathbf{F}_{k-1,k}^T + \mathbf{Q}_{k-1} \qquad (6)$$

$$\mathbf{P}_{xy,k} = \mathbf{P}_k^-\,\mathbf{H}_k^T \qquad (7)$$

$$\mathbf{P}_{yy,k} = \mathbf{H}_k\,\mathbf{P}_k^-\,\mathbf{H}_k^T. \qquad (8)$$

The Kalman gain can been then computed with Eq. (4), the a posteriori state estimate $\hat{\mathbf{x}}_k$ with Eq. (3) and the a posteriori error covariance $\mathbf{P}_k$ with Eq. (5).

If the observation operator H_k in Eq. (2) is nonlinear (a scenario analysed in this work), SKF is not applicable and a different algorithm has to be adopted to linearize H_k. If one adopts the EKF linearizing strategy, the Kalman gain and the a posteriori estimate of the covariance matrix take the following forms:

$$\mathbf{K}_k = \mathbf{P}_k^-\,\mathbf{C}_k^T\left(\mathbf{C}_k\,\mathbf{P}_k^-\,\mathbf{C}_k^T + \mathbf{D}_k\,\mathbf{R}_k\,\mathbf{D}_k^T\right)^{-1} \qquad (9)$$

$$\mathbf{P}_k = (\mathbf{I} - \mathbf{K}_k\,\mathbf{C}_k)\,\mathbf{P}_k^- \qquad (10)$$

where $\mathbf{C}_k = \partial H(\mathbf{x}, \eta_k)/\partial \mathbf{x}|_{\hat{\mathbf{x}}_k^-}$ is the Jacobian matrix of H_k with respect to the state vector $\mathbf{x}$ computed at the a priori estimate $\hat{\mathbf{x}}_k^-$, and $\mathbf{D}_k = \partial H(\hat{\mathbf{x}}_k^-, \eta)/\partial \eta|_{\overline{\eta}=0}$ is the Jacobian matrix of H_k with respect to the noise vector η at the mean value $\overline{\eta} = 0$.

2.2   Unscented Kalman Filter (UKF)

The Unscented Kalman Filter (UKF) belongs to a wider group of approaches known as Sigma Point Kalman Filters (van der Merwe, 2004). The UKF is based on the Unscented Transformation, originally introduced by Julier and Uhlman (1997, 2004) as an effective method for capturing the nonlinear propagation of the first two moments of the state distribution through a minimal set of deterministically chosen sample points.

For a state vector of dimension N, the UKF, in its basic mode, obtains a set of sigma points, consisting of $2N+1$ vectors and their associated weights, $\mathbf{S} = \left\{ \mathcal{X}_i, \mu_i^{(j)};\ i = 0 \dots 2N;\ j \in (m, c) \right\}$, completely capturing the actual mean and covariance of the random variable $\mathbf{x}$. The vectors are weighted according to the respective mean (m) and covariance (c) weights $\mu_i^{(j)}$. A selection of sigma points fulfilling this requirement is defined as follows:

$$\mathcal{X}_0 = \hat{\mathbf{x}};\ \mathcal{X}_i = \hat{\mathbf{x}} + \left(\sqrt{\gamma\,\mathbf{P}}\right)_i,\ i = 1, \dots, N;$$

$$\mathcal{X}_i = \hat{\mathbf{x}} - \left(\sqrt{\gamma\,\mathbf{P}}\right)_i,\ i = N+1, \dots, 2N$$

$$\mu_0^{(m)} = \frac{\gamma - N}{\gamma};\ \mu_0^{(c)} = \frac{\gamma - N}{\gamma} + \left(1 - \rho^2 + \beta\right);$$

$$\mu_i^{(m)} = \mu_i^{(c)} = \frac{1}{2\gamma},\ i = 1, \dots, 2N. \qquad (11)$$

The parameter γ controls the spread of the states around the mean, and it is calculated as $\gamma = \rho^2(N + \kappa)$, with $\kappa \geq 0$ to ensure semi-positive definiteness of the covariance matrix and

$0 \leq \rho \leq 1$. A good default choice is $\kappa = 0$ and ρ small enough to limit the spread of the sample states. The parameter β is introduced as a second control on the magnitude of the covariance weights. Details about the proper choice of κ, ρ and β can be found in van der Merwe (2004). The symbol $\left(\sqrt{\gamma\,\mathbf{P}}\right)_i$ is the ith column (or row) of the root square matrix $\gamma\,\mathbf{P}$, which can be regularly computed by Cholesky decomposition (e.g. Press et al., 1992).

The $2N+1$ sigma point vectors are assembled in the following matrix:

$$\mathcal{X}_{k-1} = \left[\hat{\mathbf{x}}_{k-1}\quad \hat{\mathbf{x}}_{k-1} + \sqrt{\gamma\,\mathbf{P}_{k-1}}\quad \hat{\mathbf{x}}_{k-1} - \sqrt{\gamma\,\mathbf{P}_{k-1}}\right]. \qquad (12)$$

As part of the prediction step, each sigma point vector is propagated through the dynamic state model:

$$\mathcal{X}_k^- = F_{k-1,k}\left(\mathcal{X}_{k-1}, \mathbf{u}_k\right). \qquad (13)$$

The a priori estimate of the state mean is computed as the weighted average of the transformed points:

$$\hat{\mathbf{x}}_k^- = \sum_{i=0}^{2N} \mu_i^{(m)}\,\mathcal{X}_{k,i}^-. \qquad (14)$$

The a priori estimate of the state covariance is computed as the weighted outer product of the transformed points plus the Gaussian noise covariance $\mathbf{Q}_k$:

$$\mathbf{P}_k^- = \sum_{i=0}^{2N} \mu_i^{(c)}\left(\mathcal{X}_{k,i}^- - \hat{\mathbf{x}}_k^-\right)\left(\mathcal{X}_{k,i}^- - \hat{\mathbf{x}}_k^-\right)^T + \mathbf{Q}_k. \qquad (15)$$

One alternative to incorporate the effect of the process noise on the observed sigma-points, is to augment the number of these sigma-points with N additional vectors derived from the matrix square root of the process noise covariance, and recalculating the various weights μ_i accordingly. We opted to redraw the states, thus keeping the dimensionality of the problem equal to $2N+1$, although this option has the drawback that it discards the odd moments information captured by the propagated original sample states (van der Merwe, 2004):

$$\mathcal{X}_k^- = \left[\hat{\mathbf{x}}_k^-\quad \hat{\mathbf{x}}_k^- + \sqrt{\gamma\,\mathbf{P}_k^-}\quad \hat{\mathbf{x}}_k^- - \sqrt{\gamma\,\mathbf{P}_k^-}\right]. \qquad (16)$$

The observation equation is applied to this set of state vectors:

$$\mathcal{Y}_k = H_k\left(\mathcal{X}_k^-\right). \qquad (17)$$

The forecast cross covariance between state predictions errors and observation predictions errors, $\mathbf{P}_{xy,k}$, and the forecast error covariance matrix of the observation predictions, $\mathbf{P}_{yy,k}$, are computed as follows:

$$y_k^- = \sum_{i=0}^{2N} \mu_i^{(m)} \mathcal{Y}_{k,i} \tag{18}$$

$$\mathbf{P}_{xy,k} = \sum_{i=0}^{2N} \mu_i^{(c)} \left(\mathcal{X}_{k,i} - \hat{x}_k^- \right) \left(\mathcal{Y}_{k,i} - \hat{y}_k^- \right)^T \tag{19}$$

$$\mathbf{P}_{yy,k} = \sum_{i=0}^{2N} \mu_i^{(c)} \left(\mathcal{Y}_{k,i} - \hat{y}_k^- \right) \left(\mathcal{Y}_{k,i} - \hat{y}_k^- \right)^T \tag{20}$$

where $\hat{y}_k^-$ is a weighted mean of the predicted measurements $\mathbf{Y}_{k,i}$.

The Kalman *gain* can be then straightforwardly computed with Eq. (4), while the a posteriori estimate of the state mean $\hat{x}_k$ and the covariance matrix $\mathbf{P}_k$ can be computed respectively with Eqs. (3) and (5).

In case of linear systems, the solution provided by UKF converges to that of SKF as the size of the sigma points ensemble increases (van der Merwe, 2004).

2.3 Ensemble Kalman Filter (EnKF)

The EnKF uses an ensemble of randomly chosen model trajectories, from which the necessary error covariances are estimated (Evensen, 2003). Similarly to the UKF, this method does not approximate the nonlinear process and observation models: it rather uses the true nonlinear models and it approximates the distribution of the state random variable. Moreover, it does not explicitly transform error information with a dynamic equation for computing the state error covariance matrix.

The EnKF propagates an ensemble of state vectors and each of these propagated vectors represents one realization of generated model replicas. Given an ensemble of L members, the dynamic model is applied to each member as follows:

$$x_{k,i}^- = F_{k-1,k} \left(x_{k-1,i}, u_k \right) + v_{k-1,i} \quad i = 1 \ldots L \tag{21}$$

where $x_{k,i}^-$ is the ith forecast ensemble member at time k and $x_{k-1,i}$ is the updated ensemble member at $k-1$. Vector $v_{k-1,i}$ is the ith column of a $N \times L$ matrix of perturbations generated according to a Gaussian distribution with zero mean and covariance $\mathbf{Q}_k$. In this study the current exogenous input vector is assumed to be unperturbed, to keep the analogy with the other two methods.

The sample mean and covariance can be evaluated according to the expressions:

$$\hat{x}_k^- = \frac{1}{L} \sum_{i=1}^{L} x_{k,i}^- \tag{22}$$

$$\mathbf{P}_k^- = \frac{1}{L-1} \sum_{i=1}^{L} \left(x_{k,i}^- - \hat{x}_k^- \right) \left(x_{k,i}^- - \hat{x}_k^- \right)^T. \tag{23}$$

In practice, the calculation of the approximate covariance $\mathbf{P}_k^-$ is not required. The Kalman gain $\mathbf{K}$ is obtained with Eq. (4) after computing the following covariances:

$$\mathbf{P}_{xy,k} = \frac{1}{L-1} \sum_{i=1}^{L} \left(x_{k,i}^- - \hat{x}_k^- \right) \left(y_{k,i}^- - H_k \left(\hat{x}_k^- \right) \right)^T \tag{24}$$

$$\mathbf{P}_{yy,k} = \frac{1}{L-1} \sum_{i=1}^{L} \left(y_{k,i}^- - H_k \left(\hat{x}_k^- \right) \right) \left(y_{k,i}^- - H_k \left(\hat{x}_k^- \right) \right)^T \tag{25}$$

where $y_{k,i}^- = H_k(x_{k,i}^-)$ represents the ith observation prediction at discrete time k.

An ensemble of L perturbed observations vectors $y_{k,i}$ is derived by summing perturbations $\eta_{k,i}(\eta_{k,i} \in N(0, \mathbf{R}_k)$, $i = 1 \ldots L$) to the nominal term y_k. The update step for the forecasted state ensemble members is then defined as follows:

$$x_{k,i} = x_{k,i}^- + \mathbf{K}_k \left(y_{k,i} - H_k \left(x_{k,i}^- \right) \right). \tag{26}$$

After the analysis ensemble is generated, it is propagated forward, and a new assimilation cycle starts.

The EnkF algorithm does not entail to explicitly update and store the state error covariance matrix, differently from the SKF and the UKF algorithms, where the state error covariance has to be explicitly updated to represent the change in forecast error covariance as an observation becomes available.

The ensemble forecast step of both UKF and EnKF algorithms can be parallelized by running each ensemble member on a separate processor of a parallel computer (or cluster). This can result in a significant computational advantage for the application of these two algorithms, whose computational effort is highly dependent on the size of the respective ensembles.

The optimal ensemble size is uncertain for the EnKF and it is generally heuristically chosen. Moreover, it is still not clear how the ensemble size should scale with the system dimension to achieve adequate estimates (Reichle et al., 2002a; Camporese et al., 2009). In case of linear systems, the solution of EnKF converges to that of SKF as the ensemble size increases (Evensen, 2003). For small ensemble size the EnKF is susceptible to systematic underestimation of the ensemble error covariance, caused by spurious long-range correlations (Houtekamer and Mitchell, 1998; Papadakis et al., 2010).

On the contrary, the UKF relies on a deterministically chosen set of samples to capture the statistical moments of the nonlinear model accurately, and the number of samples is univocally defined by the system dimension. However, the identification of this set of sample requires the calculation of the matrix square root of the state covariance matrix (see Eq. 11), which can be a computationally intensive process.

3 Soil water transport model

The soil water dynamics along the vertical direction is modelled using the Richards equation (Jury et al., 1991) in the h-based form:

$$C(h) \frac{\partial h}{\partial t} = \frac{\partial \left[K(h) \left(\frac{\partial h}{\partial z} + 1 \right) \right]}{\partial z} \tag{27}$$

where t is the time, z denotes the position along vertical axis (with upward orientation and zero reference value at the surface), h [L] is the matric pressure head [L], $K(h)$ [L T^{-1}] is the hydraulic conductivity function, and $C(h)$ [L^{-1}] is the differential water capacity function, obtained from the derivative $C(h) = d\theta(h)/dh$ of the water retention function $\theta(h)$ [L^3 L^{-3}].

The water retention and hydraulic conductivity functions are modelled according to the van Genuchten-Mualem model (van Genuchten, 1980):

$$\theta(h) = \theta_r + (\theta_s - \theta_r) \left[1 + |\alpha h|^n \right]^{-m} \tag{28}$$

$$K(\theta) = K_s \left(\frac{\theta - \theta_r}{\theta_s - \theta_r} \right)^\lambda \left\{ 1 - \left[1 - \left(\frac{\theta - \theta_r}{\theta_s - \theta_r} \right)^{1/m} \right]^m \right\}^2 \tag{29}$$

where θ_s [L^3 L^{-3}] is the saturated soil water content, θ_r [L^3 L^{-3}] is the residual soil water content, K_s [L T^{-1}] is the saturated hydraulic conductivity, while $\alpha > 0$ [L^{-1}], $n > 1$ [$-$], m [$-$] and λ [$-$] are empirical parameters. Following a common assumption, parameter m [$-$] is defined by the relation $m = 1 - 1/n$ and λ [$-$] is set equal to 0.5.

Equation (27), combined with the boundary conditions, can be solved by adopting a numerical scheme, which is formally equivalent to a state-space representation of the system model discretised in the time domain, as the one described by Eq. (1). Depending on the type of numerical scheme employed, the system model can be linear or nonlinear.

For this specific system, $v(t)$ represents the state noise affecting the dynamic behaviour. According to Katul et al. (1993) the zero mean state noise assumption describes reasonably well the dynamic characteristics of the soil water flow in field conditions.

The measurement model could be reduced to a simple linear relation if matric pressure head is directly measured at given soil depths. If soil water content is measured, the observation equation is described by a nonlinear model corresponding to the soil water retention function (see Eq. 28). Different observation equations are required to assimilate other sources of measurements, such as those originated by near-surface remote sensing.

Below we illustrate three numerical schemes largely employed for integrating the Richards equation.

3.1 Explicit finite difference scheme (EX)

The Explicit finite difference scheme (EX) is the most basic numerical technique for solving differential equations. However, it may suffer from instability, which may make the method inappropriate or impractical. This method, also recognised as forward Euler finite difference scheme, is the one employed by Walker et al. (2001).

This numerical scheme provides an estimate of the matric head h_k^i of the ith node at the kth time-step as function of all other quantities at the preceding time-step according to the following discrete form of Eq. (27):

$$h_k^i = \left(\frac{\Delta t_{k-1}}{C_{k-1}^i} \frac{K_{k-1}^{i-1/2}}{\Delta z^i \Delta z^u}; \ 1 - \frac{\Delta t_{k-1}}{C_{k-1}^i} \frac{\frac{K_{k-1}^{i-1/2}}{\Delta z^u} + \frac{K_{k-1}^{i+1/2}}{\Delta z^l}}{\Delta z^i}; \right.$$

$$\left. \frac{\Delta t_{k-1}}{C_{k-1}^i} \frac{K_{k-1}^{i+1/2}}{\Delta z^i \Delta z^l} \right) \begin{pmatrix} h_{k-1}^{i-1} \\ h_{k-1}^i \\ h_{k-1}^{i+1} \end{pmatrix} + \frac{\Delta t_{k-1}}{C_{k-1}^i} \frac{K_{k-1}^{i-1} - K_{k-1}^{i+1}}{2 \Delta z^i} \tag{30}$$

The subscript i for the node number is increasing downward. The soil column is divided in compartments of finite thickness Δz^i. All nodes, including the top and bottom nodes, are in the centre of the soil compartments, with $\Delta z^u = z^{i-1} - z^i$ and $\Delta z^l = z^i - z^{i+1}$. This represents a small difference with respect to the work of Walker et al. (2001), who considered nodes at the compartment extremes, positive upwards. $K_k^{i-1/2}$ and $K_k^{i+1/2}$ denote respectively the upward and the downward spatial averages of the hydraulic conductivity computed as arithmetic means. Δt_{k-1} indicates the time interval $\Delta t_{k-1} = t_k - t_{k-1}$.

For nodes $i = 1$ and $i = N$ the discrete forms at time-step k are:

$$h_k^1 = \left(\frac{\Delta t_{k-1}}{C_{k-1}^2} \frac{K_{k-1}^{1\frac{1}{2}}}{\Delta z^2 \Delta z^l}; \ 1 - \frac{\Delta t_{k-1}}{C_{k-1}^2} \frac{\frac{K_{k-1}^{\frac{1}{2}}}{\Delta z^l} + \frac{K_{k-1}^{2\frac{1}{2}}}{(z^2 - z^3)}}{\Delta z^2}; \right.$$

$$\left. \frac{\Delta t_{k-1}}{C_{k-1}^2} \frac{K_{k-1}^{2\frac{1}{2}}}{\Delta z^2 (z^2 - z^3)} \right) \begin{pmatrix} h_{k-1}^1 \\ h_{k-1}^2 \\ h_{k-1}^3 \end{pmatrix}$$

$$+ \frac{\Delta t_{k-1}}{C_{k-1}^2} \frac{K_{k-1}^1 - K_{k-1}^3}{2 \Delta z^2} - \Delta z^l \left(\frac{q_{top}}{K_{k-1}^{\frac{1}{2}}} + 1 \right) \tag{31}$$

$$h_k^N = \left(\frac{\Delta t_{k-1}}{C_{k-1}^{N-1}} \frac{K_{k-1}^{N-1-\frac{1}{2}}}{\Delta z^{N-1} (z^{N-2} - z^{N-1})}; \ 1 - \frac{\Delta t_{k-1}}{C_{k-1}^{N-1}} \right.$$

$$\left. \frac{\frac{K_{k-1}^{N-1-\frac{1}{2}}}{(z^{N-2} - z^{N-1})} + \frac{K_{k-1}^{N-\frac{1}{2}}}{\Delta z^u}}{\Delta z^{N-1}}; \ \frac{\Delta t_{k-1}}{C_{k-1}^{N-1}} \frac{K_{k-1}^{N-\frac{1}{2}}}{\Delta z^{N-1} \Delta z^u} \right)$$

$$\begin{pmatrix} h_{k-1}^{N-2} \\ h_{k-1}^{N-1} \\ h_{k-1}^N \end{pmatrix} + \frac{\Delta t_{k-1}}{C_{k-1}^{N-1}} \frac{K_{k-1}^{N-2} - K_{k-1}^N}{2 \Delta z^{N-1}} + \Delta z^u \left(\frac{q_{bot}}{K_{k-1}^{N-\frac{1}{2}} + 1} \right) \tag{32}$$

The symbols q_{top} and q_{bot} indicate the top and bottom boundary conditions, herein assumed to be of Neumann type.

The forecasting equation of the system state x_k, coinciding with the matric pressure head ($x_k^i = h_k^i$), can be obtained by combining the discrete equations written for all N nodes in the following linear state-space form:

$$\hat{x}_k^- = \mathbf{A}_{k-1}\hat{x}_{k-1} + g_{k-1} \tag{33}$$

where $\mathbf{A}_{k-1}$ is the matrix obtained by assembling the terms multiplying the state vector at time-step $k-1$ of Eqs. (30)–(32) and g_{k-1} results from the combination of the terms on the right end of Eqs. (30)–(32).

According to this formulation, the system covariance is updated by the expression:

$$\mathbf{P}_k^- = \mathbf{A}_{k-1}\mathbf{P}_{k-1}\mathbf{A}_{k-1}^T + \mathbf{Q}_{k-1}. \tag{34}$$

Once the a priori state mean, $\hat{x}_{k-1}^-$, and state covariance, $\mathbf{P}_{k-1}^-$, have been computed using Eqs. (33)–(34), the results of Eqs. (7)–(8) can be used for determining $\mathbf{K}$ in Eq. (4), and then obtaining the a posteriori estimates $\hat{x}_k$ and $\mathbf{P}_k$ by means of Eqs. (3) and (5), respectively.

3.2 Crank-Nicolson finite difference scheme (CN)

The Crank-Nicolson implicit finite difference scheme (CN) has been widely implemented for solving the Richards equation (e.g. Haverkamp et al., 1977; Santini, 1980; Romano et al., 1998). The CN scheme is numerically stable, but sensitive to spurious oscillations when the ratio of the time step to the square of the space step is large.

The discrete representation of this relationship for intermediary nodes yields:

$$\left(\frac{-K_{k-1}^{i-1/2}}{2\,\Delta z^i\,\Delta z^u}; \frac{C_{k-1}^i}{\Delta t_{k-1}} + \frac{\frac{K_{k-1}^{i-1/2}}{\Delta z^u} + \frac{K_{k-1}^{i+1/2}}{\Delta z^l}}{2\,\Delta z^i}; \frac{-K_{k-1}^{i+1/2}}{2\,\Delta z^i\,\Delta z^l} \right)$$

$$\left(\begin{array}{c} h_k^{i-1} \\ h_k^i \\ h_k^{i+1} \end{array} \right) = \left(\frac{K_{k-1}^{i-1/2}}{2\,\Delta z^i\,\Delta z^u}; \frac{C_{k-1}^i}{\Delta t_{k-1}} - \frac{\frac{K_{k-1}^{i-1/2}}{\Delta z^u} + \frac{K_{k-1}^{i+1/2}}{\Delta z^l}}{2\,\Delta z^i}; \right.$$

$$\left. \frac{K_{k-1}^{i+1/2}}{2\,\Delta z^i\,\Delta z^l} \right) \left(\begin{array}{c} h_{k-1}^{i-1} \\ h_{k-1}^i \\ h_{k-1}^{i+1} \end{array} \right) + \frac{K_{k-1}^{i-1} - K_{k-1}^{i+1}}{2\,\Delta z^i}. \tag{35}$$

For nodes $i = 1$ and $i = N$ the discrete forms at time-step $k+1$ are:

$$\left(\frac{C_{k-1}^1}{\Delta t_{k-1}} + \frac{K_{k-1}^{\frac{1}{2}}}{2\,\Delta z^1\,\Delta z^l}; \frac{-K_{k-1}^{1\frac{1}{2}}}{\Delta z^1\,\Delta z^l} \right) \left(\begin{array}{c} h_k^1 \\ h_k^2 \end{array} \right)$$

$$= \left(\frac{C_{k-1}^1}{\Delta t_{k-1}} - \frac{K_{k-1}^{\frac{1}{2}}}{2\,\Delta z^1\,\Delta z^l}; \frac{K_{k-1}^{1\frac{1}{2}}}{\Delta z^1\,\Delta z^l} \right) \left(\begin{array}{c} h_{k-1}^1 \\ h_{k-1}^2 \end{array} \right) + \frac{q_{top} - K_{k-1}^{1\frac{1}{2}}}{\Delta z^1} \tag{36}$$

$$\left(\frac{-K_{k-1}^{N-1/2}}{\Delta z^N\,\Delta z^u}; \frac{C_{k-1}^N}{\Delta t_{k-1}} + \frac{K_{k-1}^{N-1/2}}{2\,\Delta z^N\,\Delta z^u} \right) \left(\begin{array}{c} h_k^{N-1} \\ h_k^N \end{array} \right)$$

$$= \left(\frac{K_{k-1}^{N-1/2}}{2\,\Delta z^N\,\Delta z^u}; \frac{C_{k-1}^N}{\Delta t_{k-1}} - \frac{K_{k-1}^{N-1/2}}{2\,\Delta z^N\,\Delta z^u} \right) \left(\begin{array}{c} h_{k-1}^{N-1} \\ h_{k-1}^N \end{array} \right) + \frac{K_{k-1}^{-1/2} - q_{bot}}{\Delta z^N}. \tag{37}$$

As in the previous algorithm, an explicit linearization of K and C is implemented, by taking their values at the previous time-step $k-1$. Then, a linear state-space representation of the dynamic system can be easily derived by combining the set of Eqs. (35)–(37) written for each node and accounting for the boundary conditions:

$$\hat{x}_k^- = \left(\mathbf{B}_{k-1}'\right)^{-1}\mathbf{A}_{k-1}'\hat{x}_{k-1} + \left(\mathbf{B}_{k-1}'\right)^{-1}g_{k-1}'. \tag{38}$$

This equation represents the dynamic state space model obtained with the CN scheme, analogously to Eq. (33) obtained with the explicit scheme. $\mathbf{A}_{k-1}'$ and $\mathbf{B}_{k-1}'$ are tri-diagonal matrices obtained by assembling the terms in the first parentheses on the right and left hand-sides of Eqs. (35)–(37), respectively. The term g_{k-1}' is a vector obtained by assembling the terms on the right end of Eqs. (35)–(37).

The a priori estimate of the covariance matrix is calculated as follows:

$$\mathbf{P}_k^- = \left(\mathbf{B}_{k-1}'\right)^{-1}\mathbf{A}_{k-1}'\mathbf{P}_{k-1}\left[\left(\mathbf{B}_{k-1}'\right)^{-1}\mathbf{A}_{k-1}'\right]^T + \mathbf{Q}_{k-1}. \tag{39}$$

As for the explicit scheme, once the a priori system state and covariance have been determined, the current values $\hat{x}_k$ and $\mathbf{P}_k$ can be straightforwardly calculated with Eqs. (3)–(8).

3.3 Nonlinear implicit finite difference scheme (NL)

The backward Euler's finite differences (implicit) nonlinear (NL) scheme guarantees numerical stability for time steps considerably larger than those employed in the CN scheme, which explains its wide application in simulation models.

In this study we adopt the NL scheme of the Richards equation introduced by Celia et al. (1990) and further implemented by the SWAP model (van Dam, 2000).

The NL scheme has to be iteratively solved, as it includes C and θ values at the current time-step k to account for the strong nonlinearity of the differential water capacity C:

$$\left(\frac{-K_{k-1}^{i-1/2}}{\Delta z^i\,\Delta z^u}; \frac{C_{k,p-1}^i}{\Delta t_{k-1}} + \frac{\frac{K_{k-1}^{i-1/2}}{\Delta z^u} + \frac{K_{k-1}^{i+1/2}}{\Delta z^l}}{\Delta z^i}; \frac{-K_{k-1}^{i+1/2}}{\Delta z^i\,\Delta z^l} \right)$$

$$\left(\begin{array}{c} h_{k,p}^{i-1} \\ h_{k,p}^i \\ h_{k,p}^{i+1} \end{array} \right) = \frac{C_{k,p-1}^i}{\Delta t_{k-1}} h_{k,p}^i + \frac{K_{k-1}^{i-1} - K_{k-1}^{i+1}}{2\,\Delta z^i} + \theta_{k-1}^i - \theta_{k,p-1}^i. \tag{40}$$

The discrete forms for nodes $i = 1$ and $i = N$ at time-step k are:

$$\left(\frac{C_{k,p-1}^1}{\Delta t_{k-1}} + \frac{K_{k-1}^{1\frac{1}{2}}}{\Delta z^1\,\Delta z^l}; \frac{-K_{k-1}^{1\frac{1}{2}}}{\Delta z^1\,\Delta z^l} \right) \left(\begin{array}{c} h_{k,p}^1 \\ h_{k,p}^2 \end{array} \right)$$

$$= \frac{C_{k,p-1}^1}{\Delta t_{k-1}} h_{k,p-1}^1 - \frac{q_{top} + K_{k-1}^{1\frac{1}{2}}}{\Delta z^1} + \theta_{k-1}^1 - \theta_{k,p-1}^1 \tag{41}$$

$$\left(\frac{-K_{k-1}^{N-1/2}}{\Delta z^N\,\Delta z^u}; \frac{C_{k,p-1}^N}{\Delta t_{k-1}} + \frac{K_{k-1}^{N-1/2}}{\Delta z^N\,\Delta z^u} \right) \left(\begin{array}{c} h_{k,p}^{N-1} \\ h_{k,p}^N \end{array} \right)$$

$$= \frac{C_{k,p-1}^N}{\Delta t_{k-1}} h_{k,p-1}^N + \frac{q_{bot} + K_{k-1}^{N-1/2}}{\Delta z^N} + \theta_{k-1}^N - \theta_{k,p-1}^N \tag{42}$$

Table 1. Summary of assimilation algorithms adopted in combination with different numerical schemes and different types of observed variables.

Observed variable	Finite difference scheme			
	EX	CN	NL	NL
h	SKF[1]	SKF[1]	UKF[3]	EnKF[4]
θ	–	EKF[2]	UKF[3]	EnKF[4]

[1] SKF = standard Kalman Filter; [2] EKF = extended Kalman Filter; [3] UKF = unscented Kalman Filter; [4] EnKF = ensemble Kalman Filter.

where index p denotes the iteration step.

The dynamic state space model then assumes the form:

$$\mathbf{B}''_{k-1,k}\hat{\boldsymbol{x}}_k^- = \mathbf{A}''_{k-1,k}\hat{\boldsymbol{x}}_{k-1} + \boldsymbol{g}''_{k-1,k}. \tag{43}$$

$\mathbf{A}''_{k-1,k}$ is a diagonal matrix, with $\mathbf{A}^{i,i}_{k-1,k} = C^i_k/\Delta t_{k-1}$, obtained by assembling the terms multiplying the state vector at time-step $k-1$ in Eqs. (40)–(42). $\boldsymbol{g}''_{k-1,k}$ is a vector obtained by assembling the three terms on the right end of Eqs. (40)–(42). $\mathbf{B}''_{k-1,k}$ is the tri-diagonal matrix obtained by assembling the terms in the first parentheses on the left hand-side of Eqs. (40)–(42).

This type of numerical scheme requires the implementation of non-standard Kalman Filters for the a priori prediction of $\hat{\boldsymbol{x}}_{k-1}^-$ and $\mathbf{P}_{k-1}^-$. Equation (43) is employed for propagating an ensemble of sample states with both the UKF and the EnKF. In the case of the UKF, the a priori system state, $\hat{\boldsymbol{x}}_{k-1}^-$, and covariance, $\mathbf{P}_{k-1}^-$, are calculated as a weighted contribution of each propagated state, according to Eqs. (14)–(15). In the case of the EnKF, the statistics of the predicted states are computed with Eqs. (22)–(23).

4 Synthetic study

A synthetic study is performed to evaluate the relative merits of different Kalman Filter algorithms for retrieving matric head profiles by assimilating near surface soil matric head or water content measurements.

As pointed out above, the types of Kalman Filter that can be applied depend on the numerical scheme employed. The Standard Kalman Filter (SKF) can be implemented with explicit (EX) and Crank-Nicolson (CN) finite difference schemes, as long as the measurement model is linear. In this study, we employ the h-based form of the Richards equation and thus a linear measurement model occurs if the measured variable is also the matric pressure head h.

Walker et al. (2001) showed the efficiency of a standard Kalman Filter (SKF) in assimilating near surface matric pressure head measurements with an explicit finite difference scheme (EX) as compared with direct insertion of the observation values. Following Walker et al. (2001), we first compare the SKF applied to a Crank-Nicolson finite difference scheme (SKF-CN), with the SKF applied to an explicit finite difference scheme (SKF-EX). Then we analyse the relative performances of the Unscented Kalman Filter based on the implicit nonlinear finite difference scheme (UKF-NL), the Ensemble Kalman Filter also implemented with the implicit numerical scheme (EnKF-NL) and the SKF-CN assimilation algorithm.

As an alternative to the UKF or the EnKF, the EKF could be also employed in conjunction with a nonlinear numerical scheme of the Richards equation. However, the EKF, based on an explicit linearization of nonlinear equations, is less efficient in state retrieving as compared with UKF (van der Merwe, 2004) and EnKF (e.g. Reichle et al., 2002b). Further discussions about the limitations and the flaws of the EKF can be found in other studies (e.g. Julier et al., 1995; van der Merwe, 2004).

If soil water content is the measured variable, the SKF-CN as such is not applicable, and a nonlinear KF is required to overcome the nonlinearity of the measurement model H_k defined by the soil water retention function. In this case we adopt an Extended approach (EKF-CN), using Eqs. (9)–(10) for computing the Kalman gain and the a posteriori estimate of the covariance matrix, respectively.

The final analysis of this work involves the comparison between the EKF-CN, the UKF-NL and the EnKF-NL algorithms, when the soil moisture content is the observation variable.

Table 1 summarizes the assimilation schemes examined in this paper.

The numerical experiment is arranged following Walker et al. (2001), to facilitate the comparison with this previous study. The essential information of the implemented numerical experiment is summarized in Table 2. Soil column depth is 100 cm, discretised in 27 nodes and the true initial matric pressure head profile is uniformly equal to -50 cm. The boundary conditions are: constant evaporative flux of 5.78×10^{-6} cm s^{-1} at the top surface and no flux at the bottom.

All assimilation scenarios are initialised with the same poor guess of the initial matric pressure head profile, assumed to be uniformly equal to -300 cm, thus 250 cm less than the true initial uniform profile.

Matric pressure head profiles are then retrieved by assimilating 4 different observation sets, consisting of hourly h data generated in top nodes, down to depths of 0.5, 1.5, 4.5 and 10 cm, respectively. These depths are slightly different from those adopted by Walker et al. (2001), because of the small differences in the soil column discretisation, as illustrated in Sect. 3.1. The NL scheme of the soil water transfer model illustrated above has been used to generate a set of soil water content and matric head profiles, representative of the true dynamic process to be retrieved. The assimilated values are true values of either matric heads or soil moisture contents,

Table 2. Parameters and conditions employed in the synthetic generation of matric pressure head profiles and for the initialization of the assimilation algorithms.

Soil depth	100 cm
Number of nodes	27
Soil hydraulic parameters	$\theta_s = 0.54\ \text{cm}^3\ \text{cm}^{-3}$
	$\theta_r = 0.2\ \text{cm}^3\ \text{cm}^{-3}$
	$\alpha = 0.008\ \text{cm}^{-1}$
	$n = 1.8\ (-)$
	$K_s = 2.9 \times 10^{-4}\ \text{cm}^{-1}$
Top evaporative flux	$5.79 \times 10^{-6}\ \text{cm s}^{-1}$
Bottom flux	$0\ \text{cm s}^{-1}$
Initial uniform h profile	$-50\ \text{cm}$
Poor guess of initial uniform h profile	$-300\ \text{cm}$
Initial state covariance matrix	$\mathbf{P}^{i,j} = \begin{cases} 10^3\ \text{cm}^2 & \text{if } i = j;\ i, j = 1 \ldots \text{n\_nodes} \\ 0 & \text{if } i \neq j \end{cases}$
Measurement noise variance matrix	$\mathbf{R}^{i,j} = \begin{cases} 0.02\, y_i & \text{if } i = j;\ i, j = 1 \ldots \text{n\_obs} \\ 0 & \text{if } i \neq j \end{cases}$

perturbed with a random error of zero mean and a standard deviation of five percent of the absolute true values.

Employing a reference true dynamic scheme generated with a numerical scheme (NL in this case) different from that employed in the assimilation process (e.g. EX or CN) is formally equivalent to introducing a non-additive model error as further source of uncertainty in the system dynamics. However, we verified that the choice of the numerical scheme to be employed for generating the reference true dynamic process is not relevant for the comparison of the retrieved state profiles, since the profiles predicted with the three different numerical schemes are very similar to each other (herein not shown for the sake of brevity). Indeed, starting from the same initial conditions, the different numerical schemes (EX, CN and NL) can predict the evolution (a priori prediction) of the state profile with negligible numerical differences, as the time stepping adopted within each numerical scheme can be adjusted to guarantee numerical stability and achieve relatively similar numerical accuracy. This has been already proved by several previous studies, also applied for simulating infiltration fronts (e.g. Haverkamp et al., 1977; Paniconi et al., 1991; Kavetski et al., 2002), which from a numerical perspective is even more challenging than simulating an evaporation experiment, as the one examined in this study.

The EnKF has been implemented with an ensemble size of 50 members. We could verify that an ensemble size greater than 50 did not add much accuracy to the data assimilation algorithm, as also shown by Camporese et al. (2009) for the same case study. Moreover, by using 50 members for the examined case study, the dimensionality of the EnKF-NL assimilation algorithm is similar to that of the UKF-NL algorithm ($2N + 1 = 55$).

The default initial state variance is set equal to 10^3 cm^2, rather than 10^6 cm^2 as assumed by Walker at al. (2001), since using an extremely high initial state variance causes practical difficulties in the implementation of the UKF and the EnKF, as discussed later.

The amount of system noise variance is implemented in a different way. Walker et al. (2001) defined it equal to a five percent of the change in system states for each time step for the diagonal elements matrix, and zero for those off-diagonal. Entekhabi et al. (1994) considered an initial diagonal matrix accounting for the five percent of the precedent state. We adopted a diagonal system noise variance accounting for the five percent of the previous a posteriori state change, in order to avoid any ambiguity in the amount of error being incorporated, given that the time steps are variable in dependence of the chosen numerical scheme. The measurement noise covariance was set equal to two percent of the observations (either matric pressure head or soil moisture content) for the diagonal elements, and zero for all other elements, as also assumed by Walker et al. (2001).

4.1 Assimilating matric pressure heads

In Fig. 1, the profiles retrieved by assimilating daily matric pressure heads with the SKF-EX, SKF-CN, UKF-NL and EnKF-NL algorithms, are respectively compared with the true profiles as well as with the "guess" profiles. The "guess" profile, also referred to as "open loop" profile, is the one obtained without assimilating any near-surface observations, i.e. the system is simply propagated from the initial uniform conditions using the known boundary conditions.

All algorithms exhibit convergence rates of the retrieved matric head profiles to the true ones faster than that shown by Entekhabi et al. (1994), who implemented the EKF with a finite element scheme of the Richards equation for an analogous synthetic study.

The retrieved profiles tend to converge to the true ones slightly faster as the observation depth increases. For clarity, Fig. 1 illustrates only the results for the extreme observation

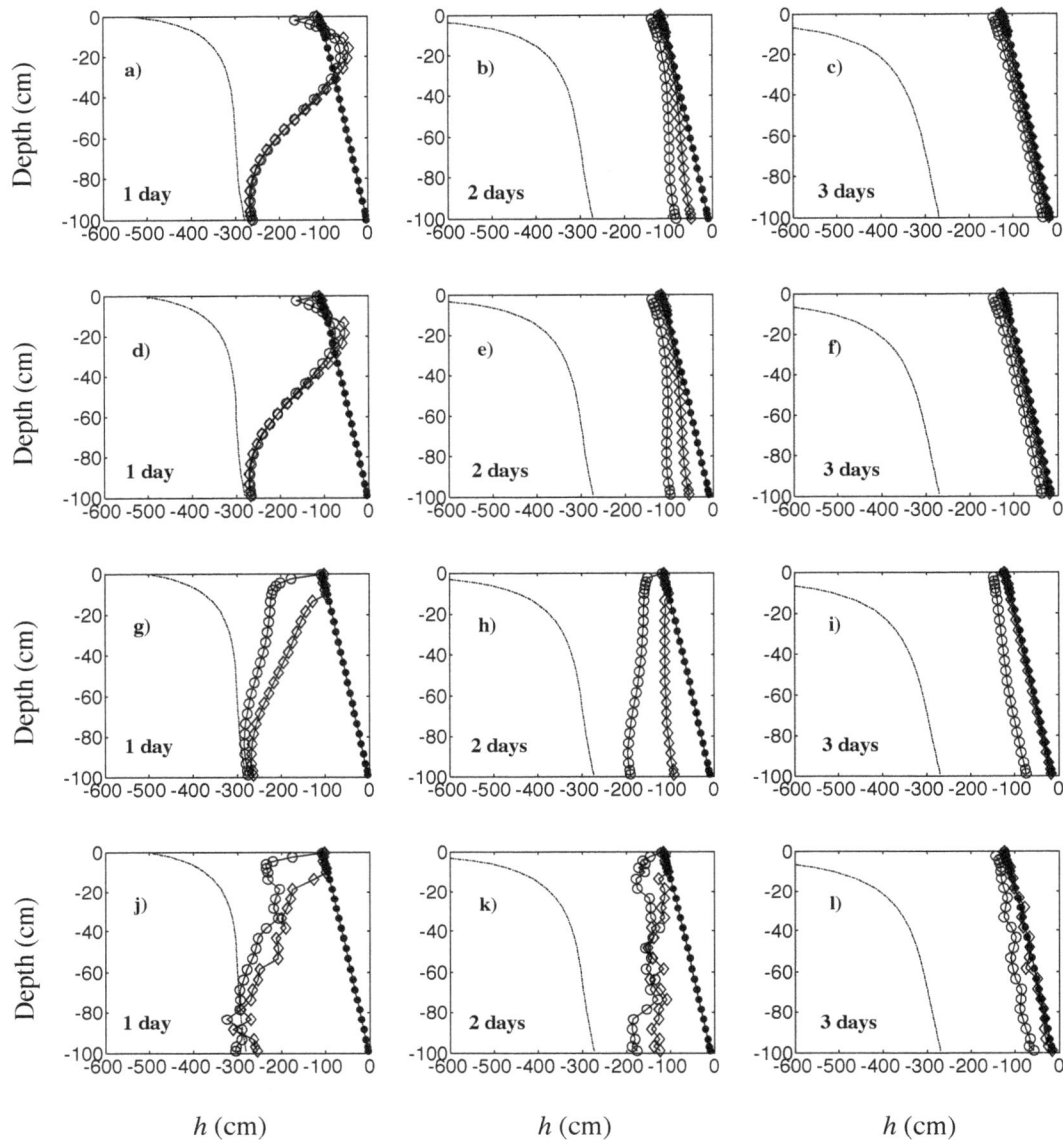

Figure 1. Retrieved profiles by assimilating daily observations of matric pressure heads involving nodes within the top 0.5 cm (open circle) and 10.5 cm (diamond) compared with the "true" profile (closed circle) and "guess" profile (dashed line), for different combinations of Kalman Filters and numerical schemes: (**a–c**) SKF-EX; (**d–f**) SKF-CN; (**g–i**) UKF-NL; (**j–l**) EnKF-NL.

depth scenarios considered in this work (i.e. observations at the nodes located within the top 0.5 and 10.5 cm, respectively), since the solutions for the other depth scenarios are bounded by those corresponding to these two extreme scenarios.

The two sets of profiles retrieved by using the SKF algorithm, respectively coupled with the explicit (Fig. 1a–c) and the Crank-Nicolson (Fig. 1d–f) numerical schemes, are practically identical. However, the explicit scheme requires time steps markedly smaller in order to guarantee numerical accuracy and stability. SKF-EX with a time step of 1 s demands about 30 times more CPU time than SKF-CN with a time step

of 60 s. The round off errors do not visibly impact the performance of the linearized approaches. Another favourable aspect of the CN numerical scheme is that it is particularly stable and thus it can be employed to assess the state profile even for a decreased precision of the model equation.

The profiles retrieved with UKF-NL and EnKF-NL are also very similar to each other, apart from the slight irregularities of the EnKF-NL profiles resulting from sample means (Eq. 22) obtained with a relatively small ensemble size. These irregularities tend to be smoothed as the ensemble size increases. Alternatively, they can be reduced by adopting specific sampling strategies both for the initial ensemble

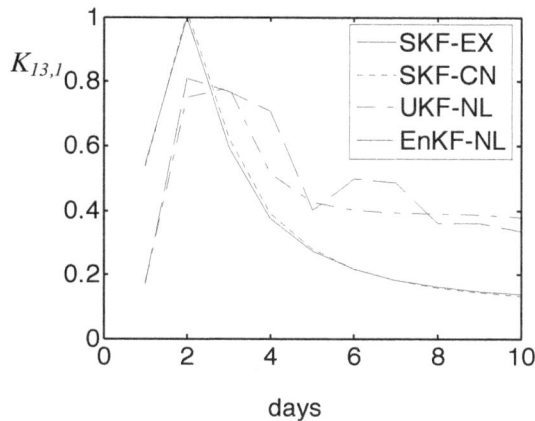

Figure 2. Evolution of the $K_{13,1}$ during the assimilation process with different assimilation algorithms.

and the measurement noise. These sampling strategies aim at generating ensembles with full rank and an improved conditioning with respect to what can be obtained with a pure random sampling, i.e. with a better representation of the error covariance matrix for a given ensemble size (Evensen, 2004).

The differences between the profiles retrieved with linear and nonlinear approaches are relatively high after the first update. UKF-NL and EnKF-NL first updates impact only slightly the medium and bottom nodes of the retrieved profile. Instead, using SKF-EX and SKF-CN a significant portion of the profile nodes is affected. These results are mainly related to differences in the prediction of the covariance $\mathbf{P}^-$ prior to the update. The first order approximation of the EX and CN schemes induces a smoothing of the a priori covariance $\mathbf{P}^-$ with respect to the NL scheme, which in turns also determines larger Kalman gain coefficients relating the state errors at middle depths to the observations at the top nodes, as illustrated in Fig. 2.

The generic Kalman gain coefficient $K_{i,j}$ is an index of the correlation between the state errors at the depth of the retrieved ith node and the observation at the depth of the assimilated jth node. Figure 2 shows the time variation of the $K_{13,1}$, i.e. the **K** coefficient associated to the retrieval of the node at a depth of 48.5 cm (at about the middle of the soil column) by assimilating the top node observation. SKF-EX and SKF-CN exhibit patterns of $K_{13,1}$ different from those of UKF-NL and EnKF-NL, with values much larger on the first day of assimilation.

One of the key differences between EnKF or UKF and SKF is in the way the error covariance is propagated: while SKF propagates a single state vector (corresponding to the mean state) and, analytically, its error covariance, EnKF and UKF propagate an ensemble of state vectors, with error covariance resulting from the distribution of the states across the ensemble. However, the differences between the profiles

retrieved after the first update cannot be ascribed to the different strategies adopted by the examined algorithms in the propagation of the error covariance. EnKF and UKF, if applied to the EX and CN numerical schemes, would yield results close to those obtained with the SKF. Actually, in the limit of many simulations (i.e. by taking a large amount of members for the EnKF and a large amount of sigma points for the UKF), the corresponding EnKF and UKF estimates (state and error covariance) would tend to those provided by the SKF, which are statistically the optimal solutions for linear systems, provided that no model errors occur, the entering noise is white and the noise covariances are known.

Moreover, the differences between the profiles retrieved after the first update cannot be ascribed to discrepancies between the a priori state profile $\hat{x}_k^-$ predicted by the different numerical schemes, since these discrepancies are negligible. Starting from the same initial poor guess, linear (EX, CN) and nonlinear numerical (NL) schemes predict the evolution of the state profile (a priori prediction) with very small differences.

In the UKF-NL and EnKF-NL algorithms, the nonlinear implicit differential scheme of the Richards equation is solved for each sample to predict its state evolution, while the update phase is activated only as the observation is available. In this numerical experiment, the NL differential scheme is resolved with an hourly time-step, thanks to its high numerical stability and accuracy, demanding a CPU time equal to almost 50 % that required for the CN scheme, which is implemented with a time step of 200 s. However, the nonlinear data assimilation approaches (UKF-NL and EnKF-NL) consume almost 50 % more CPU time than the SKF-CN, because of the cost of the ensemble propagation. The EnKF-NL is just slightly more efficient than the UKF-NL in terms of computational costs.

The effects of the system noise variance and of the initial state covariance deserve some comments.

Walker (1999) found that the performance of the retrieval algorithm is not particularly sensitive to the system noise variance. However, this result should be interpreted keeping in mind that Walker (1999) assumed the system noise variance to be equal to five percent of the state change. Instead, in the present study, the system noise variance has a more important role, as it is assumed to be equal to five percent of the previous state, similarly to Entekhabi et al. (1994).

As also shown by Walker (1999) in a sensitivity analysis for the same experimental setup, the convergence time tends to decrease as one takes higher initial state covariance values. With a value of 10^4 cm^2 applied to the diagonal elements of the initial state covariance, the retrieved profiles perfectly coincide with the true ones on third day of assimilation, regardless of the observation depth. Instead, as shown by Fig. 1, with a value of 10^3 cm^2 applied to the diagonal elements of the initial state covariance, perfect coincidence is achieved on the third day of assimilation only with an observation depth of 10.5 cm.

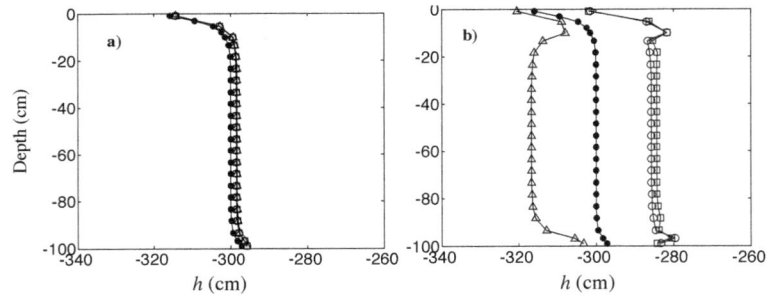

Figure 3. UKF-NL a priori estimates of the state mean after 1 h, computed with $\rho = 0.05$ (open circle), $\rho = 0.3$ (open square) and $\rho = 0.8$ (triangle), and adopting initial variances respectively equal to **(a)** 10^3 cm^2 and **(b)** 10^4 cm^2 on the diagonal elements of the state covariance matrix. The figure also shows the corresponding central state profiles (closed circle) propagated through the dynamic model on the same simulation time.

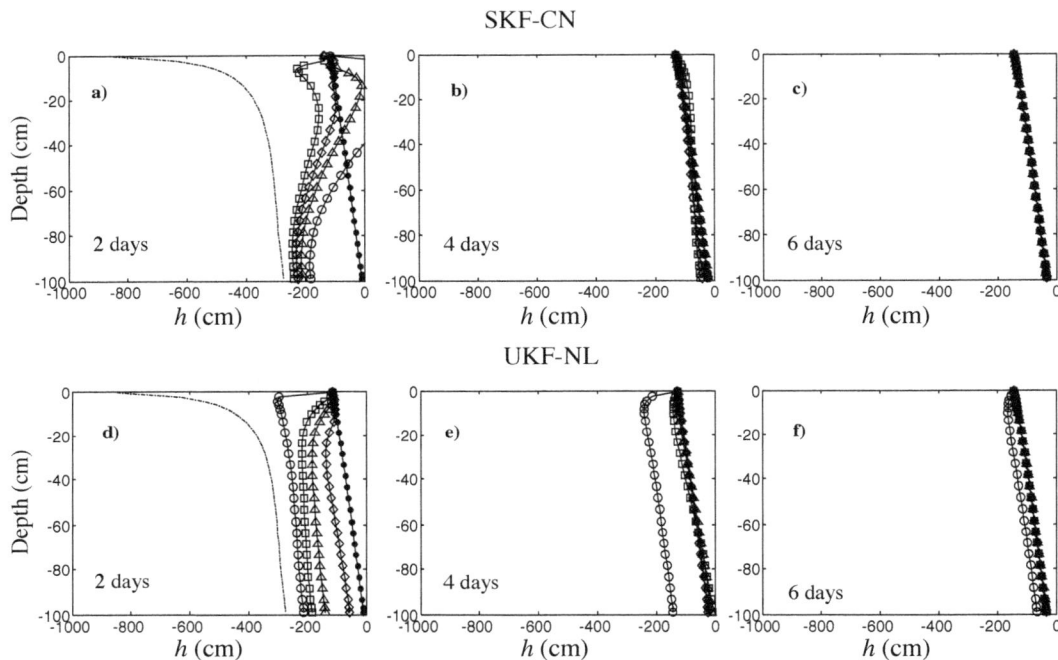

Figure 4. Retrieved profiles after **(a, d)** 2 days; **(b, e)** 4 days, and **(c, f)** 6 days using SKF-CN and UKF-NL, respectively, by assimilating matric pressure head observations every two days involving the nodes within the top 0.5 cm (open circle), 1.5 cm (square), 4.5 cm (triangle) and 10.5 cm (diamond), as compared with the "true" profile (closed circle) and "guess" profile (dashed line).

The choice of high initial state variance can actually determine some difficulties in the implementation of the UKF and EnKF algorithms. In both these algorithms, the distribution of the sample states logically depends on the magnitude of the variance. Taking a very large initial variance, without any correlation structure, could lead to sample profiles which are physically improbable, and consequently the stability of the assimilation process could degenerate. In case of the UKF, this issue can be overcome by shrinking the sample state distribution around the mean state with the scaling parameter ρ, which controls the weights attributed to the sample state distribution (Eq. 11). Figure 3 shows different

UKF-NL a priori estimates of the state mean $\hat{x}_k^-$, computed after 1 h with Eq. (14), by adopting parameter ρ respectively equal to 0.05, 0.3 and 0.8, and initial variances respectively equal to 10^3 and 10^4 cm^2 for the diagonal elements of the state covariance matrix. As a reference state profile for the comparative analysis of the different a priori estimates, Fig. 3 shows the corresponding central state profile (X_0 of Eq. 11) propagated through the dynamic model (Eq. 13) on the same simulation time. The a priori mean state estimation using the smaller initial state variance value is practically insensitive to the value of ρ, while this prior mean is highly affected by ρ when using the higher initial state variance. Taking $\rho = 0.8$, with a uniform initial profile -300 cm and state variance of 10^4 cm^2, leads to a set of sample profiles exhibiting positive matric heads for perturbed nodes. All positive matric heads were changed to zero before propagating the state through the dynamic model, thus favouring a shift of the predicted prior mean state towards smaller matric heads as compared with the central state. With ρ equal to 0.05 and 0.3 the saturated region is not sampled, but the predicted prior means are shifted towards larger matric heads with respect to the central state.

In practical applications, particularly those where the frequency of the observations is small, the evolution of the magnitudes of the state variance is unpredictable. Thus, the coefficients selected for designing the sigma point sampling strategy could turn out to be inadequate during the assimilation process. To overcome this issue, some other UKF applications adapt the value of the coefficients to guarantee the physical coherence of the sample states.

Figure 4 shows the profiles retrieved by assimilating matric head observations once every two days with the SKF-CN and the UKF-NL. In both cases the initial state covariance employed has been set equal to 10^3 cm^2. To guarantee a performance of the UKF-NL comparable with that of SKF-CN, the parameter ρ has been specifically tuned. Results show that the SKF-CN is able to retrieve the true profile already at the fourth day, i.e. at the second assimilation, practically for any observation depths. However, the profile retrieved with an observation depth of 0.5 cm exhibits positive matric heads in the upper 40 cm on the second day of assimilation. Satisfactory results with the SKF have also been obtained (not shown here) by assimilating observations once every three days and four days, except for the case when the observations are limited to the top node. SKF failed only for larger observation time-intervals, as the extremely negative pressure heads of the predicted states at the top nodes altered the singularity of the matrix operators. To achieve results similar to those obtained with SKF, specific tunings of the assimilation scheme are required with the UKF.

However, an aspect favouring the application of the UKF or the EnKF is the possibility to straightforwardly implement nonlinear state variable transformations. When the states are very far from the observations, the filtering process imposes severe gradients in the profiles, causing the estimation of temporary meaningless profiles. This could be partially avoided by making a transformation of the state (e.g. logarithmic), which implies at the same time a nonlinear transformation of the dynamic equation. This transformation, scarcely affecting results (not shown) in case of UKF or EnKF, should be treated by linearizing an already linearized equation in the case of an Extended Kalman Filter, therefore with a drastic reduction of the efficiency of the assimilation algorithm.

4.2   Assimilating of soil water content observations

The analysis has also been extended to examine the case of assimilating soil water content observations, instead of matric pressure heads.

The CN numerical scheme has been coupled with the EKF in order to handle the nonlinearity of the soil water retention function. Thus, the profiles retrieved with the EKF-CN algorithm have been compared with those retrieved with the UKF-NL and the EnKF-NL algorithms. An approach alternative to EKF-CN could have been to combine the SKF-CN algorithm for estimating the a priori mean state and covariance matrix of the state distribution, and apply the UKF just for the update phase when the soil moisture content observation becomes available, in order to address the nonlinearity of the water retention curve.

The EKF-CN algorithm entails the Jacobian matrices of the observation equation for computing the Kalman gain and the a posteriori error covariance, according to Eqs. (9)–(10). In this case study, matrix $\mathbf{D}_k$ is the identity matrix, while the non-zero coefficients of $\mathbf{C}_k$ coincide with the specific water capacity values computed at the of the observation nodes with respect to the a priori estimate $\hat{\boldsymbol{x}}_k^-$. Provided that the VGM model equations are continuous over the entire range of matric heads, the specific water capacity can be analytically derived. Thus the computation of the Kalman gain (Eq. 9) and of the a posteriori error covariance (Eq. 10) is very easy to be implemented and does not represent a significant increase of the overall computational load. Ultimately, the computational effort in the implementation of EKF-CN is not significantly different from that of SKF-CN for the examined system.

Figure 5 is designed to compare the matric head profiles retrieved with the EKF-CN, UKF-NL and EnKF-NL, analogously to Fig. 1, but assimilating soil water content observations instead of matric heads. The performances of the EnKF-NL and the UKF-NL algorithms are very similar to each other in terms of accuracy of the retrieved profiles, analogously to what observed with the same algorithms applied for assimilating matric pressure heads. The EnKF-NL is just slightly less onerous in terms of computational costs. Moreover, the profiles retrieved with the EKF-CN algorithm are markedly different from those obtained with the NL-based ones, particularly on the first day of assimilation. The EKF-CN algorithm requires a CPU time about twice smaller than the EnKF-NL and UKF-NL algorithms, because more time is required to ensure the convergence of the iterative numerical schemes implemented in the EnKF-NL and UKF-NL algorithms for each ensemble member.

The effect of the observation depth is more evident when taking soil water content instead of matric head as observation variable. For an observation depth of 10.5 cm (diamonds in Figs. 1 and 5), the convergence rate of the retrieved profiles towards the true ones by assimilating soil water content is very similar to that obtained by assimilating pressure

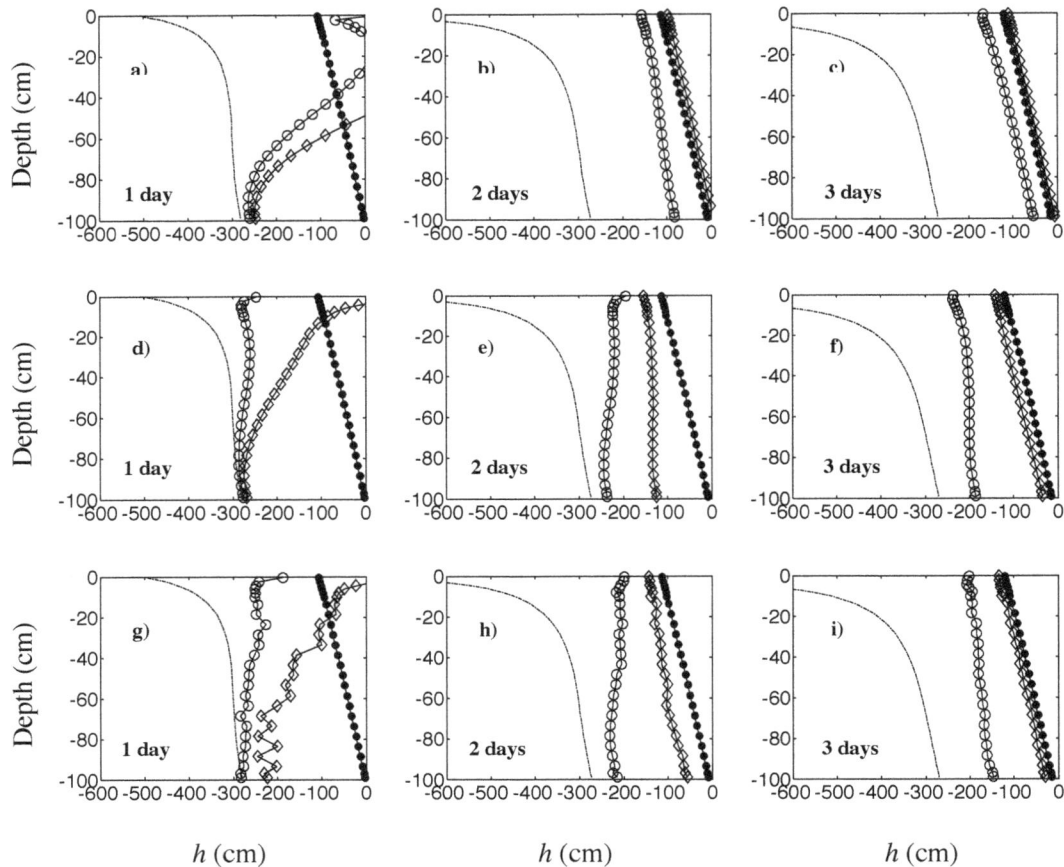

Figure 5. Retrieved profiles by assimilating daily observations of soil moisture contents involving nodes within the top 0.5 cm (open circle) and 10.5 cm (diamond) compared with the "true" profile (closed circle) and "guess" profile (dashed line), for different combinations of Kalman Filters and numerical schemes: (**a–c**) EKF-CN; (**d–f**) UKF-NL; (**g–i**) EnKF-NL.

heads. The profiles retrieved with an observation depth of 0.5 cm (open circles in Fig. 5) converge towards the true ones much more slowly than those retrieved with an observation depth of 10.5 cm, particularly for the UKF-NL and EnKF-NL schemes, for which the convergence is far from being reached even on the third day.

The differences between the matric head profiles retrieved by assimilating soil water contents instead of matric heads are entirely related to the nonlinearity of the observation equation, represented by the water retention function, which produces a deformation of the Kalman gain matrix with respect to that derived by assimilating matric heads, influencing both the state and the error covariance updates. This effect can be more easily interpreted by comparing the state profiles respectively retrieved with EKF and SKF, which show differences that are directly related to the linearization of the observation equation (first order approximation) by means of the specific water capacity.

The effect of this first order approximation is expected to be dependent on the soil properties. Coarser textured soils exhibit large gradients in the specific water capacity functions

for near-saturated conditions and thus are more vulnerable to biased predictions of the linearized observation equation. This effect can be exacerbated in dynamic processes involving infiltration, which impose limited correlation between nodes located at the opposite ends of the infiltration front, particularly in applications with real data, where model errors due to the uncertainty in the soil parameters are unavoidable.

It is important to point out that the results obtained by assimilating soil water content values, are subjected to the assumption that the parameters defining the soil water retention at the observation points coincide with those employed for simulating the soil water dynamics along the entire soil column, which is considered homogenous. However, in realistic circumstances, the retrieval algorithm should account for the model simplifications and system heterogeneity, particularly for the large spatial variability of the soil hydraulic properties (Pringle et al., 2007; Chirico et al., 2010). The "optimal" parameters defining the soil water retention at the observation points are in principle different from the "optimal" parameters defining the soil water dynamics along the soil column.

These "optimal" soil hydraulic parameters are to be considered as effective values at the scale of the observation or of the modelled system, respectively (Vereecken et al., 2007).

5   Conclusions

When designing a Kalman Filter algorithm for assimilating near surface data into the Richards equation, it would be desirable to choose the Kalman Filter considering the numerical scheme employed for solving the Richards equation and the type of assimilated variable.

It is well known that the nonlinear (NL) numerical scheme is computationally more efficient than the linearised Crank-Nicolson (CN) numerical scheme and for this reason the NL scheme is preferred by the most popular soil water dynamics simulation models. However, in some circumstances the larger computational effort required for the application of linear schemes is compensated by applying SKF analytic estimators for the propagation of the first two moments of the state profiles, with minor computational costs than the corresponding non-standard KF algorithms, without losing accuracy of the predictions.

In this study we verified this possibility by assimilating surface matric head or soil water content observations for retrieving matric head profiles of a homogeneous soil column subjected to a surface evaporation flux.

When assimilating matric heads, the standard Kalman Filter (SKF) combined with a Crank–Nicolson (CN) linear numerical scheme of the h-based form of the Richards equation, provides estimates of the matric head profiles converging faster to the true solution and with minor computational costs than the ensemble Kalman filter (EnKF) or the unscented Kalman Filter (UKF) with a NL scheme.

The SKF-CN algorithm takes advantage of both the stability of the CN numerical scheme and the linearity of the operators in the dynamic system model. An explicit numerical scheme, although being also linear, is unfeasible for practical applications, as it demands computational time-steps of the order of a few seconds.

The SKF-CN becomes unfeasible for large scale applications, due to the computational issues attached to an explicit formulation of the covariance propagation using the SKF (Reichle et al., 2002a; Reichle, 2008; Camporese et al., 2009). However, even large scale applications can often be reduced to a set of independent low-dimensional spatial systems. Reichle and Koster (2003) compared the results of an Ensemble Kalman Filter (EnKF) applied to a 1-D model to those obtained with EnKF applied to a 3-D model to estimate soil moisture in the root zone, finding that, although the 3-D approach produced more accurate results, in particular for intermediate assimilated data volumes, the 1-D EnKF satisfactorily performed with a lower demand of computational time. Using either a low volume of assimilated data or a high number of them, the differences were not very pronounced.

The relative gain of using the 1-D EnKF over the straight model (with no assimilation) can be larger than the relative gain of using the 3-D EnKF over the 1-D EnKF, at least for those circumstances where horizontal error correlations have a marginal role (Hoeben and Troch, 2000; Reichle and Koster, 2003).

In order to make the SKF-CN applicable, the form of the Richards equation should be chosen according to the assimilated variable, so that the type of variable describing the observations is equal to that describing the states or it is at least a linear transformation of it. However, this strategy is not always possible, as for example it occurs when the assimilation algorithm has to be implemented with closed on-hand model software, such as HYDRUS (Vogel et al., 1996), with a predefined nonlinear numerical scheme.

Another case is when the observed variable is soil water content while the Richards equation has to be in the h-based form in order to handle both saturated and unsaturated flows in the vadose zone, thus imposing a nonlinear observation equation in the state-space description of the dynamic system. In this case, the SKF can be replaced with the Extended Kalman Filter to handle the nonlinearity of the observation equation. With regards to the numerical evaporation test examined in this study, we could verify that, when assimilating surface soil moistures for retrieving matric head profiles, the Extended Kalman Filter combined with the Crank-Nicolson (EKF-CN) also provides state profiles converging faster to the true profiles than the UKF-NL or EnKF-NL numerical schemes, with less computational efforts. However, we also argued that the linearization of the observation equation embedded in EKF does not ensure accurate solutions for all circumstances, particularly for coarser textured soils and when sharp space-time gradients of the state vector are involved in the retrieving process.

We also compared the EnKF with the UKF, both as potential alternatives to the EKF, whose flaws and limitations for nonlinear models have been largely discussed in the literature (e.g. Julier et al., 1995; van der Merwe, 2004). The UKF has been less commonly applied in hydrological studies and, to our knowledge, previous studies illustrating the implementation of the UKF with a numerical scheme of the Richards equation have not been published.

Differently from the SKF, both UKF and EnKF sampling strategies can lead to state profiles with low physical meaning, particularly with large state covariances.

The computational efforts for implementing EnKF-NL and UKF-NL are directly linked to the corresponding sizes of the state ensemble propagated during the retrieving process. For the UKF, the size of the propagated ensemble is deterministically set equal to $2N + 1$, where N is the system dimension. The optimal ensemble size for EnKF is not known a priori and it has been often calibrated, making it dependent on both the type of examined process and the system dimension (Reichle et al., 2002a). However, UKF also requires the calibration of a scaling parameter affecting the spread of the sample state distribution around the central state, which can play a critical role in the overall retrieving process.

At least for the numerical evaporation test examined in this study, as far as the system dimension is almost half the ensemble size involved in EnKF, the two algorithms are almost equivalent in terms of computational costs. The UKF can be even more competitive than the EnKF, if the dimensionality on the nonlinear problem is much smaller than 50, which is the ensemble size generally required to gain accurate solutions in this type of applications (Camporese et al., 2009).

While dealing with different types of filters applied to non-linear systems, it is important to keep in mind that the physical model affects the ensemble generation and the uncertainty analysis, which can be relevant for skewed ensemble distributions (Drécourt, 2004), as observed by Reichle et al. (2002a) for very dry or wet soil conditions. Ultimately, as these Authors uphold, the "best" approach for a given data assimilation problem will be application dependent.

This study focussed on retrieving state profiles, while assuming that parameters are known. However, in most practical circumstances, significant uncertainties arise from the identification of the soil hydraulic parameters. The following two companion papers (Medina et al., 2013a, b) explore the capability of a dual Kalman Filter approach for simultaneous retrieval of states and parameters in the Richards equation, by examining synthetic and experimental data, respectively.

Acknowledgements. This study has been supported by P.O.N. project "AQUATEC – New technologies of control, treatment, and maintenance for the solution of water emergency". Hanoi Medina has been also supported by the Abdus Salam International Centre for Theoretical Physics (ICTP), where he has been appointed as Junior Associate. The Editor and the anonymous reviewers contributed considerably to improve the manuscript with fruitful comments.

Edited by: J. Vrugt

References

Camporese, M., Paniconi, C., Putti, M., and Salandin, P.: Ensemble Kalman filter data assimilation for a process based catchment scale model of surface and subsurface flow, Water Resour. Res., 45, W10421, doi:10.1029/2008WR007031, 2009.

Celia, M. A., Bououtas, E. T., and Zarba, R. L.: A general mass-conservative numerical solution for the unsaturated flow equation, Water Resour. Res., 26, 1483–1496, 1990.

Chirico, G. B., Medina, H., and Romano, N.: Functional evaluation of PTF prediction uncertainty: An application at hillslope scale, Geoderma, 155, 193–202, 2010.

Clark, M., Rupp, D., Woods, R., Zheng, X., Ibbitt, R., Slater, A., Schmidt, J., and Uddstrom, M.: Hydrological data assimilation with the ensemble Kalman filter: use of streamflow observations to update states in a distributed hydrological model, Adv. Water Resour., 31, 1309–1324, 2008.

Das, N. N. and Mohanty, B. P.: Root zone soil moisture assessment using remote sensing and vadose zone modeling, Vadose Zone J., 5, 296–307, 2006.

De Lannoy, G. J. M., Reichle, R. H., Houser, P. R., Pauwels, V. R. N., and Verhoest, N. E. C.: Correcting for forecast bias in soil moisture assimilation with the ensemble Kalman filter, Water Resour. Res., 43, W09410, doi:10.1029/2006WR005449, 2007.

Drécourt, J. P.: Data assimilation in hydrological modeling, Ph. D. Thesis, Environment & Resources DTU, Technical University of Denmark, Kongens Lyngby, Denmark, 2004.

Dunne, S. and Entekhabi, D.: Land surface state and flux estimation using the ensemble Kalman smoother during the Southern Great Plains 1997 field experiment, Water Resour. Res., W01407, doi:10.1029/2005WR004334, 2006.

Entekhabi, D., Nakamura, H., and Njoku, E. G.: Solving the inverse problem for soil moisture and temperature profiles by sequential assimilation of multifrequency remote sensed observations, IEEE T. Geosci. Remote, 32, 438–448, 1994.

Entekhabi, D., Rodríguez-Iturbe, I., and Castelli, F.: Mutual interaction of soil moisture state and atmospheric processes, J. Hydrol., 184, 3–17, 1996.

Evensen, G.: Sequential data assimilation with a nonlinear quasi-geostrophic model using Monte Carlo methods to forecast error statistics, J. Geophys. Res., 99, 10143–10162, 1994.

Evensen, G.: The ensemble Kalman filter: Theoretical formulation and practical implementation, Ocean Dynam., 53, 343–367, 2003.

Evensen, G.: Sampling strategies and square root analysis schemes for the EnKF, Ocean Dynam., 54, 539–560, 2004.

Haverkamp, R. M., Vauclin, M., Tourna, J., Wierenga, P. J., and Vachaud, G.: A comparison of numerical simulation models for one-dimensional infiltration, Soil Sci. Soc. Am. J., 41, 285–294, 1977.

Heathman, G. C., Starks, P. J., Ahuja, L. R., and Jackson, T. J.: Assimilation of surface soil moisture to estimate profile soil water content, J. Hydrol., 279, 1–17, 2003.

Hoeben, R. and Troch P. A.: Assimilation of active microwave observation data for soil moisture profile estimation, Water Resour. Res., 36, 2805–2819, doi:10.1029/2000WR900100, 2000.

Houtekamer, P. L. and Mitchell, H.: Data assimilation using an ensemble Kalman filter technique, Mon. Weather Rev., 126, 796–811, 1998.

Huang, C., Li, X., Lu, L., and Gu, J.: Experiments of one-dimensional soil moisture assimilation system based on ensemble Kalman filter, Remote Sens. Environ., 112, 888–900, doi:10.1016/j.rse.2007.06.026, 2007.

Julier, S. J. and Uhlmann, J. K.: A New Extension of the Kalman Filter to Nonlinear Systems, Proc. SPIE, 3068, 182–193, 1997.

Julier, S. J. and Uhlmann, J. K.: Unscented filtering and nonlinear estimation, Proc. IEEE, 92, 401–422, 2004.

Julier, S. J., Uhlmann, J. K., and Durrant-Whyte, H. F.: A new approach for filtering nonlinear systems, in: The Proceedings of the American Control Conference, Seattle WA, USA, 1628–1632, 1995.

Jury, W. A., Gardner, W. R., and Gardner, W. H.: Soil Physics, 5th Edn., John Wiley, New York, 1991.

Kalman, R. E.: A New Approach to Linear Filtering and Prediction Problems, ASME J. Basic Eng., 82D, 35–45, 1960.

Katul, G. G., Wendroth, O., Parlange, M. B., Puente, C. E., Folegatti, M. V., and Nielsen, D. R.: Estimation of in situ hydraulic conductivity function from nonlinear filtering theory, Water Resour. Res., 29, 1063–1070, 1993.

Kavetski, D., Binning, P., and Sloan, S. W.: Noniterative time stepping schemes with adaptive truncation error control for the solution of Richards equation, Water Resour. Res., 38, 1211, doi:10.1029/2001WR000720, 2002.

Luo, X. and Moroz, I. M.: Ensemble Kalman filter with the unscented transform, Physica D, 238, 549–562, 2009.

Luo, X. and Moroz, I. M.: Reply to "Comment on 'Ensemble Kalman filter with the unscented transform'", Physica D, 239, 1662–1664, 2010.

Matgen, P., Montanari, M., Hostache, R., Pfister, L., Hoffmann, L., Plaza, D., Pauwels, V. R. N., De Lannoy, G. J. M., De Keyser, R., and Savenije, H. H. G.: Towards the sequential assimilation of SAR-derived water stages into hydraulic models using the Particle Filter: proof of concept, Hydrol. Earth Syst. Sci., 14, 1773–1785, doi:10.5194/hess-14-1773-2010, 2010.

McLaughlin, D. B.: An integrated approach to hydrologic data assimilation: Interpolation, smoothing, and filtering, Adv. Water Resour., 25, 1275–1286, 2002.

Medina, H., Romano, N., and Chirico, G. B.: Kalman filters for assimilating near-surface observations into the Richards equation – Part 2: A dual filter approach for simultaneous retrieval of states and parameters, Hydrol. Earth Syst. Sci., 18, 2521–2541, doi:10.5194/hess-18-2521-2014, 2014.

Medina, H., Romano, N., and Chirico, G. B.: Kalman filters for assimilating near-surface observations into the Richards equation – Part 3: Retrieving states and parameters from laboratory evaporation experiments, Hydrol. Earth Syst. Sci., 18, 2543–2557, doi:10.5194/hess-18-2543-2014, 2014.

Milly, P. C. and Dunne, K. A.: Sensitivity of the global water cycle to the water-holding capacity of land, J. Climate, 7, 506–526, 1994.

Moradkhani, H., Sorooshian, S., Gupta, H. V., and Houser, P.: Dual state-parameter estimation of hydrological models using Ensemble Kalman filter, Adv. Water Resour., 28, 135–147, 2005.

Paniconi, C., Aldama, A. A., and Wood, E. F.: Numerical evaluation of iterative and noniterative methods for the solution of the nonlinear Richards equation, Water Resour. Res., 27, 1147–1163, doi:10.1029/91WR00334, 1991.

Papadakis, N., Mémin, E., Cuzol, A., and Gengembre, N.: Data assimilation with the weighted ensemble Kalman filter, Tellus A, 62, 673–697, 2010.

Pipunic, R. C., Walker, J. P., and Western, A.: Assimilation of remotely sensed data for improved latent and sensible heat flux prediction: A comparative synthetic study, Remote Sens. Environ., 112, 1295–1305, 2008.

Press, W. H., Teukolsky, S. A., Vetterling, W. T., and Flannery, B. P.: Numerical Recipes in C: The Art of Scientific Computing, 2nd Edn., Cambridge University Press, 1992.

Pringle, M. J., Romano, N., Minasny, B., Chirico, G. B., and Lark, R. M.: Spatial evaluation of pedotransfer functions using wavelet analysis, J. Hydrol., 333, 182–198, 2007.

Reichle, R. H.: Data assimilation methods in the Earth sciences, Adv. Water Resour., 31, 1411–1418, 2008.

Reichle, R. H. and Koster, R. D.: Assessing the impact of horizontal error correlations in background fields on soil moisture estimation, J. Hydrometeorol., 4, 1229–1242, 2003.

Reichle, R. H., McLaughlin, D. B., and Entekhabi, D.: Hydrologic data assimilation with the ensemble Kalman filter, Mon. Weather Rev., 130, 103–114, 2002a.

Reichle, R. H., Walker, J. P., Koster, R. D., and Houser, P. R.: Extended versus ensemble Kalman filtering for land data assimilation, J. Hydrometeorol., 3, 728–740, 2002b.

Reichle, R. H., Koster, R. D., Liu, P., Mahanama, S. P. P., Njoku, E. G., and Owe, M.: Comparison and assimilation of global soil moisture retrievals from the Advanced Microwave Scanning Radiometer for the Earth Observing System (AMSR-E) and the Scanning Multichannel Microwave Radiometer (SMMR), J. Geophys. Res.-Atmos., 112, D09108, doi:10.1029/2006JD008033, 2007.

Rodriguez-Iturbe, I. and Porporato, A.: Ecohydrology of Water-Controlled Ecosystems: Soil Moisture and Plant Dynamics, Cambridge University Press, 2005.

Romano, N., Brunone, B., and Santini, A.: Numerical analysis of one-dimensional unsaturated flow in layered soils, Adv. Water Resour., 21, 315–324, 1998.

Santini, A.: Model for simulating soil water dynamics considering root extraction, Proceedings of CCE Seminaires sur l'Irrigation Localisee – Agrimed Sorrento, Italy, 1980.

van Dam, J. C.: Field-scale water flow and solute transport. SWAP model concepts, parameter estimation and case studies, Doctoral Thesis Wageningen University, Wageningen, 2000.

van der Merwe, R.: Sigma-Point Kalman Filters for Probabilistic Inference in Dynamic State-Space Models, Ph. D. dissertation, University of Washington, USA, 2004.

van Genuchten, M. Th.: A closed-form equation for predicting the hydraulic conductivity of unsaturated soils, Soil Sci. Soc. Am. J., 44, 892–898, 1980.

Vereecken, H., Kasteel, R., Vanderborght, J., and Harter T.: Upscaling hydraulic properties and soil water flow processes in heterogeneous soils, Vadose Zone J., 6, 1–28, doi:10.2136/vzj2006.0055, 2007.

Vereecken, H., Huisman, J. A., Bogena, H., Vanderborght, J., Vrugt, J. A., and Hopmans, J. W.: On the value of soil moisture measurements in vadose zone hydrology: A review, Water Resour. Res., 44, W00D06, doi:10.1029/2008WR006829, 2008.

Vogel, T., Huang, K., Zhang, R., and van Genuchten, M. T.: The HYDRUS code for simulating one-dimensional water flow, solute transport and heat movement in variably-saturated media, Research Report No. 140, US Salinity Laboratory, Agricultural Research Service, USDA, Riverside, CA, 1996.

Vrugt, J. A., Bouten, W., and Weerts, A. H.: Information content of data for identifying soil hydraulic parameters from outflow experiments, Soil Sci. Soc. Am. J., 65, 19–27, 2001.

Vrugt, J. A., Bouten, W., Gupta, H. V., and Hopmans, J. W.: Toward improved identifiability of soil hydraulic parameters: On the selection of a suitable parametric model, Vadose Zone J., 2, 98–113, 2003.

Walker, J. P.: Estimating soil moisture pro?le dynamics from near-surface soil moisture measurements and standard meteorological data, Ph D. thesis, The University of Newcastle, Callaghan, New South Wales, Australia, 766 pp., 1999.

Walker, J. P., Willgoose, G. R., and Kalma, J. D.: One-dimensional soil moisture profile retrieval by assimilation of near-surface observations: a comparison of retrieval algorithms, Adv. Water Resour., 24, 631–650, 2001.

Weerts, A. H. and El Serafy, G. Y.: Particle filtering and ensemble Kalman filtering for state updating with hydrological conceptual rainfall-runoff models, Water Resour. Res., 42, 1–17, doi:10.1029/2005WR004093, 2006.

Xie, X. and Zhang, D.: Data assimilation for distributed hydrological catchment modeling via ensemble Kalman filter, Adv. Water Resour, 33, 678–690, 2010.

7

Kalman filters for assimilating near-surface observations into the Richards equation – Part 2: A dual filter approach for simultaneous retrieval of states and parameters

H. Medina[1], N. Romano[2], and G. B. Chirico[2]

[1]Department of Basic Sciences, Agrarian University of Havana, Havana, Cuba
[2]Department of Agricultural Engineering, University of Naples Federico II, Naples, Italy

Correspondence to: G. B. Chirico (gchirico@unina.it)

Abstract. This study presents a dual Kalman filter (DSUKF – dual standard-unscented Kalman filter) for retrieving states and parameters controlling the soil water dynamics in a homogeneous soil column, by assimilating near-surface state observations. The DSUKF couples a standard Kalman filter for retrieving the states of a linear solver of the Richards equation, and an unscented Kalman filter for retrieving the parameters of the soil hydraulic functions, which are defined according to the van Genuchten–Mualem closed-form model. The accuracy and the computational expense of the DSUKF are compared with those of the dual ensemble Kalman filter (DEnKF) implemented with a nonlinear solver of the Richards equation. Both the DSUKF and the DEnKF are applied with two alternative state-space formulations of the Richards equation, respectively differentiated by the type of variable employed for representing the states: either the soil water content (θ) or the soil water matric pressure head (h). The comparison analyses are conducted with reference to synthetic time series of the true states, noise corrupted observations, and synthetic time series of the meteorological forcing. The performance of the retrieval algorithms are examined accounting for the effects exerted on the output by the input parameters, the observation depth and assimilation frequency, as well as by the relationship between retrieved states and assimilated variables. The uncertainty of the states retrieved with DSUKF is considerably reduced, for any initial wrong parameterization, with similar accuracy but less computational effort than the DEnKF, when this is implemented with ensembles of 25 members. For ensemble sizes of the same order of those involved in the DSUKF, the DEnKF fails to provide reliable posterior estimates of states and parameters. The retrieval performance of the soil hydraulic parameters is strongly affected by several factors, such as the initial guess of the unknown parameters, the wet or dry range of the retrieved states, the boundary conditions, as well as the form (h-based or θ-based) of the state-space formulation. Several analyses are reported to show that the identifiability of the saturated hydraulic conductivity is hindered by the strong correlation with other parameters of the soil hydraulic functions defined according to the van Genuchten–Mualem closed-form model.

1 Introduction

Accurate determination of the water dynamics in the vadose zone is crucial for the success of many hydrological, climatic and environmental studies. The significant increase in the availability of hydrologic data sets can definitely provide extensive opportunities for reducing the uncertainty associated to the detection of the spatial and temporal variability of soil moisture. However, it also calls for more robust methods to merge new available observations and uncertain model predictions appropriately.

A crucial aspect in the application of these methods is the proper specification of the model parameters, as a function of the variables characterizing the state of the water in the soil (e.g. Heathman et al., 2003; de Lannoy et al., 2007; Vereecken et al., 2008). Parameterization of soil hydraulic properties is considered one of the main challenges in the current land surface modelling efforts (Zhu and Mohanty, 2004), particularly because hydraulic properties exhibit large spatial variability at all scales of interest, making it extremely difficult to capture hydrological behaviour at one particular scale (Pringle et al., 2007; Chirico et al., 2010).

A prerequisite to properly handle the marked variability of soil hydraulic properties in large-scale applications is the use of efficient calibration methods in terms of time and storage. Traditional methods for parameter identification generally optimize an objective function from a historical batch of data, and hence require a set of historical data to be kept in storage and processed all together, with limited flexibility to account for new available measurements (Moradkhani et al., 2005). In addition, a common problem with most of these traditional inverse methods is stability and convergence (Yeh, 1986; Abbaspour et al., 1997). Therefore, several attempts have been made to develop and apply calibration methods that circumvent these drawbacks.

Considerable progress has been achieved in the development and application of sequential data assimilation (DA) techniques. As recursive data-processing algorithms, DA methods do not require all past information to be stored; they continuously update the variables under scrutiny in the model, when new measurements become available, to improve the model forecast and evaluate the forecast accuracy (McLaughlin, 2002; Vrugt et al., 2005; Reichle, 2008). Although sequential estimation is typically applied only to the state variables, some algorithms, belonging to the family of the dual estimation methods, have been designed to simultaneously estimate model states and parameters as part of the assimilation process. This family of algorithms includes joint and dual filtering as well as expectation maximization (EM) approaches (e.g. Moradkhani et al., 2005; Liu and Gupta, 2007). EM methods have been commonly designed for offline applications, but sequential EM methods have also been proposed (Wan and Nelson, 2001).

In the dual filtering approach, a separate state-space representation is used for the states and the parameters, while in the joint approach the unknown system states and parameters are concatenated into a single higher-dimensional joint state vector. In principle, the joint approach should provide better estimates than the dual approach, because it explicitly accounts for the cross-covariance between state and parameter estimates. However, the estimation process can lead to unstable results because of complex interactions between states and parameters in nonlinear dynamic systems (Moradkhani et al., 2005; Liu and Gupta, 2007).

Some modern approaches extend the traditional parameter estimation paradigm toward a more explicit incorporation of structural data errors. Since the performance in hydrological modelling is also affected by errors in model structure and input data, model adjustment through time variation of parameters together with state variables can result in a limited understanding about the overall uncertainty (Clark and Vrugt, 2006). Nevertheless, the diagnostic analysis of a model can

be a difficult task and it is only possible after the model has been parameterized (Spaaks and Bouten, 2013). According to Renard et al. (2010), none of the current approaches appears entirely satisfactory and the optimal methodology for handling structural errors is still to be established.

Common sequential DA methods are based on standard Kalman filtering (SKF), from the innovative work of Kalman (1960). SKF became a widely used technique to merge information optimally from different sources and model predictions in linear systems (e.g. McLaughlin, 2002; Vereecken et al., 2008). Variations of the SKF algorithm have been developed to make it applicable to the sequential probabilistic inference problem within nonlinear dynamic systems, such as the extended Kalman filter (EKF) (Jazwinski, 1970), the commonly used ensemble Kalman filter (EnKF) (Evensen, 1994, 2003), and the unscented Kalman filter (UKF) (Julier et al., 1995; van de Merwe, 2004).

A fundamental difference between the Kalman filter and variational methods is that the former explicitly evolves the covariance matrix without interruption, while variational methods do not propagate error covariance information from one assimilation interval to the next (Reichle, 2008). In addition, the Kalman filter provides an analytical solution of the a posteriori state mean, while variational methods rely on numerical methods, which are considered more feasible for applications where the dimension of the state vector is very large, as with weather forecasting models (Reichle, 2008).

Kalman filter applications in hydrology (Reichle et al., 2002; Reichle and Koster, 2003; Reichle, 2008; Camporese et al., 2009) favour the use of EnKF, relying on the propagation of a random ensemble of the retrieving variable. The EnKF is an advantageous approach for highly dimensional applications, mainly because, by means of a comparably small ensemble of model trajectories, it captures the relevant parts of the error structure (Reichle, 2008). This method also facilitates the treatment of errors in model dynamics and parameters (Reichle and Koster, 2003; Moradkhani et al., 2005) and it is easily scalable. Nevertheless, the EnKF estimation based on small ensemble sizes, can be affected by spurious modes and large biases even if the ensemble mean and covariance are correct (Luo and Moroz, 2009; Lei and Baehr, 2013). Moreover, the optimal ensemble size for the EnKF is uncertain and is generally chosen on the basis of a heuristic evaluation.

The sampling strategy of the EnKF could be a drawback in large-scale applications where a small variation of the ensemble size has an important impact on the computational demand (e.g. Kumar et al., 2008). The implementation of these large-scale assimilation systems is often described as a collection of independent low dimensional assimilation problems (Crow and Wood, 2003; Reichle and Koster, 2003; Kumar et al., 2008).

Chirico et al. (2014) show that the SKF, coupled with a Crank–Nicolson numerical scheme, can be an efficient choice for low dimensional applications, because it provides retrieval performances similar to those obtained with non-linear schemes, but with less computational effort, also circumventing some issues which may arise with the sampling strategies used by the UKF and the EnKF. However, the UKF can be more flexible and computationally efficient than the EnKF in problems with low degrees of freedom, because it relies on ensemble sizes equal to twice the number of degrees of freedom plus one.

Few attempts have been made to retrieve soil water state profiles and soil hydraulic parameters simultaneously by assimilating near-surface observations with Kalman filters (e.g. Qin et al., 2009; Yang et al., 2009; Montzka et al., 2011). Tian et al. (2008) used a dual UKF for reproducing the temporal evolution of daily soil moisture under freezing conditions by assimilating satellite observations. Lü et al. (2011) developed a dual Kalman filter for estimating the root zone soil moisture using a model based on the Richards equation, by combining the EKF to update the state variables, with an optimization algorithm for retrieving parameters of soil hydraulic functions defined according to the van Genuchten–Mualem (VGM) relations (van Genuchten, 1980). Monztka et al. (2011) performed a joint approach retrieving soil moisture and VGM parameters, but using a particle filter algorithm.

Moving from the result of Chirico et al. (2014), we hypothesize that the combination of SKF applied to a linearized numerical representation of the Richards equation, and UKF applied to handle the intrinsic nonlinearities between hydraulic parameters and soil water states, could provide a suitable strategy for optimizing the prediction of the state dynamics.

The first objective of this study is to illustrate the feasibility of using a deterministic dual filter approach to perform simultaneous retrieval of soil moisture profiles and VGM parameters, with similar accuracy but reduced computational expense, as compared with ensemble Kalman filters, based on the assimilation of near-surface observations in a one-dimensional Richards' equation. The analysis is based on a synthetic test assuming uncertain observations and a poor guess of the initial states. A small structural error is also involved by implementing a different numerical solver of the Richards equation in the assimilation algorithm from that employed for generating the reference synthetic data.

The dual Kalman filter (hereafter referred to as DSUKF – dual standard-unscented Kalman filter) is designed by coupling the SKF approach for retrieving the states with the UKF for retrieving soil hydraulic parameters. For comparative purposes, the simultaneous retrieval of states and parameters is also performed using the dual ensemble Kalman filter (DEnKF), following the framework described by Moradkhani et al. (2005). Interested readers are referred to this work, widely cited by the hydrological data assimilation community.

A second objective is to compare the potential advantages and limitations of an h-based or a θ-based form of the Richards equation in the retrieval algorithm, also account-ing for different initial guesses of the parameters, observation depths, assimilation frequencies as well as the type of near-surface observations (h or θ).

2   Model and methods

2.1   Governing equation

As in the vast majority of applications in this realm, we describe the vertical movement of water under isothermal conditions in a rigid, homogeneous, variably saturated porous medium using the Richards equation (Jury et al., 1991). The following two equations represent the Richards equation in the h-based and in θ-based forms, respectively:

$$\frac{\partial \theta}{\partial t} = C(h)\frac{\partial h}{\partial t} = \frac{\partial \left[K(h)\left(\frac{\partial h}{\partial z} - 1\right)\right]}{\partial z}, \tag{1}$$

$$\frac{\partial \theta}{\partial t} = \frac{\partial \left[D(\theta)\frac{\partial \theta}{\partial z} - K(\theta)\right]}{\partial z}, \tag{2}$$

where t is time and z is soil depth taken as positive downward with $z = 0$ at the top of the profile, $C(h) = d\theta/dh$ [1/L] is the specific water capacity of the soil at matric pressure head, h, obtained by differentiating the function $\theta(h)$, and $D(\theta) = K(\theta)/C(\theta)$ [L^2/T] represents the unsaturated diffusivity.

For an efficient numerical solution of the model, it is convenient to describe the soil hydraulic properties using closed-form analytical relationships. The following non-hysteretic VGM equations (van Genuchten, 1980) are widely used in soil hydrology:

$$\theta(h) = \theta_r + (\theta_s - \theta_r)\left[1 + |\alpha h|^n\right]^{-m}, \tag{3}$$

$$K(\theta) = K_s S_e^\lambda \left[1 - \left(1 - S_e^{1/m}\right)^m\right]^2, \tag{4}$$

where θ_s is the saturated soil water content, θ_r is the residual soil water content, $S_e = (\theta - \theta_r)/(\theta_s - \theta_r)$ is the effective saturation, K_s is the saturated hydraulic conductivity, and α [L^{-1}], n (-), m(-) and λ(-) are empirical scale and shape parameters. A common assumption, also adopted in this work, is to set $\lambda = 0.5$ and pose $m = 1 - 1/n$.

2.2   Numerical formulation of the model

Chirico et al. (2014) showed that the implementation of the filtering approach upon a linearized Crank–Nicolson finite difference scheme (CN) can be an efficient algorithm for one-dimensional problems. The differentiation of Eq. (1) for intermediate nodes according to the CN scheme, leads to the expression

$$\left(\frac{-K_{k-1}^{i-1/2}}{2\Delta z^i \Delta z^u}; \frac{C_{k-1}^i}{\Delta t_{k-1}} + \frac{\frac{K_{k-1}^{i-1/2}}{\Delta z^u} + \frac{K_{k-1}^{i+1/2}}{\Delta z^l}}{2\Delta z^i}; \frac{-K_{k-1}^{i+1/2}}{2\Delta z^i \Delta z^l}\right) \begin{pmatrix} h_k^{i-1} \\ h_k^i \\ h_k^{i+1} \end{pmatrix}$$

$$= \left(\frac{K_{k-1}^{i-1/2}}{2\Delta z^i \Delta z^u} ; \frac{C_{k-1}^i}{\Delta t_{k-1}} - \frac{\frac{K_{k-1}^{i-1/2}}{\Delta z^u} + \frac{K_{k-1}^{i+1/2}}{\Delta z^l}}{2\Delta z^i} ; \frac{K_{k-1}^{i+1/2}}{2\Delta z^i \Delta z^l} \right)$$

$$\left(\begin{array}{c} h_{k-1}^{i-1} \\ h_{k-1}^i \\ h_{k-1}^{i+1} \end{array} \right) + \frac{K_{k-1}^{i-1} - K_{k-1}^{i+1}}{2\Delta z^i}, \tag{5}$$

where superscript i is the node number (increasing downward), subscript k is the time level, and $\Delta t_k = t_{k+1} - t_k$. The soil column is divided into compartments of thickness Δz^i. All nodes, including the top and bottom node, are in the centre of the soil compartments, with $\Delta z^u = z^i - z^{i-1}$ and $\Delta z^l = z^{i+1} - z^i$. The spatial averages of K are calculated as arithmetic means.

Assuming flux boundary conditions, the differential equations at the top and bottom nodes respectively are

$$\left(\frac{C_{k-1}^1}{\Delta t_{k-1}} + \frac{K_{k-1}^{1\frac{1}{2}}}{2\Delta z^1 \Delta z^l} ; \frac{-K_{k-1}^{1\frac{1}{2}}}{\Delta z^1 \Delta z^l} \right) \left(\begin{array}{c} h_k^1 \\ h_k^2 \end{array} \right) \tag{6}$$

$$= \left(\frac{C_{k-1}^1}{\Delta t_{k-1}} - \frac{K_{k-1}^{1\frac{1}{2}}}{2\Delta z^1 \Delta z^l} ; \frac{K_{k-1}^{1\frac{1}{2}}}{\Delta z^1 \Delta z^l} \right) \left(\begin{array}{c} h_{k-1}^1 \\ h_{k-1}^2 \end{array} \right)$$

$$+ \frac{q_{\text{top}} - K_{k-1}^{1\frac{1}{2}}}{\Delta z^1},$$

$$\left(\frac{-K_{k-1}^{N-1/2}}{\Delta z^N \Delta z^u} ; \frac{C_{k-1}^N}{\Delta t_{k-1}} + \frac{K_{k-1}^{N-1/2}}{2\Delta z^N \Delta z^u} \right) \left(\begin{array}{c} h_k^{N-1} \\ h_k^N \end{array} \right) \tag{7}$$

$$= \left(\frac{K_{k-1}^{N-1/2}}{2\Delta z^N \Delta z^u} ; \frac{C_{k-1}^N}{\Delta t_{k-1}} - \frac{K_{k-1}^{N-1/2}}{2\Delta z^N \Delta z^u} \right) \left(\begin{array}{c} h_{k-1}^{N-1} \\ h_{k-1}^N \end{array} \right)$$

$$+ \frac{K_{k-1}^{N-1/2} - q_{\text{bot}}}{\Delta z^N},$$

where q_{top} and q_{bot} are the fluxes at the top and bottom of the soil profile, respectively.

The analogous differential expressions of the Richards equation in the θ form (Eq. 2) can be obtained from Eqs. (5)–(7) by simply removing the soil water capacity (C) and by substituting h with θ, the hydraulic conductivity (K) of the dependent terms with the diffusivity (D), while keeping the independent terms on the right-hand side unchanged.

In the numerical scheme, the explicit linearization of K and C (or D) is implemented by taking their values at the previous time step $k - 1$. A linear state-space representation of the dynamic system can then be easily derived by combining

the set of Eqs. (5)–(7) written for each node and accounting for the boundary conditions:

$$\mathbf{B}_{k-1} x_k = \mathbf{A}_{k-1} x_{k-1} + f_{k-1}, \tag{8}$$

where x represents the state vector (i.e. either soil water contents or matric heads in the soil profile), while $\mathbf{A}_{k-1}$ and $\mathbf{B}_{k-1}$ are tridiagonal matrices obtained by assembling the terms in the first parenthesis on the right- and left-hand side of Eqs. (5)–(7), respectively. The term f_{k-1} is a vector obtained by assembling the terms on the right-hand side of the state variable at time step $k - 1$. More explicitly, Eq. (8) becomes

$$x_k = \mathbf{F}_{k-1} x_{k-1} + g_{k-1}, \tag{9}$$

where $\mathbf{F} = \mathbf{B}^{-1} \mathbf{A}$ and $g = \mathbf{B}^{-1} f$.

2.3 The dual standard-unscented Kalman filter (DSUKF) formulation

The dual filter approach has been implemented with most of the variants of the Kalman filter, applied to both linear and nonlinear problems, i.e. the SKF (Todini et al., 1976), the EKF (Nelson, 2000; Wan and Nelson, 2001), the EnKF (Moradkhani et al., 2005) and the UKF (Wan and van der Merwe, 2001; van der Merwe, 2004).

In this section we illustrate the (DSUKF) formulation, where the SKF is implemented for retrieving the states of a linear system, while the UKF is applied to handle the marked nonlinearities between states and parameters.

At every time step k, the posterior parameter estimate at time $k - 1$ is used in the state filter, while the current estimate of the states is used in the parameter filter. In the most general case, the set of system equations for the states can be written as follows:

$$x_k = F_{k-1,k} \left(x_{k-1}, u_k, \hat{w}_{k-1} \right) + v_{k-1}, \tag{10}$$

$$y_k = H_k \left(x_k, \hat{w}_{k-1} \right) + \eta_k. \tag{11}$$

In a Bayesian framework, they represent a prior distribution over the states. Equation (10) allows inferring the transition probability density of the states, while Eq. (11) determines the probability density of the observations given the prior states. The set of system equations for the parameters can be written as

$$w_k = w_{k-1} + \xi_{k-1}, \tag{12}$$

$$y_k = H_k \left(F_{k-1,k} \left(\hat{x}_{k-1}, u_k, w_k \right), w_k \right) + \varsigma_k, \tag{13}$$

representing a prior distribution over an artificial time-dependent random variable that emulates model parameters.

In the equations above, u_k is the exogenous input assumed to be known at instant t_k; v_{k-1} accounts for a simplified representation of the model errors, assumed to be a zero-mean Gaussian process noise with covariance $\mathbf{Q}_{k-1}$, while η_k is the zero mean and temporally uncorrelated observation or measurement noise with covariance $\mathbf{R}_k$, corrupting the observation of the states. The state transition density $p(x_k|x_{k-1}, u_k, w_{k-1})$ is fully specified by $F_{k-1,k}$ and the process noise distribution $p(v_{k-1})$, whereas H_k and the observation noise distribution $p(\eta_k)$ fully specify the observation likelihood $p(y_k|x_k, w_k)$.

$F_{k-1,k}$ and H_k are parameterized via the parameter vector w_k, whose evolution is artificially set up in a way similar to that employed for the state variables (Moradkhani et al., 2005) by means of a stationary process with identity state transition matrix. $\xi_{k-1} \in N(0, \mathbf{Q}_{w,k-1})$ is the noise driving parameter updating, and $\varsigma_k = \eta_k + \xi_k$ is the noise corrupting the observation equation relative to the parameters, with zero mean and covariance $\mathbf{R}_{w,k}$. The upper symbol "^" denotes the density mean of the variable.

2.3.1 UKF algorithm for parameter retrieval

The UKF, like the EnKF, is based on a strategy for the selection of the sample points, which aims to capture the posterior true mean and covariance of the retrieved variable, after the sample points are propagated through the true nonlinear system. States or parameters are still represented by a Gaussian random variable. However, in the UKF this is not specified by an ensemble of randomly chosen points, like in the EnKF, rather by using a minimal set of deterministically chosen sample points.

Considering $\hat{w}$ and $\mathbf{P}_w$, respectively, as mean and covariance of the parameter vector w to be retrieved, having the dimension equal to N_{par}, the UKF selects a set of sigma points $\mathbf{S}_i = \{\mu_i, \mathcal{W}_i, i = 0 \ldots 2N_{par}\}$, consisting of $2N_{par} + 1$ vectors $\mathcal{W}_i$ and their associated weights μ_i, completely capturing the actual mean and covariance of the random variable w. A selection of sigma points fulfilling this requirement is defined as follows:

$$\mathcal{W}_0 = \hat{w}; \mathcal{W}_i = \hat{w} + \left(\sqrt{\gamma \mathbf{P}_w}\right)_i, i = 1, \ldots, N_{par}; \mathcal{W}_i$$
$$= \hat{w} - \left(\sqrt{\gamma \mathbf{P}_w}\right)_i, i = N_{par} + 1, \ldots, 2N_{par}, \quad (14)$$

$$\mu_0^{(m)} = \frac{\gamma - N_{par}}{\gamma}; \mu_0^{(c)} = \frac{\gamma - N_{par}}{\gamma} + \left(1 - \rho^2 + \beta\right); \mu_i^{(m)}$$
$$= \mu_i^{(c)} = \frac{1}{2(\gamma)}, i = 1, \ldots, 2N_{par}. \quad (15)$$

Weight values for calculating the mean and the covariance are distinguished by the upper indexes m and c, respectively. The other parameters are defined as follows: $\gamma = \rho^2(N_{par} + \kappa)$, where ρ is a factor employed to expand or to shrink the sample state distribution around the mean; κ is a scaling parameter; β affects the weights of the points when calculating the covariance. Details about the proper choice of ρ, β and κ can be found in the work of van der Merwe (2004). The term $\left(\sqrt{\gamma \mathbf{P}_w}\right)_i$ is the ith column (or row) of the root square matrix $\gamma \mathbf{P}_w$, calculated by Cholesky decomposition (Press et al., 1992).

The evolution of the parameter mean and covariance during each time step is computed as follows

$$\hat{w}_k^- = \hat{w}_{k-1} \quad (16)$$

$$\mathbf{P}_{w,k}^- = \mathbf{P}_{w,k-1} + \mathbf{Q}_{w,k-1}. \quad (17)$$

The artificial noise covariance $\mathbf{Q}_w$ is computed as follows (Wan and Nelson, 1997; Nelson, 2000; van der Merwe, 2004):

$$\mathbf{Q}_{w,k} = \left(\lambda_{RLS}^{-1} - 1\right)\mathbf{P}_{w,k}. \quad (18)$$

The parameter $\lambda_{RLS} \in (0, 1]$ is considered a forgetting factor, as defined in the recursive least-squares (RLS) algorithm. Nelson (2000) showed that setting $\lambda_{RLS} < 1$ (i.e. the prior covariance is larger than the posterior covariance) provides an approximate exponentially decaying weight on past data. By setting $\lambda_{RLS} = 1$ (i.e. no process noise for the parameters is considered) all past data are equally weighted to obtain the current dynamics.

Whenever measurements are available, new sample states are created by substituting the a priori parameter mean, $\hat{w}_k^-$, and covariance, $\mathbf{P}_{w,k}^-$, in Eq. (14). In principle the weights, μ_i, do not change during the simulation.

The set of $2N_{par} + 1$ parameter vectors $\mathcal{W}_k$ is propagated across the model, and the observation equation, using as states the a posteriori mean at $k - 1$, $\hat{x}_{k-1}$, is expressed as follows:

$$\mathcal{Y}_k = H_k\left(F_{k-1,k}\left(\hat{x}_{k-1}, u_k, v_{k-1}, \mathcal{W}_k\right)\right). \quad (19)$$

$\mathcal{Y}_k$ also represents a set of $2N_{par} + 1$ vectors, each having N_{obs} elements.

The Kalman gain employed for modifying the parameter trajectories is obtained as follows:

$$\mathbf{K}_{w,k} = \mathbf{P}_{wy,k}\left(\mathbf{P}_{yy,k}^w + \mathbf{R}_{w,k}\right)^{-1} = \mathbf{P}_{wy,k}\left(\mathbf{P}_{\upsilon\upsilon,k}^w\right)^{-1}. \quad (20)$$

$\mathbf{P}_{wy,k}$ is computed by the following weighted outer product:

$$\mathbf{P}_{wy,k} = \sum_{i=0}^{2N_{par}} \mu_i^{(c)}\left(\mathcal{W}_{k,i} - \hat{w}_k^-\right)\left(\mathcal{Y}_{k,i} - \hat{y}_{w,k}^-\right)^T. \quad (21)$$

$\hat{y}_{w,k}^-$ is a weighted average of the predicted measurements $\mathcal{Y}_{k,i}$:

$$\hat{y}_{w,k}^- = \sum_{i=0}^{2N_{par}} \mu_i^{(m)}\mathcal{Y}_{k,i}. \quad (22)$$

$\mathbf{P}_{yy,k}^w$ is given by

$$\mathbf{P}_{yy,k}^w = \sum_{i=0}^{2N_{par}} \mu_i^{(c)}\left(\mathcal{Y}_{i,k} - \hat{y}_{w,k}^-\right)\left(\mathcal{Y}_{i,k} - \hat{y}_{w,k}^-\right)^T. \quad (23)$$

The measurement noise covariance $\mathbf{R}_{w,k}$ is assumed to be a constant diagonal matrix following the basic implementation of the dual UKF proposed by van der Merwe (2004), previously applied by Wan and Nelson (1997), Nelson (2000) and Wan and Nelson (2001) in the context of a dual EKF. This assumption leads to a recursive prediction error algorithm that minimizes a simplified cost function with respect to the parameters. This prediction error algorithm, while questionable from a theoretical perspective, has been shown to be quite useful (Wan and Nelson, 2001). As part of a dual UKF application, Gove and Hollinger (2006) showed that by setting the measurement noise covariance equal to the identity matrix, the overall trajectory of the retrieved parameter was very similar to that obtained considering the actual measurement errors, and attributed this behaviour to the robustness of the filter with respect to changes in parameter measurement variance components.

The parameter mean is updated according to the standard Kalman filter equation:

$$\hat{\boldsymbol{w}}_k = \hat{\boldsymbol{w}}_{k-1}^- + \mathbf{K}_{w,k}\left(\boldsymbol{y}_k - \hat{\boldsymbol{y}}_{w,k}^-\right). \tag{24}$$

The parameter covariance is updated as follows:

$$\mathbf{P}_{w,k} = \mathbf{P}_{w,k}^- - \mathbf{K}_{w,k}\mathbf{P}_{\upsilon\upsilon,k}^w\left(\mathbf{K}_{w,k}\right)^T. \tag{25}$$

$\mathbf{P}_{\upsilon\upsilon,k}^w$ (see also Eq. 19) represents the covariance of $\boldsymbol{y}_k - \hat{\boldsymbol{y}}_{w,k}^-$. Here we opt for the expression also employed by Julier and Uhlmann (2004), van der Merwe (2004) and Tian et al. (2008), given the nonlinearity between parameters and observations.

2.4 Algorithm for parameter sampling

Given the marked differences in the range of variation of the VGM parameters, a variable transformation is required to guarantee operational stability. Bounding parameters by means of a function of reference values and a variable correction term ensures that the model behaves reliably. Moradkhani et al. (2005) and Montzka et al. (2011) also applied some strategies to limit the overdispersion of parameter sampling.

Given that w_i is the true value of the ith parameter, the parameter estimation system makes use of the following variable transformation:

$$w_i = w_{i_{\min}} + \left(w_{i_{\max}} - w_{i_{\min}}\right)\boldsymbol{g}(\delta w_i), \tag{26}$$

where $w_{\min}$ and $w_{\max}$ represent user-defined nominal values, indicating the minimum and maximum values of the parameter, respectively, while the correction term, δw, which is the actual variable under estimation, is expressed as an independent term of a nonlinear sigmoidal function $\boldsymbol{g}(\delta w)$. This function $\boldsymbol{g}(\delta w)$, termed a "squashing function" by van der Merwe (2004), limits the absolute magnitude of iterative parameter adjustment, further preventing the divergence of the parameter estimations. Therefore, the parameters are not estimated directly, rather "correction terms" are estimated.

A preliminary analysis has shown that the approach is not very sensitive to the type of sigmoidal function and that the following relationship performed well for all of the circumstances examined:

$$\boldsymbol{g}(\delta w_i) = \frac{\delta w_i}{2(1 + |\delta w_i|)} + 0.5. \tag{27}$$

Note that $\lim_{\delta w_i \to -\infty} \boldsymbol{g}(\delta w_i) = 0$ and $\lim_{\delta w_i \to \infty} \boldsymbol{g}(\delta w_i) = 1$, in which cases $w_i = w_{i_{\min}}$ and $w_i = w_{i_{\max}}$, respectively.

3 Synthetic experimental framework

We explore the performance of the proposed dual Kalman filter with a synthetic study. The main advantage of testing the algorithm with a synthetic study is that, by knowing the true system, the results are not overshadowed by other sources of uncertainty: a fundamental aspect that should be addressed prior to evaluating algorithm performance with real data, as in the study presented by Medina et al. (2014).

3.1 Model implementation and synthetic data generation

We simulate the vertical movement of water in a homogeneous and variably saturated soil column of 100 cm. The hydraulic properties of the homogeneous soil column are identified by using the VGM parameters reported in the papers by Entekhabi et al. (1994) and Walker et al. (2001): $\theta_{s_T} = 0.54$; $\theta_{r_T} = 0.2$; $K_{s_T} = 0.00029\,\mathrm{cm\,s^{-1}}$, $\alpha_T = 0.008\,\mathrm{cm^{-1}}$ and $n_T = 1.8$, where the subscript "T" indicates the "true" values, i.e. those employed to produce the reference synthetic simulations. However, different boundary conditions have been set so as to make the synthetic study more representative from a practical perspective:

- the top boundary condition is the result of a combination of a stochastically generated daily series of rainfall plus a constant evaporation rate of $2.35\,\mathrm{mm\,d^{-1}}$;

- the bottom boundary condition is set by a zero gradient of the matric head, also known as "free drainage" condition, which also implies that this condition is affected by uncertainty in the identification of the unsaturated hydraulic conductivities.

The inclusion of a rainfall pattern allows evaluating of the dual filter performance during continuous wetting and drying processes taking place in the soil profile. In mathematical terms, the higher variability in the flux at the soil surface entails a higher temporal variability of the correlations between adjacent states, thus affecting the potential ability of the filter to adjust the soil profile. This makes the synthetic study a more representative stress test of the overall retrieval process

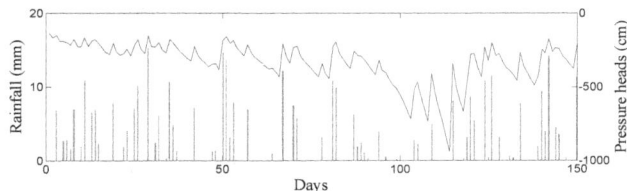

Figure 1. Rainfall pattern (bar plot) and synthetically generated "true" matric pressure head values at 5 cm depth (solid line).

than the case with a constant top boundary condition, which is the one applied by Entekhabi et al. (1994) and Walker et al. (2001).

Daily rainfall is obtained by stochastically sampling a Poisson probability distribution of the occurrence of daily events with an exponential distribution of the rainfall depth. The bar plot in Fig. 1 illustrates the synthetic daily rainfall time series for a period of 150 days.

The soil column is discretized by 27 nodes with variable node spacing; this according to van Dam's (2000) suggestion that accurate computation of soil water fluxes at the top boundary requires that the distance between the nodes close to the soil surface has to be in the order of a few centimetres. A similar criterion is followed for the bottom compartments, given the flux condition adopted for the bottom boundary.

Subsequently, time series of synthetic "true" matric head and soil moisture profiles are generated for 150 days, by setting the initial profile matric head uniformly equal to −50 cm, and by employing the nonlinear numerical scheme illustrated by Chirico et al. (2014). Figure 1 also shows the time series of the generated matric pressure head values at 5 cm depth.

3.2   Retrieval modes

The synthetic study involves the retrieval of states and parameters by assimilating near-surface observations into the Richards equation, according to three different retrieval modes:

- the h-h retrieval mode, indicating that matric head is used as both observed and state variable, with the h-based form of the Richards equation;

- the θ-θ retrieval mode, indicating that soil water content is used as both the observed and state variable, with the θ-based form of the Richards equation;

- the θ-h retrieval mode, indicating that soil water content is used as the observed variable, while matric head is used as the state variable, with the h-based form of the Richards equation.

Preliminary analyses showed that the h-h mode is prone to volatility in the parameter solution, for relatively abrupt changes in the original state variable, generating either very large or very small values. When the solution is very close to the extreme values, the parameter estimation filter sometimes loses its tracking ability, and the algorithm becomes unstable.

A successful strategy is to work with log transformed matric heads only for the parameter filter, without the need to make any change in state relationships. This alternative can be implemented straightforwardly in a nonlinear KF as UKF or EnKF, where the covariance matrices are not propagated analytically. Hence, in the h-h mode the parameter equations are not directly applied with log transformed predicted measurements and observations.

As illustrated below, the retrieval algorithm using soil moisture as a state variable is permanently stable, albeit at the expense of a slightly slower convergence speed, as compared with the case of h as a state variable. As stated by Walker et al. (2001), the soil moisture transformation not only reduces the differences between model predictions and observations, but also the numerical values of the gradients along the soil profile.

3.3   Reference scenarios

The performance of the proposed dual Kalman filter approach is evaluated with respect to reference scenarios, given by implementing the three retrieval modes introduced in the previous section, with different assimilation depths, assimilation frequencies and initial guessed parameter sets.

We simulate the assimilation of observed variables at three alternative observation depths (OD): 2, 5 and 10 cm. Escorihuela et al. (2010) found 2 cm to be the most effective soil moisture sampling depth by L-band radiometry. Nevertheless, L-band sensors receive their signal from approximately the top 5 cm, on average (Kerr, 2007). A depth of 10 cm represents the maximum observation depth that can probably be explored with the current remote sensing technology (e.g. Nichols et al., 2011).

We also examine three alternative assimilation frequencies (AF): 1, 1/3 and 1/5 d^{-1}. Daily assimilation frequency accounts for future L-band missions or a combination of different remote sensors, whilst 3 days is the minimum time interval of SMOS spaceborne platforms (Kerr et al., 2010). One observation every 5 days represents a more common remote sensing time frequency.

The assimilation scenarios with the h-based form of the Richards equation are initialized with an initial matric pressure head profile assumed to be uniformly equal to −100 cm. The assimilation scenarios with the θ-based form of the Richards equation are initialized with an initial soil water content profile uniformly equal to −0.47 cm^3 cm^{-3}, which corresponds to the soil water content at $h = -100$ cm, according to the true water retention function.

The retrieved parameters are K_s, α and n of the VGM analytical model. We assume parameters θ_s and θ_r to be known, as they can be easily measured or estimated by indirect methods, i.e. with pedotransfer functions (e.g. Chirico et al., 2007).

We considered six very dissimilar sets of initial values for the parameters K_s, α, and n, to evaluate the role exerted by different initial guesses on the performance of the retrieval process. These initial values were identified by employing the six possible permutations of the values $-1, 0$ and 1 as correction terms δw_i in Eq. (27), and subsequently in Eq. (26). The initial matrix of the normalized correction terms associated with the soil hydraulic parameters is set to be diagonal, with non-zero entries equal to 0.01, following Nelson (2000).

Table 1 shows the resulting initial values of the parameters and the corresponding $w_{\min}$ and $w_{\max}-w_{\min}$. Notice that the limit values of our parameters, $w_{\min}$ and $w_{\max}$, cover practically the whole spectrum of values reported by Carsel and Parrish (1988) for the 12 major soil textural groups, except for some sandy soils.

On the whole, 162 reference scenarios are examined, made by three retrieval modes, three observation depths, three assimilation frequencies and six initial parameter sets.

3.4 Comparative performance analyses

The DSUKF performance in retrieving the state profiles is analysed by comparing it with the DEnKF, the SKF and the "open loop" solution.

The DEnKF is implemented following the framework described by Moradkhani et al. (2005), coupled with the fully implicit numerical representation of the Richards equation described by Chirico et al. (2014). The DEnKF is run with ensembles of 12 members (hereafter referred to as DEnKF(12)) and 50 members (hereafter referred to as DEnKF(50)). Reichle and Koster (2003) found that the uncertainty in soil moisture retrieved with an EnKF applied to a one-dimensional problem, is consistently reduced with an ensemble of 12 members. Camporese et al. (2009) suggest a minimum of about 50 realizations for ensuring a suitable level of accuracy in analogous applications.

The SKF is implemented with a linearized Crank–Nicolson finite difference scheme of the Richards equation (Chirico et al., 2014), with time-independent initial guessed parameters.

The "open loop" solution is obtained without assimilating any near-surface observations, i.e. the system is simply propagated from the initial uniform conditions and the time-independent initial guessed parameters, using the known boundary conditions.

For quantitatively evaluating the performance of the retrieval algorithms, the normalized root mean square error (RMSE) between predicted and synthetic data (SD) state profiles is calculated as follows:

$$\text{RMSE}_j = \frac{1}{\sigma_{\text{SD}}}\sqrt{\sum_{i=1}^{N_{\text{nod}}}\left(x_{i,j}^{\text{P}} - x_{i,j}^{\text{SD}}\right)^2/(N_{\text{nod}}-1)}, \qquad (28)$$

where $x_{i,j}^{\text{P}}$ and $x_{i,j}^{\text{SD}}$ represent the predicted and SD state value at node i and time j, respectively, and σ_{SD} is the standard deviation of the SD state series, with $N_{\text{nod}} = 27$. Normalization is carried out to enable the comparison between θ-based and h-based retrieval processes.

The average RMSE of the last 15 days of simulation (hereafter simply referred to as RMSE) is taken as accuracy index of the retrieved state profiles, since RMSE values in the last 15 days are not affected by the poor guess of the initial condition. Under the assumption of "free drainage" at the bottom boundary, the dynamic evolution of the system's states, conditional upon a specific set of parameters, rapidly loses its dependence on the initial state values, and the effect of the poor guess of the initial state disappears after a few weeks.

The performance in state retrieval is thus analysed by comparing the RMSE of 648 experiments, resulting from the application of four retrieval algorithms (DSUKF, DEnKF(50), DEnKF(12) and SKF) to the 162 reference scenarios outlined in the previous section. In addition, another 12 experiments are undertaken for the "open loop" solutions, resulting from the application of the h-based and θ-based forms of the Richards equation with the six sets of parameters.

The reference scenarios, but only in the h-h and θ-θ retrieval modes, are also employed for assessing the capability of DSUKF to identify the unknown parameters. The effect of dealing with a nonlinear observation operator in the θ-h mode is examined by comparing the parameters retrieved in the h-h and in the θ-h retrieval modes, but using an hourly assimilation frequency.

Further insights into parameter identifiability are gained by comparing the performances of the retrieval algorithms in estimating the states when only one or two parameters are uncertain.

The effect of parameter uncertainty in retrieving states and parameters is also assessed by applying the DSUKF and the DEnKF with a large number of initial parameter combinations in the h-h and the θ-θ modes. Similarly to Moradkhani et al. (2005), 500 random sets of initial parameters are chosen by sampling them from uniform distributions within the respective limit values listed in Table 1. These 500 sets are mapped into the space of the correction terms (Eq. 25), prior to running the assimilation algorithms. The DEnKF is implemented with ensembles of 12, 25, 35 and 50 members, in order to obtain a comprehensive survey of the tradeoff between computational expense and accuracy of the two retrieval algorithms. This complementary bootstrapping analysis provides us with the posterior parameter uncertainty and the associated probability distribution of the errors of the estimated states. It involves 5000 additional experiments, resulting from the combination of five retrieval algorithms (DSUKF, DEnKF(50), DEnKF(35), DEnKF(25) and DEnKF(12)), 500 random sets of initial parameters, and two retrieval modes (h-h and θ-θ), while accounting for just one assimilation resolution (i.e. AF = 1 d^{-1} and OD = 10 cm).

Table 1. Values of the true parameters K_s, α and n, the minimum, w_{min}, and the range, $w_{max}-w_{min}$, used to constrain their distribution, and the resulting six sets of input values considered during the assimilation process.

Parameter	True	w_{min}	$w_{max}-w_{min}$	S1	S2	S3	S4	S5	S6
K_s ($\times 10^{-4}$ cm s^{-1})	2.9	0.1	6.0	4.6	4.6	3.1	3.1	1.6	1.6
α ($\times 10^{-2}$ cm^{-1})	0.8	0.1	5.0	2.6	1.35	3.85	1.35	2.6	3.85
n (–)	1.8	1.1	2.0	1.6	2.1	1.6	2.6	2.6	2.1

Table 2 provides an overview of the numerical experiments undertaken for assessing the relative performance of DUSKF.

3.5 Setting system and noise covariances

The covariance matrices of the added process and measurement noises ($\mathbf{Q}$, $\mathbf{Q_w}$, $\mathbf{R}$ and $\mathbf{R}_w$) and the initial system covariance matrices ($\mathbf{P}_0$ and $\mathbf{P}_{w,0}$) are set to be diagonal for all cases. The initial state covariance matrix accounts for a standard deviation of 32 % of the initial state value, denoting a sufficiently high, yet realistic, error, with no correlation between nodes. This corresponds to an initial covariance of 10^3 cm^2 using the h-based form of the Richards equation and 0.023 using the θ-based form. The response to a variation of the initial system covariance in a dual Kalman filter framework is less predictable than in a state Kalman filter application, where an increase in the error covariance regularly drives the system to converge faster to the true regime, provided that the variance of the observation error is lower.

Similarly to Camporese et al. (2009), a standard deviation of 1.4 % of the observed value was given to the observation noise, while a standard deviation of 2.24 % was given to the system noise. These values respectively correspond to a covariance of 2 and 5 % of the initial state value in the h-based form, as also adopted by Walker et al. (2001).

The system noise covariance was added every hour as a means of normalizing the incorporated error with respect to the time step, i.e. to make the incorporated error independent of the adopted time step.

We set $b = 2.0$ and $k = 0$ for the deterministic sampling within the UKF (see Eq. 14), as suggested by van der Merwe (2004). The ensemble size is equal to seven, i.e. twice the number of retrieved parameters (N_{par}) plus one. Through sensitivity experiments, we chose $\rho = 0.3$ using the h-h and θ-h modes, and $\rho = 0.8$ using the θ-θ mode.

The forgetting factor coefficient λ_{RLS}, affecting the artificial parameter noise covariance $\mathbf{Q}_w$ (Eq. 17), was set equal to 0.995 in the h-based form and to 0.9999 in the θ-based form, while the diagonal entries of the artificial observation noise covariance $\mathbf{R}_w$ were set equal to 0.5 and 10^{-5}, respectively. Given that these coefficients are subjectively chosen and have a major effect on parameter updates, we also examined some scenarios accounting for different combinations of λ_{RLS} and $\mathbf{R}_w$ values, as listed in Table 3 for both retrieval modes.

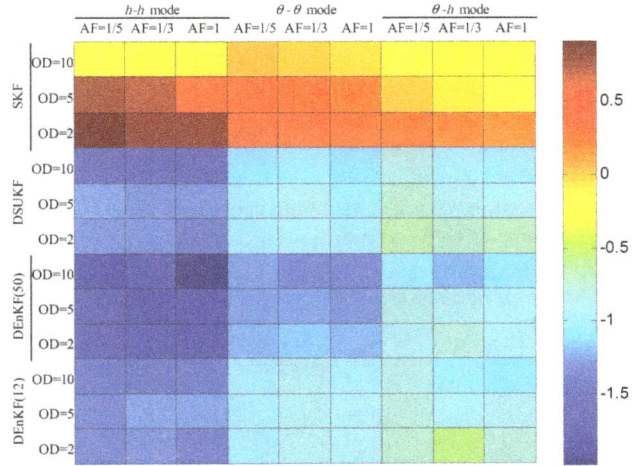

Figure 2. The colour scale indicates the logarithm of the ratio between the RMSE of SKF, DSUKF, DEnKF(50) and DEnKF(12), and the corresponding RMSE of the open loop solution, for three retrieval modes (h-h, θ-θ and θ-h), three assimilation frequencies (AF = 1, 1/3 and 1/5 d^{-1}) and three observation depths (OD = 2, 5 and 10 cm). RMSE values have been averaged among the six initial guessed parameters sets (S1–S6).

4 Results

4.1 State retrieval

The coloured grid depicted in Fig. 2 provides a comprehensive representation of the relative performances of the Kalman-based retrieval algorithms with respect to the reference scenarios. The colour scale indicates the logarithmic of the ratio between the RMSE of the examined assimilation algorithm and the RMSE of the open loop solution, both averaged among the six parameter sets. The average RMSE values retrieved in the h-h and in the θ-h modes are divided by the average RMSE open loop value obtained with the h-based form. The average RMSE values retrieved in the θ-θ mode are divided by the value obtained with the θ-based form. Thus, looking at the different retrieval modes, only the cells of the grid referring to the h-h and the θ-h modes can be directly compared.

The SKF method, involving only state retrieval, gives estimations that can be considerably poorer than the open loop solutions. Instead, both the proposed and the established dual

Table 2. Summary of numerical experiments involved in the performance analyses.

	Algorithms	Mode/form	Sets	AF (d^{-1})	OD (cm)	No. of Experiments
Reference scenarios analysis	DSUKF, DEnKF(50), DEnKF(12), SKF	h-h θ-θ θ-h	S1–S6	[1 1/3 1/5]	[2 5 10]	$4 \times 3 \times 6 \times 3 \times 3 = 648$
	Open loop	h-, θ-based	S1–S6	–	–	$2 \times 6 = 12$
Bootstrapping analysis	DSUKF, DEnKF(50), DEnKF(35), DEnKF(25), DEnKF(12)	h-h θ-θ	500 random sets	1	10	$5 \times 2 \times 500 = 5000$

Table 3. Scenarios for assessing the effect of the artificial noise variance in parameter state space equations.

h-h retrieval mode			θ-θ retrieval mode		
Scenario	λ_{RLS}	$\mathbf{R}_w$	Scenario	λ_{RLS}	$\mathbf{R}_w$
1	0.975	10^{-4}	6	0.975	10^{-5}
2	0.95	10^{-3}	7	0.995	10^{-5}
3	0.975	10^{-3}	8	0.9999	10^{-5}
4	0.999	10^{-3}	9	0.9999	10^{-3}
5	0.975	10^{-2}	10	0.9999	10^{-4}

methods improve the open loop estimations independently on the adopted scenario.

The RMSE values of the DSUKF estimations are in all cases higher than those obtained using DEnKF(50), but generally lower than the ones of the DEnKF(12). The observation depth and the assimilation frequency affect the retrieval performance of the dual filters more in the θ-θ and in the θ-h modes than in the h-h mode.

Figure 2 clearly shows that the retrieval performance with the θ-h mode is much poorer than that one obtained with the h-h mode. This occurrence confirms that a nonlinear observation operator in the θ-h mode has a relevant detrimental effect on the state-retrieving performance. Therefore, later in this section we specifically focus on the comparison of the h-h and the θ-θ retrievals, by showing additional results.

Figure 3 depicts the absolute RMSE values, in the h-h mode and the θ-θ mode, for two extreme assimilation scenarios: one with OD = 10 cm and AF = 1 d^{-1}, the other with OD = 2 cm and AF = 1/5 d^{-1}. The RMSE of the open loop simulations obtained with θ-θ mode is lower than those relevant to the h-h mode, showing the lower impact that parameter uncertainty exerts on the soil water content uncertainty with respect to the matric head uncertainty. Although the initial conditions in terms of h or θ are consistent between them according to the "true" soil water retention function, they actually have a different impact on the corresponding open loop

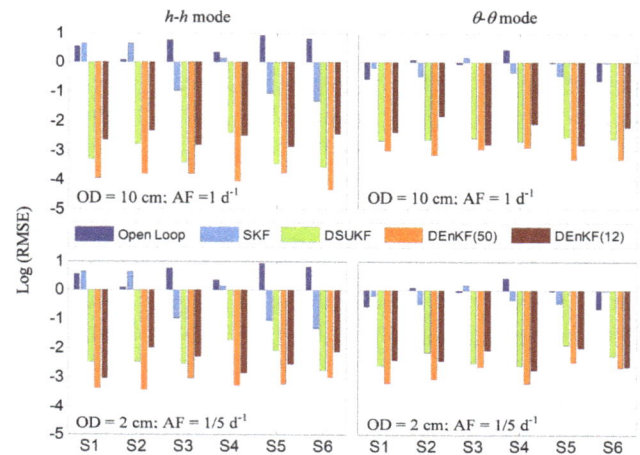

Figure 3. Logarithm of RMSE of the state profiles computed for open loop simulations, SKF, DSUKF, DEnKF(50) and DEnKF(12), in the h-h (left column) and in the θ-θ (right column) modes.

simulation errors because of the poor guess in the parameters. In addition, these results may also be biased somehow because the soil moisture space is constrained, whereas the matric head space is not and it is theoretically infinite.

As mentioned above, the SKF algorithm not only fails to considerably reduce the state uncertainty, but it could even provide worse results than the open loop solution. That is the case of sets S1 and S2 using matric heads and S1, S3, and S6 using soil moisture, all of them with $n < 2.6$. In general, larger RMSE values of the open loop simulation correspond to smaller RMSEs of the SKF approach. This behaviour is more pronounced using matric heads, partly because larger errors in h are predominantly associated with poor guessing of α, which produces a shift of the entire matric head profile that can be more easily corrected by state retrieval with SKF.

The DSUKF considerably reduces the state uncertainty in all cases. For OD = 10 cm and AF = 1 d^{-1}, the error statistics with the h-h and the θ-θ modes are about 40 and 13 times lower than those with the open loop simulations, respectively.

For OD = 2 cm and AF = 1/5 d^{-1}, the RMSE is about 18 and 9 times lower than with the corresponding open loop solutions. The RMSE with the h-h mode is always lower than that with the θ-θ mode, except for the case with the initial parameter set S4.

The RMSEs of DSUKF are about 2.5 and 1.65 larger than using DEnKF(50), respectively for the h-h mode and the θ-θ modes, almost irrespective of the time and space resolutions of the assimilated observation. DEnKF(50) also appears to be less vulnerable to the anomalies affecting the DSUKF algorithm implemented with the matric heads, but its computational time is almost seven times larger than that required for DSUKF. However, with OD = 10 cm and AF = 1 d^{-1}, the DSUKF errors are approximately 1.6 and 1.4 times lower, respectively, than those of DEnKF(12) .

For OD = 2 cm and AF = 1/5 d^{-1}, the DEnKF(12) performed similarly to the DSUKF in terms of RMSE, but it is subjected to some numerical artifacts, in particular for S4, due to the relatively large sampling errors associated with the small size of the ensemble. In addition, the computational time required for DSUKF is about 1.7 times smaller than that required for DEnKF(12).

Figure 4 shows the states retrieved using the h-h and θ-θ retrieval modes after 5, 10, 20, 50, 100 and 150 days, considering the minimum assimilation frequency of the near-surface observations (AF = 1/5 d^{-1}), together with the open loop profiles.

The two sets of open loop profiles simply reflect the same model predictions with two different representations of the system states, which are reciprocally related by means of the water retention function parameterized according to the corresponding guessed parameters. Compared with the corresponding true profiles, the open loop matric head profiles are biased toward larger matric heads, while the open loop water content profiles are mainly shifted toward lower content values.

The DSUKF, with both retrieval modes, considerably reduces the uncertainty of the states despite the initial wrong parameterization. At the 50th day, a good match is observed between the estimated and "true" profiles , almost irrespective of the initial guessed parameter set.

Figure 5 depicts the ratios of the mean RMSE within each group of parameter sets, computed during the last 15 days for different observation depths and assimilation frequencies. In general, the RMSE exhibits limited sensitivity to the observation depths under the adopted range (Fig. 5a, c). In the case of the θ-θ retrieval mode, the ratio between OD = 2 cm and OD = 5 cm is even smaller than one both for AF = 1/3 and AF = 1/5 d^{-1}, which should be due to the stochastic nature of the simulations. Increasing AF from 1/5 to 1/3 d^{-1} does not appreciably improve the error statistics. Only the transition from AF = 1/3 d^{-1} to daily assimilations consistently reduces the RMSE, particularly with matric heads (Fig. 5b, d). With the DEnKF(50) (Fig. 3), the RMSE for OD = 2 cm and AF = 1/5 d^{-1}, compared with that assuming OD = 10 cm

and AF = 1 d^{-1}, is about 2.1 times larger with the h-h mode and 1.3 times larger with the θ-θ mode, consistent with the results obtained with the DSUKF.

The ratio of the mean RMSE computed with the θ-θ retrieval mode to that obtained with the h-h mode is about 1.7 considering AF = 1 d^{-1}, while it is close to one for the other two frequencies. The values reported in Fig. 3 for DEnKF(50) show that the RMSE with the θ-θ mode is 2.4 times higher than with the h-h mode, for OD = 10 cm and AF = 1 d^{-1}, and 1.5 times higher for OD = 2 cm and AF = 1/5 d^{-1}. However, the average statistics are strongly affected by the high errors obtained for S4 and S5.

Given the stochastic nature of the problem, we examine the probabilistic distribution of the error of the estimated states by applying the DSUKF and the DEnKF with 500 random sets of initial parameters, as illustrated in Sect. 3.4. Figure 6 shows cumulative probability distributions of the RMSE obtained with DSUKF, DEnKF(12), DEnKF(25), DEnKF(35), DEnKF(50), assuming OD = 10 cm and AF = 1 d^{-1}.

Using both retrieval modes, DEnKF(50) and DEnKF(35) outperform DSUKF in terms of accuracy. The error statistics using DEnKF(35) are almost half those obtained with DSUKF when retrieving pressure heads (h-h mode). The accuracy of the DSUKF is found to be comparable to that of DEnKF(25), whose implementation demands a computational time 3.3 times larger than that required by the proposed method. The 5th and 25th percentiles of DEnKF(25) are lower than those of DSUKF with both retrieval modes. The median of the RMSE distribution with DEnKF(25) is also lower, but only for the h-h mode. However, the 75th and 95th percentiles almost redouble when the ensemble size is reduced from 35 to 25 members, leading to an increase in the skewness of the error distribution, and hence to an appreciable decrease in the accuracy of the DEnKF. The DEnKF(12) exhibits RMSE percentiles higher than those of the DSUKF, except for the 5th percentile, whose RMSE is slightly smaller in the θ-θ mode and it is equal in the h-h mode. However, it should be pointed out that the DEnKF, unlike the DSUKF, is not affected by any structural error. Indeed, DEnKF is implemented with the same nonlinear numerical solver of the Richards equation used for generating the synthetic "true" data, while DUSKF is implemented with a CN scheme, as illustrated in Sect. 2.2. As shown by Luo and Moroz (2009), high sampling errors, resulting from a small ensemble size, produce high biases and spurious modes. These sampling errors make DEnKF(12) unfeasible, due to their detrimental impact on both precision and accuracy. For a considerable number of simulations, most of them with initial n values close to 3, DEnKF(12) fails to propagate correctly the first and second moments of the states and parameters.

A favourable aspect of DEnKF is that the computational time can be reduced by the use of parallel computing. Kumar et al. (2008), for example, found that the execution time of the EnKF with 12 members, decreased about threefold when using four processors instead of one. In contrast,

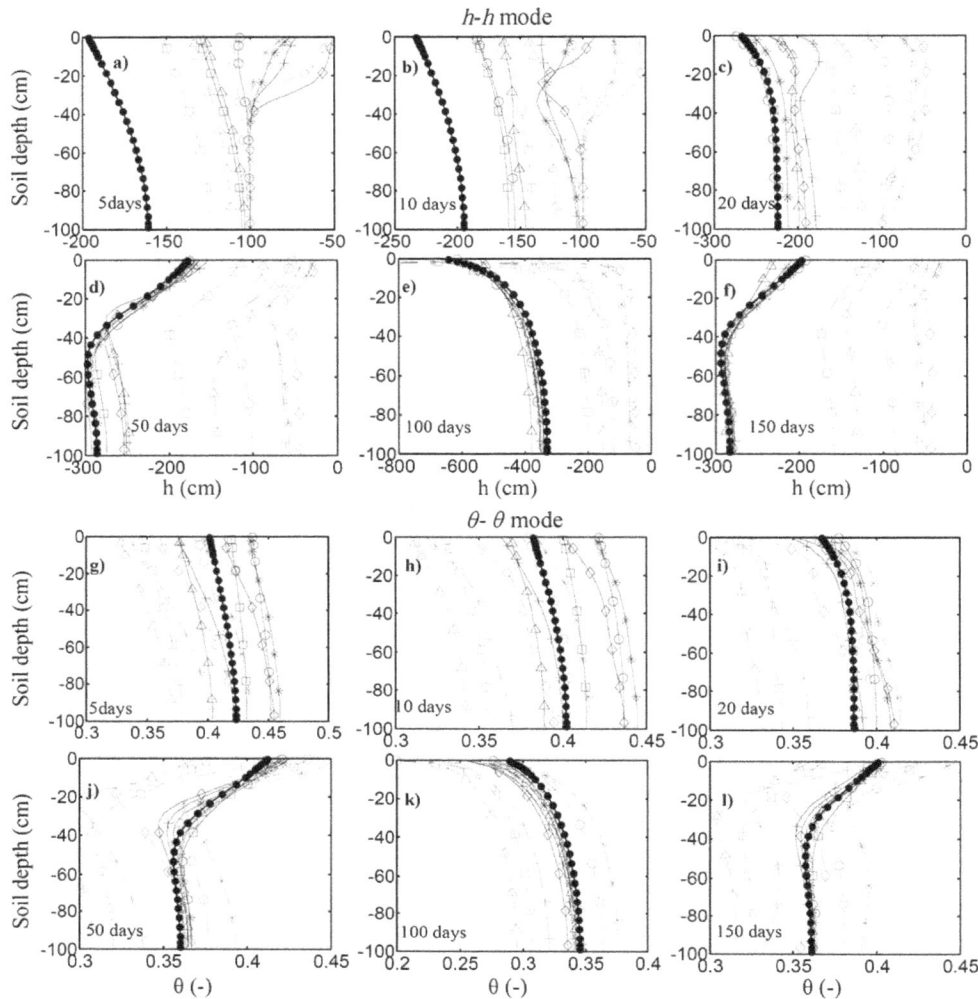

Figure 4. State profiles (solid bold lines, non-filled symbols) retrieved with DSUKF using the h-h **(a–f)** and θ-θ **(g–l)** modes, with OD = 2 cm and AF = 1/5 d^{-1}, after the following days: **(a, g)** 5; **(b, h)** 10; **(c, i)** 20; **(d, j)** 50; **(e, k)** 100 and **(f, l)** 150. The corresponding open loop simulations are also depicted (dash-dot gray lines, non-filled symbols). Comparisons account for the six sets of initial guessed parameters: S1($\bigcirc$), S2($\square$), S3($\ast$), S4($\triangle$), S5($+$) and S6($\diamondsuit$). The dotted lines with filled circles represent the true profiles.

DSUKF is not completely scalable with available computational resources.

4.2 Parameter identifiability

Figure 7 shows the retrieved parameters K_s, α and n, using both the h-h and θ-θ retrieval modes. These graphs depict the evolving patterns with the two alternative resolutions (OD = 10 cm, AF = 1 d^{-1}, and OD = 2 cm, AF = 1/5 d^{-1}).

The "true" value of α is rapidly identified during the retrieval process in both modes, but in particular using matric heads. Parameter α is also the least affected by the initial guess of the parameters under scrutiny. Indeed, since parameter α acts as a scaling factor of the state values in the soil hydraulic property functions, its retrieval is highly sensitive to the convergence rate of the first moment of the state vec-

tor. Vrugt et al. (2001, 2002) found that most of the information on α is embedded in soil water content observations just beyond the air entry value of the soil. Accordingly, in the present study, the identifiability of α is probably favoured by the relatively wet states explored in the initial stage of the synthetic experiment.

The identifiability of parameter α is seemingly also related to the relative position of the observations in the soil profile, depending on the type of simulated process. Ritter et al. (2004) performed a sensitivity analysis of three state variables (soil moisture, matric head and bottom flux) to the VGM parameters, using a soil profile with four soil horizons, and found that the average sensitivity of parameter α was higher than that of parameter n by about a factor of 2, particularly for the uppermost horizon. For the deeper horizons,

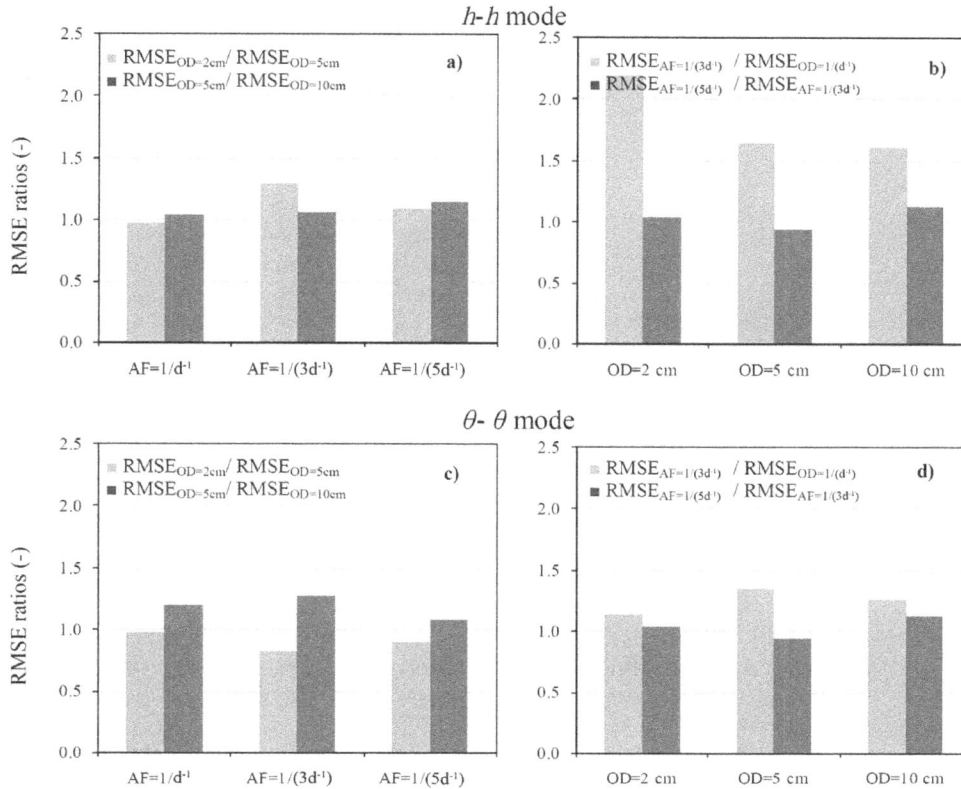

Figure 5. Ratios of the average RMSE, involving the last 15 days of simulations and the six initial guessed parameter sets, computed between contiguously sampled (**a, c**) observation depths (OD) and (**b, d**) assimilation frequencies (AF), with the DSUKF algorithm.

instead, the sensitivity to n was almost three times higher than that of parameter α. This is an interesting aspect, particularly for the issues related to near-surface observations.

Convergence toward the true n is more delayed as compared with α. A close inspection of the time series of the retrieved parameters reveals that the convergence of n for the h-based form is mainly driven by the relatively abrupt reductions in soil moisture on about the 50th, 65th, 80th, 100th and 110th days. These gradients generally induce pronounced shifts on the updated n values for sets S4 and S5, with an initial $n = 2.6$ (see Fig. 7a), while more moderate shifts for sets S1, S3, and S6, which provide systematic underestimations of n. When parameter retrieving does not account for the logarithmic transformation of the matric heads, the sharp decrease in the state variable, taking place on the 110th day, induces in some cases a failure in the retrieval algorithm. As shown by Vrugt et al. (2001, 2002), most of the information on n is embedded in observations whose matric heads are located well beyond the inflection point of the soil water retention function.

Using the h-based form of the Richards equation, we generally observe relatively large differences between the evolving patterns stably adopting $n < 2$ or $n > 2$, after the "erratic" first few updates. This behaviour deserves further attention in future studies. We ascribe these differences to the change in the shape of the soil water capacity, $C(h)$, and of the hydraulic conductivity, $K(h)$, functions, both linked to the governing equation (Eq. 1), when n changes from $n < 2$ to $n > 2$ near saturation, as addressed by Vogel et al. (2001).

The n values retrieved in the θ-θ mode exhibit low sensitivity to the cited sharp soil moisture gradients. The reduction of the posterior uncertainty in this mode is clearly lower than in the h-h mode for daily assimilations (Fig. 7c), and very small for assimilations every 5 days (Fig. 7d). The convergence is more greatly affected by decreasing AF, although the corresponding RMSE values of the retrieved state profiles appear to be rather insensitive to it. The overall performance is affected by the slower convergence of S5. Nevertheless, even for the low-resolution scenario, the method always provides convergent solutions.

However, the saturated hydraulic conductivity, K_s, was not correctly identified in the course of the 150 days of simulation. According to Wöhling and Vrugt (2011), in situ measurements of soil water dynamics contain insufficient information to warrant a reliable estimation of the soil hydraulic properties. The poor performance in terms of retrieved K_s seems to be a confirmation of their results. To solve this problem these authors suggest considering soil moisture and

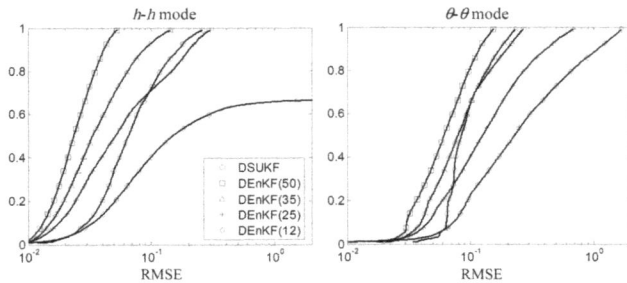

Figure 6. Cumulative probability distribution of the RMSE of DSUKF and DEnKF with 50, 35, 25 and 12 ensemble members for both the *h-h* and *θ-θ* modes.

matric head data simultaneously as part of the statistical inference of the soil hydraulic parameters.

One question arising is whether parameter K_s is necessarily more prone to problems of identifiability than parameters α and n, given their role in the VGM relationships, or if this is solely a result of the adopted experimental conditions. The limited variability of the observations being assimilated is definitely a factor that can affect a proper identification of K_s. Several authors highlighted the limitations for a successful estimation of VGM parameters, as imposed by the narrow variability of naturally occurring boundary conditions (Vrugt et al., 2001, 2002, 2003; Scharnagl et al., 2011). A wide range of soil moisture states is required to constrain the soil hydraulic functions reliably. Moreover, the use of a single metric also conspires against the desired identifiability, as pointed out by Vrugt et al. (2013).

Another possible reason is the fact that the soil water retention parameters also feature in the hydraulic conductivity function, thus enhancing the occurrence of high correlations among the model parameters. A strong correlation is found between retrieved parameters n and K_s. It is known that this strong interdependence also affects the performance of the VGM model. Especially for certain soil types, Romano and Santini (1999) showed that more successful inverse modelling results can be achieved by decoupling the hydraulic conductivity function from the water retention function. As a strategy to reduce the relative uncertainty, Scharnagl et al. (2011) suggested that the parameter K_s should be assessed soon after rainfall events, when soil moisture redistributes more rapidly in the entire soil profile, being essentially driven by gravity.

Table 4 provides further insights into parameter identifiability. We illustrate the performance of the adopted approach when the simulations involve only one or two uncertain parameters, as compared with the original method considering three uncertain parameters. We compare the average RMSE of the results obtained with the six sets of parameters within the last 15 days, respectively using open loop simulations, the state retrieval algorithm SKF, and the DSUKF, both

with OD = 10 cm and AF = 1 d^{-1}, and with OD = 2 cm and AF = 1/5 d^{-1}.

Under the conditions adopted for this experiment, the open loop simulations highlight that the uncertainty in the individual parameters, in particular that of α and n, has a very different impact on the overall uncertainty, depending on the adopted retrieval mode. The retrieval of matric heads is mainly affected by the uncertainty in parameter α, with an RMSE (1.788) more than five times higher than that with n and K_s. The soil moisture estimations are preponderantly influenced by the uncertainty in parameter n, whose RMSE (1.059) is more than three times higher than those computed considering the other two parameters uncertain. In both retrieval modes, K_s uncertainty has only a limited impact.

When considering two uncertain parameters, the influence of those pairs involving uncertain values of α in the *h-h* mode and n in the *θ-θ* mode is also predominant. In both cases the assumption of uncertain α and n parameters gives rise to the highest RMSE. As can be seen in some cases, the uncertainty of one of these dominant parameters could cause a detrimental effect, similar to that provoked by the combined uncertainty of two or even all three parameters. This result is clear evidence of the marked correlation between them. Again, simple state retrieving always provides poorer results.

Careful inspection of the DSUKF behaviour provides further insights into the non-identifiability of K_s. When we consider only the uncertainty of K_s, the method correctly converges to the true value. This can be inferred from the considerable reduction of the RMSE when using the DSUKF for daily assimilations (0.035 for the *h-h* mode and 0.040 for *θ-θ*), as compared with the analogous statistics for the open loop solution (0.3 and 0.334, respectively). However, when considering two unknown parameters, we observe (not shown here for the sake of brevity) that of the two pairs including an unknown K_s, the one with the most explanatory parameter (θ for matric heads and n for soil moisture) still fails to reach the convergence of K_s. For example, when using matric heads and we consider K_s and α uncertain, the method still fails to find the correct K_s, while when considering K_s and n uncertain, all parameter sets tend to converge to the same solution.

We have verified that the states exhibit a cross-covariance with K_s two or three orders of magnitude lower than that formed with the other two dominant parameters. Thus, the Kalman gain scarcely affects the prior K_s values consistently. We have also verified that both the DEnKF(12) and the DEnKF(50) come across the same problem. However, the DEnKF with 500 ensemble members converges to the true value. Nevertheless, even in this case, minimal deviations of α with respect to the true value cause some shifts of the evolving K_s around the true value. The problem of the identifiability of K_s appears principally related to the unavailability of an effective cross-covariance between states and K_s, which can suitably reflect the effect of the uncertainty of K_s (at least

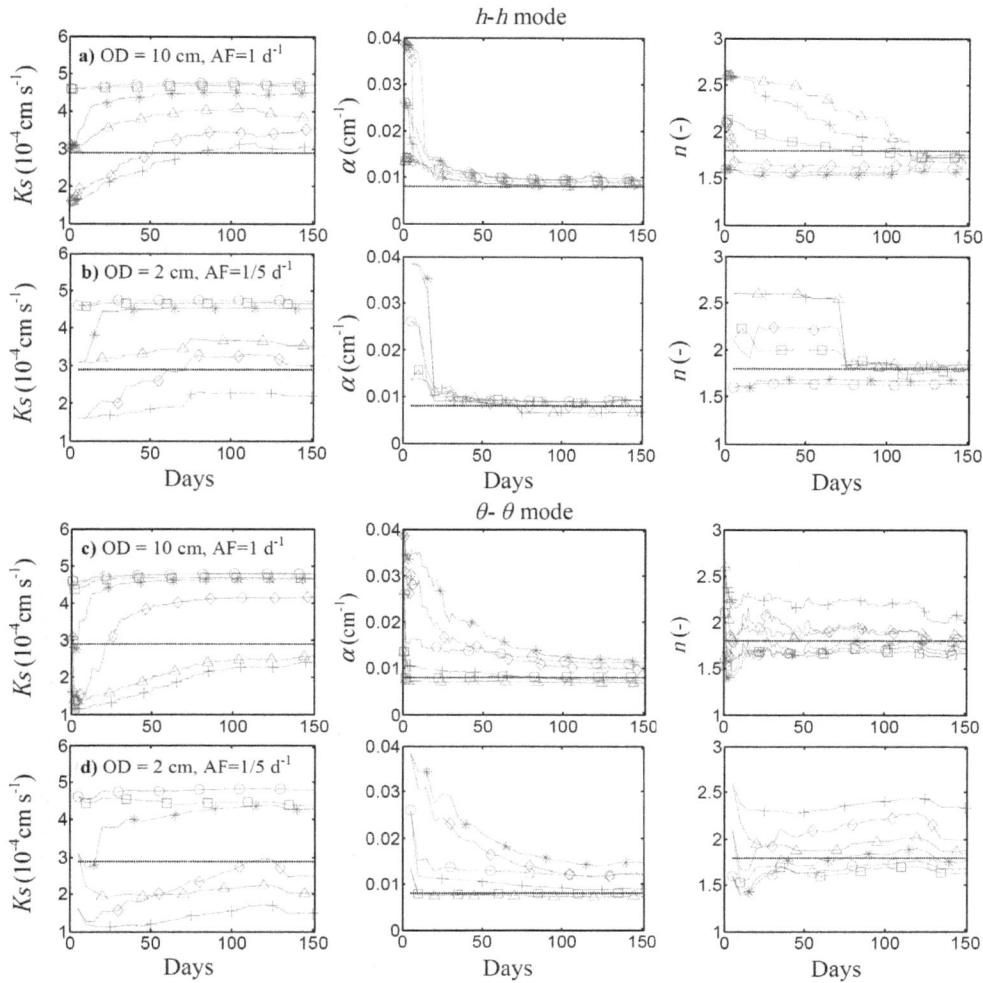

Figure 7. VGM parameters K_s, α and n retrieved with the DSUKF algorithm, using the *h-h* (**a–b**) and the θ-θ (**c–d**) retrieval modes, with the following assimilation scenarios: (**a, c**) OD = 10 cm and AF = 1 d^{-1}; (**b, d**) OD = 2 cm and AF = 1/5 d^{-1}. Comparisons account for the six pondered sets of initial parameters: S1($\bigcirc$), S2($\square$), S3($*$), S4($\triangle$), S5($+$) and S6($\diamond$). The dotted line indicates the true value.

Table 4. Average RMSE of the state profiles estimated during the last 15 days of simulations with the six sets of initial parameters (S1–S6), considering either one, two or all three uncertain parameters. The results refer to the open loop simulations, the SKF (involving only state retrieving) with OD = 10 cm and AF = 1 d^{-1}, the DSUKF with OD = 10 cm and AF = 1 d^{-1}, and the DSUKF with OD = 2 cm and AF = 1/5 d^{-1}.

Uncertain parameters	*h-h* mode				θ-θ mode			
	Open loop	SKF	DSUKF OD = 10, AF = 1 d^{-1}	DSUKF OD = 2, AF = 1/5 d^{-1}	Open loop	SKF	DSUKF OD = 10, AF = 1 d^{-1}	DSUKF OD = 2, AF = 1/5 d^{-1}
K_s	0.300	0.174	0.035	0.049	0.334	0.204	0.040	0.070
α	1.788	1.187	0.040	0.065	0.283	0.621	0.058	0.072
n	0.348	0.388	0.038	0.062	1.059	0.345	0.040	0.035
K_s, α	1.777	1.097	0.049	0.081	0.465	0.627	0.071	0.094
K_s, n	0.473	0.514	0.046	0.078	0.888	0.250	0.064	0.093
α, n	2.225	2.467	0.043	0.093	1.066	0.676	0.060	0.085
K_s, α, n	1.866	0.993	0.047	0.102	0.946	0.829	0.073	0.104

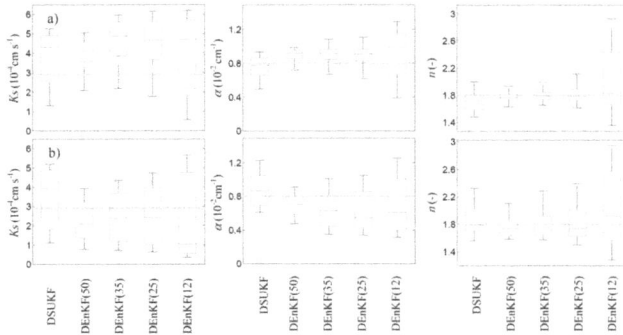

Figure 8. Box plots representing the posterior uncertainty of the parameters by performing DSUKF and DEnKF for 500 randomly chosen sets of initial parameters considering (**a**) the h-h (**b**) and θ-θ modes.

within the temporal extent of this experiment), making the model poorly sensitive to the errors of this parameter.

The box plots in Fig. 8 illustrate the posterior uncertainty of the parameters estimated with the DSUKF and the DEnKF, considering the 500 randomly chosen sets of initial parameters described in the previous section. The uncertainty ranges of the proposed method are roughly comparable to those obtained by Moradkhani et al. (2005) within the first 150 days of simulation. As emphasized, both methods fail to identify parameter K_s correctly. The decrease in accuracy by assuming only 12 ensemble members, especially during matric head retrieving, is also considerable. Both accuracy and precision of the estimated n values are superior using DEnKF under the h-h mode, except for ensembles of 12 members (Fig. 8a). However, for the same scheme, the median of parameter α provided by the DSUKF is less biased than that provided by the nonlinear approach, which overestimates the true value. The performance of the DSUKF is comparable to that of the DEnKF(25), although the interquartile ranges of α and n are slightly smaller with the latter method. The uncertainty regions of the estimated parameters in the h-h mode are narrower than in the θ-θ mode (Fig. 8b). In this latter case, the DEnKF(25) provides wider uncertainty bounds and less accurate estimates of parameters α and n than the DSUKF.

It could be argued that, given the scope of the present study, a classical calibration method could also provide similar results. For a more reliable analysis of the pros and cons of the DSUKF, we also implemented gradient iterative algorithms based on the Levenberg–Marquardt algorithm (Kool and Parker, 1988), considering the six sets of parameters S1–S6 in the h-h mode. The algorithms were implemented by exploiting the solvers embedded in the Matlab® optimization toolbox.

A major drawback of this technique is its computational time, which is about 30 times larger than that of the DSUKF. About 1 week was needed to generate the a posteriori distri-

bution from the 500 sets of initial parameters. The performance was also poorer in terms of identifiability. The retrieved n values varied between a minimum of 1.64 for S2 and a maximum of 2.23 for S5, while α varied between 0.0055 for S1 and 0.012 cm^{-1} for S6. K_s remains clearly not identified. The identifiability problems of traditional approaches like this have been documented in the literature (e.g. Kool et al., 1987; Romano and Santini, 1999; van Dam, 2000). Finally, a third difficulty is that these variational methods, although they can be easily implemented, demand some expertise for suitably tuning the parameters involved in the numerical solvers (e.g. tolerance threshold values applied in the numerical algorithm).

4.3 Influence of the type of observed variables with respect to the selected state variables

The analyses in the previous section focused on the performance of the h-h and θ-θ retrieval modes, i.e. when observed and retrieved variables are of the same type. This allows implementing a linear observation equation (Eq. 2), with a standard Kalman filter for state retrievals. Nevertheless, part of the study also focused on the relation between the type of assimilated data and the h-based form or θ-based form of the state equation.

In principle, the numerical algorithm can be structured to assimilate soil moisture observations (or some information linked to it) in the h-based form of the Richards equation by dealing with a nonlinear observation equation, above referred to as the θ-h retrieval mode. This issue can be frequent, given the structure of many widely used simulation models as well as the type of information provided by current remote sensing techniques and ground-based sensors.

At this point, it is important to note that the inversion of the observation variable, i.e. converting soil water contents to matric pressure heads by means of a water retention function with guessed (wrong) parameters, would be a serious mistake, because the observations would be significantly biased, incorporating an unpredictable error in the retrieval algorithm. By contrast, a nonlinear relationship for transforming an exogenous observation variable (such as soil surface temperature from thermal infrared remote sensing) in soil moisture can be directly employed prior to applying the observation operator H_k in Eq. (11).

The effect of dealing with a nonlinear observation operator within the retrieval algorithm is illustrated in Fig. 9. Figure 9a shows the retrieved parameters using an hourly assimilation frequency, with the h-h retrieval mode. Figure 9b shows the analogous results, but with the θ-h retrieval mode, i.e. by assimilating soil moisture observations and using matric heads as state variables. This last case requires the VGM analytical model to be used as a (nonlinear) observation equation for mapping the predicted matric heads into the soil moisture space.

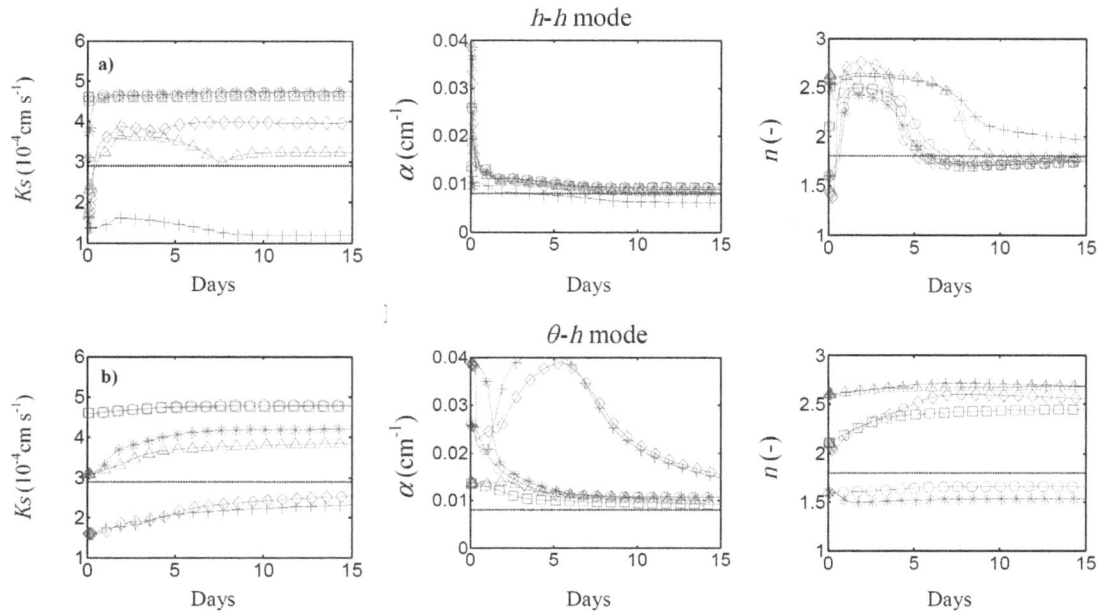

Figure 9. VGM parameters K_s, α and n retrieved with the DSUKF algorithm, using (**a**) the h-h and (**b**) θ-h retrieval modes by assimilating observations every hour, with observation depth OD $= 10$ cm. Comparisons account for the six pondered sets of initial parameters: S1(○), S2(□), S3(∗), S4(△), S5(+) and S6(◇).

The unscented algorithm is also employed for dealing with the nonlinearity of the observation equation, similarly to what is done for retrieving the parameters. During each time step we propagate the a priori mean and covariance of the matric heads using the SKF, as currently done in the h-h and θ-θ retrieval modes. When the observation vector becomes available for assimilation, we sample the predicted state variable around the mean, following the UKF precepts, using the estimated a priori covariance, $\mathbf{P}_k^-$. The sample state vectors are propagated through the VGM expression using the a posteriori mean of the parameters at time $k-1$. Then the cross-covariance between predicted states and predicted measurements, $\mathbf{P}_{xy,k}$, and the auto-covariance of these predicted measurements, $\mathbf{P}_{yy,k}$, are estimated analogously to what is done for the parameters (Eqs. 20, 22, respectively). The Kalman gain is then estimated as $\mathbf{K}_k = \mathbf{P}_{xy,k}\left(\mathbf{P}_{\upsilon\upsilon,k}\right)^{-1}$, from which the estimations of the a posteriori state mean and covariance are straightforward.

Such linearization clearly incorporates a certain amount of error, which affects the overall identifiability of the unknown parameters. Even the identifiability of parameter α is affected by this assimilation strategy. With low assimilation frequencies the algorithm is subjected to persistent failures.

According to such results, while applying the DSUKF, the state variable and observation variable should be preferably of the same type, either in the h-based form or in θ-based form, to avoid the need to linearize the observation equation (Eq. 11) with respect to the states.

Finally, it is useful to see that in an extended Kalman filter framework, the non-zero entries of the linearized observation operator H_k would correspond to the hydraulic capacities $C(h)$, evaluated in the prior state values $\hat{\mathbf{x}}_k^-$, at the observation nodes. This provides an idea of the unpredictability of the uncertainty due to the linearization process, as this is strongly influenced by the soil properties.

4.4 Influence of the initial covariance matrices

The DSUKF algorithm, like its analogous approaches, requires initial values for the state covariance, $\mathbf{P}$, and the parameter covariance, $\mathbf{P}_w$. The effects of the initial state covariance matrix, $\mathbf{P}$, and of the noise covariance matrices $\mathbf{Q}$ and $\mathbf{R}$ on the assimilation scheme are clear and have been widely examined (see for example Walker, 1999; Nelson, 2000). The values that should be used for the initial parameter covariance $\mathbf{P}_w$ and the artificial noise covariances $\mathbf{Q}_w$ and $\mathbf{R}_w$, are less clear and depend on several factors (Nelson, 2000).

An initial value 10^{-2} for the diagonal entries of $\mathbf{P}_{w,0}$ performed well for most of the cases, with both soil water contents and matric heads as state variables, as found by Nelson (2000), who also employed normalized parameterization. Once the normalized $\mathbf{P}_w$ is fixed, the values of the noise covariances depend on the variance of the data, and hence of the state variable.

The effect of evaluating different scenarios accounting for the variability of the parameter noise covariances is illustrated in Fig. 10, showing the daily retrievals of parameter n, using OD $= 10$ cm. Figure 10a and b account for scenarios

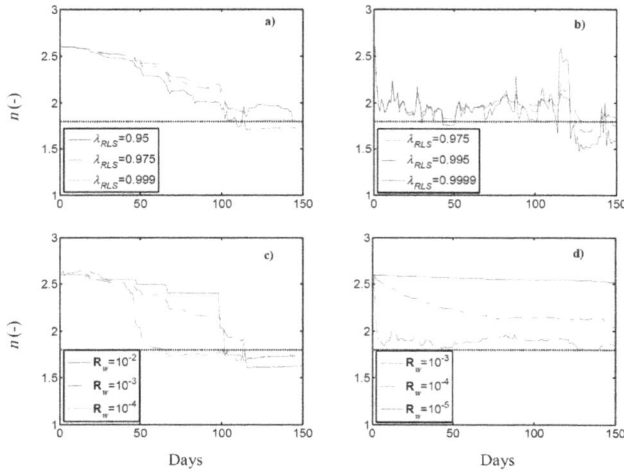

Figure 10. Daily retrieval of parameter n with the DSUKF algorithm, considering different values of the coefficients λ_{RLS} (see Eq. 25) and of the artificial observation noise covariances, according to the scenarios of Table 3: (**a**) scenarios 2–4; (**b**) scenarios 6–8; (**c**) 1, 3 and 5; (**d**) scenarios 8–10. Simulations involve the initial parameters set S4(Δ) and an observation depth OD $= 10$ cm.

from 2 to 4 and from 6 to 8, respectively, considering three values of the forgetting factor (λ_{RLS}) for the h-h mode and three values for the θ-θ mode. Assuming $\lambda_{RLS} = 1$ entails no process noise $\mathbf{Q}_w$ for the parameters, while a small λ_{RLS} incorporates a significant noise in the retrieval process.

The prediction error covariance $\mathbf{Q}_w$ is a key variable, having effects on parameter retrieving for longer time intervals, and it is decisive in convergence and tracking. For the adopted h-based form, where parameter retrieving considers the log transformation of the states, $\lambda_{RLS} = 0.975$ is appreciated as a fair value. However, when assuming the original (i.e. non-transformed) state values, we observe that a similar value of λ_{RLS} compromises the stability in some cases, due to the added volatility. Setting $\lambda_{RLS} = 0.95$ definitively adds too much error to the estimated parameters. Using the θ-θ mode, a value of around $\lambda_{RLS} = 0.9999$ produces good results (Fig. 10b), while $\lambda_{RLS} = 0.995$ improves the convergence, although it affects the stability in a few cases. An intermediate value between 0.995 and 0.9999 could be a good choice.

Van der Merwe (2000) and Wan and van der Merwe (2001) suggested two other options on how to choose the matrix $\mathbf{Q}_w$. One is to set $\mathbf{Q}_w$ equal to an arbitrary "fixed" diagonal value, which may then be annealed toward zero as training continues. Another choice is to apply a Robbins–Monro stochastic approximation scheme for estimating the innovations (see Wan and van der Merwe, 2001 and van der Merwe, 2004, for details about the applied expression). Our preliminary analysis suggests that the most efficient approach is to use the method currently adopted, i.e. the forgetting factor, although further research is required to address this issue.

The effect of the artificial observation covariance $\mathbf{R}_w$ for each retrieval mode can be evaluated by comparing the results obtained with scenarios 1, 3 and 5 (Fig. 10c), and 8, 9 and 10 (Fig. 10d). As stated by Nelson (2000), $\mathbf{R}_w$ acts as a scaling term, determining the relative influence of the initial covariance $\mathbf{P}_{w,0}$ on later covariance matrices $\mathbf{P}_{w,k}$. For a prefixed $\mathbf{P}_{w,0}$, a large $\mathbf{R}_w$ produces a more stable (i.e, lower variance) behaviour, but this produces significantly biased estimates of w_k for small time steps. A very small $\mathbf{R}_w$ exposes the algorithm to retrieve parameters toward the corresponding limiting values, undermining its stability and convergence.

Using the h-h mode (Fig. 10c), $\mathbf{R}_w = 10^{-3}$ has been found appropriate for many of the examined cases. Again, $\mathbf{R}_w = 10^{-4}$ improves the convergence but at the price of lower stability. When retrieving soil moisture (Fig. 10d), $\mathbf{R}_w$ has been set equal to 10^{-5}. $\mathbf{R}_w = 10^{-3}$ is too large, since with this value the Kalman gain $\mathbf{K}_w$ adopts a relatively low value, thus allowing for less variability in the estimations. $\mathbf{R}_w = 10^{-4}$ is also a suitable value; instead $\mathbf{R}_w = 10^{-6}$ persistently induces the collapse of the system.

In general, higher $\mathbf{R}_w$ values are required for a larger variability of the retrieved variable involved. As a general guideline, we observe a roughly linear relationship between the log of the initial state value and the log of the adopted $\mathbf{R}_w$, fulfilling the relationship $\mathbf{R}_w = 4.23 \times 10^{-5} x_0^{2.04}$, x_0 being the value of the initial states. That said, the selection of proper values for λ_{RLS} and $\mathbf{R}_w$ deserves more attention in further studies, because they act as important stability factors, particularly when using matric pressure heads.

5 Conclusions

This study presented a DSUKF formulation for the simultaneous retrieval of states and parameters controlling the soil water dynamics in a homogeneous soil column, by assimilating near-surface observations into the Richards equation. The proposed approach takes advantage of the standard Kalman filter applied to a linear numerical scheme of the Richards equation for straightforward retrieval of the states within a linear system of small dimension, and of the unscented Kalman filter for retrieving a small number of soil hydraulic parameters defined according to the nonlinear VGM relations. A transformation of variables is used to deal with the physical constraints and the marked differences in the range of variability of the VGM parameters, thus improving the operational stability.

The unscented approach deals with the nonlinearity of the model with respect to the parameters, without the need to perform any analytical differentiation, thus making the computational implementation simple and of general applicability, i.e. independent of the analytic equations employed.

By means of a synthetic experiment, we showed that the DSUKF, with an ensemble of seven sigma points of the parameter space, provides predictions with an accuracy similar to that provided by a dual ensemble Kalman filter with an ensemble size of 25 members (DEnKF(25)), but with a computational time three times smaller. The DEnKF guarantees more accurate states and parameter predictions than the DSUKF, with an ensemble of 35 or more members, but at the cost of a large increase in computational effort. The study also demonstrated that DEnKF(12), whose computational expense is slightly larger than that of the DSUKF, does not ensure a substantial reduction in the posterior uncertainty of states and parameters due to the relatively large sampling errors involved in this scheme.

Except for the case with 12 ensemble members, the DEnKF provided less biased estimates of parameter n. Instead, with the DSUKF, we achieved less biased estimates of parameter α. The performance in reducing the uncertainty of K_s was poor, both for the DSUKF and the DEnKF, even with 50 ensemble members.

The dual Kalman filter approach is able to retrieve states close to true values, even for observations at very shallow depths, i.e. using observation depths of 2 cm and an assimilation frequency of one every 5 days. Comparison between parameter initialization, observation depth and assimilation frequency, evidenced that the latter has the most dominant effect on the evolving errors.

The problem associated with the choice of either the h-based form or θ-based form of the Richards equation in the dual Kalman filter algorithm was explored. The former scheme is generally preferred for practical applications, particularly when having to deal with both saturated and unsaturated flows. The matric head retrieval algorithm outperformed that using soil water content in terms of state convergence and final accuracy. However, to avoid some stability problems in this mode, a log transformation of the predicted model observations and measurements was needed prior to applying the parameter equations.

The assimilation of near-surface soil moisture observations recalls some considerations about the system sensitivity to the VGM parameters, at least when the system is initialized with wet conditions. The identifiability of parameter α is markedly higher than that of n, particularly when using matric pressure head as state variable. The impact of the uncertainty of parameters α and n depends on whether the state and observed variables coincide with the soil water content (θ-θ retrieval mode) or with the matric pressure head (h-h retrieval mode). The retrieved matric head profiles are markedly influenced by the uncertainty of parameter α, while the retrieved soil moisture profiles are preponderantly influenced by the uncertainty of parameter n.

The identifiability of the saturated hydraulic conductivity K_s is very poor in all cases, unless K_s is the only uncertain parameter. This limitation in the identifiability of K_s can be ascribed to the strong correlation between the parameters

(not shown for the sake of brevity), and particularly between n and K_s, which in turn also causes the cross covariance between states and K_s to become too poor to make the model sensitive enough to the errors of this parameter. The limited identifiability of K_s in a dual framework was also experienced with the DEnKF. A large ensemble size (in the order of 500) is required to achieve the convergence of K_s, but the solution remains vulnerable to slight oscillations of the other two parameters (for instance of α in the h-h retrieval mode). These results suggest the advisability of employing other analytical models for the soil hydraulic functions, representing the hydraulic conductivity decoupled from the retention function.

By examining different combinations of retrieved and assimilated variables, the study also highlights some other issues to be considered beyond the efficiency of the dual Kalman filter approach. In this synthetic experiment, the performance of the overall retrieval process is significantly dampened when adopting soil moisture as the observed variable and matric pressure head as the state variable, even when observations are assimilated hourly.

Finally, we point out that the implementation of a dual (or a joint) exercise for state and parameter estimation demands more caution, as compared with a standard KF approach, particularly with respect to the initialization of the covariances, not only of the parameters, but even of the states. This can be seen as the price to be paid for achieving more accurate results.

Acknowledgements. This study was supported by the P.O.N. project AQUATEC – New technologies of control, treatment, and maintenance for the solution of water emergencies. Hanoi Medina was also supported by the Abdus Salam International Centre for Theoretical Physics (ICTP), where he has been appointed as Junior Associate. The Editor and the anonymous reviewers contributed considerably to improve the manuscript with fruitful comments.

Edited by: J. Vrugt

References

Abbaspour, K. C., van Genuchten, M. T., Schulin, R., and Schläppi, E.: A sequential uncertainty domain inverse procedure for estimating subsurface flow and transport parameters, Water Resour. Res., 33, 1879–1892, 1997.

Camporese, M., Paniconi, C., Putti, M., and Salandin, P.: Ensemble Kalman filter data assimilation for a process based catchment scale model of surface and subsurface flow, Water Resour. Res., 45, W10421, doi:10.1029/2008WR007031, 2009.

Carsel, R. F. and Parrish, R. S.: Developing joint probability distributions of soil water characteristics, Water Resour. Res., 24, 755–769, 1988.

Chirico, G. B., Medina, H., and Romano, N.: Uncertainty in predicting soil hydraulic properties at the hillslope scale with indirect methods, J. Hydrol., 334, 405–422, 2007.

Chirico, G. B., Medina, H., and Romano, N.: Functional evaluation of PTF prediction uncertainty: An application at hillslope scale, Geoderma, 155, 193–202, 2010.

Chirico, G. B., Medina, H., and Romano, N.: Kalman filters for assimilating near-surface observations into the Richards equation – Part 1: Retrieving state profiles with linear and nonlinear numerical schemes, Hydrol. Earth Syst. Sci., 18, 2503–2520, doi:10.5194/hess-18-2503-2014, 2014.

Clark, M. P. and Vrugt, J. A: Unraveling uncertainties in hydrologic model calibration: Addressing the problem of compensatory parameters, Geophys. Res. Lett., 33, L06406, doi:10.1029/2005GL025604, 2006.

Crow, W. and Wood, E.: The assimilation of remotely sensed soil brightness temperature imagery into a land surface model using ensemble Kalman filtering: a case study based on ESTAR measurements during SGP97, Adv. Water Resour., 26, 137–149, 2003.

De Lannoy, G. J. M., Reichle, R. H., Houser, P. R., Pauwels, V. R. N., and Verhoest N. E. C.: Correcting for forecast bias in soil moisture assimilation with the ensemble Kalman filter, Water Resour. Res., 43, W09410, doi:10.1029/2006WR005449, 2007.

Entekhabi, D., Nakamura, H., and Njoku, E. G.: Solving the inverse problem for soil moisture and temperature profiles by sequential assimilation of multifrequency remote sensed observations, IEEE Trans. Geosci. Remote Sensing, 32, 438–448, 1994.

Escorihuela, M. J., Chanzy, A., Wigneron, J. P., and Kerr, Y. H.: Effective soil moisture sampling depth of L-band radiometry: A case study, Remote Sens. Environ., 114, 995–1001, 2010.

Evensen, G.: Sequential data assimilation with a nonlinear quasi-geostrophic model using Monte Carlo methods to forecast error statistics, J. Geophys. Res., 99, 10143–10162, doi:10.1029/94JC00572, 1994.

Evensen, G.: The ensemble Kalman filter: Theoretical formulation and practical implementation, Ocean Dynam., 53, 343–367, 2003.

Gove, J. H. and Hollinger, D. Y.: Application of a dual unscented Kalman filter for simultaneous state and parameter estimation in problems of surface-atmosphere exchange, J. Geophys. Res., 111, D08S07, doi:10.1029/2005JD006021, 2006.

Heathman, G. C., Starks, P. J., Ahuja, L. R., and Jackson, T. J.: Assimilation of surface soil moisture to estimate profile soil water content, J. Hydrol., 279, 1–17, 2003.

Jazwinski, A.: Stochastic Processes and Filtering Theory, Academic Press, New York, 1970.

Julier, S. J. and Uhlmann, J. K.: Unscented Filtering and Nonlinear Estimation, Proc. IEEE 92, 401–422, 2004.

Julier, S. J., Uhlmann, J. K., and Durrant-Whyte, H. F.: A new approach for filtering nonlinear systems, in: The Proceedings of the American Control Conference, 1628–1632, Seattle WA, USA, 1995.

Jury, W. A., Gardner, W. R., and Gardner, W. H.: Soil Physics, 5th Edn., John Wiley, New York, 1991.

Kalman, R. E.: A New Approach to Linear Filtering and Prediction Problems, ASME J. Basic Eng., 82D, 35–45, 1960.

Kerr, Y. H.: Soil moisture from space. Where are we?, Hydrogeol. J., 15, 117–120, 2007.

Kerr, Y. H., Waldteufel, P., Wigneron, J. P., Delwart, S., Cabot, F., Boutin, J., Escorihuela, M. J., Font, J., Reul, N., Gruhier, C., Ju-

glea, S. E., Drinkwater, M. R., Hahne, A., Martin-Neira, M., and Mecklenburg, S.: The SMOS Mission: New Tool for Monitoring Key Elements of the Global Water Cycle, Proc. IEEE, 98, 666–687, 2010.

Kool, J. B. and Parker J. C.: Analysis of the inverse problem for transient unsaturated flow, Water Resour. Res., 24, 817–830, doi:10.1029/WR024i006p00817, 1988.

Kool, J. B., Parker J. C., and van Genuchten, M. T.: Parameter estimation for unsaturated flow and transport models – A review, J. Hydrol., 91, 255–293, 1987.

Kumar, S. V., Reichle, R. H., Peters-Lidard, C. D., Koster, R. D., Zhan, X., Crow, W. T., Eylander J. B., and Houser, P. R.: A land surface data assimilation framework using the land information system: description and applications, Adv. Water Resour., 31, 1419–1432, 2008.

Lei, M. and Baehr, C.: Unscented/ensemble transform-based variational filter, Physica D, 246, 1–14, 2013.

Liu, Y. and Gupta, H. V.: Uncertainty in hydrologic modeling: Toward an integrated data assimilation framework, Water Resour. Res., 43, W07401, doi:10.1029/2006WR005756, 2007.

Lü, H., Yu, Z., Zhu, Y., Drake, S., Hao, Z., and Sudicky, E. A.: Dual state-parameter estimation of root zone soil moisture by optimal parameter estimation and extended Kalman filter data assimilation, Adv. in Water Res., 34, 395–406, 2011.

Luo, X. and Moroz, I. M.: Ensemble Kalman filter with the unscented transform, Physica D, 238, 549–562, 2009.

McLaughlin, D. B.: An integrated approach to hydrologic data assimilation: Interpolation, smoothing, and filtering, Adv. Water Resour., 25, 1275–1286, 2002.

Medina, H., Chirico, G. B., and Romano, N.: Kalman filters for assimilating near-surface observations into the Richards equation – Part 3: Retrieving states and parameters from laboratory evaporation experiments, Hydrol. Earth Syst. Sci., 18, 2543–2557, doi:10.5194/hess-18-2543-2014, 2014.

Montzka, C., Moradkhani, H., Weihermuller, L., Hendricks-Franssen, H. J., Canty, M., and Vereecken, H.: Hydraulic parameter estimation by remotely-sensed top soil moisture observations with the particle filter, J. Hydrol., 399, 410–421, 2011.

Moradkhani, H., Sorooshian, S., Gupta, H. V., and Houser, P.: Dual state-parameter estimation of hydrological models using Ensemble Kalman filter, Adv. Water Resour., 28, 135–147, 2005.

Nelson, A. T.: Nonlinear estimation and modeling of noisy time-series by dual Kalman filter methods, PhD thesis, Oregon Graduate Institute of Science and Technology, 281 pp., 2000.

Nichols, S., Zhang, Y., and Ahmad, A.: Review and evaluation of remote sensing methods for soil-moisture estimation, SPIE Rev., 2, 028001, doi:10.1117/1.3534910, 2011.

Press, W. H., Teukolsky, S. A., Vetterling, W. T., and Flannery, B. P.: Numerical Recipes in C: The Art of Scientific Computing, 2 Edn., Cambridge University Press, 1992.

Pringle, M. J., Romano, N., Minasny, B., Chirico, G. B., and Lark, R. M.: Spatial evaluation of pedotransfer functions using wavelet analysis, J. Hydrol., 333, 182–198, 2007.

Qin, J., Liang, S. L., Yang, K., Kaihotsu, I., Liu, R. G., and Koike, T.: Simultaneous estimation of both soil moisture and model parameters using particle filtering method through the assimilation of microwave signal, J. Geophys. Res.-Atmos., 114, D15103, doi:10.1029/2008JD011358, 2009.

Reichle, R. H.: Data assimilation methods in the Earth sciences, Adv. Water Resour., 31, 1411–1418, 2008.

Reichle, R. H. and Koster, R. D.: Assessing the impact of horizontal error correlations in background fields on soil moisture estimation, J. Hydrometeorology, 4, 1229–1242, 2003.

Reichle, R. H., McLaughlin, D. B., and Entekhabi, D.: Hydrologic data assimilation with the ensemble Kalman filter, Mon. Weather Rev., 130, 103–114, 2002.

Renard, B., Kavetski, D., Kuczera, G., Thyer, M., and Franks, S. W.: Understanding predictive uncertainty in hydrologic modeling: The challenge of identifying input and structural errors, Water Resour. Res., 46, W05521, doi:10.1029/2009WR008328, 2010.

Ritter, A., Muñoz-Carpena, R., Regalado, C. M., Vanclooster, M., and Lambot, S.: Analysis of alternative measurement strategies for the inverse optimization of the hydraulic properties of a volcanic soil, J. Hydrol., 295, 124–139, 2004.

Romano, N. and Santini, A.: Determining soil hydraulic functions from evaporation experiments by a parameter estimation approach: Experimental verifications and numerical studies, Water Resour. Res., 35, 3343–3359, 1999.

Scharnagl, B., Vrugt, J. A., Vereecken, H., and Herbst, M.: Inverse modelling of in situ soil water dynamics: investigating the effect of different prior distributions of the soil hydraulic parameters, Hydrol. Earth Syst. Sci., 15, 3043–3059, doi:10.5194/hess-15-3043-2011, 2011.

Spaaks, J. H. and Bouten, W.: Resolving structural errors in a spatially distributed hydrologic model using ensemble Kalman filter state updates, Hydrol. Earth Syst. Sci., 17, 3455–3472, doi:10.5194/hess-17-3455-2013, 2013.

Tian, X., Xie, Z., and Dai, A.: A land surface soil moisture data assimilation system based on the dual-UKF method and the Community Land Model, J. Geophys. Res., 113, D14127, doi:10.1029/2007JD009650, 2008.

van Dam, J. C.: Field-scale water flow and solute transport. SWAP model concepts, parameter estimation, and case studies, PhD thesis, Wageningen University, Wageningen, The Netherlands, 167p., 2000.

van der Merwe, R.: Sigma-Point Kalman Filters for Probabilistic Inference in Dynamic State-Space Models, PhD dissertation, University of Washington, USA, 2004.

van Genuchten, M. Th.: A closed-form equation for predicting the hydraulic conductivity of unsaturated soils, Soil Sci. Soc. Am. J., 44, 892–898, 1980.

Vereecken, H., Huisman, J. A., Bogena, H., Vanderborght, J., Vrugt, J. A., and Hopmans, J. W.: On the value of soil moisture measurements in vadose zone hydrology: A review, Water Resour. Res., 44, W00D06, doi:10.1029/2008WR006829, 2008.

Vogel, T., van Genuchten, M. T., and Cislerova, M.: Effect of the shape of soil hydraulic functions near saturation on variably-saturated flow predictions, Adv. Water Resour., 24, 133–144, 2001.

Vrugt, J. A., Bouten, W., and Weerts, A. H.: Information content of data for identifying soil hydraulic parameters from outflow experiments, Soil Sci. Soc. Am. J., 65, 19–27, 2001.

Vrugt, J. A., W. Bouten, H. V. Gupta, and Sorooshian, S: Toward improved identifiability of hydrologic model parameters: the information content of experimental data, Water Resour. Res., 38, 1312, doi:10.1029/2001WR001118, 2002.

Vrugt, J. A., Bouten, W., Gupta, H. V., and Hopmans J. W.: Toward improved identifiability of soil hydraulic parameters: On the selection of a suitable parametric model, Vadose Zone J., 2, 98–113, 2003.

Vrugt, J. A., Diks, C. G. H., Gupta, H. V., Bouten W., and Verstraten, J. M.: Improved treatment of uncertainty in hydrologic modeling: Combining the strengths of global optimization and data assimilation, Water Resour. Res., 41, W01017, doi:10.1029/2004WR003059, 2005.

Vrugt, J. A., ter Braak, C. J. F., Diks, C. G. H., and Schoups, G.: Hydrologic data assimilation using particle Markov chain Monte Carlo simulation: Theory, concepts and applications, Adv. Water Resour., 51, 457–478, 2013.

Walker, J. P.: Estimating soil moisture profile dynamics from near-surface soil moisture measurements and standard meteorological data. Ph.D. thesis, The University of Newcastle, Callaghan, New South Wales, Australia, 766 pp., 1999.

Walker, J. P., Willgoose G. R., and Kalma, J. D.: One-dimensional soil moisture profile retrieval by assimilation of near-surface observations: a comparison of retrieval algorithms, Adv. Water Resour., 24, 631–650, 2001.

Wan, E. A. and Nelson, A. T.: Neural Dual Extended Kalman Filtering: Applications in Speech Enhancement and Monaural Blind Signal Separation, in: Proceedings of the IEEE Signal Processing Society Neural Networks for Signal Processing Workshop, New York, 466–475, 1997.

Wan, E. A. and Nelson, A. T.: Dual Extended Kalman Filter Methods, in Kalman Filtering and Neural Networks, edited by: Haykin, S., 221–280, John Wiley, Hoboken, N. J., 2001.

Wan, E., van der Merwe, R., and Nelson, A.: Dual Estimation and the Unscented Transformation, in: Neural Information Processing Systems 12, edited by: Solla, S. A., Leen, T. K., and Müller, K. R., MIT Press, 666–672, 2000.

Wöhling, T. and Vrugt, J. A.: Multiresponse multilayer vadose zone model calibration using Markov chain Monte Carlo simulation and field water retention data, Water Resour. Res., 47, W04510, doi:10.1029/2010WR009265, 2011.

Yang, K., Koike, T., Kaihotsu, I., and Qin, J.: Validation of a dual-pass microwave land data assimilation system for estimating surface soil moisture in semiarid regions, J. Hydrometeorol., 10, 780–793, 2009.

Yeh, W. W. G.: Review of parameter identification procedures in ground water hydrology: The inverse problem, Water Resour. Res., 22, 95–108, 1986.

Zhu, J. and Mohanty, B. P.: Soil hydraulic parameter upscaling for steady flow with root water uptake, Vadose Zone J., 3, 1464–1470, 2004.

Kalman filters for assimilating near-surface observations into the Richards equation – Part 3: Retrieving states and parameters from laboratory evaporation experiments

H. Medina[1] **, N. Romano**[2] **, and G. B. Chirico**[2]

[1]Department of Basic Sciences, Agrarian University of Havana, Havana, Cuba
[2]Department of Agricultural Engineering, University of Naples Federico II, Naples, Italy

Correspondence to: G. B. Chirico (gchirico@unina.it)

Abstract. The purpose of this work is to evaluate the performance of a dual Kalman filter procedure in retrieving states and parameters of a one-dimensional soil water budget model based on the Richards equation, by assimilating near-surface soil water content values during evaporation experiments carried out under laboratory conditions. The experimental data set consists of simultaneously measured evaporation rates, soil water content and matric potential profiles. The parameters identified by assimilating the data measured at 1 and 2 cm soil depths are in very good agreement with those obtained by exploiting the observations carried out in the entire soil profiles. A reasonably good correspondence has been found between the parameter values obtained from the proposed assimilation technique and those identified by applying a non-sequential parameter estimation method. The dual Kalman filter also performs well in retrieving the water state in the porous system. Bias and accuracy of the predicted state profiles are affected by observation depth changes, particularly for the experiments involving low state vertical gradients. The assimilation procedure proved flexible and very stable in both experimental cases, independently from the selected initial conditions and the involved uncertainty.

1 Introduction

With the unprecedented availability of soil moisture information, new opportunities emerge to improve the accuracy in hydrologic predictions. Nonetheless, the effective use of this information entails the implementation of techniques adequately addressing the different sources of uncertainties embedded in the simulation process (Hoeben and Troch, 2000; Vrugt et al., 2005; McLaughlin, 2002; Vereecken et al., 2008), in particular those arising from the parameterization of the soil hydraulic properties, i.e. the soil water retention and hydraulic conductivity functions, which are fundamental for reliably modelling soil water dynamics in the vadose zone.

Several studies have pointed out the potential of data assimilation (DA) techniques, in particular those based on sequential algorithms as the Kalman filter (Kalman, 1960), to improve the forecast offered by a hydrological model by using near-surface soil water content observations derived from remote sensing or ground-based networks. As defined by Liu and Gupta (2007), DA aims at providing consistent estimates of the dynamical behaviour of a system by merging the information available in imperfect models and uncertain data in an optimal way. Nonetheless, the majority of soil hydrology studies deal with the use of near-surface soil moisture information to retrieve soil moisture profiles in the unsaturated zone while assuming the soil hydraulic properties as known attributes (e.g. Entekhabi et al., 1994; Walker et al., 2001, 2004; Dunne and Entekhabi, 2005). Poorly identified parameters usually generate error and bias, which should be faced using bias-correction strategies (De Lannoy et al., 2007a, b) or, more generally, a model calibration.

Classical algorithms for model calibration, either the traditional approaches accounting for deterministic methods, or those based on the standard Bayes' law, calibrate model parameters by implementing an optimization procedure that minimizes long-term prediction errors for a given historical data set. This type of algorithms assumes time-invariant parameters and thus does not make any attempt to include information from new observations (Moradkhani et al, 2005a; Liu and Gupta, 2007; Vrugt et al., 2013), thus hindering the identification of time-varying errors associated with the several components of the uncertainty. In addition, the applicability of such algorithms is limited in those regions where the lack of historical data makes the optimization procedure less practicable (Thiemann et al., 2001).

Therefore an increasing number of hydrological DA applications has being aimed at exploring the capabilities of DA methods to improve the accuracy of modelled physical phenomena and processes by recursively retrieving both states and parameters. An interesting discussion about pros and cons of the common approaches dealing with simultaneous state and parameter estimation in hydrological applications was provided by Liu and Gupta (2007). Todini et al. (1978) provided significant insights about the simultaneous state–parameter estimation in pioneering hydrological applications. Other important studies have been carried out by Vrugt et al. (2005), who combined the ensemble Kalman filter (EnKF) to update model states recursively with an optimization technique for parameter estimation. Moradkhani et al. (2005a, b) provided a framework for dual state–parameter estimation using the EnKF and particle filtering (PF).

One controvertible issue is which of the two approaches, namely the dual or the joint Kalman filtering, performs better for the simultaneous retrieval of states and parameters. In the dual filter approach, two separate filters – one for the state space and the other for the parameter space – are implemented. Instead, the joint filter methods predict the evolution of the joint probability distribution of states and parameters, which are combined in an augmented state vector. Theoretically, the joint approach is more robust because it accounts for the cross-covariance between the states and parameter estimates (van der Merwe, 2004). However, several studies highlighted the instability and intractability of this approach as a result of the complex interactions between states and parameters in a nonlinear dynamic system (Moradkhani et al., 2005; Liu and Gupta, 2007). Medina et al. (2014) have shown that the decoupled state-space representation in the dual KF provides higher flexibility since different Kalman filters can be selected for the states and parameters, respectively, according to the specific characteristics of both the observation and system equations.

A few studies have explored the possibility of simultaneously retrieving soil moisture profiles and soil hydraulic parameters, by assimilating surface soil moisture observations (e.g. Qin et al., 2009; Yang et al., 2009). Lü et al. (2011) applied a direct insertion method for assimilating surface soil moisture within the Richards equation, coupled with a particle swarm optimization algorithm, for identifying the optimal saturated hydraulic conductivity. Monztka et al. (2011) and Plaza et al. (2012) performed recursive state and parameter

retrieval for the soil moisture estimation, but using a particle filter algorithm, which is another sequential method widely applied in the most recent studies (Vrugt et al., 2013).

Medina et al. (2014) performed a synthetic numerical study to evaluate a dual Kalman filter (named DSUKF) for real-time simultaneous prediction of soil water content or matric pressure head profiles and soil hydraulic parameters, by assimilating near-surface information in a one-dimensional Richards equation. In this approach, a standard Kalman filter is implemented with a Crank–Nicolson numerical scheme to retrieve soil state profiles, while an unscented Kalman filter (UKF) (Julier and Uhlman, 1997, 2004; van der Merwe 2004) is implemented for retrieving soil hydraulic parameters. The UKF is based on a statistical linearization of the nonlinear operators (unscented transformation), without the need of performing any analytic differentiation.

One distinguishing issue of this approach concerns the adopted linearized forward state model. The noniterative integration arising from the linearized Crank–Nicolson formulation represents a simplification of the solution of the Richards equation. However, the time stepping adopted within the numerical scheme can be adjusted to reduce the numerical error to negligible values (e.g. Haverkamp et al., 1977). Paniconi et al. (1991) show that the linearization of the Richards equation, when applied to second-order time steeping formulations as the Crank–Nicolson, reduces the accuracy to first order. Nevertheless, further improvements can be easily incorporated, as for instance reported by Kavetski et al. (2002), who demonstrated that the reliability and the efficiency of these noniterative linearized schemes can be increased by adaptive truncation error control, and can be even more efficient than analogous time stepping schemes with iterative solvers.

Medina et al. (2014) provided valuable insights about the identifiability of the parameters featuring in the van Genuchten–Mualem soil hydraulic analytical relations (van Genuchten, 1980) by means of dual Kalman filters. This study also discussed the impact exerted by parameter initialization, observation depth, and assimilation frequency on the overall DSUKF retrieval performance, as well as the influence of the formulation of the Richards differential equation (either h-based or θ-based form). The performance of the adopted approach is compared with that using the dual ensemble KF with ensemble sizes equal to 12 (DEnKF(12)) and 50 (DEnKF(50)), respectively. Reichle and Koster (2003) found that 12 ensembles consistently reduced the uncertainty of the soil moisture estimates during a one-dimensional EnKF state retrieval application. Camporese et al. (2009) suggest a minimum of about 50 realizations for ensuring a suitable level of accuracy in analogous applications. The results show that the root mean square error (RMSE) using DEnKF(50) is about 1.5 times lower, but at cost of seven times more computational central processing unit (CPU) time. However, DEnKF(12) is outperformed in terms of both uncertainty reduction and computational efficiency.

The advantage of referring to synthetic values instead of using measured values of the variables is that the former are known a priori, and thus the assessment of the algorithm performance is facilitated. On the other hand, synthetic studies greatly simplify the inherent complexity of real-world applications, where uncertainties arise from several complexities of the system at hand, such as the soil heterogeneity and measuring errors. Nevertheless, one important question concerning the validity of the approach is whether it is able to cope with real data suitably.

This paper aims at evaluating the DSUKF performance by exploiting data measured during evaporation experiments carried out on undisturbed soil cores under laboratory conditions. The strength of the evaporation experiment is that the data are gathered during a transient flow that is very close to natural processes occurring in real soils, thereby providing a highly representative hydraulic response of the soil under study. The evaporation tests selected for this study are two of those employed by Romano and Santini (1999) for evaluating a parameter optimization method developed to determine the unsaturated hydraulic properties of different soil types. The method consists of a non-sequential inversion procedure, also implemented with a Crank–Nicolson numerical scheme, but using matric pressure head data instead of soil water content data. The evaporation experiments carried out by Romano and Santini (1999) have been selected mainly because soil water content values were also measured with a relatively high vertical resolution using the gamma ray attenuation method, thus providing a valuable experimental data set to test the DSUKF approach developed by Medina et al. (2014).

The performance of the DSUKF approach is herein examined, accounting for the effects of the observation depth, the assimilation frequency, and the parameter initialization, on both state and parameter retrieval processes.

2 The dual Kalman filter formulation

A detailed description of the algorithm employed for the separate state-space representation used to retrieve states and parameters is in the paper by Medina et al. (2014). At each time step, the current estimate of the parameters is used in the state filter, and the current estimate of the states is used in the parameter filter. For the sake of effectiveness, the state and parameter filter equations employed in the assimilation algorithm are herein summarized. In the most general case, the set of system equations can be written as

$$x_k = F_{k-1,k}\left(x_{k-1}, u_k, \hat{w}_{k-1}\right) + v_{k-1} \tag{1}$$

$$y_k = H_k\left(x_k, \hat{w}_{k-1}\right) + \eta_k \tag{2}$$

for the state vector x at instant k, and

$$w_k = w_{k-1} + \xi_{k-1} \tag{3}$$

$$y_k = H_k\left(F_{k-1,k}\left(\hat{x}_{k-1}, u_k, w_k\right), w_k\right) + \varsigma_k \tag{4}$$

for the parameter vector w. In the equation above, F is the state transition function, H the observation function, v_{k-1} the zero mean Gaussian process noise with covariance Q_{k-1}, η_k the zero mean observation or measurement noise with covariance R_k, and u_k represents an exogenous input to the system. The parameter update is artificially set up by means of a stationary process with an identity state transition matrix. $\xi_{k-1} \in N\left(0, Q_{w,k-1}\right)$ is the noise driving the parameter updating, and $\varsigma_k \in N\left(0, R_{w,k}\right)$ is the noise corrupting the observation equation relative to the parameters, both artificially settled. The caret symbol ("^") denotes the density mean of the variable.

2.1 The standard Kalman filter formulation for linear state retrieval

The linear algorithm for the state retrieval is summarized in the following three phases.

I. Initialization:

$$\hat{x}_0 = E\left[x_0\right] \tag{5}$$

$$\mathbf{P}_0 = E\left[\left(x_0 - \hat{x}_0\right)\left(x_0 - \hat{x}_0\right)^T\right] \tag{6}$$

$$\mathbf{Q}_0 = E\left[\left(v_0 - \bar{v}_0\right)\left(v_0 - \bar{v}_0\right)^T\right] \tag{7}$$

$$\mathbf{R}_0 = E\left[\left(\eta_0 - \bar{\eta}_0\right)\left(\eta_0 - \bar{\eta}_0\right)^T\right] \tag{8}$$

Subscript "0" indicates initial values.

II. Prediction phase is carried out by computing the state mean and covariances respectively as follows:

$$\hat{x}_k^- = \mathbf{F}_{k-1,k}\left(\hat{w}_{k-1}\right)\hat{x}_{k-1} + g_{k-1,k}\left(u_k, \hat{w}_{k-1}\right) \tag{9}$$

$$\mathbf{P}_k^- = \mathbf{F}_{k-1,k}\,\mathbf{P}_{k-1},\,\mathbf{F}_{k-1,k}^T + \mathbf{Q}_k, \tag{10}$$

where subscript k indicates the time step; $\hat{x}_k^-$ and $\mathbf{P}_k^-$ represent the a priori predictions of the state mean and covariance, respectively; $\mathbf{F}_{k-1,k}$ is the linear state transition matrix of operator F in Eq. (1), which depends nonlinearly on the parameters w_{k-1}; $g_{k-1,k}$ is a vector accounting for the effect of the exogenous input u_k, and it is also nonlinearly dependent on the parameters w_{k-1}.

III. Correction phase is carried out for updating estimates with the last observation:

$$\mathbf{K}_k = \mathbf{P}_k^- \mathbf{H}_k^T \left(\mathbf{H}_k, \mathbf{P}_k^- \mathbf{H}_k^T + \mathbf{R}_k \right)^{-1}, \tag{11}$$

$$\hat{\mathbf{x}}_k = \left(\hat{\mathbf{x}}_k^- \right) + \mathbf{K}_k \left(\mathbf{y}_k - \mathbf{H}_k \left(\hat{\mathbf{x}}_k^- \right) \right), \tag{12}$$

$$\mathbf{P}_k = \mathbf{P}_k^- \mathbf{K}_k \mathbf{H}_k \mathbf{P}_k^-, \tag{13}$$

where $\mathbf{K}_k$ is the Kalman gain, expressing the ratio of the expected cross-covariance matrix of the process prediction error and the observation prediction error, $\mathbf{y}_k$ represents the actual measurements, and $\hat{\mathbf{x}}_k$ and $\mathbf{P}_k$ represent the states posterior density mean and covariance, respectively. $\mathbf{H}_k$ stands for the linear observation operator (Eq. 2) in matrix form.

2.2 The unscented Kalman filter (UKF) formulation for parameter estimation

In the UKF formulation, the distribution of the parameters is represented by a Gaussian random variable, being specified using a minimal set of sample points, so-called sigma points, which are deterministically selected to capture the true mean and covariance of the variable completely and, when propagated through the true nonlinear system, to capture the posterior mean and covariance accurately up to the second order for any nonlinearity (van Der Merwe, 2004; Chirico et al., 2014; Medina et al., 2014).

The algorithm for the dynamic retrieval of the unknown parameters can be summarized in the following four phases.

I. Initialization of the parameter vector, $\hat{\mathbf{w}}_0$, and covariance, $\mathbf{P}_{w,0}$:

$$\hat{\mathbf{w}}_0 = E[\mathbf{w}] \tag{14}$$

$$\mathbf{P}_{w,0} = E\left[(\mathbf{w} - \hat{\mathbf{w}}_0)(\mathbf{w} - \hat{\mathbf{w}}_0)^T \right] \tag{15}$$

II. Time update phase, in order to get the corresponding a priori predictions:

$$\hat{\mathbf{w}}_k^- = \hat{\mathbf{w}}_{k-1} \tag{16}$$

$$\mathbf{P}_{w,k}^- = \mathbf{P}_{w,k-1} + \mathbf{Q}_{w,k-1}, \tag{17}$$

where $\mathbf{Q}_{w,k} = \left(\lambda_{\mathrm{RLS}}^{-1} - 1 \right) \mathbf{P}_{w,k}$, being $\lambda_{\mathrm{RLS}} \in (0, 1]$ a forgetting factor as defined in the recursive least-squares (RLS) algorithm. This relationship for $\mathbf{Q}_{w,k}$ is chosen on the basis of the results illustrated in Medina et al. (2014), where alternative expressions have been compared.

III. Computation of the sigma points $\mathcal{W}_{k,i}$ for the measuring update:

$$\mathcal{W}_{k,i} = \left[\hat{\mathbf{w}}_k^-, i = 0 \quad \hat{\mathbf{w}}_k^- + \left(\sqrt{\gamma \mathbf{P}_{w,k}^-} \right)_i, i = 1, \ldots \right.$$
$$\left. N_{\mathrm{par}} \quad \hat{\mathbf{w}}_k^- - \left(\sqrt{\gamma \mathbf{P}_{w,k}^-} \right)_i, i = N_{\mathrm{par}} + 1, \ldots 2N_{\mathrm{par}} \right], \tag{18}$$

where N_{par} is equal to the number of parameters to be retrieved, while γ is a coefficient scaling the sampled parameter distribution around the mean parameter vector.

IV. Measuring update equations:

$$\mathcal{Y}_k = H_k \left(F_{k-1,k} \left(\hat{\mathbf{x}}_{k-1}, \mathbf{u}_k, \mathcal{W}_k \right) \right) \tag{19}$$

Then mean, $\hat{\mathbf{y}}_{w,k}^-$, cross-covariance, $\mathbf{P}_{wy,k}$, and covariance, $\mathbf{P}_{yy,k}^w$, of the transformed sigma points $\mathcal{Y}_k$ are respectively calculated as follows:

$$\hat{\mathbf{y}}_{w,k}^- = \sum_{i=0}^{2N_{\mathrm{par}}} \mu_i^{(m)} \mathcal{Y}_{k,i}, \tag{20}$$

$$\mathbf{P}_{wy,k} = \sum_{i=0}^{2N_{\mathrm{par}}} \mu_i^{(c)} \left(\mathcal{W}_{k,i} - \hat{\mathbf{w}}_k^- \right) \left(\mathcal{Y}_{k,i} - \hat{\mathbf{y}}_{w,k}^- \right)^T, \tag{21}$$

$$\mathbf{P}_{yy,k}^w = \sum_{i=0}^{2N_{\mathrm{par}}} \mu_i^{(c)} \left(\mathcal{Y}_{k,i} - \hat{\mathbf{y}}_{w,k}^- \right) \left(\mathcal{Y}_{k,i} - \hat{\mathbf{y}}_{w,k}^- \right)^T, \tag{22}$$

where μ_i are the weights related to the sigma point i, conditioned to $\sum_{i=0}^{2N_{\mathrm{par}}} \mu_i = 1$. Weight values for calculating the mean and the covariance are indicated by the superscripts m and c, respectively.

The Kalman gain, $\mathbf{K}_{w,k}$, and the posterior parameter mean, $\hat{\mathbf{w}}_k$, and covariance, $\mathbf{P}_{w,k}$, are calculated respectively as follows:

$$\mathbf{K}_{w,k} = \mathbf{P}_{wy,k} \left(\mathbf{P}_{yy,k}^w + \mathbf{R}_{w,k} \right)^{-1} = \mathbf{P}_{wy,k} \left(\mathbf{P}_{vv,k}^w \right)^{-1}, \tag{23}$$

$$\hat{\mathbf{w}}_k = \hat{\mathbf{w}}_k^- + \mathbf{K}_{w,k} \left(\mathbf{y}_k - \hat{\mathbf{y}}_{w,k}^- \right), \tag{24}$$

$$\mathbf{P}_{w,k} = \mathbf{P}_{w,k}^- - \mathbf{K}_{w,k} \mathbf{P}_{vv,k}^w \left(\mathbf{K}_{w,k} \right)^T, \tag{25}$$

where $\mathbf{P}_{vv,k}^w$ represents the covariance of $\mathbf{y}_k - \hat{\mathbf{y}}_{w,k}^-$. More details about the UKF implementation can be found in Medina et al. (2014).

3 Materials and methods

3.1 Soil water flow governing equation

Water movement in a vertical soil column, modelled as a homogeneous, variably saturated porous medium under isothermal conditions, is simulated using the θ-based form of the Richards equation:

$$\frac{\partial \theta}{\partial t} = \frac{\partial}{\partial z} \left[\left(D(\theta) \frac{\partial \theta}{\partial z} - K(\theta) \right) \right], \tag{26}$$

where t is time, z soil depth taken positive downward, with $z = 0$ at the top of the profile, θ the soil water content [L^3 L^{-3}], $D(\theta) = K(\theta)\frac{dh}{d\theta}$ [L^2/T] is the unsaturated diffusivity, with K [L/T] being the unsaturated hydraulic conductivity and h the soil water matric pressure head [L].

The constitutive relationships characterizing the soil hydraulic properties are the van Genuchten–Mualem (VGM) parametric relations (van Genuchten, 1980):

$$\theta(h) = \theta_r + \theta_s - \theta_r \left[1 + |\alpha h|^n\right]^{-m}, \tag{27}$$

$$K(\theta) = K_S S_e^{\lambda}\left[1 - \left(1 - S_e^{1/m}\right)^m\right]^2, \tag{28}$$

where θ_s is the saturated soil water content, θ_r the residual soil water content, $S_e = (\theta - \theta_r)/(\theta_s - \theta_r)$ the effective saturation, and K_s the saturated hydraulic conductivity, whereas α [L^{-1}], n (-), m (-) and λ (-) are empirical parameters. A common assumption, also adopted in this work, is to consider $\lambda = 0.5$ and $m = 1 - 1/n$.

3.2 Crank–Nicolson finite difference scheme

The numerical solution of the θ-based form of the Richards equation (Eq. 26) is implemented according to the Crank–Nicolson finite difference scheme, with an explicit linearization of both the soil hydraulic conductivity K and the diffusivity D, which takes on the following form for the intermediate nodes of the soil column:

$$\left(-\frac{D_k^{i-1/2}}{2\Delta z^i \Delta z^u}; \frac{1}{\Delta t_k} + \frac{\frac{D_k^{i-1/2}}{\Delta z^u} + \frac{D_k^{i+1/2}}{\Delta z^l}}{2\Delta z^i}; -\frac{D_k^{i+1/2}}{2\Delta z^i \Delta z^l}\right)\left(\begin{array}{c}\theta_{k+1}^{i-1}\\\theta_{k+1}^{i}\\\theta_{k+1}^{i+1}\end{array}\right)$$

$$= \left(\frac{D_k^{i-1/2}}{2\Delta z^i \Delta z^u}; \frac{1}{\Delta t_k} - \frac{\frac{D_k^{i-1/2}}{\Delta z^u} + \frac{D_k^{i+1/2}}{\Delta z^l}}{2\Delta z^i}; \frac{D_k^{i+1/2}}{2\Delta z^i \Delta z^l}\right)$$

$$\left(\begin{array}{c}\theta_k^{i-1}\\\theta_k^{i}\\\theta_k^{i+1}\end{array}\right) + \frac{K_k^{i-1} - K_k^{i+1}}{2\Delta z^i}, \tag{29}$$

where subscript i is the node number (increasing downward), superscript k the time step, and $\Delta t^k = t^{k+1} - t^k$. All the nodes, including the top and bottom nodes, are located in the centre of each soil compartment in which the soil column is discretized, with $\Delta z^u = z^i - z^{i-1}$, $\Delta z^l = z^{i+1} - z^i$ and Δz^i the compartment thickness (cm). The spatial averages of K are calculated as arithmetic means.

Flux conditions imposed at the upper and lower boundaries are expressed, respectively, by the following equations:

$$\left(\frac{1}{\Delta t_k} + \frac{D_k^{1+1/2}}{2\Delta z^1 \Delta z^l}; -\frac{D_k^{1+1/2}}{2\Delta z^1 \Delta z^l}\right)\left(\begin{array}{c}\theta_{k+1}^{1}\\\theta_{k+1}^{2}\end{array}\right) \tag{30}$$

$$= \left(\frac{1}{\Delta t_k} - \frac{D_{1+1/2}^{k}}{2\Delta z^1 \Delta z^l}; \frac{D_{1+1/2}^{k}}{2\Delta z^1 \Delta z^l}\right)\left(\begin{array}{c}\theta_k^{1}\\\theta_k^{2}\end{array}\right) + \frac{q_{top} - K_k^{1+1/2}}{\Delta z^1}$$

$$\left(-\frac{D_k^{n-1/2}}{2\Delta z^n \Delta z^u}; \frac{1}{\Delta t_k} + \frac{D_k^{n-1/2}}{2\Delta z^n \Delta z^u}\right)\left(\begin{array}{c}\theta_{k+1}^{n-1}\\\theta_{k+1}^{n}\end{array}\right) \tag{31}$$

$$= \left(\frac{D_k^{n-1/2}}{2\Delta z^n \Delta z^u}; \frac{1}{\Delta t_k} - \frac{D_k^{n-1/2}}{2\Delta z^n \Delta z^u}\right)\left(\begin{array}{c}\theta_k^{n-1}\\\theta_k^{n}\end{array}\right) + \frac{K_k^{n-1/2} - q_{bot}}{\Delta z^n}$$

being q_{top} and q_{bot} the fluxes at the top and the bottom of the soil profile, respectively.

In the numerical scheme, an explicit linearization of K and D is implemented by taking their values at the previous time step k-1. Then, a linear state-space representation of the dynamic system can be easily derived by combining the set of Eqs. (29)–(31) written for each node and accounting for the boundary conditions:

$$B_{k-1}\theta_k = A_{k-1}\theta_{k-1} + f_{k-1}, \tag{32}$$

where A_{k-1} and B_{k-1} are tri-diagonal matrices obtained by assembling the terms in the first pair of parentheses on the right- and left-hand side of Eqs. (29)–(31), respectively. The term f_{k-1} is a vector obtained by assembling the terms on the right-hand side of the state variable at time step k-1. More explicitly, Eq. (32) becomes

$$\theta_k = F_{k-1}\theta_{k-1} + g_{k-1} \tag{33}$$

by making $\mathbf{F}\mathbf{B}^{-1}\,\mathbf{A}$ and $g = \mathbf{B}^{-1}\,f$, which correspond to the analogous terms in Eqs. (9)–(10).

3.3 Experimental data set

We used data published on an earlier study since it was judged to be helpful to explore advanced methods about a common problem already mentioned in the literature. We used the data reported in the paper by Romano and Santini (1999) and collected during evaporation tests on the two undisturbed soil core named as GA3 and GB1. Each of these two soil cores had an inner diameter of 8.0 cm and a length of 12.0 cm. In order to provide a clear view of the soil water dynamic process, against which the performance of DSUKF has been evaluated, a brief description of the evaporation tests is herein provided, while further details can be found in Romano and Santini (1999).

The undisturbed soil core, after being completely saturated from the bottom, is induced to a state of hydrostatic equilibrium with the matric pressure head value at the bottom end almost at zero. The sample cylinder is then completely sealed at the bottom and positioned on a plate, supported by a strain-gauge load cell measuring the soil sample weight, while a small fan is positioned near the top. Tensiometers connected to pressure transducers are inserted at various depths to monitor the soil water pressure head. The evaporation experiment is carried out until the formation of air bubbles causes the breakdown of the hydraulic connection between the last working tensiometer and the corresponding pressure transducer. Tensiometers were inserted at the following three soil

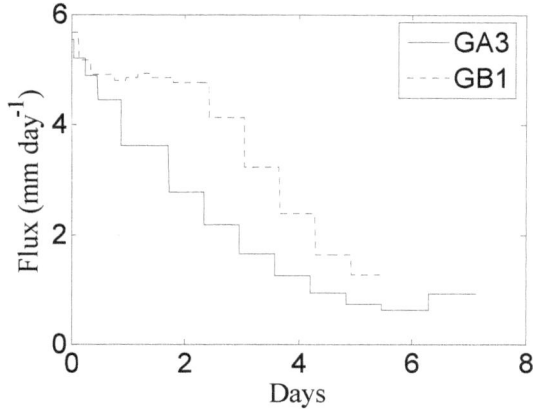

Figure 1. Evaporation flux at the soil surface calculated from the water balance of each soil sample between consecutive measurements of the soil profiles.

depths: 3, 6, and 9 cm. Additionally, soil water content profiles during the experiment were measured with a gamma ray attenuation device, with a vertical resolution of 1.0 cm.

Table 1 lists the basic physical properties of the two soil samples together with the VGM model parameters α, n and Ks, which have been estimated by Romano and Santini (1999) by applying a non-sequential parameter estimation method and are employed as reference values in this study.

Soil water content profiles, with a time update of 600 s, have been built by polynomial interpolation of the gamma ray measurements at all soil depths and taken as "true" state values.

The assimilation algorithm samples the observation values from the interpolated soil water content profiles. The evaporation rate at the soil surface is estimated by applying a water balance equation between two consecutive measurements of the soil water content profiles, under the assumption of a constant evaporation flux during the measurement interval. This approach provides an approximate temporal pattern of the upper boundary fluxes, with step changes, as depicted in Fig. 1.

The duration of the assimilated evaporation process is 170 h for GA3 and 131 h for GA1, which approximately correspond to 7 and 5.5 days, respectively.

4   DSUKF implementation

Similarly to the synthetic study of Medina et al. (2014), the saturated (θ_s) and the residual (θ_r) soil water contents are assumed to be known, as these parameters can be easily determined by direct or indirect methods (e.g. Pringle et al., 2007; Chirico et al., 2010). As reported in the paper by Romano and Santini (1999), the values of parameter θ_s are fixed to 0.31 cm³ cm⁻³ for GA3 and 0.35 cm³ cm⁻³ for GB1, re-

spectively. The values of parameter θ_r are instead defined according to the values suggested by Carsel and Parrish (1988) for soils of the same textural class: $\theta_r = 0.067$ cm³ cm⁻³ for a silt loam soil as GA3, and $\theta_r = 0.08$ cm³ cm⁻³ for a loam soil as GB1, which are both slightly smaller than the air-dried values assumed by Romano and Santini (1999) (see Table 1).

The unscented Kalman filter is thus implemented to retrieve the remaining parameters K_S, α and n. As shown in Medina et al. (2014), a variable transformation is applied to constrain the retrieved parameter values to a certain physically meaningful range and thus to guarantee operational stability. Considering that w_i is the true value of the ith parameter, the parameter estimation procedure makes use of the following variable transformation:

$$w_i = w_{i_{min}} + \left(w_{i_{max}} - w_{i_{min}}\right) s(\delta w_i), \qquad (34)$$

where w_{min} and w_{max} represent user-defined nominal values constraining the minimum and maximum values of the parameter, respectively; the correction terms δw_i are the actual variables under estimation and are expressed as independent terms of a nonlinear sigmoidal function $s(\delta w)$. The sigmoidal function, designed to limit the absolute magnitude of the estimated adjustment, is defined as follows:

$$s(\delta w_i) = \frac{\delta w_i}{2\left(1 + |\delta w_i|\right)} + 0.5. \qquad (35)$$

Table 2 summarizes the initial conditions employed for the state variables and the covariance matrices, as well as the examined observation depths and assimilation frequencies.

Uniform profiles have been assumed as initial state condition, with values $\theta_0 = 0.28$ cm³ cm⁻³ for soil core GA3 and $\theta_0 = 0.31$ cm³ cm⁻³ for soil core GB1. These are approximately average values between saturation and the soil water content measured at the soil surface at the beginning of the experiments.

Our evaluations considered three different observation depths (OD): 1, 2, and 12 cm. Escorihuela et al. (2010) found that an OD of 2 cm is the most effective soil moisture sampling depth for L-band radiometry, even in comparison with smaller depths. The observation depth of OD = 12 cm, which means the assimilation of the entire soil core, has been included as benchmark performance for OD = 1 cm and OD = 2 cm in the parameter identification. Two assimilation frequencies have been used: a finer frequency with assimilations every 2 h (AF = 1/2 h⁻¹), and a coarser one with assimilations every 12 h (AF = 1/12 h⁻¹).

The initial covariance matrices are all diagonals and are defined similarly to Medina et al. (2014): the initial state covariance matrix is set to a value of 1000 % of the mean state profile on the diagonal elements; the initial matrix of the normalized correction terms associated with the soil hydraulic parameters is assigned to a value of 0.01 on the diagonal elements; the diagonal of the observation noise autocovariance matrix is updated using the 2 % of the observed state vector,

Table 1. Physical and soil hydraulic properties of the two soil samples employed for the evaporation experiments.

Soil Sample	Texture	ρb, g cm^{-3}	θ_s	θ_r*	α** (10^{-2} cm^{-1})	n**	K_s** (10^{-4} cm s^{-1})
GA3	Silty loam	1.592	0.310	0.080	1.75	2.27	0.222
GB1	Loam	1.572	0.348	0.120	1.67	2.70	1.56 5

* Air-dried value. ** Estimated by inversion method (Romano and Santini, 1999).

Table 2. Values adopted for the initialization and implementation of the retrieval algorithm.

Input variable	Soil sample	
	GA3	GB1
Initial state variable	0.28 cm^3 cm^{-3}	0.31 cm^3 cm^{-3}
Observation depths (ODs)	1, 2 and 12 cm	
Assimilation frequency (AF)	2 and 12 h	
State covariance matrices		
Initial state covariance matrix $P_0^{i,i}$; $i = 1 \ldots N_{\mathrm{nod}}$	0.8 cm^6 cm^{-6}	
Process noise updating $Q_0^{i,i}$; $i = 1 \ldots N_{\mathrm{nod}}$	0.05 x$_i$ cm^6 cm^{-6}	
Observation noise updating $R_0^{i,i}$; $i = 1 \ldots N_{\mathrm{obs}}$	0.02 y$_i$ cm^6 cm^{-6}	
Parameter covariance matrices		
Initial normalized correction terms matrix $P_{w,0}^{i,i}$; $i = 1 \ldots N_{\mathrm{par}}$	0.01	
Forgetting factor λ_{RLS} (linked to Q_w, Eq. 17)	0.9999	
Artificial noise covariance $R_w^{i,i}$; $i = 1 \ldots N_{\mathrm{obs}}$	1.0×10^{-5}	

N_{nod} is the number of nodes (states); N_{par} is the number of parameters under scrutiny; N_{obs} is the number of observations; x and y represent the state and the observation vectors, respectively.

while the system noise covariance is updated assuming the 5 % of the profile state vector.

The limiting and initialization values of parameters K_s, α and n are also identical to those employed by Medina et al. (2014). Table 3 summarizes the minimum value ($w_{\min}$) and the prescribed range ($w_{\max}$-$w_{\min}$) of each parameter, as well the resulting values of the sets of initial parameters. The six sets of initial parameters are employed for evaluating the influence of the initial condition on the performance of the parameter retrieval algorithm.

For quantitatively evaluating the performance of the involved schemes, the mean error (ME) and the root mean square error (RMSE) between retrieved and true state profiles are computed as follows:

$$ME_j = \frac{1}{N} \sum_{i=1}^{N} \left(x_{i,j}^{\mathrm{guess}} - x_{i,j}^{\mathrm{true}} \right), \tag{36}$$

$$RMSE_j = \left[\frac{1}{N-1} \sum_{i=1}^{N} \left(x_{i,j}^{\mathrm{guess}} - x_{i,j}^{\mathrm{true}} \right)^2 \right]^{1/2}, \tag{37}$$

where $x_{i,j}^{\mathrm{guess}}$ and $x_{i,j}^{\mathrm{true}}$ represent the guess and true state value at node i and time j, respectively.

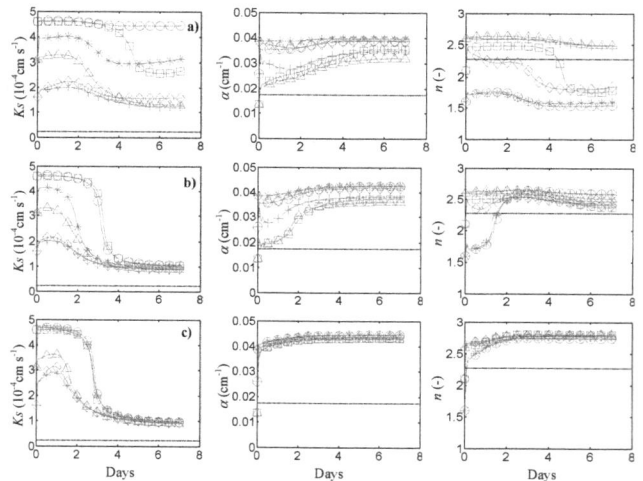

Figure 2. Retrieved VGM parameters K_s, α and n for GA3 soil sample with assimilation frequency AF $= 1/2$ h^{-1}, and observation depth (**a**) OD $= 1$ cm, (**b**) OD $= 2$ cm and (**c**) OD $= 12$ cm. Comparisons account for the six pondered sets of initial parameters S1($\bigcirc$), S2($\square$), S3($*$), S4($\triangle$), S5($+$) and S6($\Diamond$). The dotted line indicates the value of the parameter found by Romano and Santini (1999).

Table 3. Values of $w_{\min}$ and $w_{\max}-w_{\min}$ used to constrain the distribution of the parameters K_s, α and n, and the resulting six sets of parameter values employed for initializing the assimilation algorithm.

Parameter	$w_{\min}$	$w_{\max}-w_{\min}$	S1	S2	S3	S4	S5	S6
K_s (cm s^{-1})	1×10^{-5}	6×10^{-4}	4.6×10^{-4}	4.6×10^{-4}	3.1×10^{-4}	3.1×10^{-4}	1.6×10^{-4}	1.6×10^{-4}
α (cm^{-1})	1×10^{-3}	5×10^{-2}	2.6×10^{-2}	1.35×10^{-2}	3.85×10^{-2}	1.35×10^{-2}	2.6×10^{-2}	3.85×10^{-2}
n (-)		1.1	2.0	1.6	2.1	1.6	2.6	2.6	2.1

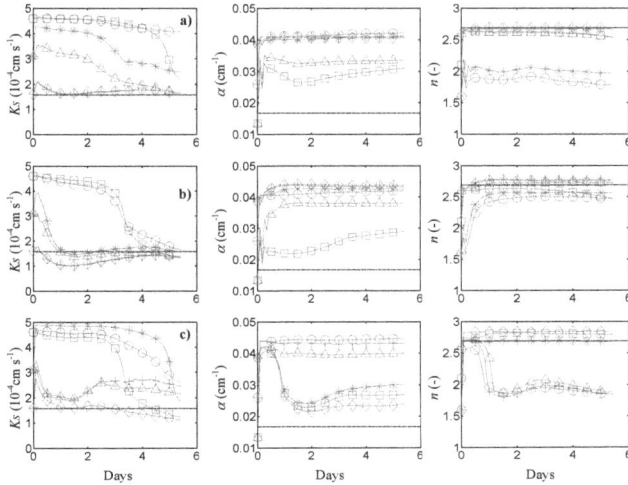

Figure 3. Retrieved VGM parameters K_s, α and n for GB1 soil sample with assimilation frequency AF $= 1/2$ h^{-1}, and observation depth **(a)** OD $= 1$ cm, **(b)** OD $= 2$ cm and **(c)** OD $= 12$ cm. Comparisons account for the six pondered sets of initial parameters S1(○), S2(□), S3(∗), S4(△), S5(+) and S6(◇). The dotted line indicates the value of the parameter found by Romano and Santini (1999).

5 Results

5.1 Parameter retrieval

Figures 2 and 3 show the temporal patterns of the retrieved VGM parameters for GA3 and GB1, respectively, obtained by assimilating the observed soil water content every 2 h within the three different observation depths (ODs): 1, 2, and 12 cm. For soil core GA3 (Fig. 2), there is a very good agreement between parameter values retrieved using OD $= 2$ cm (Fig. 2b) and OD $= 12$ cm (Fig. 2c). It is particularly interesting to note the consistency in the convergence of K_s, which in general is the less identifiable parameter (Medina et al., 2014). The final retrieved K_s value is approximately 1×10^{-4} cm s^{-1}, higher than both the value estimated by Romano and Santini (1999) for the VGM hydraulic model (0.222×10^{-4} cm s^{-1}) and the value directly estimated with the falling head method (0.345×10^{-4} cm s^{-1}). However, the final retrieved K_s is in the range of values obtained by Romano and Santini (1999) with analytical models of the soil hydraulic properties other than the VGM. When us-

ing OD $= 1$ cm, parameter n converges to two main values: one approximately 2.5 and the other one approximately 1.7 (Fig. 2a). For OD $= 2$ cm and OD $= 12$ cm, the convergence patterns are less dependent on the initial set of parameter values. For OD $= 2$ cm, n converges to a value approximately of 2.5, while for OD $= 12$ cm it converges to 2.7, thus in both cases higher than the reference value of 2.27.

For soil core GB1 (Fig. 3), the retrieval process provides values of K_s and n that converge toward those identified by Romano and Santini (1999) ($K_s = 1.565 \times 10^{-4}$ cm s^{-1} and $n = 2.7$, respectively). Using OD $= 2$ cm, the means of K_s and n are found practically equal to those reported by Romano and Santini (1999). The parameters retrieved by assimilating the entire profile (OD $= 12$ cm, Fig. 3c) follow two distinct patterns. The parameter values retrieved with the initial sets S2, S3, and S6 follow patterns fairly close to the values found by Romano and Santini (1999), with mean K_s, α, and n of approximately 1.5×10^{-4} cm s^{-1}, 0.028 cm^{-1} and 2.8, respectively. The parameter values retrieved with the initial sets S1, S4, and S5 follow patterns with average values $K_S = 2.4 \times 10^{-4}$ cm s^{-1}, $\alpha = 0.043$ cm^{-1} and $n = 1.8$, which are relatively close to the values reported by Carsel and Parrish (1988) for loam soils: $K_s = 5.0 \times 10^{-4}$ cm s^{-1}, $\alpha = 0.036$ cm^{-1} and $n = 1.56$. The convergence patterns obtained with OD $= 1$ cm and 2 cm (Fig. 3a and b, respectively) reflect rather well these two alternative parameter space solutions.

Compared with K_s and n, parameter α is much less affected by the observation depth and the initial parameterization. Preliminary sensitivity analyses (not presented here for the sake of brevity) have highlighted that higher initial soil water content values favour the parameter identifiability and the increase of the convergence rate of α, as a result of a relatively higher amount of information of α retrievable for soil water states close to the air entry value (Vrugt et al., 2001, 2002; Medina et al., 2014). In both experiments GA3 and GB1, parameter α converges predominantly toward a value of approximately 0.04 cm^{-1}, which is notably higher than those estimated by Romano and Santini (1999).

This relative inconsistency can be justified considering that the assimilation algorithm is implemented by exploiting the soil water content as an observation variable, whilst Romano and Santini (1999) employed pressure head values measured at three depths (3.0, 6.0, and 9.0 cm) to estimate

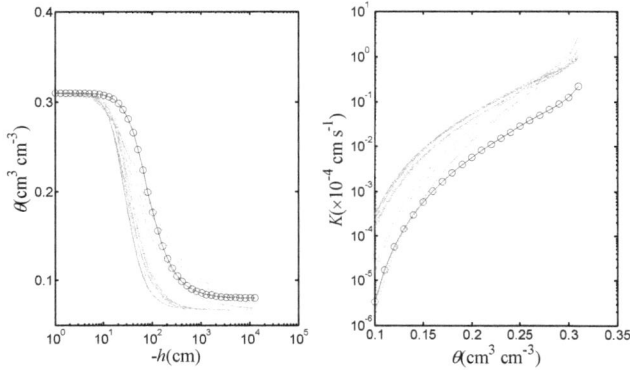

Figure 4. Comparison of the 18 (corresponding to 3 observation depths times 6 initial parameter sets) soil water retention curves $\theta(h)$, and hydraulic conductivity functions, $K(\theta)$, using the converging parameters for GA3 (grey solid lines). The solid lines with markers indicate the corresponding functions defined with the parameters found by Romano and Santini (1999). Note that in this study $\theta_r = 0.067$ cm^3 cm^{-3}, while Romano and Santini (1999) assumed $\theta_r = 0.08$ cm^3 cm^{-3}.

Figure 5. Comparison of the 18 (corresponding to 3 observation depths times 6 initial parameter sets) soil water retention curves θ (h), and hydraulic conductivity functions, $K(\theta)$, using the converging parameters for GB1 (grey solid lines). The solid lines with markers indicate the corresponding functions defined with the parameters found by Romano and Santini (1999). Note that in this study $\theta_s = 0.35$ cm^3 cm^{-3} and $\theta_r = 0.08$ cm^3 cm^{-3}, while Romano and Santini (1999) assumed $\theta_s = 0.348$ cm^3 cm^{-3} and $\theta_r = 0.12$ cm^3 cm^{-3}.

the soil hydraulic parameters with a non-sequential inverse method. Indeed, parameter α acts as a scaling factor of the pressure head values with respect to the soil moistures in the VGM model, and its identifiability through an inverse method can be highly affected by the type of information employed (e.g. Šimůnek and van Genuchten, 1996; Ritter et al., 2004; Wöhling and Vrugt, 2011). As discussed later, the observed discrepancies can also be partly attributed to some inaccuracy in the identification of the final solution due to the high correlation between the van Genuchten parameters. The narrow range covered by the state variables in the considered experiment can have a negative impact on the final results. Several authors have emphasized some difficulties in the identification of the VGM parameters as a consequence of the narrow variability of naturally occurring boundary conditions (Scharnagl et al., 2011; Vrugt et al., 2001, 2002).

With reference to soil core GA3, Fig. 4 shows the comparisons between the soil water retention and hydraulic conductivity functions obtained using the 18 retrieved parameter vectors, and those functions described by the optimized parameter values of Romano and Santini (1999). These vectors are obtained through 18 combinations of 3 different observation depths and 6 initial parameter sets. The denser groups of water retention and hydraulic conductivity curves, corresponding to the solutions using OD $= 2$ cm and 12 cm, have a slope similar to the corresponding reference curves of Romano and Santini (1999), but are shifted toward higher values of h and K. The groups of curves obtained using OD $= 1$ cm have a different slope, and they match fairly well the reference curves only in the dry range.

Similarly to Fig. 4, Fig. 5 compares the soil hydraulic functions obtained in this study for soil core GB1 and those

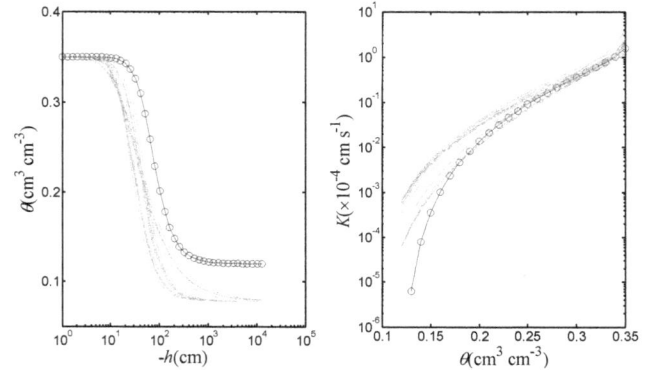

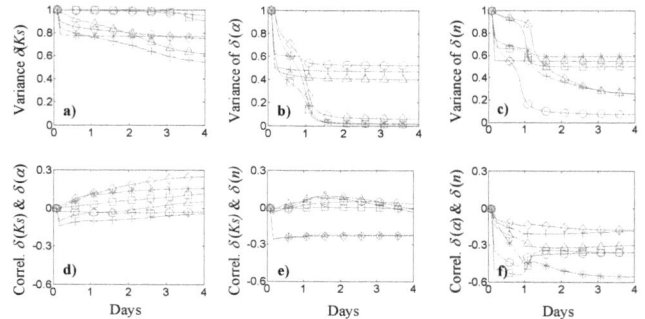

Figure 6. Evolving variances of the correction terms (**a**) δ (K_s), (**b**) δ (α) and (**c**) δ (n) associated with the VGM parameters and correlations (**d–e**) between these terms during the first 4 days using the GA3 data series, with assimilation frequency AF $= 1/2$ h^{-1} and observation depth OD $= 12$ cm.

optimized by Romano and Santini (1999). Again, our estimated soil water retention curves are shifted with respect to the reference curve, except in the dry range of the graph because of the difference between the residual soil moisture of the cited study (0.12 cm^3 cm^{-3}) and the value assumed in this study (0.078 cm^3 cm^{-3}). The estimated hydraulic conductivities match very well the reference curve in the wet range, but depart in the dry one, also as result of the difference in the residual soil water contents.

Some difficulty in correctly identifying the solution is definitely related to the multivariate correlation structure induced in the parametric distribution. This question is evident from Fig. 6, which illustrates the behaviour of the evolving variance and correlation of the correction terms associated with the estimated parameters for GA3 (see also Eqs. 34 and

35) assuming OD = 12 cm. This is roughly the pattern observed for both experiments, independently from the adopted observation depths. The temporal reduction of $\delta(K_s)$ variance (Fig. 6a) is small compared with that of $\delta(\alpha)$ and $\delta(n)$ (Fig. 6b and c). Moreover, $\delta(K_s)$ predominantly exhibits a positive correlation with $\delta(\alpha)$ and a negative correlation with $\delta(n)$ (Fig. 6d and e), while the correlation is always negative between $\delta(\alpha)$ and $\delta(n)$. The signs of these correlation values are consistent with those found by Romano and Santini (1999), although they examined the actual parameter values rather than a nonlinear transformation of them, as done in this study. It is also important to point out that the correlation structure influences the retrieved parameters in different ways, depending on the initial conditions and on the type of retrieval algorithm employed, which can be either sequential, such as in the present study, or non-sequential such as that employed by Romano and Santini (1999).

Finally, a closer inspection of the performance of the assimilation algorithm with relatively low assimilation frequencies is obtained from Fig. 7, which depicts the temporal patterns of the retrieved parameters using $AF = 1/12\,h^{-1}$ and OD = 1 cm, thus involving only 15 assimilation events for GA3 and 11 for GB1. Nevertheless, there is a good agreement between these patterns and the analogous patterns using $AF = 1/2\,h^{-1}$ (Figs. 2a and 3a). Also the covariance and correlation structure follow the general trends previously described and the effect of the negative correlation between α and n can be visually perceived. For example, it is worth noting that the lowest convergent value of n in Fig. 7b, obtained with the parameter initializations S1 and S3, corresponds to the highest convergent value of α.

In summary, all these outcomes illustrate the performance of the proposed approach in terms of parameter retrieval, in a real scenario, characterized by a limited amount of assimilated observations and by observation errors embedded as part of the experimental data. In general, by using both OD = 1 cm and OD = 2 cm, it is possible to identify efficiently the sets of parameters similar to those obtained by assimilating the entire soil moisture profile. Although, the performance using OD = 2 cm was found markedly better. The retrieval for GB1 was clearly affected by the initial parameterization; instead the response of GB3 was practically insensitive to this factor, except for OD = 1 cm.

It is important to note that the physically constrained nature of soil water content, in this study chosen as state variable, precludes the simulation of saturated conditions and entails mathematical shortcomings for the UKF sampling strategy. In fact, the retrieval algorithm can in principle sample parametric solutions involving a "wrong" saturation (i.e. giving place to state values being higher than θ_s) or "wrong" dry conditions (i.e. giving place to state values being smaller than θ_r). Notice that these limitations are also attributable to the ensemble Kalman filter, which also involves parameter sampling around a mean state vector. When these meaningless values are sampled, the algorithm simply changes them

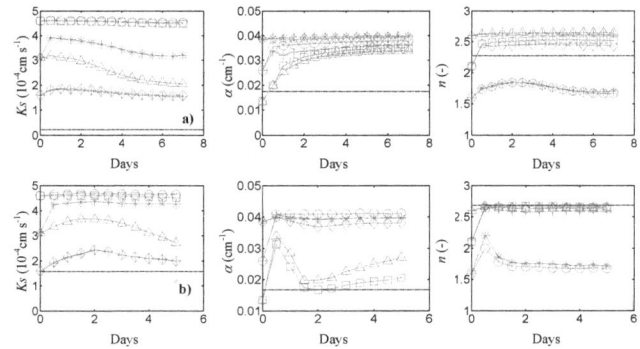

Figure 7. Retrieved VGM parameters K_s, α and n using (**a**) GA3 and (**b**) GB1 experimental data, with assimilation frequency $AF = 1/12\,h^{-1}$ and observation depth OD = 1 cm. Comparisons account for the six pondered sets of initial parameters S1($\bigcirc$), S2($\square$), S3($*$), S4($\triangle$), S5($+$) and S6($\lozenge$). The dotted line indicates the value of the parameter found by Romano and Santini (1999).

to keep the state vector solutions within the valid range. Aimed to provide a general solution circumventing this issue, several alternative strategies have been unsuccessfully pondered. One of them was to use adaptive coefficients, scaling the sigma point distribution in the unscented approach (see Eq. 17), in order to shrink the deterministic sampling of the parameter around the mean. However, this demands a high computational cost, and provides temporarily biased retrieved parameters, affecting also tracking and convergence. It was also pondered the use of a "temporarily adaptive" θ_s (or in principle θ_r), i.e. making θ_s as the maximum soil water content value whenever at least one state value exceeds the adopted actual value. However, this gives place to an irreversible state biasing. Even including θ_s as an additional unknown parameter to be retrieved would not avoid this issue, but would rather make it more evident.

5.2   State retrieval

Figures 8 and 9 depict the retrieved states after 1, 4 and 7 days for GA3 and 1, 3 and 5 days for GB1, by assimilating observations every 2 h, within the three observation depths examined (1, 2 and 12 cm). The results show that the dual filter algorithm is generally able to retrieve the true state profiles with a relatively low dependence from the identified parametric array. For soil core GA3 and using OD = 1 cm, the differences between retrieved and measured soil moistures are still large after 7 days (Fig. 8c). Nevertheless, the variability between the six involved simulations is barely noticeable, indicating that the initial parameterization has a low weight at this stage of the assimilation process. Using OD = 2 cm, a very good match between retrieved and measured profiles is found already at the fourth day (Fig. 8e).

In the case of the GB1 experiment, the retrieval process is forced to deal with a relatively larger soil heterogeneity,

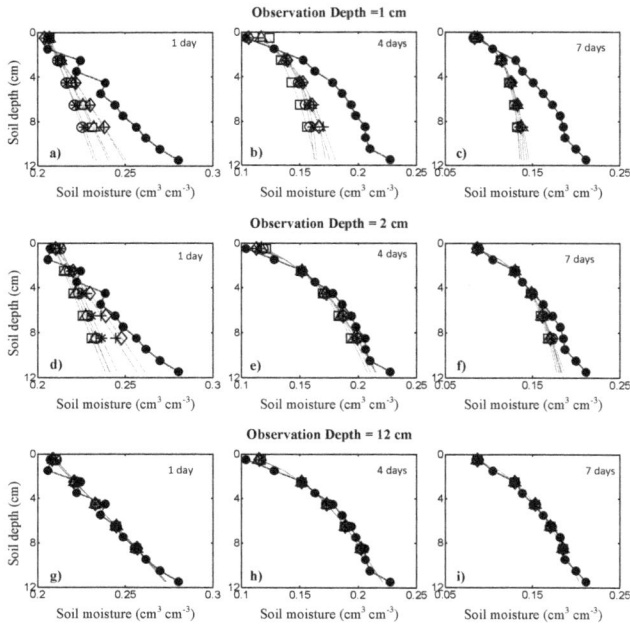

Figure 8. Retrieved soil moisture profiles (solid lines) using the GA3 data series after 1 day (**a, d, g**), 4 days (**b, e, h**) and 7 days (**c, f, i**), with assimilation frequency AF = 1/2 h^{-1} and observation depths: (**a–c**) OD = 1 cm, (**d–f**) OD = 2 cm, and (**g–i**) OD = 12 cm. Comparisons account for the six pondered sets of initial parameters S1(○), S2(□), S3(∗), S4(△), S5(+) and S6(◇). The dotted line with solid circles represents the measured profile.

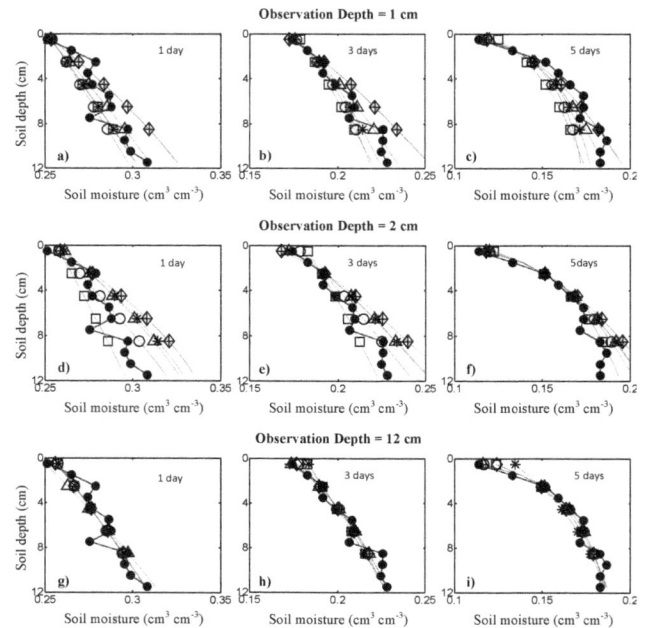

Figure 9. Retrieved soil moisture profiles (solid lines) using the GB1 data series after 1 day (**a, d, g**), 4 days (**b, e, h**) and 7 days (**c, f, i**), with assimilation frequency AF = 1/2 h^{-1} and observation depths: (**a–c**) OD = 1 cm, (**d–f**) OD = 2 cm, and (**g–i**) OD = 12 cm. Comparisons account for the six pondered sets of initial parameters S1(○), S2(□), S3(∗), S4(△), S5(+) and S6(◇). The dotted line with solid circles represents the measured profile.

as shown by the irregular scatter of the soil water content data points along the vertical direction (Fig. 9). This vertical variability makes the retrieval process more sensitive to the parameter initialization, allowing for a wider spectrum of probable soil moisture profiles. However, this spectrum of probable soil moisture profiles well represents the soil moisture "anomalies" and the differences between the retrieved soil moisture profiles on the fifth day is very small.

The comparison between the retrieval performances achieved with the observation depths OD = 1 cm and OD = 2 cm provides contrasting results for the two experiments. For soil core GA3, the differences between the soil water content profiles retrieved with OD = 1 cm and OD = 2 cm are still significant on the fourth and seventh day, whereas for soil core GB1 the profiles are similar at both the third and the fifth day. This feature can be more clearly appreciated by visual inspection of Fig. 10 showing the temporal evolution of the ME and RMSE indices for both these experiments. As the initial soil water content is close to the profile mean value, the errors at the beginning of the simulation are relatively small. Nonetheless, it is important to point out that the evolving pattern of these statistics, as the overall performance of the retrieval algorithm, is scarcely affected by the initial soil water content.

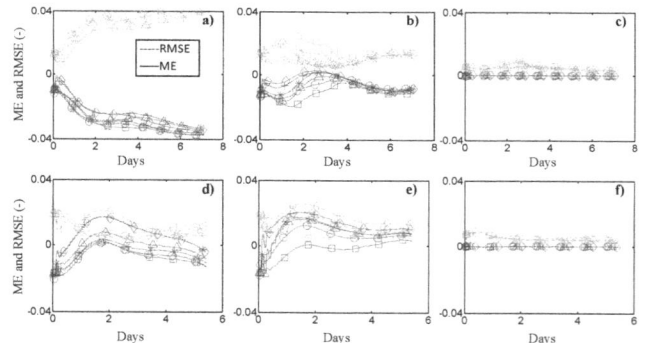

Figure 10. Evolving mean errors (MEs) and root mean square errors (RMSEs) between predicted and measured profiles using GA3 (**a–c**) and GB1 (**d–f**) experimental data series, with assimilation frequency A F = 1/2 h^{-1}, and observation depths: (**a, d**) OD = 1 cm, (**b, e**) OD = 2 cm, and (**c, f**) OD = 12 cm. The analysis accounts for the six pondered sets of initial parameters S1(○), S2(□), S3(∗), S4(△), S5(+) and S6(◇).

Figure 10a shows that relatively high errors occur for GA3 with OD = 1 cm, with increasing RMSE and ME (in absolute terms) values up to the second day and after the fourth day. A main cause of this relatively poor performance of the approach using GA3 with OD = 1 cm is that water fluxes during the last stage of this experiment are very small (see Fig. 1),

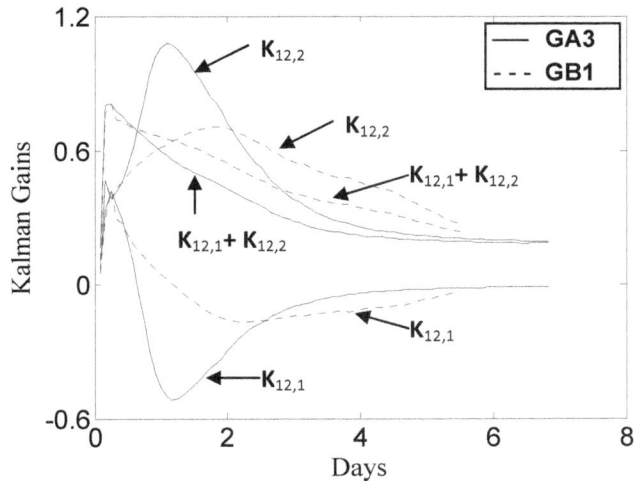

Figure 11. Evolving Kalman gain coefficients $K_{12,1}$, $K_{12,2}$ and the sum $K_{12,1} + K_{12,2}$ for GA3 and GB1 using observation depths of 1cm and the parameter initialization S6. $K_{12,1}$ and $K_{12,2}$ describe the influence of the first and second observation nodes, respectively, on node 12 (the bottom one).

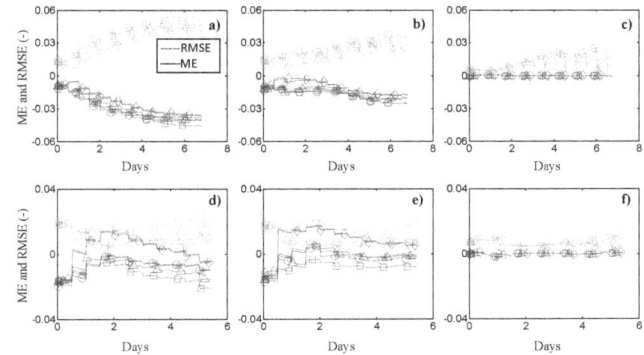

Figure 12. Evolving mean errors (MEs) and root mean square errors (RMSEs) between predicted and measured profiles using GA3 (**a–c**) and GB1 (**d–f**) experimental data series, with assimilation frequency AF $= 1/12\,\mathrm{h}^{-1}$, and observation depths: (**a, d**) OD $= 1$ cm, (**b, e**) OD $= 2$ cm, and (**c, f**) OD $= 12$ cm. The analysis accounts for the six pondered sets of initial parameters S1($\bigcirc$), S2 ($\square$), S3 ($*$), S4 ($\triangle$), S5 ($+$) and S6 ($\lozenge$).

which implies that the soil water content gradients are very small at the surface and, in turn, poor information is provided by the very top observation nodes to the lower nodes, about the ongoing process. For the same reasons, a small increase of RMSE and ME is also noticeable after the fourth day with OD $= 2$ cm.

To see how these physical constraints impact on the dynamics of the states retrieval process, Fig. 11 shows the evolving Kalman gain coefficients $K_{12,1}$, $K_{12,2}$ and $K_{12,1} + K_{12,2}$ for both data series using OD $= 1$ cm and parameter initialization S6 (the parameter initialization has a limited impact on this aspect). Provided that the Kalman gain for OD $= 1$ cm is a matrix of 12 rows (equal to the nodes of the soil profile) and 2 columns (equal to the observed nodes), $K_{12,1}$ and $K_{12,2}$ describe how the value retrieved at the twelfth node is influenced by the observations assimilated from the first and second nodes, respectively.

Usually, a value of $K_{i,j}$ close to one indicates a high effect of the assimilated node j on the retrieved node i, whilst a value close to zero indicates a neglecting effect. Given the correspondence between states and observations, then involving an observation operator **H** with ones in the diagonal elements and zero in the off-diagonal elements, a negative value of $K_{i,j}$ manifests the negative cross-covariance between the states i and j. The sum $K_{12,1} + K_{12,2}$ gives an idea of the combined effect of the two observation nodes, provided that the differences between model predictions and observations are similar.

$K_{12,1}$ for GA3 is almost zero after approximately 4 days. The sum $K_{12,1} + K_{12,2}$ after this moment is constant and essentially equal to $K_{12,2}$. In other words, the assimilation algorithm is still acting on the bottom node, but just using

the information provided by the second node, while the top node is barely contributing to the retrieved value. The term $K_{12,1} + K_{12,2}$ obtained for GB1 is higher than that obtained for GA1 practically during the entire assimilation period. In the case of GB1, the assimilation of the top node, by means of the $K_{12,1}$ coefficient, slightly influences the retrieval value of the bottom node, almost till the end of the experiment.

The negative values of $K_{12,1}$ seem to be favoured by the adopted experimental settings. At the beginning of the experiment, the guess states transits from a constant matric potential profile to a distribution similar to a constant head profile; hence, the bottom node moves to a wetter range, while the evaporative process drives an opposite trend on the top node. The zero flux condition at the bottom profile favours the relatively large duration of this phase of the process. During these stages the correlation between the first and the bottom nodes use to be markedly negative. According to Fig. 11 this phase lasts approximately 1 and 2 days for GA3 and GB1, respectively. When this trend changes, with the bottom nodes also moving to a drier range (although much more slowly than the top node), the estimated $K_{12,1}$ increases (i.e. it gets less negative values).

Notice that GB1 Kalman gain coefficients tend to converge toward those of the GA3 experiment after the fourth day. In fact, as it occurs for GA3, the ME and RMSE values for GB1 using OD $= 1$ cm also slightly increase at the end of the experiment (Fig. 10d). More precisely, the GB1 flux at its last stage is about 1 mm day^{-1}, and it is similar to that found for GA3 from about the fourth day (see Fig. 1).

Thus, the dual filter approach performs well in all cases, except for GA3 using OD $= 1$ cm, where the state retrieval process is relatively slow mainly due to the characteristics of the experiment, beside other factors such as the diminishing state covariance as the experiment advances, the narrow

Table 4. Mean error (ME) and root mean square error (RMSE) between predicted and measured soil moisture profiles for GA3 and GB1 data series at the end of the evaporation tests. Values account for the six sets of initial parameters (S1–S6), two observation depths, (OD = 1 and OD = 2 cm) and two assimilation frequencies (AF = $1/2\,h^{-1}$ and AF = $1/12\,h^{-1}$).

| | | GA3 | | | | GB1 | | | |
| | | OD = 1 cm | | OD = 2 cm | | OD = 1 cm | | OD = 2 cm | |
AF	Initial set	ME	RMSE	ME	RMSE	ME	RMSE	ME	RMSE
$1/2\,h^{-1}$	S1	−0.0334	0.0393	−0.0137	0.0193	−0.0129	0.0145	0.0061	0.0080
	S2	−0.0344	0.0405	−0.0151	0.0209	−0.0126	0.0144	0.0032	0.0046
	S3	−0.0310	0.0366	−0.0123	0.0176	−0.0106	0.0118	0.0074	0.0096
	S4	−0.0319	0.0376	−0.0132	0.0188	−0.0083	0.0096	0.0076	0.0099
	S5	−0.0307	0.0362	−0.0107	0.0157	−0.0041	0.0063	0.0101	0.0132
	S6	−0.0311	0.0367	−0.0118	0.0170	−0.0042	0.0064	0.0100	0.0131
$1/12\,h^{-1}$	S1	−0.0427	0.0491	−0.0213	0.0275	−0.0096	0.0118	−0.0028	0.0073
	S2	−0.0402	0.0461	−0.0219	0.0281	−0.0211	0.0278	−0.0079	0.0201
	S3	−0.0334	0.0382	−0.0158	0.0210	−0.0101	0.0124	−0.0009	0.0056
	S4	−0.0504	0.0588	−0.0250	0.0342	−0.0155	0.0201	−0.0017	0.0112
	S5	−0.0501	0.0589	−0.0259	0.0360	−0.0046	0.0088	0.0056	0.0090
	S6	−0.0283	0.0325	−0.0148	0.0200	−0.0053	0.0095	0.0053	0.0089

range of the soil moisture content values covered by the top node, which also implies small differences between model predictions and observations. As a demonstration of the relative efficiency of the proposed approach, ME and MAE values decrease when the analysis of the errors obtained for GA3 using OD = 1 cm is limited to the top five nodes.

Analogously to Fig. 10, Fig. 12 depicts the temporal evolution of the ME and RMSE values, but using AF = $1/12\,h^{-1}$. The reduced assimilation frequency gives mainly place to an increase of the ME and RMSE for GA3, while a higher dependence of these statistics from the initial parameter sets for GB1. The error temporal patterns evidence that the profile retrieval process follows trends similar to those observed for the higher resolution, but with a slower convergence rate. It is interesting to observe that when assimilating the entire profile for GA3, the algorithm predicts the correct average soil water content (as the ME is almost null), but the retrieved profiles present deviations from the observed values along the soil column, as testified by the increasing RMSE. Unfortunately, the analysis is limited in time by the short duration of the experimental data series.

Table 4 summarizes the ME and RMSE values computed for OD = 1 cm and 2 cm and AF = $1/2\,h^{-1}$ and $1/12\,h^{-1}$ at the end of the each simulation. The approach is able to provide good results within the limited time conceded by the duration of the experiments. The average ME with OD = 1 cm is larger than that obtained with OD = 2 cm, by 2.25 times for GA3 and by 1.3 times for GB1. Similar results occur for the average RMSE values. Instead, AF = $1/12\,h^{-1}$ produces higher errors than AF = $1/2\,h^{-1}$ by 1.45 times for GA3, while by 1.4 times for GB1 with OD=1 cm and 0.34 times for GB1 with OD = 2 cm. The results are generally more sensible to the observation depths than to the assimilation frequency, due to the small state vertical gradients, particularly for GA3.

6 Conclusions

This study has shown the potential of the dual Kalman filter approach in retrieving both parameters and states simultaneously. The performances of the proposed approach have been evaluated with data obtained from evaporation experiments carried out in the laboratory on two different soil cores.

The dual Kalman filter approach is based on a standard Kalman filter for retrieving state values and an unscented Kalman filter for retrieving the parameters of the soil hydraulic property functions. The approach adopts a linearized numerical scheme of the θ-based form of the Richards equation, based on the Crank–Nicolson finite differences, granting the linearity of both the state and the observation equations and thus enabling a direct optimal retrieval of the first and second moment of the states with a standard Kalman filter.

By assimilating soil moisture observations up to soil depths of 1 and 2 cm, the approach allows properly identifying a set of parameters that is in a very good agreement with that one obtained by assimilating the entire observed profiles. The retrieved parameters are also in a reasonably good agreement with the parameters found by Romano and Santini (1999), particularly for K_s and n in the case of the GB1 experiment. The retrieved parameter α is larger than that estimated by Romano and Santini (1999), as result of the different type of information employed for estimating the parameters.

The method also shows a good performance in terms of state retrieval and proved to be able to deal even with the somewhat larger heterogeneities of soil core GB1. The prediction performance proved to be more sensitive to the observation depths than to the assimilation frequency, when changing the observation depths from 1 to 2 cm, while the assimilation frequency from $1/2\,\mathrm{h}^{-1}$ to $1/12\,\mathrm{h}^{-1}$.

It is important to note the marked flexibility and stability of the approach, independently from the errors associated with the initial states and parameter sets. Further work is needed to investigate whether this approach is also able to cope with two- or three-dimensional problems of soil water flux efficiently, by undertaking more complex assimilation processes.

Acknowledgements. The laboratory evaporation tests have been conducted in the Soil Hydrology Laboratory of the Department of Agricultural Engineering – University of Napoli Federico II, under the supervision of prof. Alessandro Santini. This study has been supported by P. O. N. project AQUATEC – New technologies of control, treatment, and maintenance for the solution of water emergency. Hanoi Medina has been also supported by the Abdus Salam International Centre for Theoretical Physics (ICTP), where he has been appointed as Junior Associate.

Edited by: N. Verhoest

References

Camporese, M., Paniconi, C., Putti, M., and Salandin, P.: Ensemble Kalman filter data assimilation for a process based catchment scale model of surface and subsurface flow, Water Resour. Res., 45, W10421, doi:10.1029/2008WR007031, 2009.

Carsel, R. F. and Parrish, R. S.: Developing joint probability distributions of soil water retention characteristics, Water Resour. Res., 24, 755–769, doi:10.1029/WR024i005p00755, 1988.

Chirico, G. B., Medina, H., and Romano, N.: Functional evaluation of PTF prediction uncertainty: An application at hillslope scale, Geoderma, 155, 193–202, 2010.

Chirico, G. B., Medina, H., and Romano, N.: Kalman filters for assimilating near-surface observations into the Richards equation – Part 1: Retrieving state profiles with linear and nonlinear numerical schemes, Hydrol. Earth Syst. Sci., 18, 2503–2520, doi:10.5194/hess-18-2503-2014, 2014.

De Lannoy, G. J. M., Houser, P. R., Pauwels, V. R. N., and Verhoest, N. E. C.: State and bias estimation for soil moisture profiles by an ensemble Kalman filter: Effect of assimilation depth and frequency, Water Resour. Res., 43, W06401, doi:10.1029/2006WR005100, 2007a.

De Lannoy, G. J. M., Reichle, R. H., Houser, P. R., Pauwels, V. R. N., and Verhoest, N. E. C.: Correcting for forecast bias in soil moisture assimilation with the ensemble Kalman filter, Water Resour. Res., 43, W09410, doi:10.1029/2006WR005449, 2007b.

Dunne, S. and Entekhabi, D.: An ensemble-based reanalysis approach to land data assimilation, Water Resour. Res., 41, W02013, doi:10.1029/2004WR003449, 2005.

Entekhabi, D., Nakamura, H., and Njoku, E. G.: Solving the inverse problem for soil moisture and temperature profiles by sequential assimilation of multifrequency remote sensed observations, IEEE Trans. Geosci. Remote Sens., 32, 438–448, 1994.

Escorihuela, M. J., Chanzy, A., Wigneron, J. P., and Kerr, Y. H.: Effective soil moisture sampling depth of L-band radiometry: A case study, Remote Sense Environ., 114, 995–1001, 2010.

Haverkamp, R. M., Vauclin, M., Tourna, J., Wierenga, P. J., and Vachaud, G.: A comparison of numerical simulation models for one-dimensional infiltration, Soil Sci. Soc. Am. J., 41, 285–294, 1977.

Hoeben, R. and Troch P. A.: Assimilation of active microwave observation data for soil moisture profile estimation, Water Resour. Res., 36, 2805–2819, doi:10.1029/2000WR900100, 2000.

Julier, S. J. and Uhlmann, J. K. A New Extension of the Kalman Filter to Nonlinear Systems, in: Proc. SPIE – Int. Soc. Opt. Eng. (USA), 3068, 182–193, 1997.

Julier, S. J. and Uhlmann, J. K.: Unscented Filtering and Nonlinear Estimation, Pro. IEEE, 92, 401–422, 2004.

Jury, W. A., Gardner, W. R., and Gardner, W. H.: Soil Physics, 5th Edn., John Wiley, New York, 1991.

Kalman, R. E.: A New Approach to Linear Filtering and Prediction Problems, ASME J. Basic Eng., 82D, 35–45, 1960.

Kavetski, D., Binning, P., and Sloan, S. W.: Noniterative time stepping schemes with adaptive truncation error control for the solution of Richards equation, Water Resour. Res., 38, 1211, doi:10.1029/2001WR000720, 2002.

Liu, Y. and Gupta H. V.: Uncertainty in hydrologic modeling: Toward an integrated data assimilation framework, Water Resour. Res., 43, W07401, doi:10.1029/2006WR005756, 2007.

Lü, H., Yu, Z., Horton, R., Zhu, Y., Wang, Z., Hao, Z. and Xiang, L.: Multi-scale assimilation of root zone soil water predictions, Hydrol. Process., 25, 3158–3172, doi:10.1002/hyp.8034, 2011.

McLaughlin, D. B.: An integrated approach to hydrologic data assimilation: Interpolation, smoothing, and filtering, Adv. Water Resour., 25, 1275–1286, 2002.

Medina, H., Romano, N., and Chirico, G. B.: Kalman filters for assimilating near-surface observations into the Richards equation – Part 2: A dual filter approach for simultaneous retrieval of states and parameters, Hydrol. Earth Syst. Sci., 18, 2521–2541, doi:10.5194/hess-18-2521-2014, 2014.

Montzka, C., Moradkhani, H., Weihermuller, L., Hendricks-Franssen, H. J., Canty, M., and Vereecken, H.: Hydraulic parameter estimation by remotely-sensed top soil moisture observations with the particle filter, J. Hydrol., 399, 410–421, 2011.

Moradkhani, H., Sorooshian, S., Gupta, H. V., and Houser, P.: Dual state–parameter estimation of hydrologic models using ensemble Kalman filter, Adv. Water Resour., 28, 135–147, 2005a.

Moradkhani, H., Hsu, K. L., Gupta, H. V., and Sorooshian, S.: Uncertainty assessment of hydrologic model states and parameters: Sequential data assimilation using the particle filter, Water Resour. Res., 41, W05012, doi:10.1029/2004WR003604, 2005b.

Paniconi, C., Aldama, A. A., and Wood, E. F.: Numerical evaluation of iterative and noniterative methods for the solution of the nonlinear Richards equation, Water Resour. Res., 27, 1147–1163, 1991.

Plaza, D. A., De Keyser, R., De Lannoy, G. J. M., Giustarini, L., Matgen, P., and Pauwels, V. R. N.: The importance of parameter resampling for soil moisture data assimilation into hydrologic models using the particle filter, Hydrol. Earth Syst. Sci., 16, 375-390, doi:10.5194/hess-16-375-2012, 2012.

Pringle, M. J., Romano, N., Minasny, B., Chirico, G. B., and Lark, R. M.: Spatial evaluation of pedotransfer functions using wavelet analysis, J. Hydrol., 333, 182–198, 2007.

Qin, J., Liang, S. L., Yang, K., Kaihotsu, I., Liu, R. G., and Koike, T.: Simultaneous estimation of both soil moisture and model parameters using particle filtering method through the assimilation of microwave signal, J. Geophys. Res. Atmos., 114, D15103, doi:10.1029/2008JD011358, 2009.

Reichle, R. H. and Koster, R. D.: Assessing the impact of horizontal error correlations in background fields on soil moisture estimation, J. Hydrometeorol., 4, 1229–1242, 2003.

Ritter, A., Muñoz-Carpena, R., Regalado, C. M., Vanclooster, M., and Lambot, S.: Analysis of alternative measurement strategies for the inverse optimization of the hydraulic properties of a volcanic soil, J. Hydrol., 295, 124–139, 2004.

Romano, N. and Santini, A.: Determining soil hydraulic functions from evaporation experiments by a parameter estimation approach: Experimental verifications and numerical studies, Water Resour. Res., 35, 3343–3359, 1999.

Scharnagl, B., Vrugt, J. A., Vereecken, H., and Herbst, M.: Inverse modelling of in situ soil water dynamics: investigating the effect of different prior distributions of the soil hydraulic parameters, Hydrol. Earth Syst. Sci., 15, 3043–3059, doi:10.5194/hess-15-3043-2011, 2011.

Šimůnek, J. and van Genuchten, M. T.: Estimating Unsaturated Soil Hydraulic Properties from Tension Disc Infiltrometer Data by Numerical Inversion, Water Resour. Res., 32,, 2683–2696, doi:10.1029/96WR01525, 1996.

Thiemann, M., Trosset, M., Gupta, H. V., and Sorooshian, S.: Bayesian recursive parameter estimation for hydrologic models, Water Resour. Res., 37, 2521–35, 2001.

Todini, E.: Mutually interactive state/parameter estimation (MISP), in Application of Kalman Filter to Hydrology, Hydraulics and Water Resources, edited by: Chiu, C.-L., University of Pittsburgh, Pittsburgh, Pa., 1978.

van der Merwe, R.: Sigma-Point Kalman Filters for Probabilistic Inference in Dynamic State-Space Models, PhD dissertation, University of Washington, USA, 2004.

van Genuchten, M. Th.: A closed-form equation for predicting the hydraulic conductivity of unsaturated soils, Soil Sci. Soc. Am. J., 44, 892–898, 1980.

Vereecken, H., Huisman, J. A., Bogena, H., Vanderborght, J., Vrugt, J. A., and Hopmans, J. W.: On the value of soil moisture measurements in vadose zone hydrology: A review, Water Resour. Res., 44, W00D06, doi:10.1029/2008WR006829, 2008.

Vrugt, J. A., Bouten, W., and Weerts, A. H.: Information content of data for identifying soil hydraulic parameters from outflow experiments, Soil Sci. Soc. Am. J., 65, 19–27, 2001.

Vrugt, J. A., Bouten, W., Gupta, H. V., and Sorooshian, S.: Toward improved identifiability of hydrologic model parameters: the information content of experimental data, Water Resour. Res., 38, 1312, doi:10.1029/2001WR001118, 2002. 2044, 2045, 2002.

Vrugt, J. A., Diks, C. G. H., Gupta, H. V., Bouten, W., and Verstraten, J. M.: Improved treatment of uncertainty in hydrologic modeling: Combining the strengths of global optimization and data assimilation, Water Resour. Res., 41, W01017, doi:10.1029/2004WR003059, 2005.

Vrugt, J. A., ter Braak, C. J. F., Diks, C. G. H., and Schoups, G.: Hydrologic data assimilation using particle Markov chain Monte Carlo simulation: Theory, concepts and applications, Adv. Water Resour., 51, 457-478, doi:10.1016/j.advwatres.2012.04.002, 2013.

Walker, J. P., Willgoose G. R., and Kalma, J. D.: One-dimensional soil moisture profile retrieval by assimilation of near-surface observations: a comparison of retrieval algorithms, Adv. Water Resour., 24, 631–650, 2001.

Walker, J. P., Houser, P. R., and Willgoose, G. R.: Active microwave remote sensing for soil moisture measurement: a field evaluation using ERS-2, Hydrol. Process., 18, 1975–1997, 2004.

Wöhling, Th. and Vrugt, J. A.: Multi-response multi-layer vadose zone model calibration using Markov chain Monte Carlo simulation and field water retention data, Water Resour. Res., 47, W04510, doi:10.1029/2010WR009265, 2011.

Yang, K., Koike, T., Kaihotsu, I., and Qin, J.: Validation of a dual-pass microwave land data assimilation system for estimating surface soil moisture in semiarid regions, J. Hydrometeorol., 10, 780–793, 2009.

9

Multiple causes of nonstationarity in the Weihe annual low-flow series

Bin Xiong[1], Lihua Xiong[1], Jie Chen[1], Chong-Yu Xu[1,2], and Lingqi Li[1]

[1]State Key Laboratory of Water Resources and Hydropower Engineering Science, Wuhan University, Wuhan 430072, P.R. China
[2]Department of Geosciences, University of Oslo, P.O. Box 1022 Blindern, 0315 Oslo, Norway

Correspondence: Lihua Xiong (xionglh@whu.edu.cn)

Abstract. Under the background of global climate change and local anthropogenic activities, multiple driving forces have introduced various nonstationary components into low-flow series. This has led to a high demand on low-flow frequency analysis that considers nonstationary conditions for modeling. In this study, through a nonstationary frequency analysis framework with the generalized linear model (GLM) to consider time-varying distribution parameters, the multiple explanatory variables were incorporated to explain the variation in low-flow distribution parameters. These variables are comprised of the three indices of human activities (HAs; i.e., population, POP; irrigation area, IAR; and gross domestic product, GDP) and the eight measuring indices of the climate and catchment conditions (i.e., total precipitation P, mean frequency of precipitation events λ, temperature T, potential evapotranspiration (EP), climate aridity index AI_{EP}, base-flow index (BFI), recession constant K and the recession-related aridity index AI_K). This framework was applied to model the annual minimum flow series of both Huaxian and Xianyang gauging stations in the Weihe River, China (also known as the Wei He River). The results from stepwise regression for the optimal explanatory variables show that the variables related to irrigation, recession, temperature and precipitation play an important role in modeling. Specifically, analysis of annual minimum 30-day flow in Huaxian shows that the nonstationary distribution model with any one of all explanatory variables is better than the one without explanatory variables, the nonstationary gamma distribution model with four optimal variables is the best model and AI_K is of the highest relative importance among these four variables, followed by IAR, BFI and AI_{EP}. We conclude that the incorporation of multiple indices related to low-flow generation permits tracing various driving forces. The established link in nonstationary analysis will be beneficial to analyze future occurrences of low-flow extremes in similar areas.

1 Introduction

Low flow is defined as the flow of water in a stream during prolonged dry weather (WMO, 2009). Yu et al. (2014) quantitatively described a low-flow event as a segment of hydrograph during a period of dry weather with discharge values below a preset (relatively small) threshold. According to WMO (2009), annual minimum flows averaged over several days can be used to measure low flows. During low-flow periods, the magnitude of river flow will greatly restrict its various functions (e.g., providing water supply for production and living, diluting waste water, ensuring navigation, meeting ecological water requirement). Therefore, the investigation of the magnitude and frequency of low flows is of primary importance for engineering design and water resources management (Smakhtin, 2001). In recent years, low flows, as an important part of river flow regime, have been attracting an increasing attention of hydrologists and ecologists in the context of the significant impacts of climate change and human activities (HAs; Bradford and Heinonen, 2008; Du et al., 2015; Kam and Sheffield, 2015; Kormos et al., 2016; Liu et al., 2015; Sadri et al., 2016). In general, under the impact of a changing environment, combinations of multiple factors, such as precipitation change, temperature change, irrigation area (IAR) change and construction of reservoirs, can drive various patterns of streamflow changes (Liu et al., 2017; Tang et al., 2016). Unfortunately, when subjected to a variety of influencing forces, low flow is more vulnerable than high flow or mean flow. Therefore, it is a pretty important issue in hydrology to identify low-flow changes, track multiple driving factors and quantify their contributions from the perspective of hydrological frequency analysis.

In hydrological analysis and design, conventional frequency analysis estimates the statistics of a hydrological time series based on recorded data with the stationary hypothesis which means that this series is "free of trends, shifts or periodicity (cyclicity)" (Salas, 1993). However, global warming and human forces have changed climate and catchment conditions in some regions. Time-varying climate and catchment conditions (TCCCs) can affect all aspects of the flow regime, i.e., changing the frequency and magnitude of floods, altering flow seasonality and modifying the characteristics of low flows. The hypothesis of stationarity has been suspected (Milly et al., 2008). If this problematic method is still used, the frequency analysis may lead to high estimation error in hydrological design. Therefore, considerable literature has introduced the concept of hydrologic nonstationarity into analysis of various hydrological variables, such as annual runoff (Arora, 2002; Jiang et al., 2015a; Xiong et al., 2014; Yang and Yang, 2013), flood (Gilroy and Mccuen, 2012; Kwon et al., 2008; Yan et al., 2017; Zhang et al., 2015), low flow (Du et al., 2015; Jiang et al., 2015b; Liu et al., 2015), precipitation (Gu et al., 2017; Mondal and Mujumdar, 2015; Villarini et al., 2010) and so on. Compared with the literature on annual runoff, floods and precipitation, the literature on the nonstationary analysis of low flow is relatively limited.

Previous hydrological literature on frequency analysis of nonstationary hydrological series mainly focuses on two aspects: development of the nonstationary method and exploration of covariates reflecting changing environments. Strupczewski et al. (2001) presented the method of time-varying moment which assumes that the hydrological variable of interest obeys a certain distribution type, but its moments change over time. The method of time-varying moment was modified to be the method of time-varying parameter values for the distribution representative of hydrologic data (Richard et al., 2002). Villarini et al. (2009) presented this method using the generalized additive models for location, scale and shape parameters (GAMLSS; Rigby and Stasinopoulos, 2005), a flexible framework to assess nonstationary time series. The time-varying parameter method can be extended to the physical covariate analysis by replacing time with any other physical covariates (Jiang et al., 2015b; Kwon et al., 2008; López and Francés, 2013; Liu et al., 2015; Villarini and Strong, 2014). For example, Jiang et al. (2015b) used reservoir index as an explanatory variable based on the time-varying copula method for bivariate frequency analysis

of nonstationary low-flow series in Hanjiang River, China. Du et al. (2015) took precipitation and air temperature as the explanatory variables to explain the inter-annual variability in low flows of the Weihe River, China (also known as the Wei He River). Liu et al. (2015) took the sea surface temperature in the Nino3 region, the Pacific Decadal Oscillation, the sunspot number (3 years ahead), the winter areal temperature and precipitation as the candidate explanatory variables to explain the inter-annual variability in low flows of Yichang station, China. Kam and Sheffield (2015) ascribed the increasing inter-annual variability of low flows over the eastern United States to the North Atlantic Oscillation and Pacific North America.

To our knowledge, compared with the nonstationary flood frequency analysis, the studies on the nonstationary frequency analysis of low-flow series are not very extensive because of incomplete knowledge of low-flow generation (Smakhtin, 2001). Most of these studies explain nonstationarity of low-flow series only by using climatic indicators or a single indicator of human activity. However, the indicators of catchment conditions (e.g., recession rate) related to physical hydrological processes have seldom been attached in nonstationary modeling of low-flow series. This lack of linking with hydrological processes makes it impossible to accurately quantify the contributions of influencing factors for the nonstationarity of low-flow series, and such a scientific demand for tracing the sources of nonstationarity of low-flow series and qualifying their contributions motivated the present study. The knowledge of low-flow generation has been increased by efforts of hydrologists, which can help develop physical covariates to address nonstationarity. Low flows generally originate from groundwater or other delayed outflows (Smakhtin, 2001; Tallaksen, 1995). Their generation relates to both an extended dry weather period (leading to a climatic water deficit) and complex hydrological processes which determine how these deficits propagate through the vegetation, soil and groundwater system to streamflow (WMO, 2009). Thus, not only climate condition drivers (e.g., potential evaporation exceeds precipitation), but also catchment condition drivers (e.g., the faster hydrologic response rate to precipitation) can cause low flows.

The significant factors such as precipitation, temperature, evapotranspiration (EP), streamflow recession, large-scale teleconnections and human forces may play important roles in influencing low-flow generation (Botter et al., 2013; Giuntoli et al., 2013; Gottschalk et al., 2013; Kormos et al., 2016; Sadri et al., 2016). Gottschalk et al. (2013) presented a derived low-flow probability distribution function with climate and catchment characteristics parameters (i.e., the mean length of dry spells λ^{-1} and recession constant of streamflow K) as its distribution parameters. Botter et al. (2013) derived a measurable index (λ^{-1}/K) which can be used for discriminating erratic river flow regimes from persistent river flow regimes. Recently, Van Loon and Laaha (2015) used climate and catchment characteristics (e.g., the duration of dry spells in precipitation and the base-flow index, BFI) to explain the duration and deficit of the hydrological drought event and offered a further understanding of low-flow generation. These studies indicated that climate and catchment conditions play an important role in producing low flows.

The goal of this study is to trace origins of nonstationarity in low flows through developing a nonstationary low-flow frequency analysis framework with the consideration of the time-varying climate and catchment conditions and human activity. In this framework, the climate and catchment conditions are quantified using the eight indices, i.e., meteorological variables (total precipitation P, mean frequency of precipitation events λ, temperature T and potential evapotranspiration), basin storage characteristics (base-flow index, recession constant K) and aridity indexes (climate aridity index AI_{EP}, the recession-related aridity index AI_K). The specific objectives of this study are (1) to find the most important index to explain the nonstationarity of low-flow series, (2) to determine the best subset of TCCCs indices and/or human activity indices (i.e., population, POP; irrigation area; and gross domestic product, GDP) for the final model through the stepwise selection method to identify nonstationary mode of low-flow series and (3) to quantify the contribution of selected explanatory variables to the nonstationarity.

This paper is organized as follows. Section 2 describes the methods. The Weihe River basin and available data sets used in this study are described in Sect. 3, followed by a presentation of the results and discussion in Sect. 4. Section 5 summarizes the main conclusions.

2 Methodology

The flowchart of how to organize the nonstationary low-flow frequency analysis framework is shown in Fig. 1. The whole process is divided into three steps. The first step is the preliminary analysis, including the graphical presentation of both explanatory variables and low-flow series, the statistical test for nonstationarity, and the correlations between each explanatory variable and each low-flow series. The second step is the single covariate analysis for the most important explanatory variable. The third step is the multiple covariate analysis for the optimal combination. We use a low-flow frequency analysis model and stepwise regression method to accomplish the last two steps. In the following subsections, first, the low-flow frequency analysis model is constructed based on the nonstationary probability distributions method, in which distribution parameters serving as response variables can vary as functions of explanatory variables. Second, the distribution types used to build the nonstationary model are outlined. Then, the candidate explanatory variables related to the time-varying climate and catchment conditions and human activity are clarified. Finally, estimation of model parameters and selection of models are illustrated.

2.1 Construction of the low-flow nonstationary frequency analysis model

Generally, a nonstationary frequency analysis model can be established based on the time-varying distribution parameters method (Du et al., 2015; López and Francés, 2013; Liu et al., 2015; Richard et al., 2002; Villarini and Strong, 2014). For the nonstationary probability distribution $f_Y\left(Y_t \mid \boldsymbol{\theta}^t\right)$, let Y_t be a random variable at time t ($t = 1, 2, \ldots, N$) and vector $\boldsymbol{\theta}^t = [\theta_1^t, \theta_2^t, \ldots, \theta_m^t]$ be the time-varying parameters. The number of parameters m in hydrological frequency analysis is generally limited to three or less. The function relationship between the kth parameter θ_k^t and the multiple explanatory variables is expressed as follows:

$$g_k\left(\theta_k^t\right) = h_k\left(x_1^t, x_2^t, \ldots, x_n^t\right), \tag{1}$$

where $x_1^t, x_2^t, \ldots, x_n^t$ are explanatory variables, n is the number of explanatory variables, $g_k(\cdot)$ is the link function which ensures the compliance with restrictions on the sample space and is usually set to natural logarithm for the given negative predictions and $h_k(\cdot)$ is the function for nonstationary modeling. The generalized linear model theory (GLM; Dobson and Barnett, 2012) is used to build function relationships between distribution parameters and their explanatory variables. In GLMs, the response relationship can be generally expressed as

$$g_k\left(\theta_k^t\right) = \alpha_{0k} + \sum_{i=1}^{n} \alpha_{ik} x_i^t, \tag{2}$$

where α_{ik} ($i = 0, 1, 2, \ldots, n, k = 1, \ldots, m$) are the GLM parameters.

In order to compare the nonstationary models constructed by various combinations of explanatory variables, Eq. (2) is modified in this study using the dimensionless method for the standard GLM parameters. The value of θ_k^t could be assumed to be equal to its mean ($\overline{\theta}_k$) when all explanatory variables are equal to their mean ($\overline{x}_i$), i.e.,

$$\theta_k^t\left(x_1^t = \overline{x}_1, x_2^t = \overline{x}_2, \ldots, x_n^t = \overline{x}_n\right) = \overline{\theta}_k. \tag{3}$$

Equation (2) is then modified as

$$
\begin{aligned}
g_k\left(\frac{\theta_k^t}{\overline{\theta}_k}\right) &= \beta_{0k} + \sum_{i=1}^{i=n} \beta_{ik} z_i^t \\
z_i^t &= \frac{x_i^t - \overline{x}_i}{s_i}, i = 1, 2, \ldots, n \\
\beta_{0k} &= g_k\left(\frac{\theta_k^t}{\overline{\theta}_k} \mid \theta_k^t = \overline{\theta}_k\right) = g_k(1),
\end{aligned}
\tag{4}
$$

where z_i^t is the normalized explanatory variable, s_i is the standard deviation of x_i^t and β_{ik} ($i = 1, 2, \ldots, n, k = 1, \ldots, m$) are the standard GLM parameters. Letting the link function $g_k(\cdot)$ be the natural logarithmic function $\ln(\cdot)$ and θ_k^t be the distribution parameter in $[\theta_1^t, \theta_2^t, \ldots, \theta_m^t]$ with the most significant change, the degree of nonstationarity in low-flow series

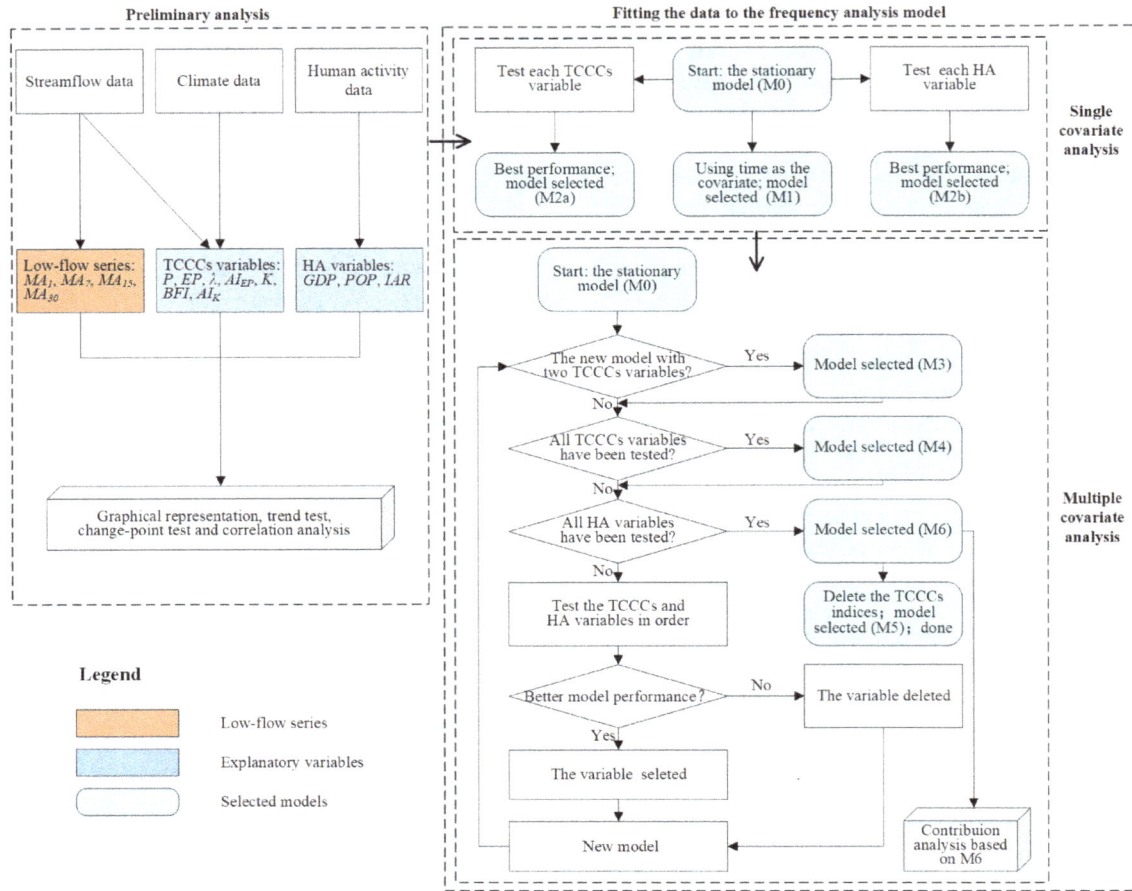

Figure 1. The framework of nonstationary low-flow frequency analysis.

can be defined as $\ln(\theta_l^t) - \ln(\overline{\theta}_l)$. Then, the contribution c_i^t of each explanatory variable x_i^t to $\ln(\theta_l^t) - \ln(\overline{\theta}_l)$ could be defined as

$$c_i^t = \beta_{il} \frac{x_i^t - \overline{x}_i}{s_i}. \tag{5}$$

2.2 Candidate distribution functions

We need to select the form of probability distribution $f_Y(\cdot)$ to determine what type of nonstationary frequency curves will be produced. Various probability distributions have been compared or suggested in modeling of low-flow series (Du et al., 2015; Hewa et al., 2007; Liu et al., 2015; Matalas, 1963; Smakhtin, 2001). An extensive overview of distribution functions for low flow is given in Tallaksen et al. (2004). Following these recommendations, we consider five distributions, i.e., Pearson type III (PIII), gamma (GA), Weibull (WEI), lognormal (LOGNO) and generalized extreme value (GEV) as candidates in this study (Table 1). In the case of Pearson type III distribution, considering that the parameter θ_3 of Pearson type III as lower bound should approach zero

and the parameter θ_3 of GEV is quite sensitive and difficult to be estimated, we assume them to be constant in this study.

2.3 Candidate explanatory variables

We look for variables $x_1^t, x_2^t, \ldots, x_n^t$ that can explain parts of the variations in distribution parameters θ^t. From the perspective of low-flow generation, the dependency between low-flow regime and both climate and catchment conditions has been presented by previous studies (Botter et al., 2013; Gottschalk et al., 2013; Van Loon and Laaha, 2015). We focus on eight measuring indices: precipitation, mean frequency of precipitation events, temperature, potential evapotranspiration, climate aridity index, base-flow index, recession constant and recession-related aridity index. These indices were chosen to incorporate time-varying climate and catchment conditions in nonstationary modeling of low-flow frequency and serve as candidate explanatory variables. Climate variables (i.e., precipitation, mean frequency of precipitation events, temperature, potential evapotranspiration and climate aridity index) are related to both water supply source and water loss and are therefore selected as candi-

Table 1. The probability density functions and moments (the mean and variance) for the candidate distributions in this study.

Distributions	Probability density function	Distribution moments
Pearson type III (PIII)	$f_Y(y \mid \theta_1, \theta_2, \theta_3) = \dfrac{(y-\theta_3)^{1/\theta_2^2 - 1}}{\Gamma(1/\theta_2^2)(\theta_1\theta_2^2)^{1/\theta_2^2}} \exp\left(-\dfrac{y-\theta_3}{\theta_1\theta_2^2}\right)$ $y > \theta_3, \theta_3 > 0, \theta_1 > 0, \theta_2 > 0$	$E[Y] = \theta_1 + \theta_3$ $\mathrm{Var}[Y] = \theta_1^2\theta_2^2$
Gamma (GA)	$f_Y(y \mid \theta_1, \theta_2) = \dfrac{(y)^{1/\theta_2^2 - 1}}{\Gamma(1/\theta_2^2)(\theta_1\theta_2^2)^{1/\theta_2^2}} \exp\left(-\dfrac{y}{\theta_1\theta_2^2}\right)$ $y > 0, \theta_1 > 0, \theta_2 > 0$	$E[Y] = \theta_1$ $\mathrm{Var}[Y] = \theta_1^2\theta_2^2$
Weibull (WEI)	$f_Y(y \mid \theta_1, \theta_2) = \left(\dfrac{\theta_2}{\theta_1}\right)\left(\dfrac{y}{\theta_1}\right)^{\theta_2 - 1} \exp\left(-\left(\dfrac{y}{\theta_1}\right)^{\theta_2}\right)$ $y > 0, \theta_1 > 0, \theta_2 > 0$	$E[Y] = \theta_1\Gamma(1 + 1/\theta_2)$ $\mathrm{Var}[Y] = \theta_1^2\left[\Gamma\left(1 + \tfrac{2}{\theta_2}\right) - \Gamma^2\left(1 + \tfrac{1}{\theta_2}\right)\right]$
Lognormal (LOGNO)	$f_Y(y \mid \theta_1, \theta_2) = \dfrac{1}{y\theta_2\sqrt{2\pi}} \exp\left\{-\dfrac{[\log(y) - \theta_1]^2}{2\theta_2^2}\right\}$ $y > 0, \theta_2 > 0$	$E[Y] = w^{1/2}e^{\theta_1}$ $\mathrm{Var}[Y] = w(w-1)e^{2\theta_1}$ $w = \exp\left(\theta_2^2\right)$
Generalized extreme value (GEV)	$f_Y(y \mid \theta_1, \theta_2, \theta_3) = \dfrac{1}{\theta_2}\left[1 + \theta_3\left(\dfrac{y-\theta_1}{\theta_2}\right)\right]^{-1/\theta_3 - 1} \exp\left\{-\left[1 + \theta_3\left(\dfrac{y-\theta_1}{\theta_2}\right)\right]^{-1/\theta_3}\right\}$ $-\infty < \theta_1 < \infty, \theta_2 > 0, -\infty < \theta_3 < \infty$	$E[Y] = \theta_1 - \dfrac{\theta_2}{\theta_3} + \dfrac{\theta_2}{\theta_3}\eta_1$ $\mathrm{Var}[Y] = \theta_2^2\left(\eta_2 - \eta_1^2\right)/\theta_3^2$ $\eta_m = \Gamma(1 - m\theta_3)$

Table 2. Description of the developed nonstationary models using time, TCCCs indices and/or HA indices as explanatory variables.

Model codes	Distribution					Description	
	GA	WEI	LOGNO	PIII	GEV	Variable category	The numbers of variables
M0	GA_M0	WEI_M0	LOGNO_M0	PIII_M0	GEV_M0	–	Zero
M1	GA_M1	WEI_M1	LOGNO_M1	PIII_M1	GEV_M1	Time	One
M2a	GA_M2a	WEI_M2a	LOGNO_M2a	PIII_M2a	GEV_M2a	TCCCs	One
M2b	GA_M2b	WEI_M2b	LOGNO_M2b	PIII_M2b	GEV_M2b	HA	One
M3	GA_M3	WEI_M3	LOGNO_M3	PIII_M3	GEV_M3	TCCCs	Two
M4	GA_M4	WEI_M4	LOGNO_M4	PIII_M4	GEV_M4	TCCCs	Identified by the stepwise selection
M5	GA_M5	WEI_M5	LOGNO_M5	PIII_M5	GEV_M5	HA	Identified by the stepwise selection
M6	GA_M6	WEI_M6	LOGNO_M6	PIII_M6	GEV_M6	TCCCs + HA	Identified by the stepwise selection

date variables. It has been shown that the base-flow index and recession constant reflect the storage and release capability of the catchments (Van Loon and Laaha, 2015). The recession-related aridity index reflects both the water supply and storage capability (Botter et al., 2013). In addition to TCCCs indices, the three indices of human activity (irrigation area, population and gross domestic product) are related to water withdrawal loss for agricultural, domestic and industrial purposes and are therefore included. The detailed reasons for selecting all indices are summarized in Table 3. The values of them at each year could be estimated from hydrometeorological data and human activity data. Annual precipitation (P) and temperature (T) are calculated directly by meteorological data. The remaining TCCCs indices need to

be estimated indirectly. Detailed estimation procedures are shown in the following subsections.

2.3.1 Annual mean frequency of precipitation events (λ)

Annual mean frequency of precipitation events is defined as an index to represent the intensity of precipitation recharge to the streamflow:

$$\lambda = \frac{1}{W}\sum_{w=1}^{w=W}\frac{N_w(A)}{t_r}, \tag{6}$$

where $N_w(A)$ is the number of daily rainfall events A (with values more than the threshold 0.5 mm) in wth windows with a length t_r; W is the number of windows.

2.3.2 Annual climate aridity index (AI_{EP})

The ratio of annual potential evaporation to precipitation, commonly known as the climate aridity index, has been used to assess the impacts of climate change on annual runoff (Arora, 2002; Jiang et al., 2015a). The climate aridity index largely reflects the climatic regimes in a region and determines runoff rates (Arora, 2002). Therefore, we choose the annual climate aridity index as a measure of time-varying climate and catchment conditions and estimate its value in a whole region using

$$AI_{EP} = \frac{EP}{P}, \tag{7}$$

where P is annual areal precipitation (mm) and EP is annual areal potential evapotranspiration (mm). The Hargreaves equation (Hargreaves and Samani, 1985) is applied to calculate EP using the R package "Evapotranspiration" (Guo, 2014).

2.3.3 Annual base-flow index (BFI)

The base-flow index (BFI) is defined as the ratio of base flow to total flow. This index has been applied to quantify catchment conditions (e.g., soil, geology and storage-related descriptors) to explain hydrological drought severity (Van Loon and Laaha, 2015). We also choose annual base-flow index as a measure of TCCCs. BFI is estimated using a hydrograph separation procedure in the R package "lfstat" (Koffler and Laaha, 2013).

2.3.4 Annual streamflow recession constant (K)

The recession constant is an important catchment characteristic index measuring the timescale of the hydrological response and reflecting water retention ability in the upstream catchment (Botter et al., 2013). Various estimation methods have been developed to extract recession segments and to parameterize characteristic recession behavior of a catchment (Hall, 1968; Sawaske and Freyberg, 2014; Tallaksen, 1995).

In this study, annual recession analysis (ARA) is performed to obtain the annual streamflow recession constant (K). In ARA, the linearized Dupuit–Boussinesq equation is used to parameterize characteristic recession behavior of a catchment and is written as

$$-\frac{dQ_t}{dt} = \frac{1}{K} Q_t, \tag{8}$$

where Q_t is the value at time t. Equation (8) is investigated by plotting data points $\frac{dQ_t}{dt}$ against Q_t of all extracted recession segments from hydrographs at each year. The criteria of recession segment extraction are based on the Manual on Low-flow Estimation and Prediction (WMO, 2009). Then, the annual recession rate (K^{-1}) is estimated as the slope of the fitted straight line of these data points with the least-squares method. We calculated K using the R package "lfstat" (Koffler and Laaha, 2013).

2.3.5 Annual recession-related aridity index (AI_K)

In this study, the recession-related aridity index is defined as the ratio of recession rate (K^{-1}) to mean precipitation frequency (λ), denoted as

$$AI_K = \frac{K^{-1}}{\lambda}. \tag{9}$$

This ratio plays an important role in controlling the river flow regime (Botter et al., 2013; Gottschalk et al., 2013) and serves as an indicator measuring the recession-related aridity degree of the streamflow in the river channel. For example, the faster recession process or lower precipitation frequency may lead to increased runoff loss or decreased precipitation supply. Consequently, the higher the value AI_K is, the more likely low-flow events occur, and vice versa.

2.4 Parameter estimation

The model parameters including $\overline{\theta}_k(k = 1, 2, \ldots, m)$ and $\beta_{ik}(i = 1, 2, \ldots, n, k = 1, \ldots, m)$ need to be estimated. $\overline{\theta}_k(k = 1, 2, \ldots, m)$ are estimated from outputs of stationary frequency analysis through the maximum likelihood method. We have

$$L\left(\overline{\theta}_1, \overline{\theta}_2, \ldots, \overline{\theta}_m\right) = \sum_{t=1}^{t=N} \ln\left[f_Y\left(y_t | \overline{\theta}_1, \overline{\theta}_2, \ldots, \overline{\theta}_m\right)\right], \tag{10}$$

where y_t is observed low flow at time t and N is the number of samples. The parameters $\beta_{ik}(i = 1, 2, \ldots, n, k = 1, \ldots, m)$ are estimated through the maximum likelihood method to produce nonstationary low-flow frequency curves:

$$L\begin{pmatrix} \beta_{11}, \ldots, \beta_{n1} \\ \ldots \\ \beta_{1m}, \ldots, \beta_{nm} \end{pmatrix} = \sum_{t=1}^{t=N} \ln \tag{11}$$

$$\left\{f_Y\left(y_t | \theta_1^t\left(z_1^t, \ldots, z_n^t | \beta_{11}, \ldots, \beta_{n1}\right), \ldots, \theta_m^t\left(z_1^t, \ldots, z_n^t | \beta_{1m}, \ldots, \beta_{nm}\right)\right)\right\}.$$

The residuals (normalized randomized quintile residuals) are used to test the goodness of fit of fitted model objects (Dunn and Symth, 1996):

$$\hat{r}_t = \Phi^{-1}\left(F_Y\left(y_t | \hat{\theta}^t\right)\right), \tag{12}$$

where $F_Y(\cdot)$ is the cumulative distribution of y_t and $\Phi^{-1}(\cdot)$ is the inverse function of the standard normal distribution. The distribution of the true residuals $\hat{r}_t$ converges to standard normal if the fitted model is correct. A worm plot (Buuren and Fredriks, 2001) is used to check whether $\hat{r}_t$ have a standard normal distribution.

2.5 Model selection

Model selection contains the selection of the type of probability distribution and the selection of the explanatory variables to explain the response variables (i.e., distribution parameters θ_1 and θ_2). In order to obtain the final optimal

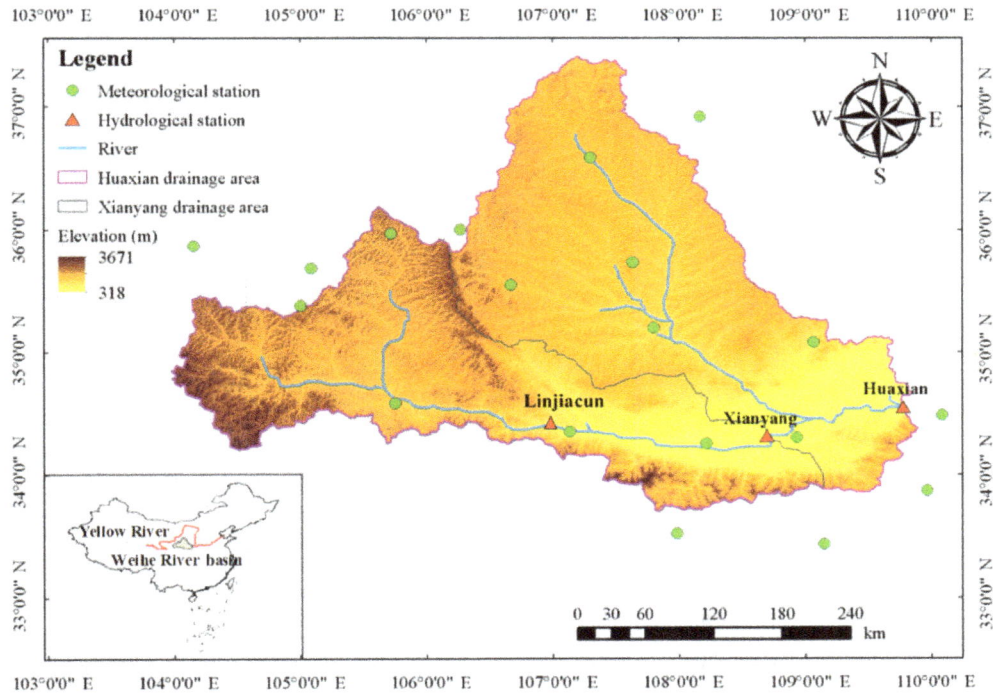

Figure 2. Location, topography, hydro-meteorological stations and river systems of the Weihe River basin.

Table 3. The summary of candidate explanatory variables and reason of selection.

Category	Name	Indices	Reason of selection (related to)	Unit
TCCCs				
	P	Precipitation	Main supply source	mm
	λ	Mean frequency of precipitation events	Water supply intensity	per day
	T	Temperature	Evaporation loss	$\circ$
	EP	Potential evapotranspiration	Evaporation loss	mm
	AI_{EP}	Climate aridity index	Degree of meteorological drought	–
	BFI	Base-flow index	Water storage capability	–
	K	Recession constant	Water storage capability	day
	AI_K	Recession-related aridity index	Both the water storage and supply capability	–
HA				
	IAR	Irrigation area	Both irrigation diversion and evaporation loss	10^6 hm^2
	POP	Population	Water withdrawal loss for agricultural, domestic and industrial purposes	10^6
	GDP	Gross domestic product	Water withdrawal loss for agricultural, domestic and industrial purposes	CNY 10^9

model, the selection of the explanatory variables for θ_1 and θ_2 is conducted by stepwise selection strategies (Stasinopoulos and Rigby, 2007; Venables, 2002): i.e., select a best subset of candidate explanatory variables for θ_1 using a forward approach (which starts with no explanatory variable in the model and tests the addition of each explanatory variable using a chosen model fit criterion); given this subset for θ_1 select another subset for θ_2 (forward). The stepwise selection strategies can get a series of stepwise models with different numbers of explanatory variables, as shown in Fig. 1. In order to detect how the number of explanatory variables influences the performance of the model for describing nonsta-

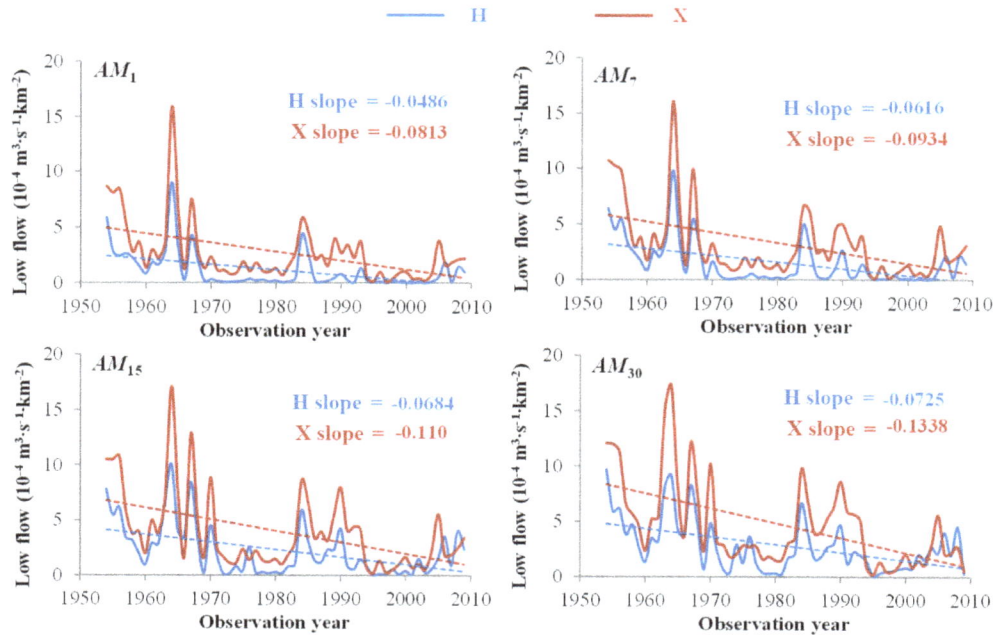

Figure 3. The annual minimum low flows and fitted trend lines in both Huaxian (H) and Xianyang (X) gauging stations.

tionarity, we investigate the eight types of stepwise models as shown in Table 2: the zero-covariate model or stationary model (M0), the time covariate model (M1), the single physical covariate model M2 (single TCCCs covariate model M2a or single HA covariate model M2b), two TCCCs covariates model (M3), the optimal TCCCs covariates model (M4), the optimal HA covariates model (M5) and the final model (M6). The model fit criterion is based on the Akaike's information criterion (AIC; Akaike, 1974) as shown by the following

$$AIC = -2ML + 2df, \qquad (13)$$

where ML is the log-likelihood in Eq. (11) and df is the number of degrees of freedom. The model with the lower AIC value was considered better.

3   Study area and data

3.1   The study area

The Weihe River, located in the southeast of the northwest Loess Plateau, is the largest tributary of the Yellow River, China. The Weihe River has a drainage area of $134\,766\,km^2$, covering the coordinates of $33°42'$–$37°20'$ N, $104°18'$–$110°37'$ E (Fig. 2). This catchment generally has a semi-arid climate, with extensive continental monsoonal influence. Average annual precipitation of the whole area over the period 1954–2009 is about 540 mm and has a wide range (400–1000 mm) in various regions. Under the significant impacts of climate change and human activities in the Weihe River basin in recent decades, the hydrological regime of the

river has changed over time (Du et al., 2015; Jiang et al., 2015a; Xiong et al., 2015).

In the Weihe basin, the impacts of agricultural irrigation on runoff have been found to be significant (Jiang et al., 2015a; Lin et al., 2012). Lin et al. (2012) mentioned that the annual runoff of the Weihe River was significantly affected by irrigation diversion of the Baoji Gorge irrigation district. The irrigated area of the Baoji Gorge irrigation district increased over time since the founding of P.R. China in 1949, and, due to one influential irrigation system project in that area, it became more than twice as large as the original irrigation area since 1971. Jiang et al. (2015a) demonstrated that, in the Weihe basin, irrigated area, as compared with the other indices, e.g., population, gross domestic product and cultivated land area, was a more suitable human explanatory variable for explaining the time-varying behavior of annual runoff. With the above background, it is important to consider the effects of human activities that mainly originate from irrigation diversion, especially for studying low-flow series in this basin. The estimations of annual recession rate (K^{-1}) by the daily streamflow data are expected to incorporate the information of impacts of water diversions on the low flows in the river channel.

3.2   Data

We used daily streamflow records (1954–2009) provided by the Hydrology Bureau of the Yellow River Conservancy Commission from both Huaxian station (with a drainage area of $106\,500\,km^2$) and Xianyang station (with a drainage area of $46\,480\,km^2$). Low-flow extreme events were selected from

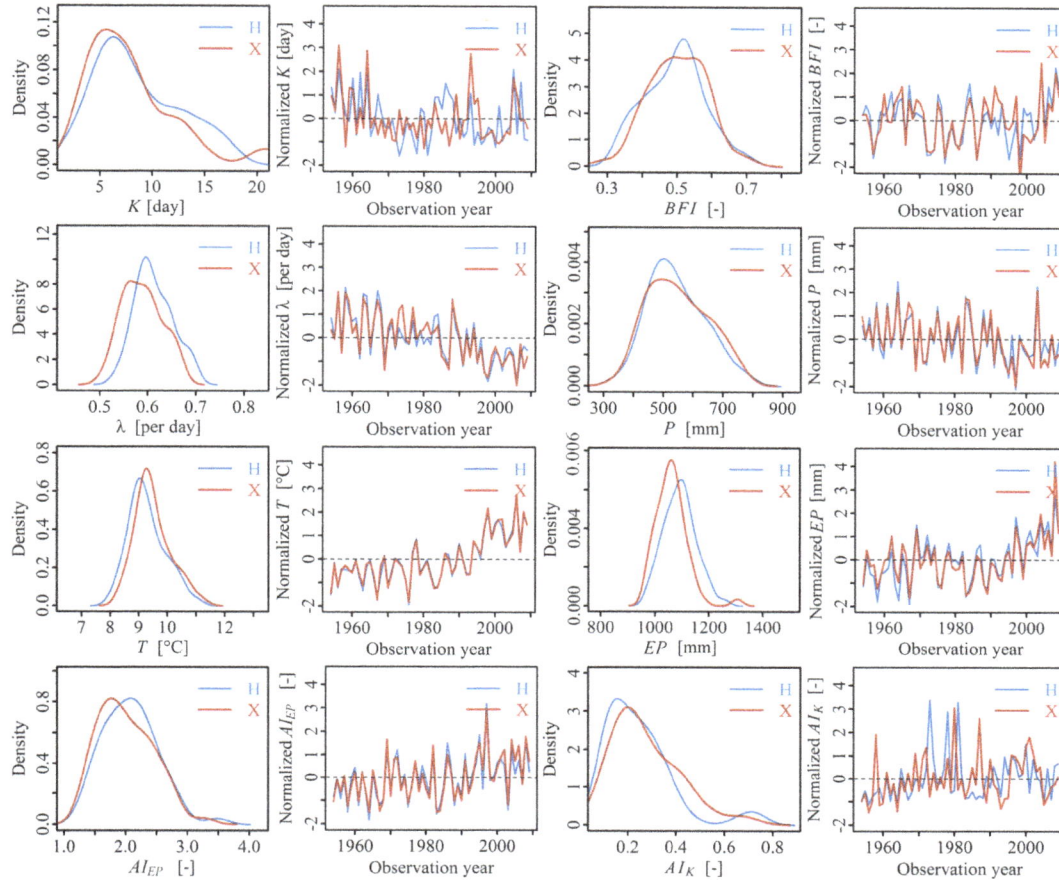

Figure 4. Frequency distributions (using the kernel density estimations) and time series processes of TCCCs variables in both Huaxian (H) and Xianyang (X) stations.

the daily streamflow series using the widely used annual minimum series method (WMO, 2009). AM_n is the annual minimum n-day flow during hydrological year beginning on 1 March. Consequently, AM_1, AM_7, AM_{15} and AM_{30} are selected as low-flow extreme events in this study. The original measure unit of streamflow data ($m^3 s^{-1}$) is converted to $10^{-4} m^3 s^{-1} km^{-2}$ for convenience of comparison of results between the Huaxian and Xianyang gauging stations

We downloaded daily total precipitation and daily mean air temperature records for 19 meteorological stations over the basin from the National Climate Center of the China Meteorological Administration (source: http://www.cma.gov.cn/). The areal average daily series of both variables above Huaxian and Xianyang stations are calculated using the Thiessen polygon method (Szolgayova et al., 2014; Thiessen, 1911). The annual average temperature (T) and annual total precipitation (P) over the period 1954–2009 are calculated for each catchment.

Human activity data (i.e., gross domestic product, population and irrigation area) were taken from annals of statistics provided by the Shaanxi Provincial Bureau of Statistics

(http://www.shaanxitj.gov.cn/) and Gansu Provincial Bureau of Statistics (source: http://www.gstj.gov.cn/).

4 Results and discussion

4.1 Identification of nonstationarity

The graphical representation and statistical test provide a preliminary analysis for low-flow nonstationarity. The graphical representations of time-series data help visualize the trends of related variables (i.e., low flow, TCCCs and HA variables), the density distributions of TCCCs variables, and the correlations between low-flow variables and these explanatory variables. In Fig. 3, four annual minimum streamflow series (AM_1, AM_7, AM_{15} and AM_{30}) in both Huaxian and Xianyang gauging stations show overall decreasing trends, as indicated by the fitted (dashed) trend lines. Compared with Huaxian, Xianyang has a larger runoff modulus (the flow per square kilometer) and a larger decrease in annual minimum streamflow series. For example, the decline slope of AM_{30} is

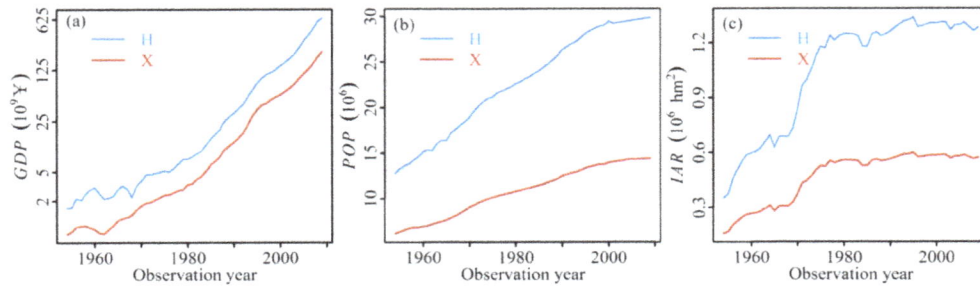

Figure 5. HA indices in both Huaxian (H) and Xianyang (X). **(a)**, **(b)** and **(c)** are for gross domestic production (GDP), population (POP) and irrigated area (IAR), respectively.

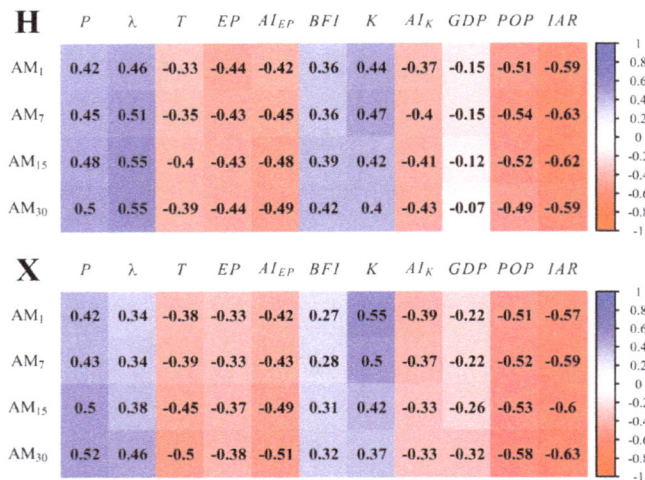

Figure 6. The Pearson correlation coefficients matrix between the annual minimum flow series and candidate explanatory variables in Huaxian (H) and Xianyang (X) stations; the darker color intensity represents a higher level of correlation (blue indicates positive correlation and red indicates negative correlations).

-0.0725 $(10^{-4}\,\mathrm{m^3\,s^{-1}\,km^{-2}\,yr^{-1}})$ in Huaxian station while Xianyang station it is -0.1338 $(10^{-4}\,\mathrm{m^3\,s^{-1}\,km^{-2}\,yr^{-1}})$.

Figure 4 shows the kernel density estimations and time processes of TCCCs variables for both Huaxian (H) and Xianyang (X) stations. The results show that these variables have different variation patterns. For example, the mean frequency of precipitation events (λ) has a decreasing trend, while temperature (T) has an increasing trend. As presented by Fig. 5, three HA variables have a significant upward trend, especially the irrigation area which is increased greatly after about 1970, suggesting that the impact of human activities in this basin has increased over time.

The significance of trends in the four annual minimum streamflow series and TCCCs variables is tested by the Mann–Kendall trend test (Kendall, 1975; Mann, 1945; Yue et al., 2002), and the change points in these series are detected by Pettitt's test (Pettitt, 1979). The results in Table 4 show that, in both Huaxian and Xianyang stations, the de-

creasing trends in all the four low-flow series ($\mathrm{AM_1}$, $\mathrm{AM_7}$, $\mathrm{AM_{15}}$ and $\mathrm{AM_{30}}$) and two explanatory variables (λ and P), as well as the increasing trends in T, EP, and $\mathrm{AI_{EP}}$ are significant at the 0.05 level (Table 4), but BFI shows no significant trends. However, K and AI_K had significantly decreasing trends only in Huaxian station (p value < 0.05). The results of change-point detection show that all low-flow series are located at 1968–1971 (p value < 0.05) except $\mathrm{AM_{30}}$ at Xianyang station whose change point is located at 1993 (p value < 0.05); for the eight candidate explanatory variables, the change points of the variables related to temperature (T, EP, $\mathrm{AI_{EP}}$) in both stations are located at 1990–1993 (p value < 0.05), the change points of the variables related to precipitation (λ, P) in both stations are close at 1984–1990 (p value ≤ 0.186) and the change points of the variables related to streamflow recession (K, AI_K) in Huaxian station are located at 1968–1971 (p value < 0.05). However, BFI in both stations and K and AI_K in Xianyang station show no significant change points.

A preliminary attribution analysis is performed using the Pearson correlation matrix to investigate the relations between the annual minimum series and eight candidate explanatory variables. Figure 6 indicates that there are significant linear correlations between the four minimum low-flow series ($\mathrm{AM_1}$, $\mathrm{AM_7}$, $\mathrm{AM_{15}}$ and $\mathrm{AM_{30}}$) and all the explanatory variables except GDP have the absolute values of Pearson correlation coefficients larger than 0.27 (p value < 0.05). These potential physical causes of nonstationarity in low flows are further considered by establishing the low-flow nonstationary model with TCCCs and HA variables in the following section.

4.2 Nonstationary frequency analysis models

4.2.1 Single covariate models

Figure 7 presents the AIC values of the four types of models (M0, M1, M2a and M2b) fitted for the low-flow series ($\mathrm{AM_1}$, $\mathrm{AM_7}$, $\mathrm{AM_{15}}$ and $\mathrm{AM_{30}}$). Some interesting results are shown as follows. First, nonstationary models (M1, M2a and M2b) have lower AIC values than stationary model (M0), which

Table 4. The results of trend test and change-point detection for both the four low-flow series and TCCCs variables in Huaxian and Xianyang.

Station	Variable	Mann–Kendall test		Pettitt's test	
		S	p value	Change point	p value
Huaxian					
	AM_1	−564	6.91E−05(***)	1968	1.34E−03(**)
	AM_7	−560	7.79E−05(***)	1968	1.44E−03(**)
	AM_{15}	−438	2.01E−03(**)	1971	4.85E−03(**)
	AM_{30}	−378	7.71E−03(**)	1971	9.96E−03(**)
	P	−292	3.97E−02(*)	1985	1.86E−01( )
	λ	−632	8.20E−06(***)	1984	3.02E−04(***)
	T	752	1.11E−07(***)	1993	8.17E−06(***)
	EP	548	1.11E−04(***)	1993	1.98E−03(**)
	AI_{EP}	384	6.79E−03(**)	1990	6.03E−02(•)
	BFI	52	7.19E−01( )	1998	3.88E−01( )
	K	−312	2.79E−02(*)	1968	8.11E−02(•)
	AI_K	376	8.04E−03(**)	1971	3.60E−02(*)
Xianyang					
	AM_1	−517	2.65E−04(***)	1968	2.2E−03(**)
	AM_7	−483	6.58E−04(***)	1970	2.5E−03(**)
	AM_{15}	−474	8.29E−04(***)	1971	2.2E−03(**)
	AM_{30}	−570	5.78E−05(***)	1993	4.5E−04(***)
	P	−414	3.51E−03(**)	1990	1.45E−02(*)
	λ	−652	4.21E−06(***)	1984	6.00E−05(***)
	T	724	3.22E−07(***)	1993	5.41E−06(***)
	EP	372	8.74E−03(**)	1993	3.01E−03(**)
	AI_{EP}	454	1.37E−03(**)	1993	8.82E−03(**)
	BFI	64	6.56E−01( )	2003	8.65E−01( )
	K	−210	1.39E−01( )	1966	2.03E−01( )
	AI_K	290	4.11E−02(*)	1968	1.63E−01( )

Significance codes: 0 "***" 0.001 "**" 0.01 "*" 0.05 "•" 0.1 " " 1

suggests that nonstationary models are worth considering. Second, for Huaxian station, irrespective of the chosen explanatory variables, the distribution type plays an important role in modeling nonstationary low-flow series. For example, PIII, GA and WEI distributions in AM_{15} and AM_{30} cases have lower AIC values than LOGNO and GEV distributions. However, for Xianyang, choosing a suitable explanatory variable may be more important than choosing a distribution type. For example, variables t, P, T, AI_{EP}, POP and IAR in most cases have lower AIC values than the other explanatory variables. Finally, in Huaxian, the lowest AIC values for modeling AM_1, AM_7, AM_{15} and AM_{30} are found in GEV_M2b_IAR, LOGNO_M2b_IAR, PIII_M2a_AI_K and GA_M2a_AI_K, respectively, while in Xianyang the lowest AIC values for modeling AM_1, AM_7, AM_{15} and AM_{30} are found in GEV_M2b_IAR, GEV_M2b_IAR, PIII_M2b_IAR and GEV_M2b_IAR, respectively. These results indicated that for explaining nonstationarity of low flow in Huaxian station, IAR is the most dominant HA variable and AI_K is the most dominant TCCCs variable, while in Xianyang the most dominant HA variable is IAR and the most dominant TCCCs

variables causing nonstationarity in AM_1, AM_7, AM_{15} and AM_{30} are K, AI_{EP}, AI_{EP} and T, respectively.

Figure 8 shows the diagnostic assessment of the GA_M2 model (with the optimal explanatory variable) for AM_{30} in both Huaxian and Xianyang stations. The centile curve plots of GA_M2 (Fig. 8a and b) show the observed values of AM_{30}, the estimated median and the areas between the 5th and 95th centile. Figure 8a shows the response relationship between AM_{30} and AI_K in Huaxian: the increase in AI_K means the smaller magnitude of low-flow events because a high value of AI_K (faster stream recession or fewer rainy days) may lead to faster water loss or less supply. In Fig. 8b, the higher values of IAR means the smaller magnitude of low-flow events, which suggests that IAR plays an important role in driving low-flow generation in Xianyang. Figure 8c and d show that the worm points are within the 95 % confidence intervals, thereby indicating a good model fit and a reasonable model construction.

(a) H

	LOGNO	PIII	GEV	WEI	GA	AIC
AM_1						
M0_none	112.4	112.9	114.3	108.6	110.9	114.3
M1_t	94.8	97.0	90.3	94.4	97.0	
M2a_P	106.9	97.9	104.5	97.2	97.9	
M2a_λ	99.9	99.1	97.8	96.9	99.1	
M2a_T	101.6	102.8	99.9	100.5	102.8	
M2a_EP	101.5	99.9	100.5	98.3	99.9	
M2a_AI_{EP}	104.8	96.8	103.0	95.8	96.8	
M2a_BFI	103.4	99.2	97.6	96.7	99.2	
M2a_K	102.5	96.7	101.5	95.4	96.7	
M2a_AI_K	103.6	95.8	101.8	95.0	95.8	
M2b_GDP	109.5	107.6	105.7	105.8	107.6	
M2b_POP	91.9	94.5	85.2	91.6	94.5	
M2b_IAR	90.3	91.5	83.8	88.1	91.5	83.8

OPTIMAL MODEL: GEV_M2b_IAR

	LOGNO	PIII	GEV	WEI	GA	AIC
AM_7						
M0_none	151.8	155.0	158.2	153.5	155.2	158.2
M1_t	134.7	138.5	137.9	139.1	140.7	
M2a_P	144.6	138.5	147.4	140.4	140.7	
M2a_λ	137.7	139.4	142.4	140.3	141.6	
M2a_T	141.6	144.6	144.4	145.2	146.7	
M2a_EP	141.6	142.6	147.0	143.7	144.8	
M2a_AI_{EP}	142.7	137.9	146.2	139.5	140.0	
M2a_BFI	141.8	141.5	146.1	142.1	143.7	
M2a_K	139.9	138.3	146.3	139.7	140.5	
M2a_AI_K	139.9	135.7	140.7	137.7	138.0	
M2b_GDP	149.0	149.7	151.7	150.6	151.9	
M2b_POP	131.7	136.0	134.0	136.5	138.2	
M2b_IAR	127.9	132.3	130.7	132.5	134.5	127.9

OPTIMAL MODEL: LOGNO_M2b_IAR

	LOGNO	PIII	GEV	WEI	GA	AIC
AM_{15}						
M0_none	211.9	203.6	222.2	209.0	209.0	222.2
M1_t	200.6	190.2	210.1	197.5	197.4	
M2a_P	200.6	187.6	210.3	195.0	194.9	
M2a_λ	197.6	188.3	204.7	195.4	195.4	
M2a_T	201.8	193.5	209.0	200.7	200.8	
M2a_EP	202.3	192.7	210.6	199.9	199.9	
M2a_AI_{EP}	199.4	187.3	209.0	194.6	194.5	
M2a_BFI	200.7	190.4	209.3	197.7	197.7	
M2a_K	199.9	190.7	208.0	198.0	198.0	
M2a_AI_K	197.5	184.2	205.7	192.3	192.2	
M2b_GDP	209.8	199.0	220.2	206.5	206.5	
M2b_POP	197.7	188.0	206.5	195.1	195.1	
M2b_IAR	192.8	184.2	206.5	191.1	191.2	184.2

OPTIMAL MODEL: PIII_M2b_AI_K

	LOGNO	PIII	GEV	WEI	GA	AIC
AM_{30}						
M0_none	241.4	237.3	244.1	236.5	236.3	244.1
M1_t	233.0	225.4	233.5	225.4	225.5	
M2a_P	229.0	220.9	228.4	220.6	220.6	
M2a_λ	228.5	221.4	227.1	221.1	221.1	
M2a_T	233.4	227.2	232.2	227.7	227.5	
M2a_EP	232.9	226.2	233.3	226.4	226.4	
M2a_AI_{EP}	228.1	220.5	227.5	220.1	220.1	
M2a_BFI	226.2	223.5	228.5	224.0	223.5	
M2a_K	230.1	224.9	232.6	225.1	225.0	
M2a_AI_K	224.5	217.5	220.8	217.7	217.4	
M2b_GDP	239.3	233.1	236.1	234.2	234.1	
M2b_POP	230.4	223.1	230.2	223.0	223.0	
M2b_IAR	224.4	218.9	222.8	218.6	218.3	217.4

OPTIMAL MODEL: GA_M2b_AI_K

(b) X

	LOGNO	PIII	GEV	WEI	GA	AIC
AM_1						
M0_none	227.2	226.5	228.0	227.2	226.3	228.0
M1_t	211.1	210.6	209.5	211.5	209.9	
M2a_P	215.6	211.9	215.6	212.9	211.5	
M2a_λ	217.2	216.1	218.6	217.7	216.5	
M2a_T	211.9	212.8	211.5	214.4	212.7	
M2a_EP	217.0	216.6	214.9	218.4	217.1	
M2a_AI_{EP}	213.6	211.3	214.1	212.3	210.8	
M2a_BFI	220.8	217.0	218.8	218.9	217.7	
M2a_K	215.4	211.4	212.9	212.2	210.7	
M2a_AI_K	218.2	214.0	217.4	215.2	214.0	
M2b_GDP	222.5	220.0	219.3	221.7	220.7	
M2b_POP	208.8	207.6	207.4	209.7	207.8	
M2b_IAR	208.0	206.1	205.1	208.2	206.3	205.1

OPTIMAL MODEL: GEV_M2b_IAR

	LOGNO	PIII	GEV	WEI	GA	AIC
AM_7						
M0_none	243.5	243.8	245.8	245.1	244.1	245.8
M1_t	227.9	228.2	228.0	229.7	227.9	
M2a_P	232.3	229.0	233.6	230.4	229.0	
M2a_λ	233.8	233.3	236.8	235.5	234.2	
M2a_T	229.1	230.1	230.5	232.2	230.3	
M2a_EP	234.4	234.0	234.7	236.4	235.0	
M2a_AI_{EP}	230.6	228.5	232.6	230.0	228.4	
M2a_BFI	235.9	234.2	236.2	236.8	235.4	
M2a_K	233.1	230.6	232.5	232.3	230.7	
M2a_AI_K	235.6	232.2	236.3	234.1	232.8	
M2b_GDP	239.1	237.7	238.7	239.8	238.7	
M2b_POP	225.5	225.5	225.8	227.9	226.0	
M2b_IAR	223.5	223.2	222.5	225.7	223.6	222.5

OPTIMAL MODEL: GEV_M2b_IAR

	LOGNO	PIII	GEV	WEI	GA	AIC
AM_{15}						
M0_none	269.9	269.3	273.3	270.0	269.3	273.3
M1_t	254.5	253.5	256.1	254.6	253.4	
M2a_P	256.2	251.6	258.9	252.4	251.4	
M2a_λ	259.3	257.4	263.4	259.0	258.1	
M2a_T	254.3	253.5	256.7	255.0	253.6	
M2a_EP	260.6	258.5	262.7	260.2	259.3	
M2a_AI_{EP}	254.7	251.3	258.2	252.1	251.0	
M2a_BFI	261.6	258.6	262.7	260.5	259.5	
M2a_K	262.0	259.0	262.9	260.8	259.8	
M2a_AI_K	263.1	259.2	264.8	261.0	260.2	
M2b_GDP	265.0	262.1	266.9	263.7	263.0	
M2b_POP	252.3	251.4	253.8	253.2	251.9	
M2b_IAR	250.3	249.5	250.5	251.4	249.9	249.5

OPTIMAL MODEL: PIII_M2b_IAR

	LOGNO	PIII	GEV	WEI	GA	AIC
AM_{30}						
M0_none	290.7	290.6	293.4	290.8	289.8	293.4
M1_t	271.6	270.4	271.5	271.8	270.1	
M2a_P	275.1	271.1	277.0	272.1	270.9	
M2a_λ	276.9	274.7	279.6	276.0	274.8	
M2a_T	270.8	270.3	271.1	272.1	270.1	
M2a_EP	280.9	278.3	282.1	279.9	278.7	
M2a_AI_{EP}	273.7	270.5	276.0	271.6	270.2	
M2a_BFI	282.3	278.8	284.1	280.3	279.3	
M2a_K	284.0	281.1	284.1	282.9	281.7	
M2a_AI_K	283.9	279.9	284.0	281.5	280.5	
M2b_GDP	283.3	279.5	283.5	281.0	280.1	
M2b_POP	269.4	268.5	269.3	270.5	268.6	
M2b_IAR	268.4	267.8	267.2	269.9	267.8	267.2

OPTIMAL MODEL: GEV_M2b_IAR

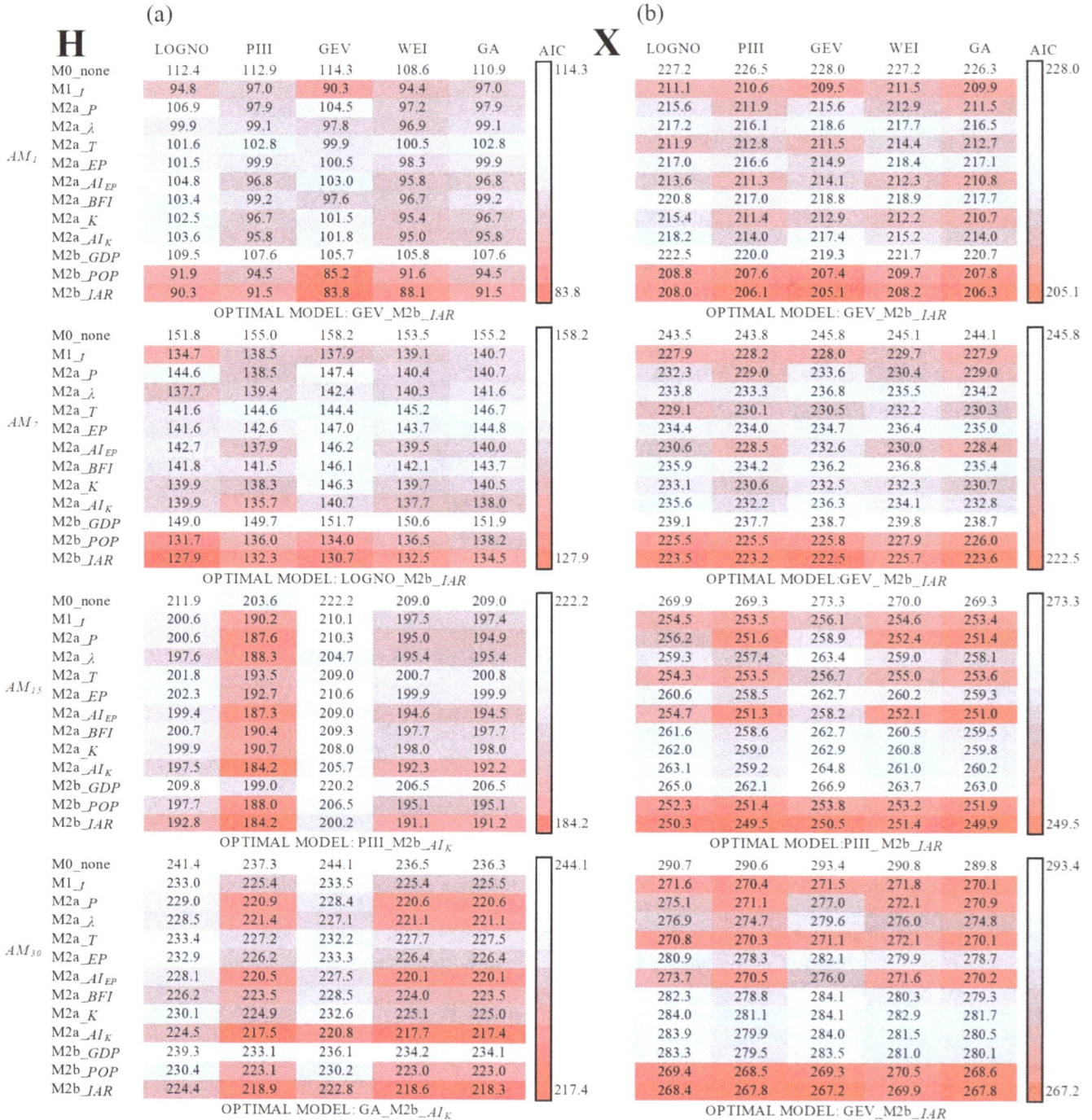

Figure 7. Comparisons among M0, M1 and M2 based on the AIC values for the four observed low-flow series in Huaxian (H) at **(a)** and Xianyang (X) at **(b)**; darker red color represents a higher goodness of fit.

4.2.2 Multiple covariate models

Figure 9 shows the AIC values of the stationary model (M0), time covariate model (M1) and physical covariate models (M2a, M2b, M3, M4, M5 and M6) for AM$_{30}$. As shown in Fig. 9, M4 (nonstationary GA distribution with the optimal TCCCs variables) has a good performance; after adding the HA variables, M6 with the lowest AIC value is attained; it can be found that the combination of multiple TCCCs variables plays a major role in changing the low flows of the Weihe River, but the influence of HA variables should not be ignored.

H   X

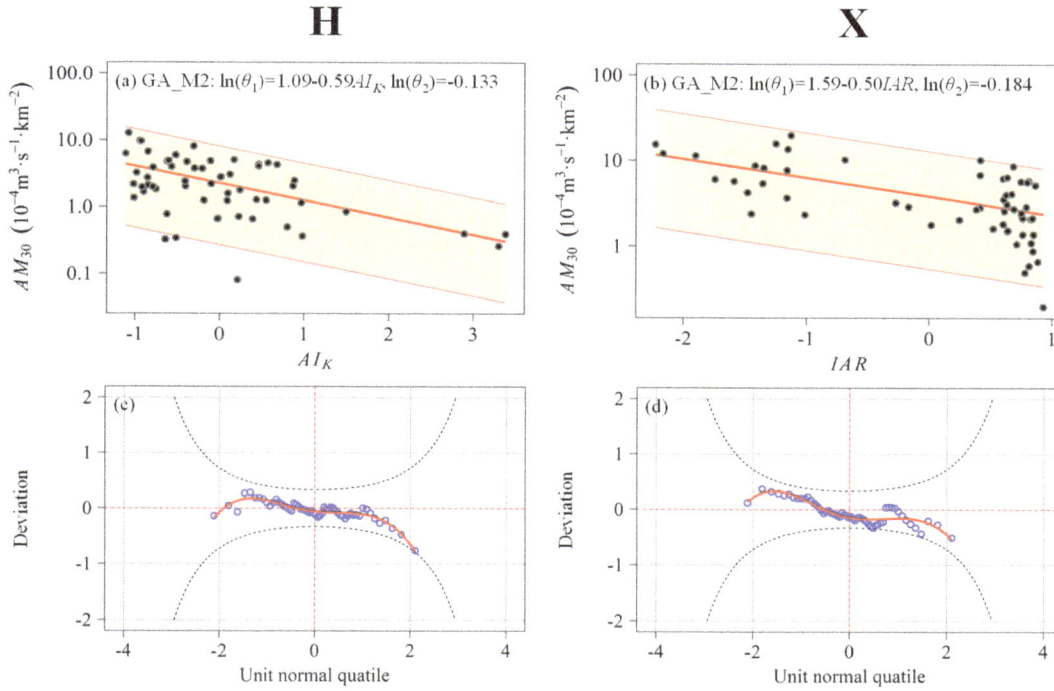

Figure 8. Performance assessments of GA_M2 for AM_{30} in Huaxian (H) on the left panel and Xianyang (X) on the right panel. **(a)** and **(b)** are the centile curves plots of GA_M2 (red lines represent the centile curves estimated by GA_M2; the 50th centile curves are indicated by thick red; the yellow-filled areas are between the 5th and 95th centile curves; the black points indicate the observed series); **(c)** and **(d)** are the worm plots of GA_M2 for the goodness-of-fit test; a reasonable model fit should have the data points fall within the 95 % confidence intervals (between the two red dashed curves).

H   X

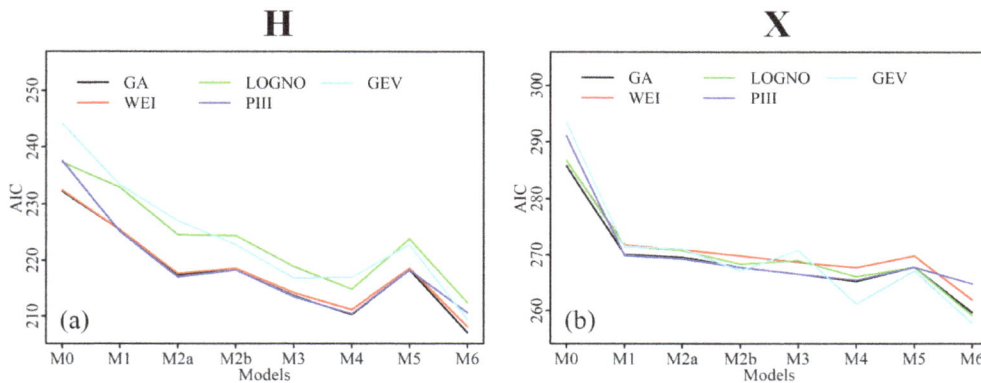

Figure 9. Comparisons of performance of stationary model (M0), time covariate model (M1) and physical covariate models (M2a, M2b, M3, M4, M5 and M6 with their corresponding optimal explanatory variables) for AM_{30} in Huaxian (H) at **(a)** and Xianyang (X) at **(b)**.

A summary of frequency analysis based on nonstationary GA distribution AM_{30} is presented in Table 5. We choose to focus on M4, M5 and M6. When only using TCCCs variables to model nonstationary low-flow frequency distribution, the results of M4 show the optimal combination of explanatory variables for all low-flow series contains more than three variables. For example, for AM_{30} of Huaxian, the optimal combination of TCCCs variables includes AI_K, BFI and AI_{EP}. When only HA variables are used, the results of M5

show IAR is important to the low flows in this area. And M4 has a better performance than M5. When using both TCCCs variables and HA variables, the results of M6 show the optimal combination contains multiple TCCCs variables and the irrigation area. For Huaxian, the optimal combination of all explanatory variables is AI_K, IAR, BFI and AI_{EP}, while for Xianyang, the optimal combination is IAR, AI_{EP} and BFI. We can also find that if two TCCCs variables are highly correlated, they do not seem to be selected as the explanatory

Table 5. The summary of frequency analysis using GA distribution for AM_{30} in Huaxian and Xianyang.

Station	Model codes	Variables	AIC	Distribution parameters		
				$\ln(\theta_1)$	$\ln(\theta_2)$	θ_3
Huaxian						
	GA_M0	–	232.3	1.09	−0.133	–
	GA_M1	t	225.5	$1.09-0.32t$	−0.133	–
	GA_M2	AI_K	217.4	$1.09-0.59AI_K$	−0.133	–
	GA_M2b	IAR	218.3	$1.09-0.47IAR$	−0.133	–
	GA_M3	AI_K, BFI	213.7	$1.09-0.50AI_K+0.32BFI$	−0.133	–
	GA_M4	AI_K, BFI, AI_{EP}	211.1	$1.09-0.40AI_K+0.32BFI-0.34AI_{EP}$	−0.133	–
	GA_M5	IAR	218.3	$1.09-0.47IAR$	−0.133	–
	GA_M6	AI_K, IAR, BFI, AI_{EP}	207.0	$1.09-0.30AI_K-0.27IAR+0.32BFI-0.23AI_{EP}$	−0.133	–
Xianyang						
	GA_M0	–	285.8	1.59	−0.184	–
	GA_M1	t	270.1	$1.59-0.48t$	−0.184	–
	GA_M2a	T	270.1	$1.59-0.50T$	−0.184	–
	GA_M2b	IAR	267.8	$1.59-0.50IAR$	−0.184	–
	GA_M3	T, P	267.1	$1.59-0.34T+0.32P$	−0.184	–
	GA_M4	T, P, BFI, K	265.4	$1.59-0.33T+0.27P+0.22BFI+0.18K$	−0.184	–
	GA_M5	IAR	267.8	$1.59-0.50IAR$	−0.184	–
	GA_M6	IAR, AI_{EP}, BFI	259.7	$1.59-0.28IAR-0.36\,AI_{EP}+0.26BFI$	$-0.184+0.23IAR$	–

H   **X**

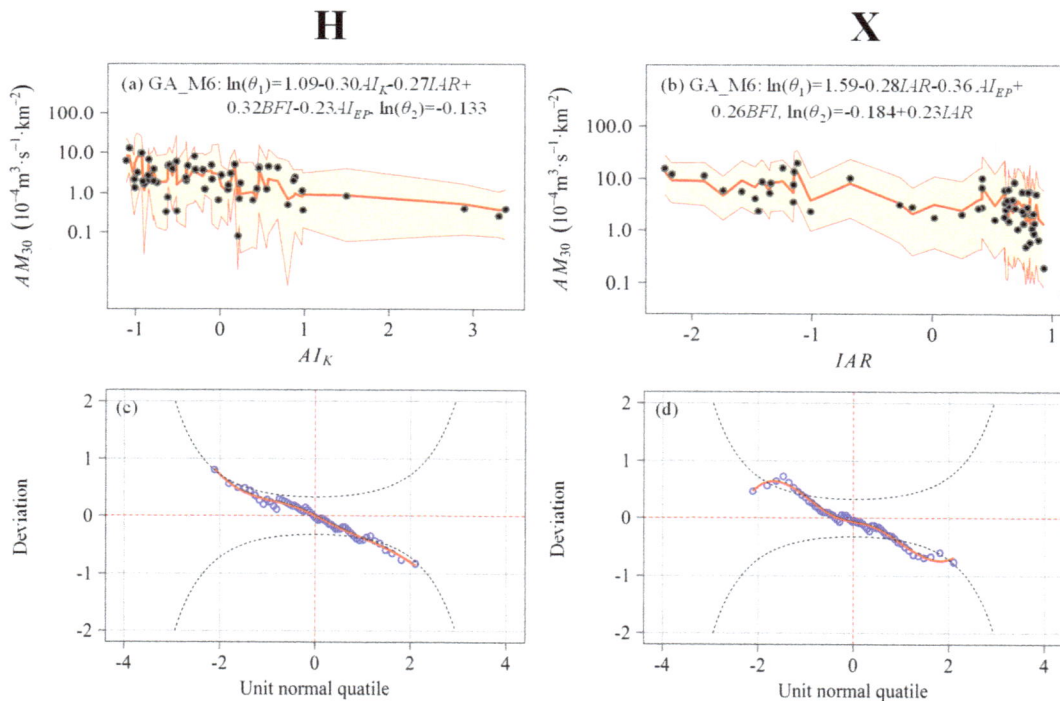

Figure 10. Performance assessments of GA_M6 for AM_{30} in Huaxian (H) on the left panel and Xianyang (X) on the right panel. **(a)** and **(b)** are the centile curves plots of GA_M6 (red lines represent the centile curves estimated by GA_M6; the 50th centile curves are indicated by thick red; the yellow-filled areas are between the 5th and 95th centile curves; the filled black points indicate the observed series); **(c)** and **(d)** are the worm plots of GA_M6 for the goodness-of-fit test; a reasonable model fit should have the data points fall within the 95 % confidence intervals (between the two red dashed curves).

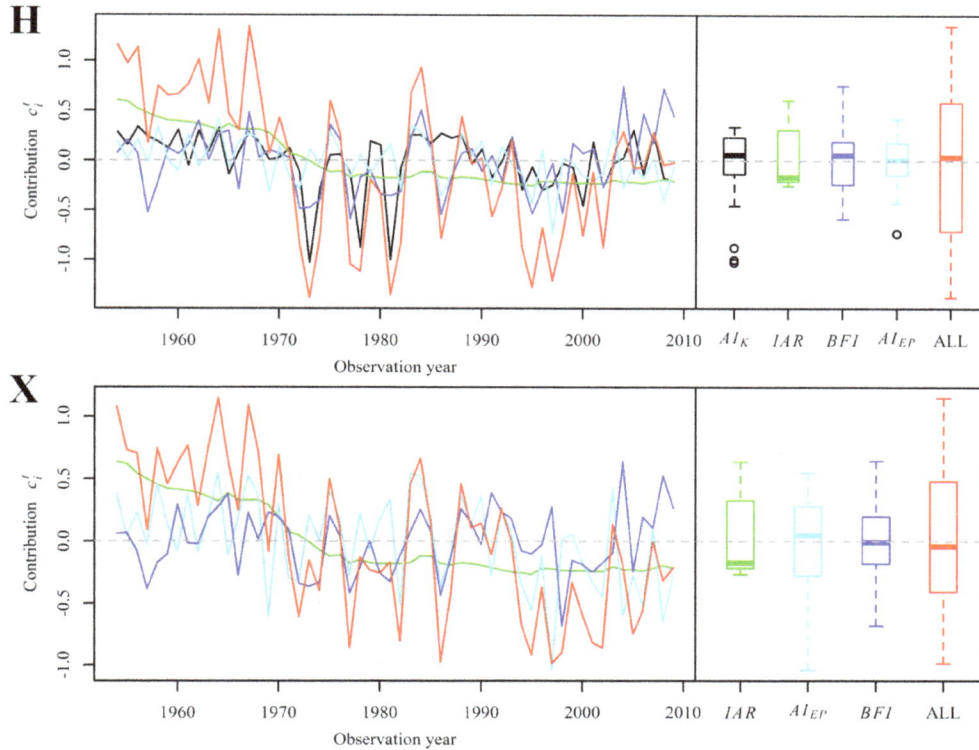

Figure 11. Contribution of selected explanatory variables to $c_i^t = \ln\left(\theta_1^t\right) - \ln\left(\overline{\theta}_1\right)$ in different periods based on GA_M6.

variables at the same time. For example, in terms of air temperature (T), evapotranspiration and the climate aridity index (AI_{EP}), only one of them will appear in the optimal combination. This suggests that multicollinearity problem in the multiple variables analysis can be reduced, which will help obtain more reliable GLM parameters for contribution analysis.

The diagnostic assessment of the GA_M6 model for AM_{30} at two stations is presented by Fig. 10. The centile curve plots of GA_M6 (Fig. 10a and b) show a more sophisticated nonstationary modeling than GA_M2 (Fig. 8). When using GA_M6 to model AM_{30} in Huaxian (Fig. 10a), similar to GA_M2, the lower low flows are found to also correspond to a higher value of AI_K, but GA_M6 is able to identify the more complex variation patterns of low flows through the incorporation of IAR, BFI and AI_{EP}. Figure 10c and d show that the data points of worm plots of GA_M6 are almost within the 95 % confidence intervals, thereby indicating an acceptable model fit and a reasonable model construction.

Figure 11 presents the contribution of each selected explanatory variable to $\ln\left(\theta_1^t\right) - \ln\left(\overline{\theta}_1\right)$ in the observation year based on GA_M6 for AM_{30} in Huaxian and Xianyang. We can find that for Huaxian, the simulation value of $\ln\left(\theta_1^t\right)$ frequently occurs below $\ln\left(\overline{\theta}_1\right)$ during the two periods of about 1970–1982 and 1993–2003, which is in accordance with the observed decrease in AM_{30} of Huaxian station during these periods. In the former period 1970–1982, both AI_K and

BFI contribute a lot of negative amount to $\ln\left(\theta_1^t\right) - \ln\left(\overline{\theta}_1\right)$, whereas, during 1993–2003, the contribution of both AI_K and BFI decreases significantly. However, IAR has almost equal negative contribution to $\ln\left(\theta_1^t\right) - \ln\left(\overline{\theta}_1\right)$ in both periods. Unlike the former three variables, the significant negative contribution of AI_{EP} is only found in 1993–2003. For AM_{30} of Xianyang, the contribution of IAR, AI_{EP} and BFI is similar to that at Huaxian station in two periods; however, AI_K is not included in the final model.

4.3 Discussion

The impacts of both human activities and climate change on low flows of the study area led to time-varying climate and catchment conditions. Nonstationary modeling for annual low-flow series using TCCCs variables and/or HA variables as explanatory variables is clearly different from either the stationary model (M0) or the time covariate model (M1). The result demonstrates that considering multiple drivers (e.g., the variability in catchment conditions), especially in such an artificially influenced river, is necessary for nonstationary modeling of annual low-flow series.

In this study area, nonstationary modeling considering TCCCs is supported by the following facts and findings. For human activities, an important milestone representative is the completion and operation of the irrigation system on the plateau in the Baoji Gorge irrigation district since 1971

(Sect. 3.1). Figure 5c shows the change in irrigation area in this basin. And the change-point detection test in Sect. 4.1 shows that significant change points of low-flow series occur exactly at around 1971. This result demonstrates that changes in AM_{30} may involve a consequence of this project. In addition to human activities, climate change also makes a considerable contribution to nonstationarity of low flows, as suggested by nonstationary modeling using TCCCs variables with stepwise analysis. Actually, climate driving pattern may strengthen after nearly 1990, which is indicated by change-point detection test of both annual mean temperature (T) and annual precipitation (P) as well as the behavior of annual low-flow series after nearly 1990. There are two faster recession periods, the 1970s and the 1990s, as shown in Fig. 4. The reasons for the faster recession are likely to be related to the above-mentioned project (e.g., the increasing diversion for irrigation) and climate change (e.g., the intensified evaporation) but also could be human alterations on catchment properties, such as vegetation cover change. In conclusion, the temporal variability in irrigation area, air temperature, precipitation (the frequency and volume of rain events) and streamflow recession should be the main driving factors of generating low-flow regimes in this basin. Overall, the causes of nonstationarity in the category for two gauging stations have no clear difference but have some differences in the relative importance. As shown in Table 5, when modeling the low-flow series of Huaxian using TCCCs variables, the optimal model (M4) preferred the variables that are related to recession process; however, for Xianyang, the preferred variables are related to temperature. The reason for this may be that, as a downstream station, Huaxian station suffers more intensive human activity, so that the importance of temperature change to the low-flow change is reduced meanwhile the importance of streamflow recession (related to the capability of water storage) change is enhanced. Ignoring the negative impacts of the errors in estimating annual recession constants (K) which are caused by insufficient data points of extracted stream segments at some wet years may lead to the propagation of high errors in annual recession analysis and accordingly affect the quality of nonstationary frequency analysis when K is used as an explanatory variable. Further study will give a more reliable estimation of K through the improvement of annual recession analysis. In addition, it should be noted that the population recorded in the annals of statistics may not be equal to the actual population living in the catchment. If the population in the annals is used as the explanatory variable, this difference may lead to the uncertainty of model parameter estimations. Nonetheless, it is the best population data so far and the explanatory variable POP is excluded in the final model (M6).

5 Conclusion

There is an increasing need to develop an effective nonstationary low-flow frequency model to deal with nonstationarities caused by climate change and time-varying anthropogenic activities. In this study, time-varying climate and catchment conditions in the Weihe River basin were measured by annual time series of the eight indices, i.e., total precipitation (P), mean frequency of precipitation events (λ), temperature (T), potential evapotranspiration, climate aridity index (AI_{EP}), base-flow index, recession constant (K) and the recession-related aridity index (AI_K). The nonstationary distribution model was developed using these eight TCCCs indices and/or three HA indices as candidate explanatory variables for frequency analysis of time-varying annual low-flow series caused by multiple drivers. The main driving forces of the decrease in low flows in the Weihe River include reduced precipitation, warming climate, increasing irrigation area and faster streamflow recession. Therefore, a complex deterioration mechanism resulting from these factors demonstrates that, in this arid and semi-arid area, the water resources could be vulnerable to adverse environmental changes, thus portending increasing water shortages. The nonstationary low-flow model considering TCCCs can provide the knowledge of low-flow generation mechanism and give a more reliable design of low flows for infrastructure and water supply.

Competing interests. The authors declare that they have no conflict of interest.

Acknowledgements. The study was financially supported by the National Natural Science Foundation of China (NSFC grants 51525902 and 51479139) and projects from State Key Laboratory of Water Resources and Hydropower Engineering Science, Wuhan University. We greatly appreciate the editor and three reviewers for their insightful comments and constructive suggestions that helped us to improve the manuscript.

Edited by: Fuqiang Tian

References

Akaike, H.: A new look at the statistical model identification, IEEE T. Automat. Contr., 19, 716–723, 1974.
Arora, V. K.: The use of the aridity index to assess climate change effect on annual runoff, J. Hydrol., 265, 164–177, 2002.

Botter, G., Basso, S., Rodriguez-Iturbe, I., and Rinaldo, A.: Resilience of river flow regimes, P. Natl. Acad. Sci. USA, 110, 12925–12930, 2013.

Bradford, M. J. and Heinonen, J. S.: Low Flows, Instream Flow Needs and Fish Ecology in Small Streams, Can. Water Resour. J., 33, 165–180, 2008.

Buuren, S. V. and Fredriks, M.: Worm plot: a simple diagnostic device for modelling growth reference curves, Stat. Med., 20, 1259–1277, 2001.

Dobson, A. J. and Barnett, A. G.: An Introduction to Generalized Linear Models, Third Edition, J. R. Stat. Soc., 11, 272–272, 2012.

Du, T., Xiong, L., Xu, C.-Y., Gippel, C. J., Guo, S., and Liu, P.: Return period and risk analysis of nonstationary low-flow series under climate change, J. Hydrol., 527, 234–250, 2015.

Dunn, P. K. and Symth, G. K.: Randomized quantile residuals, J. Comput. Graph. Stat., 5, 236–244, 1996.

Gilroy, K. L. and Mccuen, R. H.: A nonstationary flood frequency analysis method to adjust for future climate change and urbanization, J. Hydrol., 414–415, 40–48, 2012.

Giuntoli, I., Renard, B., Vidal, J. P., and Bard, A.: Low flows in France and their relationship to large-scale climate indices, J. Hydrol., 482, 105–118, 2013.

Gottschalk, L., Yu, K.-X., Leblois, E., and Xiong, L.: Statistics of low flow: Theoretical derivation of the distribution of minimum streamflow series, J. Hydrol., 481, 204–219, 2013.

Gu, X., Zhang, Q., Singh, V. P., and Shi, P.: Changes in magnitude and frequency of heavy precipitation across China and its potential links to summer temperature, J. Hydrol., 547, 718–731, 2017.

Guo, D.: An R Package for Implementing Multiple Evapotranspiration Formulations, International Environmental Modelling and Software Society, in: Proceedings of the 7th International Congress on Environmental Modelling and Software, edited by: Ames, D. P., Quinn, N. W. T., Rizzoli, A. E., 15–19 June, San Diego, California, USA, ISBN-13: 978-88-9035-744-2, 2014.

Hall, F. R.: Base flow recessions: A review, Water Resour. Res., 4, 973–983, 1968.

Hargreaves, G. H. and Samani, Z. A.: Reference Crop Evapotranspiration From Temperature, Appl. Eng. Agric., 1, 96–99 1985.

Hewa, G. A., Wang, Q. J., McMahon, T. A., Nathan, R. J., and Peel, M. C.: Generalized extreme value distribution fitted by LH moments for low-flow frequency analysis, Water Resour. Res., 43, 227–228, 2007.

Jiang, C., Xiong, L., Wang, D., Liu, P., Guo, S., and Xu, C.-Y.: Separating the impacts of climate change and human activities on runoff using the Budyko-type equations with time-varying parameters, J. Hydrol., 522, 326–338, 2015a.

Jiang, C., Xiong, L., Xu, C.-Y., and Guo, S.: Bivariate frequency analysis of nonstationary low – series based on the time – copula, Hydrol. Process., 29, 1521–1534, 2015b.

Kam, J. and Sheffield, J.: Changes in the low flow regime over the eastern United States (1962–2011): variability, trends, and attributions, Climatic Change, 135, 639–653, 2015.

Kendall, M. G.: Rank Correlation Methods, Griffin, London, 1975.

Koffler, D. and Laaha, G.: LFSTAT – Low-Flow Analysis in R, Egu General Assembly, Vienna, Austria, 15, available at: https: //cran.r-project.org/web/packages/lfstat/index.html (last access: 15 March 2017), 2013.

Kormos, P. R., Luce, C. H., Wenger, S. J., and Berghuijs, W. R.: Trends and sensitivities of low streamflow extremes to discharge timing and magnitude in Pacific Northwest mountain streams, Water Resour. Res., 52, 4990–5007, 2016.

Kwon, H.-H., Brown, C., and Lall, U.: Climate informed flood frequency analysis and prediction in Montana using hierarchical Bayesian modeling, Geophys. Res. Lett., 35, L05404, https://doi.org/10.1029/2007GL032220, 2008.

López, J. and Francés, F.: Non-stationary flood frequency analysis in continental Spanish rivers, using climate and reservoir indices as external covariates, Hydrol. Earth Syst. Sci., 17, 3189–3203, https://doi.org/10.5194/hess-17-3189-2013, 2013.

Lin, Q. C., Huai-En, L. I., and Xi-Jun, W. U.: Impact of Water Diversion of Baojixia Irrigation Area to the Weihe River Runoff, Yellow River, 34, 106–108, 2012.

Liu, D., Guo, S., Lian, Y., Xiong, L., and Chen, X.: Climate-informed low-flow frequency analysis using nonstationary modelling, Hydrol. Process., 29, 2112–2124, 2015.

Liu, J., Zhang, Q., Singh, V. P., and Shi, P.: Contribution of multiple climatic variables and human activities to streamflow changes across China, J. Hydrol., 545, 145–162 2017.

Mann, H. B.: Nonparametric Tests Against Trend, Econometrica, 13, 245–259, 1945.

Matalas, N. C.: Probability distribution of low flows, U.S. Geological Survey professional Paper, 434-A, 1963.

Milly, P. C. D., Betancourt, J., Falkenmark, M., Hirsch, R. M., Kundzewicz, Z. W., Lettenmaier, D. P., and Stouffer, R. J.: Stationarity Is Dead: Whither Water Management?, Science, 319, 573–574, 2008.

Mondal, A. and Mujumdar, P. P.: Modeling non-stationarity in intensity, duration and frequency of extreme rainfall over India, J. Hydrol., 521, 217–231, 2015.

Pettitt, A. N.: A Non-Parametric Approach to the Change-Point Problem, J. R. Stat. Soc., 28, 126–135, 1979.

Richard, W. K., Marc, B. P., and Philippe, N.: Statistics of extremes in hydrology, Adv. Water Resour., 25, 1287–1304, 2002.

Rigby, R. A. and Stasinopoulos, D. M.: Generalized additive models for location, scale and shape, Appl. Statist., 54, 507–554, 2005.

Sadri, S., Kam, J., and Sheffield, J.: Nonstationarity of low flows and their timing in the eastern United States, Hydrol. Earth Syst. Sci., 20, 633–649, https://doi.org/10.5194/hess-20-633-2016, 2016.

Salas, J. D.: Analysis and modeling of hydrologic time series, Handbook of Hydrology, McGraw Hill, NewYork, Chapter 19, 1–72, 1993.

Sawaske, S. R. and Freyberg, D. L.: An analysis of trends in baseflow recession and low-flows in rain-dominated coastal streams of the pacific coast, J. Hydrol., 519, 599–610, 2014.

Smakhtin, V. U.: Low flow hydrology – a review, J. Hydrol., 240, 147–186, 2001.

Stasinopoulos, D. M. and Rigby, R. A.: Generalized additive models for location scale and shape (GAMLSS) in R, J. Stat. Softw., 23, https://doi.org/10.18637/jss.v023.i07, 2007.

Strupczewski, W. G., Singh, V. P., and Feluch, W.: Non-stationary approach to at-site flood frequency modeling I. Maximum likelihood estimation, J. Hydrol., 248, 123–142, 2001.

Szolgayova, E., Parajka, J., Blöschl, G., and Bucher, C.: Long term variability of the Danube River flow and its relation to precipitation and air temperature, J. Hydrol., 519, 871–880, 2014.

Tallaksen, L. M.: A review of baseflow recession analysis, J. Hydrol., 165, 349–370, 1995.

Tallaksen, L. M., Madsen, H., and Hisdal, H.: Hydrological Drought- Processes and Estimation Methods for Streamflow and Groundwater, Elsevier B.V., the Netherlands, 2004.

Tang, Y., Xi, S., Chen, X., and Lian, Y.: Quantification of Multiple Climate Change and Human Activity Impact Factors on Flood Regimes in the Pearl River Delta of China, Adv. Meteorol., 2016, 1–11, https://doi.org/10.1155/2016/3928920, 2016.

Thiessen, A. H.: Precipitation averages for large areas, Mon. Weather Rev., 39, 1082–1084, 1911.

Van Loon, A. F. and Laaha, G.: Hydrological drought severity explained by climate and catchment characteristics, J. Hydrol., 526, 3–14, 2015.

Venables, W. N. and Ripley, B. D.: Modern Applied Statistics with S, Springer, 4. edition, New York, 2002.

Villarini, G. and Strong, A.: Roles of climate and agricultural practices in discharge changes in an agricultural watershed in Iowa, Agriculture, Ecosystems & Environment, 188, 204–211, 2014.

Villarini, G., Smith, J. A., Serinaldi, F., Bales, J., Bates, P. D., and Krajewski, W. F.: Flood frequency analysis for nonstationary annual peak records in an urban drainage basin, Adv. Water Resour., 32, 1255–1266, 2009.

Villarini, G., Smith, J. A., and Napolitano, F.: Nonstationary modeling of a long record of rainfall and temperature over Rome, Adv. Water Resour., 33, 1256–1267, 2010.

WMO: Mannual on Low-fow Estimation and Prediction, WMO-No. 1029, Switzerland, 2009.

Xiong, L., Jiang, C., and Du, T.: Statistical attribution analysis of the nonstationarity of the annual runoff series of the Weihe River, Water Sci. Technol., 70, 939–946, 2014.

Xiong, L., Du, T., Xu, C.-Y., Guo, S., Jiang, C., and Gippel, C. J.: Non-Stationary Annual Maximum Flood Frequency Analysis Using the Norming Constants Method to Consider Non-Stationarity in the Annual Daily Flow Series, Water Resour. Manag., 29, 3615–3633, 2015.

Yan, L., Xiong, L., Liu, D., Hu, T., and Xu, C.-Y.: Frequency analysis of nonstationary annual maximum flood series using the time – varying two – component mixture distributions, Hydrol. Process., 31, 69–89, 2017.

Yang, H. and Yang, D.: Evaluating attribution of annual runoff change: according to climate elasticity derived using Budyko hypothesis, Egu General Assembly, 15, 14029, 2013.

Yu, K.-X., Xiong, L., and Gottschalk, L.: Derivation of low flow distribution functions using copulas, J. Hydrol., 508, 273–288, 2014.

Yue, S., Pilon, P., and Cavadias, G.: Power of the Mann–Kendall and Spearman's rho tests for detecting monotonic trends in hydrological series, J. Hydrol., 259, 254–271, 2002.

Zhang, Q., Gu, X., Singh, V. P., Xiao, M., and Chen, X.: Evaluation of flood frequency under non-stationarity resulting from climate indices and reservoir indices in the East River basin, China, J. Hydrol., 527, 565–575, 2015.

10

Cosmic-ray neutron transport at a forest field site: the sensitivity to various environmental conditions with focus on biomass and canopy interception

Mie Andreasen[1], **Karsten H. Jensen**[1], **Darin Desilets**[2], **Marek Zreda**[3], **Heye R. Bogena**[4], and **Majken C. Looms**[1]

[1]Department of Geosciences and Natural Resource Management, University of Copenhagen, Copenhagen, Denmark
[2]Hydroinnova LLC, Albuquerque, New Mexico, USA
[3]Department of Hydrology and Water Resources, University of Arizona, Arizona, USA
[4]Agrosphere IBG-3, Forschungszentrum Jülich GmbH, Jülich, Germany

Correspondence to: Mie Andreasen (mie.andreasen@ign.ku.dk)

Abstract. Cosmic-ray neutron intensity is inversely correlated to all hydrogen present in the upper decimeters of the subsurface and the first few hectometers of the atmosphere above the ground surface. This correlation forms the base of the cosmic-ray neutron soil moisture estimation method. The method is, however, complicated by the fact that several hydrogen pools other than soil moisture affect the neutron intensity. In order to improve the cosmic-ray neutron soil moisture estimation method and explore the potential for additional applications, knowledge about the environmental effect on cosmic-ray neutron intensity is essential (e.g., the effect of vegetation, litter layer and soil type). In this study the environmental effect is examined by performing a sensitivity analysis using neutron transport modeling. We use a neutron transport model with various representations of the forest and different parameters describing the subsurface to match measured height profiles and time series of thermal and epithermal neutron intensities at a field site in Denmark. Overall, modeled thermal and epithermal neutron intensities are in satisfactory agreement with measurements; however, the choice of forest canopy conceptualization is found to be significant. Modeling results show that the effect of canopy interception, soil chemistry and dry bulk density of litter and mineral soil on neutron intensity is small. On the other hand, the neutron intensity decreases significantly with added litter-layer thickness, especially for epithermal neutron energies. Forest biomass also has a significant influence on the neutron intensity height profiles at the exam-

ined field site, altering both the shape of the profiles and the ground-level thermal-to-epithermal neutron ratio. This ratio increases with increasing amounts of biomass, and was confirmed by measurements from three sites representing agricultural, heathland and forest land cover. A much smaller effect of canopy interception on the ground-level thermal-to-epithermal neutron ratio was modeled. Overall, the results suggest a potential to use ground-level thermal-to-epithermal neutron ratios to discriminate the effect of different hydrogen contributions on the neutron signal.

1 Introduction

Soil moisture plays an important role in water and energy exchanges at the ground–atmosphere interface, but is difficult to measure at the intermediate spatial scale (hectometers). The cosmic-ray method has been developed to circumvent the shortcomings of existing measurement procedures for soil moisture detection at this scale (e.g., Zreda et al., 2008 and Franz et al., 2012). The cosmic-ray neutron intensity (eV range) at the ground surface is a product of the elemental composition and density of the surrounding air and soil matrix. Hydrogen is an essential element controlling neutron transport because of its physical properties and often relatively high concentration close to the land surface. As a result, neutron intensity is inversely correlated with the hydrogen content of the surrounding hectometers of air and

Table 1. Dynamics of different hydrogen pools.

	Static (yearly)	Quasi-static (sub-yearly)	Dynamic (daily)
Soil moisture			×
Tree roots		×	
Soil organic matter		×	
Water in soil minerals	×		
Vegetation (cellulose, water)		×	×
Snow		×	×
Puddles			×
Open water (river, sea, lake)		×	
Canopy-intercepted water			×
Buildings/roads	×		
Atmospheric water vapor			×

top decimeters of the ground (Zreda et al., 2008). Since soil moisture often forms the major dynamic pool of hydrogen within the footprint of the detector, neutron intensity measurements have been found to be suitable for soil moisture estimation.

Nonetheless, cosmic-ray neutron intensity detection also holds a potential for estimating the remaining pools of hydrogen (Zreda et al., 2008; Desilets et al., 2010). Hydrogen is stored statically in water in soil minerals and buildings/roads, quasi-statically in above- and belowground biomass, soil organic matter, snow and lakes/streams, or dynamically in soil water, atmospheric water vapor and canopy-intercepted precipitation (see Table 1).

To date, studies have primarily aimed to advance the cosmic-ray neutron method for soil moisture estimation by determining correction models to remove the effect of other influencing pools of hydrogen.

Rosolem et al. (2013) examined the effect of atmospheric water vapor on the neutron intensity (with energies 10–100 eV; 1 eV = 1.6×10^{-19} J) using neutron transport modeling and presented a method to rescale the measured neutron intensity to reference conditions. This correction for changes in atmospheric water vapor has become a standard procedure for the preparation of cosmic-ray neutron data along with corrections for temporal variations in barometric pressure and incoming cosmic radiation (Zreda et al., 2012).

Several studies have focused on improving the N_0 calibration parameter used for soil moisture estimation not only at forest field sites but also at high-yielding crop field sites such as maize. Bogena et al. (2013) demonstrated the importance of including the litter layer in the calibration for cosmic-ray neutron soil moisture estimation at field locations with a significant litter layer. Furthermore, the N_0 calibration parameter obtained from field measurements was found to decrease with increasing biomass (Rivera Villarreyes et al., 2011; Hornbuckle et al., 2012; Hawdon et al., 2014; Baatz et al., 2015). In order to account for this effect Baatz et al. (2015) defined a N_0-based correction model to remove the effect of biomass on the neutron intensity signal. A sim-

ilar correcting approach to improve the cosmic-ray neutron soil moisture estimation method by removing the influence of biomass and snow was presented by Tian et al. (2016). However, the study distinguishes itself by considering the ratio of the neutron intensity measured by the bare detector and the moderated detector instead of the effect on the N_0 parameter. Iwema et al. (2015) and Heidbüchel et al. (2016) applied the N_0 calibration function and obtained improved cosmic-ray neutron soil moisture estimates by performing more than one calibration campaign per field site and defining a site-specific calibration function. Heidbüchel et al. (2016) speculated that the curve shape of the standard N_0 calibration function is insufficient at the studied forest field site because of the presence of a litter layer and spatially heterogeneous soil moisture conditions within the neutron detector footprint. A different approach was presented by Franz et al. (2013b). Here a universal calibration function was proposed where separate estimates of the various hydrogen pools were included for cosmic-ray neutron soil moisture estimation.

Few studies have explored the potential of using the cosmic-ray neutron method for applications other than soil moisture. Desilets et al. (2010) distinguished snow and rain events using measurements of two neutron energy bands, and Sigouin and Si (2016) reported an inverse relationship between snow water equivalent and the neutron intensity measured using the moderated detector. Franz et al. (2013a) demonstrated an approach to isolate the effect of vegetation on the neutron intensity signal and estimated area average biomass water equivalent in agreement with independent measurements. The signals of biomass and canopy interception on neutron intensity, measured using the moderated detector, have been investigated by Baroni and Oswald (2015). They accounted the higher soil moisture estimated using the cosmic-ray neutron method compared to the up-scaled soil moisture measured at point scale to be the impact of canopy interception and biomass. The two pools of hydrogen were then separated in accordance to their dynamics.

The ability to separate the signals of the different hydrogen pools on the neutron intensity is valuable both for the advancement of the cosmic-ray neutron soil moisture estimation method and for the potential of additional applications. The potential of determining canopy interception and biomass from the cosmic-ray neutron intensity is of interest as they represent essential hydrological and ecological variables. Both are difficult and expensive to measure continuously at larger scales.

Canopy interception is for some climatic and environmental settings an important variable to include in water balance studies, as well as in hydrological and climatological modeling. For the forest site studied here the canopy interception loss was found to be 31–34 % of the gross precipitation (Ringgaard et al., 2014). A common method to estimate canopy interception is by subtracting the precipitation measured at ground level below canopy (throughfall) from precipitation measured above the forest canopy (gross precipitation) using standard precipitation gauges. However, the spatial scale of measurement is small and is not representative of larger areas as the canopy interception is highly heterogeneous. In order to obtain a representative measure of canopy interception, multiple throughfall stations must be installed. This is labor intensive and measurement uncertainties are significant. Precipitation underestimation due to wind turbulence, wetting loss and forest debris plugging the measurement gauge at the forest floor are sources of uncertainty (Dunkerley, 2000).

The forest biomass represents an important resource for timber industry and renewable energy. Furthermore, forest modifies the weather through the mechanisms and feedbacks related to evapotranspiration, surface albedo and roughness. Carbon sequestration by afforestation and an effective forest management is a widely used method to decrease the concentration of carbon dioxide in the atmosphere and thereby attenuate the greenhouse effect (Lal, 2008). The carbon sequestration in vegetation can be quantified by monitoring the growth of biomass over time. The most conventional and accurate method to estimate forest biomass is the use of allometric models describing the relationship between the biomass of a specific tree species and easily measurable tree parameters, such as tree height and tree diameter at breast height (Jenkins et al., 2003). However, this approach is time consuming and labor intensive because numerous trees have to be surveyed to obtain accurate and representative results (Popescu, 2007). Remote sensing technology offers alternative methods to estimate biomass as high correlations are found between spectral bands and vegetation parameters. One method providing high-resolution maps is airborne light detection and ranging (lidar) technology (Boudreau et al., 2008). The lidar system is installed in small aircrafts and digitizes the first and last return of near-infrared laser recordings. The canopy height can be obtained at decimeter gridsize scale and the biomass can be estimated from regression models. Instruments and aircraft surveys are expensive, and measurements of tree growth will often be at a coarse temporal resolution.

This study is an initial step towards reaching the overall objective of improving the cosmic-ray neutron soil moisture estimation method, especially at field locations with several pools of hydrogen. Furthermore, we wish to investigate the potential of biomass and canopy interception estimation using the cosmic-ray neutron intensity measurements. Here, the aim is to address this goal using only cosmic-ray neutron intensity measurements and not auxiliary information (e.g., biomass measurements using allometric models and tree surveys).

Previous studies examining the effect of hydrogen on cosmic-ray neutron intensity has for most cases considered a single neutron energy range (neutron intensity measured using the moderated neutron detector) at a single height level (typically 1.5 m above the ground). Thermal and epithermal neutrons are both sensitive to hydrogen. However, they are characterized by very different physical properties and reaction patterns resulting in different height profiles, as well as unique responses to environmental settings at the immediate ground–atmosphere interface. For this reason, thermal and epithermal neutron intensity at multiple height levels above the ground surface are considered in this study as the combination may provide additional information. Furthermore, neutron transport modeling sets the basis for this study. Neutron transport modeling of specific sites is limited and has only been performed for non-vegetated field sites (Franz et al., 2013b; Andreasen et al., 2016). In this context, forest sites are especially complex to conceptualize as the number of free parameters is relatively high (e.g., biomass, litter, soil chemistry, interception and the structure of the forest). Here, we first focus on modeling a forest field site. The model is developed from measured soil and vegetation parameters at the specific locality. The modeled neutron intensity profiles are evaluated against profile measurements, and time series of neutron intensity measurements at two heights. Following, the environmental impact on thermal and epithermal neutron intensities are identified and quantified by applying a sensitivity analysis. The environmental impact refers to the effect of the specific properties and settings of the field site on neutron transport. This includes vegetation, litter, soil composition and layers, and canopy interception. For the sensitivity analysis, one component at the time is changed in the model and the sensitivity of the component is quantified by calculating the change in the neutron intensity relative to a reference model. Measurements at an agricultural field site with no biomass and at a heather field site with a smaller amount of biomass are used to underpin the influence of certain environmental variables (e.g., biomass, litter layer).

To our knowledge this is the first study based on both measurements and modeling, which provides a quantitative analysis of the potential of using the cosmic-ray technique for estimation of interception and biomass.

2 Method

2.1 Terminology and neutron energies

The energy of a neutron determines the probability of the neutron interacting with other elements (cross section) and the type of interaction (i.e., absorbing or scattering). Overall, an important threshold for the behavior of low-energy neutrons is present at energies somewhere below 0.5 eV. The specific energy ranges of thermal, epithermal and fast neutrons are ambiguous. For the purpose of this paper the following terminology for neutron energies is used:

- thermal: energy range 0–0.5 eV;

- epithermal: energies above 0.5 eV;

- fast: energy range 10–1000 eV.

When modeling neutron transport for hydrological applications, it is common to consider fast energy ranges (10–100 or 10–1000 eV) (Desilets et al., 2010; Desilets and Zreda, 2013; Rosolem et al., 2013; Franz et al., 2013b; Köhli et al., 2015), whereas measurements using standard soil moisture neutron detectors are sensitive to the entire epithermal energy range (Andreasen et al., 2016). Here, the term epithermal neutrons will be used for both measured neutrons of energies above 0.5 eV and modeled neutrons of energies 10–1000 eV.

The probability of absorption reactions is greater for thermal neutrons, while the probability of scattering reactions is greater for neutrons of epithermal energies. For this reason thermal and epithermal neutron height profiles are very different at the ground–atmosphere interface. The epithermal neutron intensity increases with height above the ground surface as the neutrons at higher elevations have been scattered less than neutrons closer to the ground surface. The production rate of thermal neutrons is high in the soil and low in the air. This is related to the high density of the soil and the low density of air. The absorption rate of thermal neutrons is significant in both the ground and in the air. In the air, this is due to the presences of nitrogen. This results in a decreasing thermal neutron intensity with height until approximately 150 m at which point the thermal neutron intensity is unaffected by the soil. Above this point the thermal neutron intensity will increase with height following a similar curve as neutrons of higher energies.

2.2 Cosmic-ray neutron detection

2.2.1 Equipment

Cosmic-ray neutron intensity was measured using the CR1000/B system from Hydroinnova LLC, Albuquerque, New Mexico. The system has two detectors that consist of tubes filled with boron-10 (enriched to 96 %) trifluoride ($^{10}BF_3$) proportional gas. The neutron detection relies on the $^{10}B(n, \alpha)^7Li$ reaction for converting thermal neutrons into charged particles (α) and then into an electronic signal. One detector is unshielded (bare detector), while the other is shielded by 25 mm of high-density polyethylene (moderated detector). These different configurations give the bare and moderated tubes different energy sensitivities.

The thermal neutron absorption cross section of ^{10}B is very high (3835 barns) (1 barn = 10^{-24} cm^2) (Sears et al., 1992). This absorption cross section decreases rapidly with increasing neutron energy following a $1/E_n^{0.5}$ law (where E_n is neutron energy) (Knoll, 2010). Therefore, the energies measured by the bare tube comprise a continuous distribution, which is heavily weighted toward thermal neutrons (< 0.5 eV), with a small proportion of epithermal neutrons also being detected (< 10 %) (Andreasen et al., 2016).

The moderated detector is more sensitive to higher neutron energies (> 0.5 eV). The purpose of the polyethylene is to slow (moderate) epithermal neutrons through interactions with hydrogen in order to increase the probability of them being captured by ^{10}B in the detector. At the same time the polyethylene attenuates the thermal neutron flux through neutron capture by hydrogen. Nonetheless, still a large proportion originates from below 0.5 eV (approximately 40 % of the thermal neutrons detected by the bare detector) (Andreasen et al., 2016).

Following Poissonian statistics (Knoll, 2010), the relative measurement uncertainty of a given neutron intensity, N, decreases with increasing neutron intensity as the standard deviation equals $N^{0.5}$.

The measured neutron intensities are corrected for variations in barometric pressure, atmospheric water vapor and incoming cosmic-ray intensity following procedures of Zreda et al. (2012) and Rosolem et al. (2013). Unfortunately, the water vapor correction of Rosolem et al. (2013) is only valid for epithermal neutron measurements. Since the development of correction methods is beyond the scope of this study, we refrained from using a vapor correction for the measured thermal neutron intensities. From preliminary modeling conducted by the authors and R. Rosolem, personal communication (2015), we believe that this missing correction will only have a minor effect on our results (Andreasen et al., 2016). Nevertheless, we suggest that future studies should investigate the effect of water vapor on thermal neutron intensities and develop appropriate correction methods.

2.2.2 Pure thermal and epithermal neutron detection

We expect thermal and epithermal neutrons to have unique responses to environmental properties and settings. Therefore, it is important to consider pure signals of thermal and epithermal neutrons, and not simply the raw neutron intensity signal measured by the bare and moderated detectors. In order to limit the epithermal and thermal neutron contribution to the bare and the moderated detectors, respectively, we use the cadmium-difference method (Knoll, 2010; Glasstone and Edlund, 1952). The thermal absorption cross section of cadmium is very high (approximately 3500 barns) for neutron energies below 0.5 eV. The cross section drops to approximately 6.5 barns at neutron energy 0.5 eV and remains low with increasing neutron energies. Thus, a cadmium-shielded neutron detector only measures neutrons of energies higher than 0.5 eV. The epithermal neutron intensity was measured from a cadmium-shielded moderated detector, while the thermal neutron intensity was calculated by subtracting the neutron intensity measured by the cadmium-shielded bare detector from the neutron intensity measured by the bare detector (unshielded). The cadmium-difference method is described in Andreasen et al. (2016) in detail.

Appropriate neutron energy correction models were applied in order to obtain pure thermal and pure epithermal neutron intensity measurements for the time periods when the cadmium-difference method was not applied (Andreasen et al., 2016). The neutron energy correction models were obtained from field campaigns applying the cadmium-difference method on bare and moderated detectors at various locations (height levels and land covers).

2.2.3   Footprint

The footprint of the bare detector is unexplained, while the footprint of the moderated detector was determined from modeling by Desilets and Zreda (2013) and Köhli et al. (2015). However, the findings of these two studies were inconsistent. Desilets and Zreda (2013) used the neutron transport code Monte Carlo N-Particle eXtended (MCNPx) and found the footprint to be nearly 600 m in diameter in dry air, while Köhli et al. (2015) using the Ultra Rapid Adaptable Neutron-Only Simulation (URANOS) estimated the footprint to be 260–480 m in diameter depending on the air humidity, soil moisture and vegetation. The potential mismatch in the footprint of the bare and the moderated detectors is a concern when combining the neutron intensity measurements. Nevertheless, the environmental conditions at the field sites are fairly homogeneous, and although the footprint might be different we assume as a first approximation that the neutron intensity measured using the bare and the moderated detectors are comparable.

2.2.4   Field measurements

Three field sites are used in this study; the primary site is Gludsted plantation, and two secondary sites are Voulund farmland and Harrild heathland. The three sites are all located within the Skjern River catchment in the western part of Denmark and represents the three major land use types (Fig. 1) of the Danish hydrological observatory (HOBE) (Jensen and Illangasekare, 2011). The sites are situated at an elevation of approximately 50–60 m above sea level on an outwash plain from the last glaciation composed of nutrient depleted sandy stratified soils. Harrild heathland is located 1 km south of Voulund farmland, both approximately 10 km west of Gludsted plantation.

Gludsted plantation forest field site ($56°04'24''$ N, $9°20'06''$ E) is situated within a coniferous forest plantation covering an area of around 3500 ha. The trees of the plantation are densely planted in rows and are in general composed of Norway spruce with small patches of Sitka spruce, Larch and Douglas fir. Within the field site area (38 ha), the trees were estimated to be up to 25 m high and the dry aboveground biomass to be around $100 \pm 46\,t\,ha^{-1}$ (1 standard deviation) using lidar images from 2006 and 2007 (Nord-Larsen and Schumacher, 2012). The dry belowground biomass was calculated to be $25\,t\,ha^{-1}$ using a root-to-shoot ratio (the weight of the roots to the weight of the aerial part of the tree) for Norway spruce of 0.25 (Levy et al., 2004). Information on the vegetation at the forest field site (e.g., tree species, ages, heights and trunk diameters) is acquired

Figure 1. Map showing the location of the three field sites; G: Gludsted plantation, V: Voulund farmland, and H: Harrild heathland. The circles represent the footprint of the neutron detector (radius = 300 m). Green color corresponds to forest, beige to agriculture and purple to heathland.

from a register managed by the Danish Nature Agency (representative of the 2012 conditions); see Table 2.

In Scandinavian forests, around 79 % of the total aboveground biomass of Norway spruce is stored within the tree trunks. The remaining 21 % is found in the branches and needles (termed foliage). A typical density of the tree trunk is $0.83\,g\,cm^{-3}$ (Serup et al., 2002). The major component of the tree biomass is cellulose ($C_6H_{10}O_5$) and represents around 55 % of the total mass, while the remaining 45 % is vegetation water (Serup et al., 2002). Based on these approximations, the wet above- and belowground biomass at the field site area are estimated to be 182 and $45\,t\,ha^{-1}$, respectively. With a leaf area index (LAI) of 4.5 and a canopy interception capacity coefficient of $0.5\,mm\,LAI^{-1}$ (Andreasen et al., 2013), the maximum storage of canopy-intercepted rain is estimated to be 2.25 mm.

Soil samples were collected within the footprint of the cosmic-ray neutron detector on 26–27 August 2013 following the procedure of Franz et al. (2012). Based on these samples the organic-rich litter layer is found to be 5–10 cm thick. The dry bulk density of the litter and mineral layer are calculated by oven-drying the soil samples (Table 2), and the soil organic matter content of the mineral soil is determined from the loss-on-ignition method (16.9 % in 10–20 cm depth and 7.6 % in 20–30 cm depth). A time series of volumetric soil moisture is calculated from cosmic-ray neutron intensity, starting in spring, 2013, using the standard N_0 method as presented in Desilets et al. (2010). Lastly, the chemical composition of the soil matrix is estimated for two random soil samples collected at 20–25 cm depth using the X-ray fluorescence (XRF) analysis (Table 3).

The element Gadolinium (Gd) can have a significant impact on thermal neutron intensity even at low concentrations due to its very high absorption cross section of 49 000 barns. The detection limit of the XRF in this study is 50 ppm for Gd.

Table 2. Average tree height, tree diameter and dry bulk density (bd_{dry}) of the litter layer and the mineral soil at Gludsted plantation field site. Tree height and diameter are representative of conditions for year 2012.

	Average	Standard deviation	Max.	Min.
Tree height* [m]	11	6	25	3
Tree diameter* [m]	0.14	0.08	0.34	0.03
Dry bulk density litter layer [$g\,cm^{-3}$]	0.34	0.29	1.09	0.09
Dry bulk density mineral soil [$g\,cm^{-3}$]	1.09	0.28	1.53	0.22

* Data obtained from the Danish Nature Agency.

Table 3. Chemical composition of major elements at Gludsted plantation determined using X-ray fluorescence analysis on soil samples collected in 0.20–0.25 m depth.

	Gludsted plantation [%]
O	52.78
Si	44.86
Al	1.54
K	0.53
Ti	0.29

The two soil samples from Gludsted plantation both have Gd concentration below the detection limit of the XRF. Inductively coupled plasma mass spectrometry (ICP-MS) detects metals and several non-metals at very small concentrations and was used to characterize the soil chemistry of a nearby field site with similar soil conditions (Salminen et al., 2005). A Gd concentration of 0.51 ppm was found at that site and we assume this value to be representative of the conditions at Gludsted plantation.

Gludsted plantation is a heavily equipped research field site with a 38 m high tower for measurements at multiple heights within the forest canopy. At Gludsted plantation, CR1000/B systems were installed at ground level (1.5 m height) and canopy level (27.5 m height) in the spring of 2013. Hourly neutron intensities have been continuously detected (Andreasen et al., 2016) except for short periods where the detectors were used for other types of measurements or during times of malfunctions. Neutron intensity profiles extending from the ground surface to 35 m height above the ground were measured at approximately 5 m increments during two field campaigns on 28–29 November 2013 and 12–14 March 2014. In order to obtain comparability between measurements and modeling pure thermal and epithermal neutron signals were estimated using neutron energy correction models on measurements from bare and moderated detectors, respectively. Both the time series and neutron height profile measurements were corrected. Additionally, during the field campaign on 12–14 March 2014 an epithermal neutron intensity profile (with no thermal contribu-

tion) was measured using a cadmium-shielded moderated detector (Andreasen et al., 2016). For the profile measurements neutron intensities were recorded at a 10 min time resolution. As the thermal neutron intensity decreases significantly with height, we choose to extend the time of measurement with the height level to maintain a low and consistent measurement uncertainty. The volumetric soil moisture estimated from the cosmic-ray neutron intensity (Zreda et al., 2008) was $0.18\,m^3\,m^{-3}$ during both field campaigns.

Voulund farmland (56°02′14″ N, 9°09′38″ E) is an agricultural field site. In 2015, the fields were cropped with spring barley. After harvest in the late summer until plowing in spring 2016 (prior to sowing) the fields were covered with stubble (around 10 cm high). A 25 cm layer of relatively organic-rich soil (4.45 % soil organic matter) is found at the top of the soil column and is a result of the cultivation practices. More information about the field site can be found in Andreasen et al. (2016). Ground-level neutron intensities were measured on 22 and 23 September 2015 at Voulund farmland (Andreasen et al., 2016). The measurements were conducted using the bare and the moderated neutron detectors normally installed at Gludsted plantation and data were logged every 10 min. In order to obtain pure thermal and epithermal neutron intensities the neutron energy correction models were applied.

Harrild heathland (56°01′33″ N, 9°09′29″ E) is a shrubland field site dominated by grasses and heather. The heathland is maintained by controlled burning; however, the field site area has not recently been burned. The organic-rich litter layer is found to be around 10 cm thick during soil sampling field campaigns at the field site. Due to podsolization a low permeable hardpan-layer hindering percolation to deeper depths is present at around 25–30 cm depth. In the period from 27 October to 16 November 2015, the ground-level thermal and epithermal neutron intensity was measured directly at Harrild heathland using the cadmium-difference method (Knoll, 2010). The cadmium-difference method was applied using two bare and one moderated detector normally installed at Gludsted plantation. The neutron intensity was integrated and recorded on an hourly basis.

2.3 Neutron transport modeling

The three-dimensional Monte Carlo N-Particle transport code version 6 (MCNP6) (Pelowitz, 2013) simulating thermal and epithermal neutrons is used to model the forest field site. The code holds libraries of measured absorption and scattering cross sections used to compute the probability of interactions between Earth elements and neutrons. The MCNP6 combines Monte Carlo N-Particle Transport code version 5 (MCNP5) and Monte Carlo N-Particle Extended Radiation Transport code (MCNPX). MCNPX has been used for most neutron transport modeling within the field of hydrology (Desilets et al., 2013; Rosolem et al., 2013; Zweck et al., 2013). However, the improved and more advanced MCNP6 has recently been introduced. This updated version provided neutron intensity profiles in better agreement with measurements at the Voulund farmland field site (Andreasen et al., 2016).

The number of particle histories released at the center of the upper boundary of the model domain is specified to obtain an uncertainty below 1 %. The released particles represent a distribution of high-energy particles typical for the spectrum of incoming cosmic-rays traveling through the atmosphere. The modeled neutron intensities are normalized per unit source particle providing relative values (Zweck et al., 2013). In order to obtain values comparable to measurements conversion factors are used (Andreasen et al., 2016). The conversion factors 3.739×10^{12} and 1.601×10^{13} are multiplied by the modeled thermal neutron fluences in the energy range of 0–0.5 eV and epithermal neutron fluences in the energy range 10–1000 eV, respectively. We stress that, the conversion factors are detector-specific as well as dependent on the horizontal area of the model setup in MCNP6.

2.3.1 The Gludsted plantation reference model

The model domain of MCNP6 is defined by cells of varying geometry, and each cell is assigned a specific chemical composition and density. The lowest 4 m of the Gludsted plantation reference model consists of subsurface layers. The chemical composition of the mineral soil is prescribed according to the chemical composition from XRF measurements: assumed Gd concentration of 0.51 ppm, wet belowground biomass (cellulose) of 45 t ha^{-1}, dry bulk density of 1.09 g cm^{-3} and volumetric soil moisture content of 0.18 m^3 m^{-3}. The litter layer is defined according to the chemical composition of cellulose, dry bulk density of 0.34 g cm^{-3} and moisture content similar to that of mineral soil. The same volumetric soil moisture was used for the whole soil column, as the volumetric soil moisture profile was unknown for the days of neutron profile measurements. The atmosphere is composed of 79 % nitrogen and 21 % oxygen by volume and extends from the forest canopy surface to the upper boundary of the model domain at approximately 2 km height. Here, an incoming spectrum

adapted to the specific level of the atmosphere is specified (Hughes and Marsden, 1966). The density of air is assumed to be 0.001165 g cm^{-3}. Throughout the domain, multiple sub-layers of varying vertical discretization cover the vertical extent of the model in order to record neutron intensities at multiple heights and depths from the ground surface. The thickness of the layers decreases with proximity to the ground surface ranging in thickness from 0.025 to 0.20 m for the subsurface layers and from 1 to 164 m for the layers above the ground surface. The neutron intensity detectors are represented by 1 m high layers extending the full lateral model domain (400 m $\times$ 400 m) and are used from the ground to 28 m height corresponding to the measured heights. Reflecting surfaces constrain the model domain. Thus, the particles reaching a model boundary will be reflected specularly back into the model domain. Wet aboveground biomass of 182 t ha^{-1} is distributed within the forest canopy layers, i.e., from the ground surface to 25 m above the ground (Table 4).

The proper way to conceptualize the forest canopy in the model setup is not obvious and the sensitivity to forest representation on neutron intensity is therefore investigated using four model setups of increasing complexity. In the first representation (model foliage; Fig. 2b) the same material composed of cellulose and air (foliage) is assigned all forest canopy layers. In order to obtain a wet aboveground biomass of 182 t ha^{-1}, a relatively low density of 0.00189 g cm^{-3} is calculated for the material. In order to allow for a forest canopy layer to be composed of multiple materials (cellulose and air) and densities (massive tree trunks and less dense foliage, air), the horizontal discretization of the forest canopy layers is reduced to smaller cells for the next tree model setups. The bole of each tree is for all three model setups represented by a cylinder with a diameter of 0.14 m, a composition of cellulose, and a density of 0.83 g cm^{-3}. A tree is placed at the center of each cell and extends from the ground surface to the top of the forest canopy layer. In the second representation (model tree trunk, air; Fig. 2c) the horizontal discretization of the forest canopy layers is set to 4.20 by 4.20 m and the remaining volume beyond the bole of the tree consist of air alone (density 0.001165 g cm^{-3}). Thus, for this model all biomass is stored in the bole of the trees and the cell size is adjusted to obtain a wet aboveground biomass of 182 t ha^{-1} resulting in 9070 trees within the model domain. In the third representation (model tree trunk, foliage; Fig. 2d) the horizontal discretization of the forest canopy layers is 4.72 by 4.72 m and the remaining volume beyond the bole of the tree is made of foliage. As previously described, the share of biomass stored in the tree trunk and the foliage is 79 and 21 %, respectively, typical of Norway spruce. The foliage material is a composite of cellulose and air and the density is the sum of the two (0.001318 g cm^{-3}). A total of 7182 trees are evenly spaced within the model domain. The fourth and most complex forest canopy conceptualization (model tree trunk, foliage; Fig. 2e) is equal to the model tree trunk,

Table 4. Forest properties used in modeling.

	Models				
	No vegetation	50 t ha^{-1}	100 t ha^{-1b}	200 t ha^{-1}	400 t ha^{-1}
Dry aboveground biomass [t ha^{-1}]	0	50	100	200	400
Wet aboveground biomass [t ha^{-1}]	0	91	182	364	727
Dry belowground biomass [t ha^{-1}]	0	12.5	25	50	100
Wet belowground biomass [t ha^{-1}]	0	23	45	91	182
Tree trunk density [g cm^{-3}]^a	–	0.83	0.83	0.83	0.83
Tree trunk radius [m]^a	–	0.07	0.07	0.07	0.07
Tree height [m]^a	–	25	25	25	25
Foliage density [g cm^{-3}]^a	–	0.00134	0.00151	0.00185	0.00255
Foliage band [m]^a	–	2.44	1.70	1.18	0.82
Sub-cell area [m]^a	–	6.67 × 6.67	4.72 × 4.72	3.34 × 3.34	2.36 × 2.36

^{a} Specific for model with forest conceptualization of model tree trunk, foliage, air. ^{b} Reference model.

foliage except that air is also included in the description of the forest canopy layers and the density of the foliage is increased to obtain the same aboveground biomass as for the other models. The foliage is specified as a 1.7 m thick band around the tree cylinder and the density of foliage material composed of air and cellulose is 0.00151 g cm^{-3}.

2.3.2 Sensitivity to environmental conditions

The sensitivity of thermal and epithermal neutron intensities to volumetric soil moisture is examined using modeling. The volumetric soil moisture in the Gludsted plantation reference model is specified to 0.18 m^3 m^{-3} and both drier and wetter soils are modeled to test the sensitivity, i.e., 0.05, 0.10, 0.25, 0.35 and 0.45 m^3 m^{-3}. The forest canopy conceptualizations of model tree trunk, foliage, air and model foliage are used.

The thermal and epithermal neutron intensity is a product of hydrogen abundance as well as elemental composition. The Gludsted plantation reference model with the complex forest conceptualization (model tree trunk, foliage, air) is used to test the sensitivity of thermal and epithermal neutron intensities to soil chemistry. It holds the most complex soil chemistry (fourth-order complexity) with multiple subsurface layers composed of measured concentrations of major elements determined by XRF, soil organic matter, gadolinium and roots (Table 3). In order to test the effect of simplifying the soil chemistry a component is excluded one at the time: (1) third-order complexity – soil organic matter is excluded; (2) second-order complexity – soil organic matter and roots are excluded; (3) first-order complexity – soil organic matter, roots and gadolinium are excluded; and (4) pure SiO$_2$ – all other components are excluded.

The sensitivity of the modeled thermal and epithermal neutron intensities to the presence of the organic litter layer is investigated using the Gludsted plantation reference model with the complex forest conceptualization (model tree trunk,

foliage, air), in which the thickness of the litter layer is set to be 10.0 cm. Sensitivity simulations are carried out for the following thicknesses of the litter layer: 0.0, 2.5, 5.0 and 7.5 cm. For all litter-layer models, the total thickness of the subsurface is kept constant at 4 m.

The materials of forest floor litter and mineral soil differ distinctly in terms of chemical composition and dry bulk density. The determination of dry bulk density of the two materials is characterized by high measurement uncertainty, especially for the litter as sampling and drying is very challenging for materials including large amounts of soil organic matter (O'Kelly, 2004). Given that the elemental composition and density of the soil matrix is relevant for the neutron intensity the sensitivity of dry bulk density on thermal and epithermal neutron intensity is examined. The dry bulk density of the Gludsted plantation reference model is 0.34 g cm^{-3} for the litter layer and 1.09 g cm^{-3} for the mineral soil. The Gludsted plantation reference model with the complex forest conceptualization (model tree trunk, foliage, air) is used to test the sensitivity applying four scenarios: (1) higher dry bulk density of the litter layer (0.50 g cm^{-3}), (2) higher dry bulk density of the mineral soil (1.60 g cm^{-3}), (3) lower dry bulk density of the litter layer (0.20 g cm^{-3}), and (4) lower dry bulk density of the mineral soil (0.60 g cm^{-3}). All values with the exception of higher dry bulk density of 1.60 g cm^{-3} for the mineral soil (standard value for quartz; soil particle density of 2.65 g cm^{-3} and a porosity of 0.40) are within the range of the measurements at the site (see Table 2).

The Gludsted plantation reference model with the complex forest conceptualization (model tree trunk, foliage, air) is used to test the sensitivity to canopy interception by increasing the density and water content of the cells described by foliage material. The forest canopy of the reference model is dry (foliage material density 0.00151 g cm^{-3}). In order to test the effect, water equivalent to 1 mm (foliage material density 0.00155 g cm^{-3}), 2 mm (foliage material den-

Vertical model conceptualization

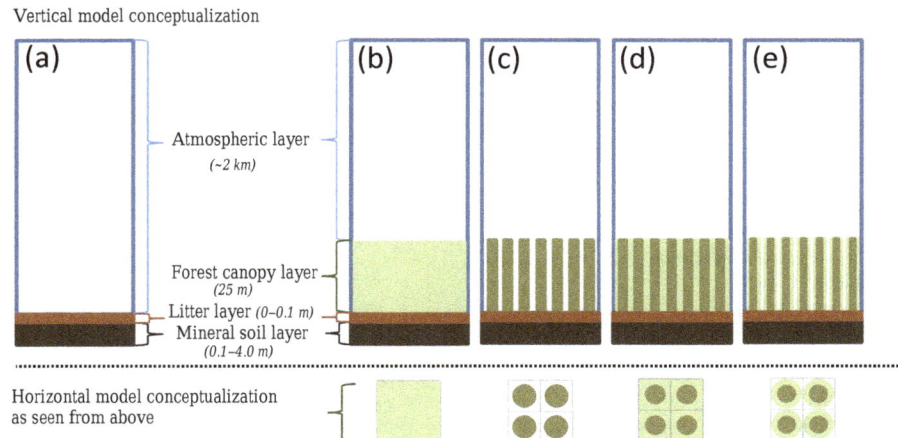

Figure 2. Model conceptualizations of forest. **(a)** no forest canopy layer (model name: No vegetation; **(b)** homogeneous foliage layer with a uniformly distributed biomass (model name: Foliage); **(c)** cylindrical tree trunks with air in between (model name: Tree trunk, Air); **(d)** cylindrical tree trunks with foliage in between (model name: Tree trunk, Foliage, Air); **(e)** cylindrical tree trunks enveloped in a foliage cover with air in between (model name: Tree trunk, Foliage, Air). The bottom four figures illustrate the forest conceptualization seen from above.

sity $0.00159\,\mathrm{g\,cm^{-3}}$) and 4 mm (foliage material density $0.00167\,\mathrm{g\,cm^{-3}}$) of canopy interception is added to the foliage volume. This changes both the wet bulk density and the atomic fraction of the foliage material.

The sensitivity to biomass is investigated using the Gludsted plantation reference model with the complex forest conceptualization (model tree trunk, foliage, air) and the simplified model setup (model foliage). The biomass of the Gludsted plantation reference model is equivalent to a dry aboveground biomass of $100\,\mathrm{t\,ha^{-1}}$ and a dry belowground biomass of $25\,\mathrm{t\,ha^{-1}}$, following the root-to-shoot ratio of 0.25 typical of Norway spruce. This distribution is used for both model setups. For the sensitivity analysis, one model without vegetation (model $0\,\mathrm{t\,ha^{-1}}$, Fig. 2a) and three models with different amounts of biomass are used (see Table 4). The forest canopy layer extending uniformly from the ground to 25 m above the ground surface is for the model with no vegetation assigned with the material composition and density of air. The amount of biomass modeled for the three remaining models is equivalent to a dry aboveground biomass of (1) $50\,\mathrm{t\,ha^{-1}}$, (2) $200\,\mathrm{t\,ha^{-1}}$ and (3) $400\,\mathrm{t\,ha^{-1}}$. The size of the cells in the forest layers and the density of the foliage material are adjusted in order to obtain the correct amount of biomass.

3 Results

3.1 Gludsted plantation

The neutron intensity profiles for Gludsted plantation are modeled using four different forest canopy conceptualizations. The model results are presented in Fig. 3 along with time series of hourly and daily ranges of thermal and epither-

mal neutron intensities collected at the Gludsted plantation during the period 2013–2015 (Andreasen et al., 2016), and measured/estimated thermal and epithermal neutron intensity profiles (November 2013 and March 2014). Note that a decrease in the epithermal neutron intensity from the ground level to 5 m above the ground surface was measured in March 2014. This is in disagreement with theory (see Sect. 2.1) and is expected to be a result of measurement uncertainties. Following the Poissonian statistics the relative uncertainty decreases with increasing neutron intensity. The relative measurement uncertainty is therefore higher for the hourly time series data than for the multi-hourly (2–12 h) and daily measurements. Accordingly, we choose to rely mostly on the daily averages of time series measurements.

Overall, time series and profile measurements provide similar results in agreement with theory. The thermal neutron intensity decreases considerably with height above ground surface and is at canopy level reduced by around 50 % compared to at the ground level. The epithermal neutron intensity increases slightly with height and is around 10–15 % higher at the canopy level compared to the ground level. Overall, a remarkable agreement between measured and modeled neutron intensities is seen in Fig. 3. We stress that no calibration of the governing physical properties in the forest model is performed and that the estimates are based on measured properties. The ground- and canopy-level thermal and epithermal neutron intensity for the four forest canopy conceptualization models are provided in Table 5. All modeled neutron intensity profiles are within the range of hourly time series measurements, and in particular the thermal neutron profiles are in agreement with measurements. The models of the more complex forest canopy conceptualizations, including a tree trunk, provide similar thermal and epithermal neu-

Table 5. Modeled ground level (1.5 m) and canopy-level (27.5 m) thermal neutron intensity and epithermal neutron intensity for the Gludsted plantation models including four different forest canopy conceptualizations (see Fig. 3).

Models	Thermal 1.5 m	Thermal 27.5 m	Epithermal 1.5 m	Epithermal 27.5 m
Foliage	573	207	681	813
Tree trunk, air	484	272	610	695
Tree trunk, foliage	536	261	619	716
Tree trunk, foliage, air	504	257	623	717

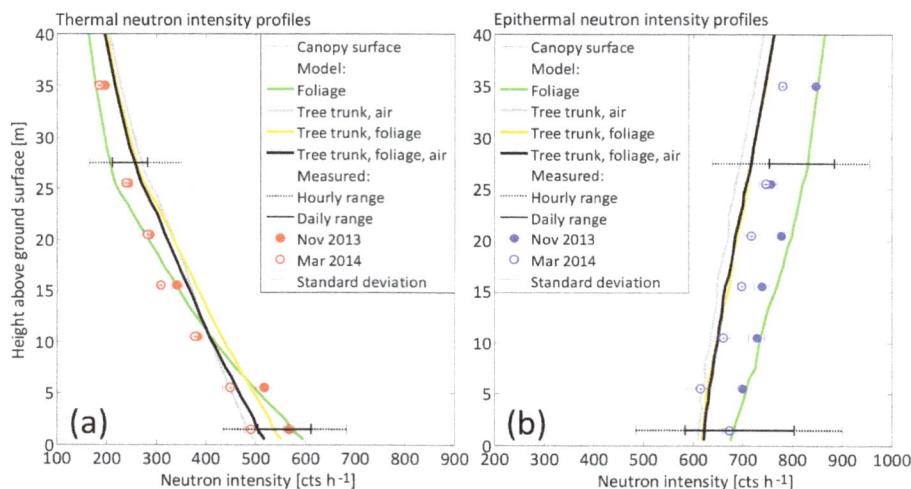

Figure 3. Measured and modeled (**a**) thermal and (**b**) epithermal neutron intensity profiles at Gludsted plantation. Hourly and daily ranges of variation of thermal and epithermal neutron intensities at ground and canopy level for the period 2013–2015. Gludsted plantation is modeled using four different forest canopy conceptualizations (see Fig. 2).

tron profiles. The ground- and canopy-level thermal neutron intensity of models with forest canopy conceptualization of model tree trunk, foliage and model tree trunk, foliage, air are within the daily ranges of the time series measurements. In contrast, the modeled epithermal neutron profiles of the more complex models are slightly underestimated and the profile slope is steeper than the measured profiles. Nevertheless, the modeled epithermal neutron intensity profile is still within the ranges of the time series of hourly measurements at both height levels. The neutron intensity profiles of the simpler forest canopy conceptualization of model foliage is less steep and is the only model providing an epithermal neutron intensity profile within the daily ranges of the time series measurements at both the ground and canopy level. All in all, then compared to the range of daily time series measurements, the best fit of the thermal measurements is found using a more complex conceptualization, while the simple foliage conceptualization matches the epithermal measurements better.

In this study, a sensitivity analysis is performed using the most complex model to examine the effect of soil moisture, soil dry bulk density and composition, litter and mineral soil-layer thickness, canopy interception and biomass on

the thermal, and epithermal neutron transport at the immediate ground–atmosphere interface. Since the most appropriate forest canopy conceptualization is not obvious from Fig. 3, the simplest forest canopy conceptualization was also used to examine the effect of soil moisture and biomass on the neutron transport.

3.2 Soil moisture

The modeled thermal and epithermal neutron intensity profiles of model tree trunk, foliage, air and model foliage using six different volumetric soil moisture, 0.05, 0.10, 0.18, 0.25, 0.35 and 0.45 $m^3\,m^{-3}$, are presented in Figs. 4 and 5, respectively. To enable comparison the measurements included in Fig. 3 are also included in Figs. 4 and 5. The sensitivity of soil moisture on thermal and epithermal neutron intensities at the ground- and canopy-level relative to the model tree trunk, foliage, air and model foliage at reference conditions (volumetric soil moisture 0.18 $m^3\,m^{-3}$) is provided in Table 6.

As expected, the thermal and epithermal neutron intensity decreases with increasing soil moisture (Table 6, Figs. 4 and 5). For both model setups, the largest changes in neutron intensity occur at the dry end of the soil moisture range and

Table 6. Sensitivity in modeled ground-level (1.5 m) and canopy-level (27.5 m) thermal neutron intensity and epithermal neutron intensity due to (1) volumetric soil moisture, (2) soil chemistry, (3) litter-layer thickness, (4) mineral soil and litter dry bulk density (bd_{dry}), (5) canopy interception and (6) biomass. The sensitivity is provided in absolute values and are relative to the simulations based on model tree trunk, foliage, air[a] and model foliage[b], (see Fig. 3 and Table 5). Values provided in parentheses specifies the direct effect of one-by-one excluding soil organic matter (third-order complexity), Gd (second-order complexity), belowground biomass (first-order complexity) and site-specific major elements soil chemistry (SiO_2).

	Models	Thermal 1.5 m	Thermal 27.5 m	Epithermal 1.5 m	Epithermal 27.5 m
Soil moisture	$0.18\,m^3\,m^{-3}$	504[a]	257[a]	623[a]	717[a]
(Fig. 4)	$0.05\,m^3\,m^{-3}$	100	47	131	109
	$0.10\,m^3\,m^{-3}$	45	20	58	50
	$0.25\,m^3\,m^{-3}$	−25	−12	−27	−23
	$0.35\,m^3\,m^{-3}$	−47	−22	−53	−45
	$0.45\,m^3\,m^{-3}$	−59	−28	−69	−59
Soil moisture	$0.18\,m^3\,m^{-3}$	573[b]	207[b]	681[b]	813[b]
(Fig. 5)	$0.05\,m^3\,m^{-3}$	119	40	142	115
	$0.10\,m^3\,m^{-3}$	56	18	68	53
	$0.25\,m^3\,m^{-3}$	−27	−9	−30	−23
	$0.35\,m^3\,m^{-3}$	−50	−16	−55	−48
	$0.45\,m^3\,m^{-3}$	−64	−21	−74	−61
Soil chemistry	Fourth-order complexity	504[a]	257[a]	623[a]	717[a]
	Third-order complexity	19 (+19)	8 (+8)	25 (+25)	14 (+14)
	Second-order complexity	18 (−1)	9 (+1)	27 (−2)	17 (+3)
	First-order complexity	22 (+4)	10 (+1)	26 (−1)	18 (+1)
	SiO_2	27 (+5)	11 (+1)	23 (−3)	19 (+1)
Litter layer	10.0 cm	504[a]	257[a]	623[a]	717[a]
(Fig. 6a)	7.5 cm	11	4	26	22
	5.0 cm	18	9	53	41
	2.5 cm	24	12	85	71
	No litter layer	22	17	131	113
Density	Gludsted plantation[a]	504[a]	257[a]	623[a]	717[a]
	Higher litter layer bd_{dry}	−7	−5	−10	−6
	Higher mineral soil bd_{dry}	15	5	17	10
	Lower litter layer bd_{dry}	7	2	14	10
	Lower mineral soil bd_{dry}	−26	−13	−22	−18
Canopy interception	Dry canopy	504[a]	257[a]	623[a]	717[a]
(Fig. 6b)	1 mm	4	−2	−3	0
	2 mm	7	−3	−5	5
	4 mm	15	−7	−5	2
Biomass	$100\,t\,ha^{-1}$	504[a]	257[a]	623[a]	717[a]
(Fig. 6c)	No vegetation	−67	−21	99	85
	$50\,t\,ha^{-1}$	−16	−8	45	33
	$200\,t\,ha^{-1}$	14	2	−70	−47
	$400\,t\,ha^{-1}$	21	2	−172	−116
Biomass	$100\,t\,ha^{-1}$	573[b]	207[b]	681[b]	813[b]
(Fig. 6d)	No vegetation	−136	29	41	−28
	$50\,t\,ha^{-1}$	0	24	13	−23
	$200\,t\,ha^{-1}$	−9	−32	−26	22
	$400\,t\,ha^{-1}$	−48	−59	−82	73

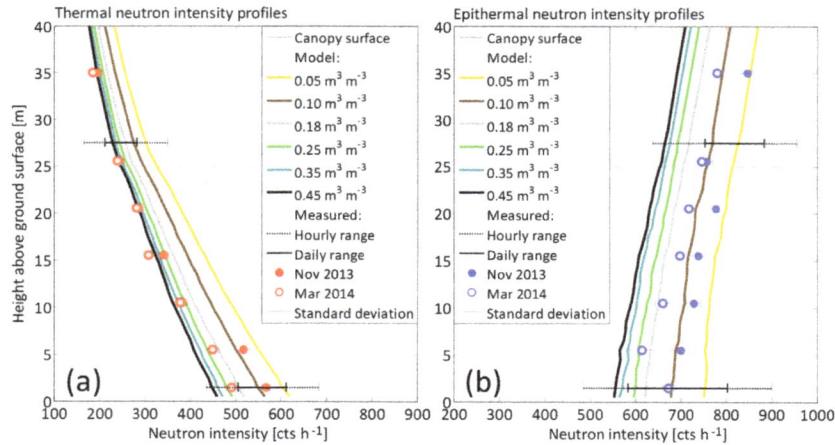

Figure 4. Sensitivity to volumetric soil moisture using model tree trunk, foliage, air. Measured and modeled **(a)** thermal and **(b)** epithermal neutron intensity profiles at Gludsted plantation. Hourly and daily ranges of variation of thermal and epithermal neutron intensities at ground and canopy level for the period 2013–2015.

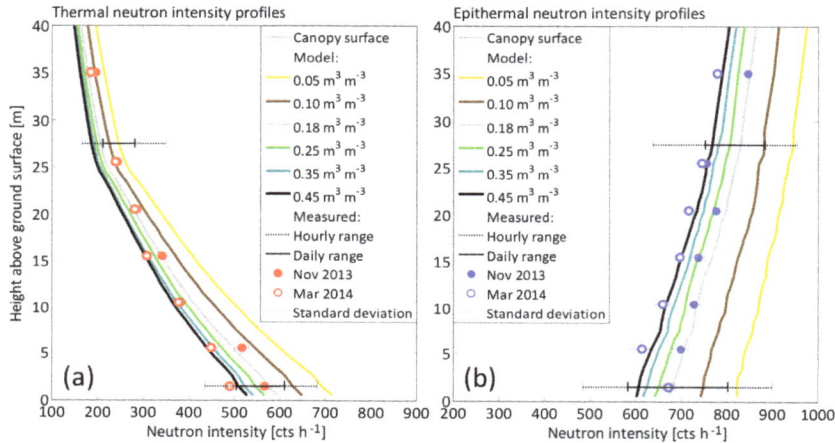

Figure 5. Sensitivity to volumetric soil moisture using model foliage. Measured and modeled **(a)** thermal and **(b)** epithermal neutron intensity profiles at Gludsted plantation. Hourly and daily ranges of variation of thermal and epithermal neutron intensities at ground and canopy level for the period 2013–2015.

for the epithermal neutrons. For model tree trunk, foliage, air (Fig. 4), only a minor decrease in the sensitivity of soil moisture on epithermal neutron intensity is observed going from ground-level to canopy-level (approximately 15 % reduction in intensity range corresponding to a volumetric soil moisture change of $0.40 \, \mathrm{m^3 \, m^{-3}}$). On the other hand, the sensitivity of the thermal neutron intensity is reduced more than 50 % (Table 6) most likely caused by the lower mean-free path length of the thermal neutrons compared to that of epithermal neutrons. The model with a simple forest canopy conceptualization provides thermal and epithermal neutron intensities slightly more sensitive to soil moisture (Fig. 5). Neutron intensity at dry and wet soil conditions is represented by the range of time series neutron intensity measurements. Overall, the modeled neutron intensities are within the measure-

ment range and the more appropriate model setup for Gludsted plantation is not obvious from the modeling results.

3.3 Subsurface properties

Thermal and epithermal neutron intensity profiles are modeled using model tree trunk, foliage, air (with fourth-order complexity) and models of decreasingly complex soil. Soil organic matter, belowground biomass, Gd and the chemical composition from XRF measurements are excluded one at the time (from third- to first-order complexity) and the final model includes a simple silica soil (SiO_2). The exact sensitivity of excluding the different components on ground- and canopy-level thermal and epithermal neutron intensity is quantified in Table 6 (see values in parentheses). Only the removal of soil organic matter (third-order complexity)

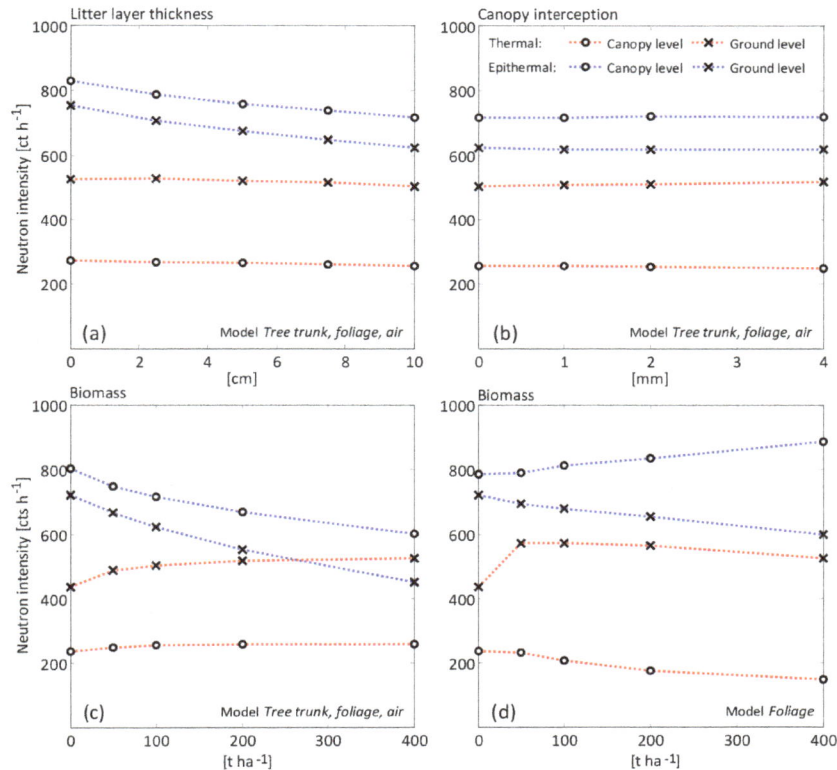

Figure 6. Sensitivity of ground- and canopy-level thermal and epithermal neutron intensity to **(a)** litter-layer thickness using model tree trunk, foliage, air, **(b)** canopy interception using model tree trunk, foliage, air and biomass using **(c)** model tree trunk, foliage, air and **(d)** model foliage, respectively.

changes the neutron intensity significantly at Gludsted plantation; i.e., an increase in the ground-level thermal and epithermal neutron intensity of $19 \, \text{cts h}^{-1}$ (cts = counts) and $25 \, \text{cts h}^{-1}$, respectively, is observed.

The thermal and epithermal neutron intensity is also modeled for a forest with litter layer of various thicknesses (Fig. 6a). The model tree trunk, foliage, air including a 10.0 cm thick litter layer is used along with forest models with litter layers of 0.0, 2.5, 5.0 and 7.5 cm thickness.

Neutron intensities are found to decrease with an increasing layer of litter, having the greatest impact on the epithermal neutron intensities (see also Table 6). Thereby, the thermal-to-epithermal neutron (t / e) ratio is altered when changing the thickness of the litter layer. This effect is most pronounced when the model without a litter layer is compared to the model with just a thin 2.5 cm thick litter layer. Since a considerable range of dry bulk density values (see Table 2) is measured within the footprint of the neutron detector, the sensitivity of neutron intensity to litter and mineral soils dry bulk density is examined using four model setups. Relative to the Gludsted plantation reference model, higher and lower values of dry bulk density are used. The first model includes a higher dry bulk density of $0.50 \, \text{g cm}^{-3}$ for the litter layer, while the second model holds a higher dry bulk den-

sity of $1.60 \, \text{g cm}^{-3}$ for the mineral soil. The third model has a low dry bulk density of $0.20 \, \text{g cm}^{-3}$ specified for the litter layer, and in the fourth model the mineral soil is described by a low dry bulk density of $0.60 \, \text{g cm}^{-3}$. The four model setups only provided slightly different thermal and epithermal neutron intensities (Table 6). Nevertheless, a reverse response of changed bulk densities is observed. A decrease in neutron intensity is obtained both by increasing the dry bulk density of the litter material and decreasing the dry bulk density of the mineral soil. Conversely, higher neutron intensities are computed by decreasing the dry bulk density of the litter material and increasing the dry bulk density of the mineral soil.

3.4 Canopy interception

The effect of canopy interception on thermal and epithermal neutron intensity is modeled using model tree trunk, foliage, air (Fig. 6b and Table 6). Except for a slight increase in ground-level thermal neutron intensities with wetting of the forest canopy, no effect of canopy interception on ground- and canopy-level thermal and epithermal neutron intensity is observed. A maximum change of approximately 3 % ($15 \, \text{cts h}^{-1}$) is observed for thermal neutron intensity at ground level going from a dry canopy to 4 mm

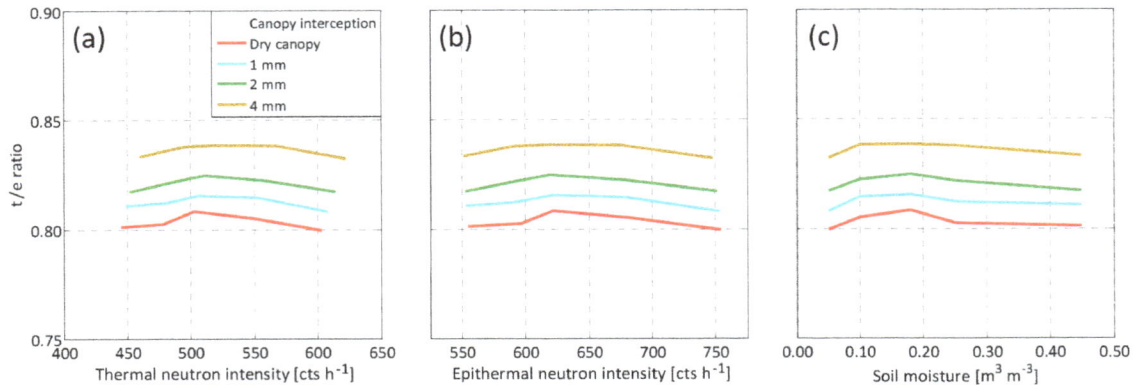

Figure 7. Modeled ground-level thermal-to-epithermal neutron intensity ratios using the model tree trunk, foliage, air for a dry forest canopy and canopy interception of 1, 2 and 4 mm plotted against modeled **(a)** ground-level thermal neutron intensity, **(b)** ground-level epithermal neutron intensity and **(c)** volumetric soil moisture.

of canopy interception. At the specific field site a maximum canopy storage capacity of 2.25 mm is expected, producing a change in observed ground-level thermal neutron intensity of approximately $7\,\mathrm{cts\,h^{-1}}$. Given an average neutron intensity of $504\,\mathrm{cts\,h^{-1}}$ of ground-level thermal neutrons with the installed detectors, an uncertainty of $22\,\mathrm{cts\,h^{-1}}$ is expected based solely on Poissonian statistics (see Sect. 2.2.1). Thus, the signal of canopy interception is within the measurement uncertainty, and cannot be identified at Gludsted plantation using the available cosmic-ray neutron measurements.

Although detection of canopy interception at Gludsted plantation is unfavorable it may still be possible at more appropriate conditions. Canopy interception modeling as described above is therefore also performed for volumetric soil moisture 0.05, 0.10, 0.25 and $0.40\,\mathrm{m^3\,m^{-3}}$. Ground level t / e ratio of the 20 model combinations are plotted against ground-level thermal neutron intensity, ground-level epithermal neutron intensity and volumetric soil moisture (Fig. 7). We chose not to include measurements in the figure because the measurement uncertainty at a relevant integration time is greater than the signal of canopy interception.

Overall, ground-level t / e ratio is found to be independent of ground-level thermal neutron intensity (Fig. 7a), ground-level epithermal neutron intensity (Fig. 7b) and volumetric soil moisture (Fig. 7c). Furthermore, the ground-level t / e ratio is found to increase with increasing canopy interception being on average 0.804 and 0.836 for a dry canopy and 4 mm of canopy interception, respectively. Overall, the same increase in ground-level t / e ratio is obtained per 1 mm additional canopy interception.

3.5 Biomass

The sensitivity to the amount of forest biomass on thermal and epithermal neutron intensity using the forest canopy conceptualization of model tree trunk, foliage, air and model foliage are presented in Fig. 6c and d, respectively. The neutron

intensity is provided for a scenario with no vegetation and models with biomass equivalent to dry aboveground biomass of 50, 100 (Gludsted plantation), 200 and $400\,\mathrm{t\,ha^{-1}}$.

Forest biomass is seen to significantly alter the thermal and epithermal neutron intensity both with regards to the differences between ground- and canopy-level neutron intensity, and ground-level t / e ratios (Fig. 6c and d). The direction and magnitude of these changes are found to be different depending on the two forest canopy conceptualizations. For the model tree trunk, foliage, air, the increase in biomass results in an increase in thermal neutron intensity, while the epithermal neutron intensity decreases (Fig. 6c). From ground level and up to an elevation of approximately 20 m the sensitivity to the amount of biomass on the neutron intensity is almost the same. From 20 m height, the sensitivity decreases with increasing elevation and for thermal neutrons the signal of biomass is almost gone at canopy level (not presented here). At canopy level, the sensitivity on epithermal neutrons is reduced, yet a strong signal remains.

Increasing the biomass in the model foliage from 0 to $50\,\mathrm{t\,ha^{-1}}$ (Fig. 6d) results in a considerable increase in ground-level thermal neutron intensity ($136\,\mathrm{cts\,h^{-1}}$, Table 6) while at canopy-level thermal neutron intensity is almost unaltered. A further increase in biomass ($>50\,\mathrm{t\,ha^{-1}}$) decreases both ground- and canopy-level thermal neutron intensities. The epithermal neutron intensity decreases at ground level and increase proportionally at canopy level with increasing amounts of biomass. The epithermal neutrons produced in the ground escape to the air and are moderated by the biomass, resulting in reduced epithermal neutron intensity with greater amounts of biomass. All models provide in accordance to theory increasing epithermal neutron intensity with height; however, the reduced steepness of the neutron height profiles with added biomass is unexplained. Oppositely to model tree trunk, foliage, air, the ground-level thermal neutron intensity decreases with added biomass.

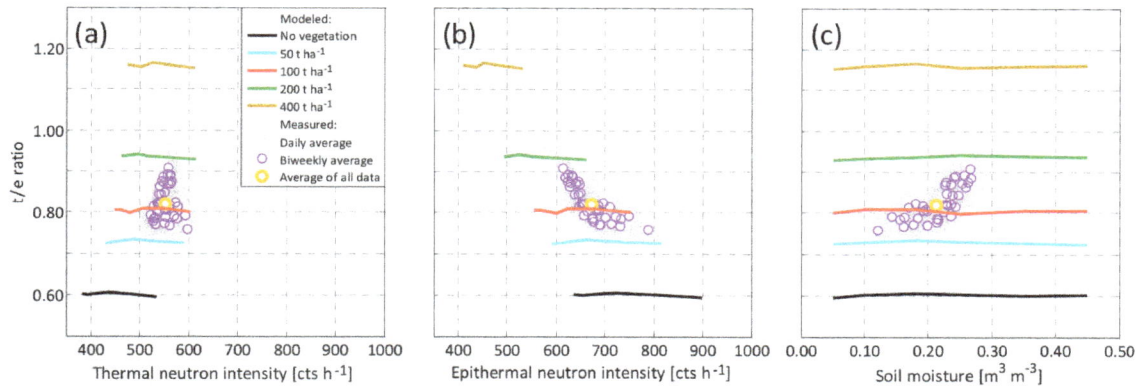

Figure 8. Neutron intensities measured at Gludsted plantation in the time period 2013–2015 and modeled using the model tree trunk, foliage, air. Ground-level thermal-to-epithermal neutron intensity ratio plotted against measured and modeled **(a)** ground-level thermal neutron intensity, **(b)** ground-level epithermal neutron intensity and **(c)** volumetric soil moisture.

As shown in Figs. 3, 6c and d, the resulting thermal and epithermal neutron intensity profiles depend highly on the chosen model setup (forest conceptualization). At this stage, we cannot determine which conceptualization is more realistic, and we therefore choose to use both conceptualizations in the further analysis. Overall, a positive correlation is found for the differences between ground- and canopy-level neutron intensity (thermal and epithermal neutron energies) and the amount of biomass (Fig. 6c and d, and Table 6). However, the model tree trunk, foliage, air and model foliage provide different relationships, and measurements and modeling are not fully in agreement. Alternatively, one can also potentially use the t / e ratio at the ground level to assess biomass. The advantage is that only one station is needed – and that at a convenient location. This would also allow for surveys of biomass estimations to be conducted from mobile cosmic-ray neutron intensity detector systems, e.g. installed in vehicles.

The measured and modeled ratios are again provided using both forest canopy conceptualization, i.e., model tree trunk, foliage, air (Fig. 8) and model foliage (Fig. 9). The ratios are plotted against (a) ground-level thermal neutron intensity, (b) ground-level epithermal neutron intensity and (c) volumetric soil moisture estimated using the N_0 method (Desilets et al., 2010). Measurements are provided as daily averages, biweekly averages and as a total average of the whole 2-year period.

The modeled ground-level t / e ratio increases with forest biomass (Figs. 8 and 9). Drying or wetting of soil change the thermal and epithermal neutron intensity proportionally and the ratios are accordingly found to be independent of changes in the ground-level thermal neutron intensity, the ground-level epithermal neutron intensity and volumetric soil moisture. However, this independence is not seen in the measurements, where the ground-level epithermal neutron intensity and soil moisture (Figs. 8c and 9c) in particular seem to impact the ratio. A fairly proportional increase in the ground-

level t / e ratio with respect to greater amounts of biomass is found when using model tree trunk, foliage, air (Figs. 8 and 10). Contrarily, when using model foliage, a more uneven increase in the ratio with increasing amounts of biomass is provided (Figs. 9 and 10). A major increase in the ground-level t / e ratio of around 0.22 appears from no vegetation to a dry aboveground biomass of $50\,t\,ha^{-1}$. However, additional amounts of biomass only increase the ground-level t / e ratio slightly. With additional $350\,t\,ha^{-1}$ biomass (from 50 to $400\,t\,ha^{-1}$ dry aboveground biomass) the t / e ratio increases by only 0.05.

Overall, a remarkable agreement is seen for the model tree trunk, foliage, air in Fig. 8 when comparing the 2-year average of the measured ratio with the modeled value of Gludsted plantation ($100\,t\,ha^{-1}$ dry aboveground biomass, Fig. 8). The biweekly averages of measurements are all within the ratios modeled for biomass of 50–$200\,t\,ha^{-1}$. For the model foliage in Fig. 9, the measured ratio is in better agreement with a lower biomass ($50\,t\,ha^{-1}$ dry aboveground biomass). The small increase in t / e ratio with increasing amounts of biomass of model foliage causes the biweekly averages of the measurements to exceed both the lower and upper boundary of ratios provided by the models of 50 and $400\,t\,ha^{-1}$ dry aboveground biomass.

4   Discussions

4.1   Neutron height profile measurements and forest conceptualization

Slightly different neutron height profiles and t / e ratios were measured during the field campaigns in November 2013 and March 2014 (Figs. 3–5). The area average soil moisture estimated using the measured cosmic-ray neutron intensity was similar for the two field campaigns. The different neutron height profiles could therefore instead be a result of dissimi-

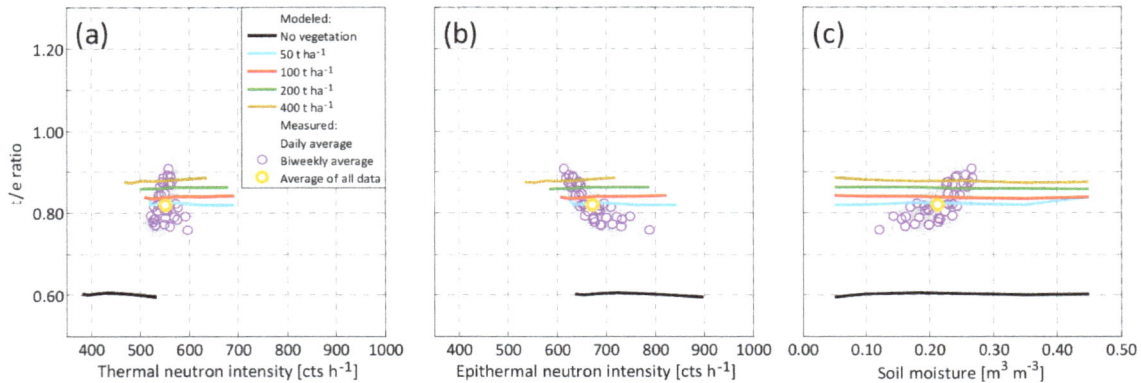

Figure 9. Neutron intensities measured at Gludsted plantation in the time period 2013–2015 and modeled using the model foliage. Ground-level thermal-to-epithermal neutron intensity ratio plotted against measured and modeled (**a**) ground-level thermal neutron intensity, (**b**) ground-level epithermal neutron intensity and (**c**) volumetric soil moisture.

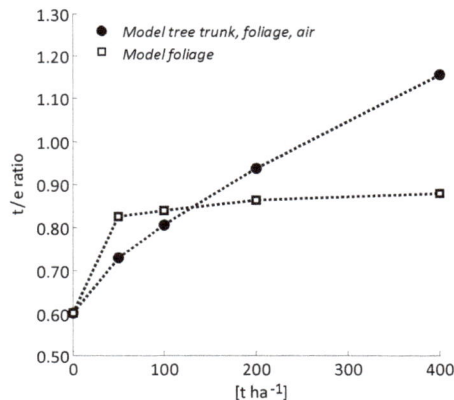

Figure 10. Ground-level thermal-to-epithermal neutron ratio plotted against biomass equivalent to dry aboveground biomass of 50, 100 (Gludsted plantation), 200 and $400\,t\,ha^{-1}$ using model tree trunk, foliage, air and model foliage, respectively.

lar soil moisture profiles or different soil moisture of the litter layer and the mineral soil. During two out of three soil sampling field campaigns, different soil moisture of the litter layer and the mineral soil was observed at Gludsted plantation (soil samples were collected at 18 locations within a circle of 200 m in radius and in 6 depths from 0 to 30 cm depth following the procedure of Franz et al., 2012). Additionally, the different neutron height profiles could also be a result of the different climate and weather conditions related to the seasons of detections (spring and fall). However, both neutron profiles are within the ranges of the daily time series measurements, and we therefore still believe that they can be used in the assessment of the modeled neutron profiles. For future studies we recommend soil sample field campaigns to be conducted on the days of neutron profile measurements.

The neutron transport at the ground–atmosphere interface was found to be sensitive to the level of complexity of the for-

est canopy conceptualization; however, the more appropriate conceptualization was not identified. Improved comparability to measurements may be obtained by advancing the forest canopy conceptualization. Currently, one tree is defined and repeated throughout the model domain. The trees are placed in rows and the same settings are applied from the ground surface to 25 m height. In order to advance the forest canopy conceptualization, trees of different heights and diameters could be included, and the placement of the trees could be more according to the actual placement of trees at the forest field site. Additionally, variability in tree trunk diameter, foliage density and volume with height above the ground surface could be implemented.

4.2 The sensitivity of neutron intensity to soil chemistry and dry bulk density

In contrary to the results obtained at Voulund farmland by Andreasen et al. (2016), the sensitivity of thermal and epithermal neutron intensity profiles to soil chemistry was found to be minor at Gludsted plantation. The soil organic matter content at Voulund farmland is smaller and the soil chemistry is, except from a few elements (added in relation to farming activities; spreading of manure and agricultural lime), similar to Gludsted plantation. Modeling shows that the sensitivity to soil chemistry at Gludsted plantation is dampened by the considerable amount of hydrogen present in the litter at the forest floor and the forest biomass (not presented here). Accordingly, the effect of litter and mineral soil dry bulk density on neutron intensity is expected to be greater at non-vegetated field site. The reverse effect of increased dry bulk density of litter and mineral soil on neutron intensity is a result of the different elemental composition of the two materials. The production rate of low-energy neutrons ($< 1\,MeV$) per incident high-energy neutron is higher for interactions with elements of higher atomic mass ($A^{2/3}$, where A is the atomic mass) (Zreda et al., 2012). Heavier elements are in particular found in mineral soil and an increase in the dry bulk density entails a higher production rate and therefore higher neutron intensity. The concentration of hydrogen is increased with an increased dry bulk density of litter material resulting in a greater moderation and absorption of neutrons, and as a consequence lower neutron intensities. To summarize, the mineral soil acts as a producer of thermal and epithermal neutrons, while the litter acts as an absorber.

4.3 The potential of cosmic-ray neutron canopy interception detection

Ground-level thermal neutron intensity was found to be sensitive to canopy interception; however, the signal is small and within the measurement uncertainty at Gludsted plantation. In order to obtain a signal-to-noise ratio of 1, either an 11 h integration time or 11 detectors similar to the installed detector are needed. However, longer integration times are not appropriate when considering Gludsted plantation as the return time of canopy interception (cycling between precipitation and evaporation) often is short (half-hourly to hourly time resolution). Although the change in the t / e ratio with wetting/drying of the forest canopy is small the canopy interception may potentially be measured using cosmic-ray neutron intensity detectors at locations with (1) a high neutron intensity level (lower latitude and/or higher altitude, (2) more sensitive neutron detectors and (3) greater amounts of canopy interception with longer residence time (e.g., snow). We suggest future studies investigating the effect of canopy interception on the neutron intensity signal to be performed at locations matching one or more of these criteria.

4.4 The sensitivity of biomass to neutron intensity

The neutron intensity depends on how many neutrons are produced, down-scattered to lower energies and absorbed. Including biomass to a system increases the concentration of hydrogen and leads to reduced neutron intensity as the moderation and absorption is intensified. Despite this, increased thermal neutron intensity is provided with greater amounts of forest biomass using model tree trunk, foliage, air (see Fig. 6c). We hypothesize that forest biomass enhances the rate of moderation more than the rate of absorption. Thus, higher thermal neutron intensity is obtained as the number of thermal neutrons generated by the moderation of epithermal neutrons exceeds the number of thermal neutrons absorbed. This behavior may be due to the large volume of air within the forest canopy. The probability of thermal neutrons to interact with elements within this space is low as the density of air is low. Overall when applying model foliage both thermal and epithermal neutron intensity decreases with added amounts of biomass (see Fig. 6d). The deviating behavior (compared to model tree trunk, foliage, air) may be due to the different elemental concentration of the forest canopy layers. Here, no space is occupied by a material of very low ele-

mental density and may lead to an increased absorption of thermal neutrons.

The discrepancy of measured and modeled ground-level t / e ratios (Figs. 8 and 9) could be related to (1) shortcomings in the model setup, i.e., a need for an even more realistic forest conceptualization, and more detailed and up-to-date forest information; a model including a sufficient representation of the field site will provide neutron height profiles and t / e ratios more representative of the real conditions; (2) discrepancy of measured and modeled energy ranges as discussed in Andreasen et al. (2016); and (3) unrepresentative biomass estimate. The 100 t ha^{-1} dry aboveground biomass was estimated using lidar images from both 2006 and 2007 and therefore not completely representative of the 2013–2015 conditions (because of tree growth). Furthermore, the biomass estimate varied considerably within the image (standard deviation = 46 t ha^{-1}), and the image coverage did not fully match the footprint of the cosmic-ray neutron intensity detector.

4.5 Cosmic-ray neutron biomass detection

The proposed possibility of estimating biomass at a hectometer scale using ground-level t / e ratios was tested. The modeled ground-level t / e ratio is compared with measurements of two additional field sites located close to Gludsted plantation. The three field sites have similar environmental settings (e.g., neutron intensity, soil chemistry), though different land covers with different amounts of biomass (stubble pasture, heathland and forest).

At Voulund farmland the ground-level t / e ratio was measured to be 0.53 and 0.58 on 22 and 23 September 2015, respectively. Only minor amounts of organic matter were present in the stubble and residual of spring barley harvested in August 2015. Additionally, the ground-level t / e ratio was determined based on modeling of bare ground and site-specific soil chemistry measured at Voulund farmland (Andreasen et al., 2016). The modeled ratio was found to be 0.56 in agreement with the measured ratios. The ratio modeled based on the non-vegetated conceptualization of Gludsted plantation was slightly higher (0.60, see Figs. 8 and 9). Here, a 10 cm thick litter layer was included in the model. The sensitivity analysis on the effect of litter layer on neutron intensity (Fig. 6a and Table 6) implies that lower ground-level t / e ratios are found at locations with a thin or no litter layer.

The ground-level t / e ratio at the Harrild heathland was measured to 0.66 during the period 27 October to 16 November 2015. The ratio is slightly higher than the non-vegetated model for Gludsted plantation. Both field sites have a considerable layer of litter, and the slightly higher t / e ratio relative to the non-vegetated Gludsted plantation may be due to biomass in the form of grasses, heather plants and bushes present at Harrild heathland. At Gludsted plantation, the ratio is 0.73 for dry aboveground biomass equivalent of 50 t ha^{-1}. Accordingly, the ratio measured at Harrild heathland is somewhere in between the ratio modeled for a non-vegetated field site and a field site with biomass equivalent to $50\,t\,ha^{-1}$ dry aboveground biomass.

Measuring ground-level t / e ratios for biomass estimation at a hectometer scale is promising as the measured ratio increases with increasing amounts of litter and biomass in correspondence to modeling. Still, ground-level t / e ratio detection at locations of known biomass should be accomplished to test the suggested relationships. We recommend a detection system with higher sensitivity to be used when a location of low neutron intensity rates (like Gludsted plantation) is surveyed, unless long periods of measurements can be conducted at each measurement location. This can be accomplished by using larger sensors, an array of several sensors and/or sensors that are more efficient, as is done in roving surveys (Chrisman and Zreda, 2013; Franz et al., 2015).

5   Conclusion

The potential of applying the cosmic-ray neutron intensity method for other purposes than soil moisture detection was explored using profile and time series measurements of neutron intensities combined with neutron transport modeling. The vegetation and subsurface layers of the forest model setup were described by average measurements and estimates. Four forest canopy conceptualizations of increasing complexity were used. Without adjusting parameters and variables, modeled thermal and epithermal neutron intensity profiles compared fairly well with measurements; however, some deviations from measurements were observed for each of the four forest canopy conceptualization models. The more appropriate forest canopy conceptualization was not obvious from the results as the best fit to thermal neutron measurements was found using the complex forest canopy conceptualization, including a tree trunk and multiple materials, while the better fit to epithermal neutron measurements was found using the simplest forest canopy conceptualization, including a homogenous layer of foliage material. A sensitivity analysis was performed to quantify the effect of the forest's governing parameters/variables on the neutron transport profiles. The sensitivity of canopy interception, dry bulk density of litter and mineral soil, and soil chemistry on neutron intensity was found to be small. The ground-level t / e neutron ratio was found to increase with increasing amounts of canopy interception and to be independent of ground-level thermal neutron intensity, ground-level epithermal neutron intensity and soil moisture. However, the increase was minor and the measurement uncertainty was found to exceed the signal of canopy interception at a timescale appropriate to detect canopy interception at Gludsted plantation (half-hourly to hourly). Neutron intensity was found to be more sensitive to litter layer, soil moisture and biomass at the forest field site. An increased litter layer at the forest floor resulted in reduced neutron intensities, particularly for epithermal neutrons. Forest biomass

was found to alter the thermal and epithermal neutron transport significantly, both in terms of the shape of the neutron profiles and the t / e neutron ratios. The response to altered amounts of biomass on thermal and epithermal neutron intensity is non-unique for the simple and complex forest conceptualization and further advancement of the forest representation is therefore necessary. Still, cosmic-ray neutron intensity detection for biomass estimation at an intermediate scale is promising. Both the difference between ground- and canopy-level thermal and epithermal neutron intensity, respectively, and the ground-level t / e ratios were found to increase with additional amounts of biomass using the simple and complex forest canopy conceptualization. The best agreement between measurements and modeling was obtained for the ground-level t / e neutron ratio using a model with a complex forest canopy conceptualization. Additionally, the modeled ratios were found to agree well with two nearby field sites with different amounts of biomass (a bare ground agricultural field and a heathland field site).

Competing interests.   The authors declare that they have no conflict of interest.

Acknowledgements. We acknowledge The Villum Foundation (www.villumfonden.dk) for funding the HOBE project (www.hobe.dk). Lars M. Rasmussen and Anton G. Thomsen (Aarhus University) are greatly thanked for the extensive help in the field. We would like to extend our gratitude to Vivian Kvist Johannsen and Johannes Schumacher from the Section for Forest, Nature and Biomass, University of Copenhagen. Finally, we also acknowledge the NMDB database (www.nmdb.eu), founded under the European Union's FP7 programme (contract no. 213007) for providing data. Jungfraujoch neutron monitor data were kindly provided by the Cosmic Ray Group, Physikalisches Institut, University of Bern, Switzerland.

Edited by: M. Weiler

References

Andreasen, M., Andreasen, L. A., Jensen, K. H., Sonnenborg, T. O., and Bircher, S.: Estimation of Regional Groundwater Recharge Using Data from a Distributed Soil Moisture Network, Vadose Zone J., 12, 1–18, doi:10.2136/vzj2013.01.0035, 2013.

Andreasen, M., Jensen, K. H., Zreda, M., Desilets, D., Bogena, H., and Looms, M. C: Modeling cosmic ray neutron field measurements, Water Resour. Res., 52, 6451–6471, doi:10.1002/2015WR018236, 2016.

Baatz, R., Bogena, H. R., Hendricks Franssen, H.-J., Huisman, J. A., Montzka, C., and Vereecken, H.: An empirical vegetation correction for soil moisture content quantification using cosmic ray probes, Water Resour. Res., 51, 2030–2046, doi:10.1002/2014WR016443, 2015.

Baroni, G. and Oswald, S. E.: A scaling approach for the assessment of biomass changes and rainfall interception using cosmic-ray neutron sensing, J. Hydrol., 525, 264–276, doi:10.1016/j.jhydrol.2015.03.053, 2015.

Bogena, H. R., Huisman, J. A., Baatz, R., Hendricks Franssen, H.-J., and Vereecken, H.: Accuracy of the cosmic-ray soil water content probe in humid forest ecosystems: The worst case scenario, Water Resour. Res., 49, 1–14, doi:10.1002/wrcr.20463, 2013.

Boudreau, J., Nelson, R. F., Margolis, H. A., Beaudoin, A., Guindon, L., and Kimes, D. S.: Regional above ground forest biomass using airborne and spaceborne LiDAR in Québec, Remote Sens. Environ., 112, 3876–3890, doi:10.1016/j.rse.2008.06.003, 2008.

Chrisman, B. and Zreda, M.: Quantifying mesoscale soil moisture with the cosmic-ray rover, Hydrol. Earth Syst. Sci., 17, 5097–5108, doi:10.5194/hess-17-5097-2013, 2013.

Desilets, D., Zreda, M., and Ferré, T. P. A.: Nature's neutron probe: Land surface hydrology at an elusive scale with cosmic rays, Water Resour. Res., 46, W11505, doi:10.1029/2009WR008726, 2010.

Desilets, D. and Zreda, M.: Footprint diameter for a cosmic-ray soil moisture probe: Theory and Monte Carlo simulations, Water Resour. Res., 49, 1–10, doi:10.1002/wrcr.20187, 2013.

Dunkerley, D.: Measuring interception loss and canopy storage in dryland vegetation: a brief review and evaluation of available research strategies, Hydrol. Process., 14, 669–678, doi:10.1002/(SICI)1099-1085(200003)14:4<669::AID-HYP965>3.0.CO;2-I, 2000.

Franz, T. E., Zreda, M., Rosolem, R., and Ferre, T. P. A.: Field Validation of a Cosmic-Ray Neutron Sensor Using a Distributed Sensor Network, Vadose Zone J., 2012, 1–10, doi:10.2136/vzj2012.0046, 2012.

Franz, T. E., Zreda, M. Rosolem, R., Hornbuckle, B. K., Irvin, S. L., Adams, H., Kolb, T. E., Zweck, C., and Shuttleworth, W. J.: Ecosystem-scale measurements of biomass water using cosmic ray neutrons, Geophys. Res. Lett., 40, 3929–3933, doi:10.1002/grl.50791, 2013a.

Franz, T. E., Zreda, M., Rosolem, R., and Ferre, T. P. A.: A universal calibration function for determination of soil moisture with cosmic-ray neutrons, Hydrol. Earth Syst. Sci., 17, 453–460, doi:10.5194/hess-17-453-2013, 2013b.

Franz, T. E., Wang, T., Avery, W., Finkenbiner, C., and Brocca, L.: Combined analysis of soil moisture measurements from roving and fixed cosmic ray neutron probes for multiscale real-time monitoring, Geophys. Res. Lett., 42, 3389–3396, doi:10.1002/2015GL063963, 2015.

Glasstone, S. and Edlund, M. C.: The elements of nuclear reactor theory, 5th Edn., Van Nostrand, New York, 416 pp., 1952.

Hawdon, A., McJannet, D., and Wallace, J.: Calibration and correction procedures for cosmic-ray neutron soil moisture probes located across Australia, Water Resour. Res., 50, 5029–5043, doi:10.1002/2013WR015138, 2014.

Heidbüchel, I., Güntner, A., and Blume, T.: Use of cosmic-ray neutron sensors for soil moisture monitoring in forests, Hydrol. Earth Syst. Sci., 20, 1269–1288, doi:10.5194/hess-20-1269-2016, 2016.

Hornbuckle, B., Irvin, S., Franz, T., Rosolem, R., and Zweck, C.: The potential of the COSMOS network to be a source of new soil moisture information for SMOS and SMAP, in: Geoscience and Remote Sensing Symposium (IGARSS), IEEE International, 1243–1246, doi:10.1109/IGARSS.2012.6351317, 2012.

Hughes, E. B. and Marsden, P. L.: Response of a standard IGY neutron monitor, J. Geophys. Res., 71, 1435–1444, doi:10.1029/JZ071i005p01435, 1966.

Iwema, J., Rosolem, R., Baatz, R., Wagener, T., and Bogena, H. R.: Investigating temporal field sampling strategies for site-specific calibration of three soil moisture-neutron intensity parameterisation methods, Hydrol. Earth Syst. Sci., 19, 3203–3216, doi:10.5194/hess-19-3203-2015, 2015.

Jenkins, J. C., Chojnacky, D. C. Heath, L. S., and Birdsey, R. A.: National-Scale Biomass Estimators for United States Tree Species, Forest Sci., 49, 12–35, 2003.

Jensen, K. H. and Illangasekare, T. H.: HOBE-A hydrological observatory in Denmark, Vadose Zone J., 10, 1–7, doi:10.2136/vzj2011.0006, 2011.

Knoll, G. F.: Radiation Detection and Measurement, Fourth Edn., John Wiley & Sons, Inc., New Jersey, 2010.

Köhli, M., Schrön, M., Zreda, M., Schmidt, U., Dietrich, P., and Zacharias, S.: Footprint characteristics revised for field-scale soil moisture monitoring with cosmic-ray neutrons, Water Resour. Res., 51, 5772–5790, doi:10.1002/2015WR017169, 2015.

Lal, R.: Carbon sequestration, Philos. T. Roy. Soc. B, 363, 815–830, doi:10.1098/rstb.2007.2185, 2008.

Levy, P. E., Hale, S. E., and Nicoll, B. C.: Biomass expansion factors and root : shoot ratios for coniferous tree species in Great Britain, Forestry, 77, 421–430, doi:10.1093/forestry/77.5.421, 2004.

Nord-Larsen, T. and Schumacher, J.: Estimation of forest resources from a country wide laser scanning survey and national forest inventory data, Remote Sens. Environ., 119, 148–157, doi:10.1016/j.rse.2011.12.022, 2012.

O'Kelly, B.: Accurate determination of moisture content of organic soils using the oven drying method, Dry. Technol., 22, 1767–1776, doi:10.1081/DRT-200025642, 2004.

Pelowitz, D. B.: MCNP6™ User's Manual, Version 1, Los Alamos National Laboratory report LA-CP-13-00634, Rev. 0, 2013.

Popescu, S. C.: Estimating biomass of individual pine trees using airborne lidar, Biomass Bioenerg., 31, 646–655, doi:10.1016/j.biombioe.2007.06.022, 2007.

Ringgaard, R., Herbst, M., and Friborg, T.: Partitioning forest evapotranspiration: Interception evaporation and the impact of canopy structure, local and regional advection, J. Hydrol., 517, 677–690, doi:10.1016/j.jhydrol.2014.06.007, 2014.

Rivera Villarreyes, C. A., Baroni, G., and Oswald, S. E.: Integral quantification of seasonal soil moisture changes in farmland by cosmic-ray neutrons, Hydrol. Earth Syst. Sci., 15, 3843–3859, doi:10.5194/hess-15-3843-2011, 2011.

Rosolem, R., Shuttleworth, W. J., Zreda, M., Franz, T. E., Zeng, X., and Kurc, S. A.: The Effect of Atmospheric Water Vapor on Neutron Count in the Cosmic-Ray Soil Moisture Observing System, J. Hydrometeorol., 14, 1659–1671, doi:10.1175/JHM-D-12-0120.1, 2013.

Salminen, R., Batista, M. J., Bidovec, M., Demetriades, A., De Vivo, B., De Vos, W., Duris, M., Gilucis, A., Gregorauskiene, V., Halamić, J., and Heitzmann, P.: Geochemical Atlas of Europe. Part 1: Background Information, Methodology and Maps. Espoo, Geological Survey of Finland, 36 figures, 362 maps, 526 pp., 2005.

Sears, V. F.: Neutron scattering lengths and cross sections, Neutron News, 3, 29–37, doi:10.1080/10448639208218770, 1992.

Serup, H. (Ed.), Falster, H., Gamborg, C., Gundersen, P., Hansen, L., Heding, N., Jakobsen, H. H., Kofman, P., Nikolaisen, L., and Thomsen, I.: Wood for Energy Production: Technology-Environment-Economy, 2nd Revised Edn., Centre for Biomass Technology, Denmark, 69 pp., 2002.

Sigouin, M. J. P. and Si, B. C.: Calibration of a non-invasive cosmic-ray probe for wide area snow water equivalent measurement, The Cryosphere, 10, 1181–1190, doi:10.5194/tc-10-1181-2016, 2016.

Tian, Z., Li, Z., Liu, G., Li, B., and Ren, T.: Soil Water Content Determination with Cosmic-ray Neutron Sensor: Correcting Aboveground Hydrogen Effects with Thermal / Fast Neutron Ratio, J. Hydrol., 540, 923–933, doi:10.1016/j.jhydrol.2016.07.004, 2016.

Zreda, M., Desilets, D., Ferré, T. P. A., and Scott, R. L.: Measuring soil moisture content non-invasively at intermediate spatial scale using cosmic-ray neutrons. Geophys. Res. Lett., 35, L21402, doi:10.1029/2008GL035655, 2008.

Zreda, M., Shuttleworth, W. J., Zeng, X., Zweck, C., Desilets, D., Franz, T., and Rosolem, R.: COSMOS: the COsmic-ray Soil Moisture Observing System, Hydrol. Earth Syst. Sci., 16, 4079–4099, doi:10.5194/hess-16-4079-2012, 2012.

Zweck, C., Zreda, M., and Desilets, D.: Snow shielding factors for cosmogenic nuclide dating inferred from Monte Carlo neutron transport simulations, Earth Planet. Sc. Lett., 379, 64–71, doi:10.1016/j.epsl.2013.07.023, 2013.

11

Hydraulic characterisation of iron-oxide-coated sand and gravel based on nuclear magnetic resonance relaxation mode analyses

Stephan Costabel[1], Christoph Weidner[2,a], Mike Müller-Petke[3], and Georg Houben[2]

[1]Federal Institute for Geosciences and Natural Resources, Wilhelmstraße 25–30, 13593 Berlin, Germany
[2]Federal Institute for Geosciences and Natural Resources, Stilleweg 2, 30655 Hannover, Germany
[3]Leibniz Institute for Applied Geophysics, Stilleweg 2, 30655 Hannover, Germany
[a]current address: North Rhine Westphalian State Agency for Nature, Environment and Consumer Protection, Leibnizstr. 10, 45659 Recklinghausen, Germany

Correspondence: Stephan Costabel (stephan.costabel@bgr.de)

Abstract. The capability of nuclear magnetic resonance (NMR) relaxometry to characterise hydraulic properties of iron-oxide-coated sand and gravel was evaluated in a laboratory study. Past studies have shown that the presence of paramagnetic iron oxides and large pores in coarse sand and gravel disturbs the otherwise linear relationship between relaxation time and pore size. Consequently, the commonly applied empirical approaches fail when deriving hydraulic quantities from NMR parameters. Recent research demonstrates that higher relaxation modes must be taken into account to relate the size of a large pore to its NMR relaxation behaviour in the presence of significant paramagnetic impurities at its pore wall. We performed NMR relaxation experiments with water-saturated natural and reworked sands and gravels, coated with natural and synthetic ferric oxides (goethite, ferrihydrite), and show that the impact of the higher relaxation modes increases significantly with increasing iron content. Since the investigated materials exhibit narrow pore size distributions, and can thus be described by a virtual bundle of capillaries with identical apparent pore radius, recently presented inversion approaches allow for estimation of a unique solution yielding the apparent capillary radius from the NMR data. We found the NMR-based apparent radii to correspond well to the effective hydraulic radii estimated from the grain size distributions of the samples for the entire range of observed iron contents. Consequently, they can be used to estimate the hydraulic conductivity using the well-known Kozeny–Carman equation without any calibration that is otherwise necessary when predicting hydraulic conductivities from NMR data. Our future research will focus on the development of relaxation time models that consider pore size distributions. Furthermore, we plan to establish a measurement system based on borehole NMR for localising iron clogging and controlling its remediation in the gravel pack of groundwater wells.

1 Introduction

Iron oxides are, due to their abundance and reactive properties, amongst the most important mineral phases in the geosphere (Cornell and Schwertmann, 2003; Colombo et al., 2014). They encompass a variety of oxides, hydroxides and oxihydroxides of predominantly ferric iron but all are referred to as iron oxides in this study for the sake of brevity. They form some of the most important commercial iron ores worldwide but also play a vital role in soils and aquifers. As weathering products, iron oxides control the conditions for soil genesis and degradation (Stumm and Sulzberger, 1991; Kappler and Straub, 2005) and the mobility of nutrients, trace metals, and contaminants (Cornell and Schwertmann, 2003; Colombo et al., 2014; Cundy et al., 2014). Particularly in many tropic and subtropic soils, the building processes of iron oxide exhibit high temporal dynamics and may change the environmental conditions within a few years, which makes it necessary to further develop measurement techniques to characterise and monitor the corresponding status of soils and aquifers.

Furthermore, iron oxides play a negative role when forming in wells and drains used for the extraction of fluids from the subsurface, e.g. in drinking water production, oil wells, dewatering of mines or bogs, landfill leachate collection systems, and geothermal energy systems (Houben, 2003a; Larroque and Franceschi, 2011; Medina et al., 2013). The formation of iron oxide incrustations negatively affects the performance of these systems by blocking the entrance openings and the pore space of gravel pack and formation (Weidner et al., 2012). The removal of such deposits is expensive and time-consuming. Their spatial distribution is often inhomogeneous (Houben and Weihe, 2010; Weidner, 2016). It is therefore imperative to identify their exact location and to characterise their degree of clogging to successfully target rehabilitation measures. Ideally, this is to be done before the incrustation gained a state at which fluid movement through the pore space is significantly hindered in order to ensure maximum chance of success of the remediation activities. Although the chemical (Stumm and Lee, 1960; Pham and Waite, 2008; Geroni and Sapsford, 2011; Larese-Casanova et al., 2012) and biological processes (Tuhela et al., 1997; Cullimore, 2000; Emerson et al., 2010) involved are well investigated, accurate methods for identifying and characterising the location and degree of in situ iron mineralisation are still not available.

Geophysical field and borehole methods have the potential to comply with this demand. Methods such as electrical resistivity tomography, electromagnetics, and ground-penetrating radar are sensitive to different phases and concentrations of iron oxides in the pore space (e.g. Van Dam et al., 2002; Atekwana and Slater, 2009; Abdel Aal et al., 2009). The same is true for the method of nuclear magnetic resonance (NMR, e.g. Bryar et al., 2000; Keating and Knight, 2007, 2008, 2010). The aim of this laboratory study is to assess the potential of NMR for identifying the location and concentration of iron oxide coatings in water-saturated porous media and the assessment of their hydraulic effects.

Geophysical applications of NMR relaxometry are used in hydrocarbon exploration, hydrogeology, and environmental and soil sciences for estimating pore liquid contents, pore size distributions, and permeability. When applied in boreholes and a laboratory setting, NMR is able to identify different pore fluid components, e.g. water and oil (e.g. Bryar and Knight, 2003; Hertzog et al., 2007), to distinguish between clay-bound, capillary-bound and mobile pore water (e.g. Prammer et al., 1996; Coates et al., 1999; Dunn et al., 2002), and to provide hydraulic and soil physical parameters (e.g. Dlugosch et al., 2013; Costabel and Yaramanci, 2011, 2013; Sucre et al., 2011; Knight et al., 2016). As a non-invasive subsurface tool, it is used for investigating the subsurface distributions of water content and hydraulic conductivity and allows for the lithological categorisation of aquifers and aquitards (e.g. Legchenko et al., 2004; Costabel et al., 2017).

NMR relaxometry for hydraulic characterisation of porous media takes advantage of the paramagnetic properties of the pore surface. The NMR measurement observes the exchange of energy between stimulated proton spins of the pore fluid and the pore walls and thereby provides a proxy for pore surface-to-volume ratios, i.e. pore sizes. However, existing approaches to estimating pore sizes and permeabilities demand material-specific calibration (Kenyon, 1997; Coates et al., 1999), which is expected to be particularly difficult for materials containing a large amount of paramagnetic species (Keating and Knight, 2007). Moreover, NMR relaxation measurements are affected by additional effects such as the occurrence of additional energy losses within the pore fluid (Bryar et al., 2000; Bryar and Knight, 2002), ferromagnetism and corresponding disturbances of the magnetic fields (Keating and Knight, 2007, 2008), and the existence of pore geometries with a high level of complexity, e.g. capillaries with angular cross sections or fractal pore surfaces (Sapoval et al., 1996; Mohnke et al., 2015; Müller-Petke et al., 2015). Different iron oxide phases can produce any of these effects and can thus significantly bias the results. Foley et al. (1996) demonstrated for instance that the amount of paramagnetic iron minerals is linearly correlated with the NMR relaxation rate for materials with otherwise identical pore space. Keating and Knight (2007, 2010) found that NMR relaxation is not only influenced by the amount but also by the specific kind of iron oxide mineral. Additional complexity might occur if paramagnetic and ferromagnetic particles accumulate inhomogeneously inside the pore space (Grunewald and Knight, 2011; Keating and Knight, 2012).

In this study, we investigate the effects of paramagnetic iron oxide coatings, particularly for coarse material. For large pores in the so-called slow diffusion regime, the otherwise linear relationship between relaxation time and pore size is disturbed because higher relaxation modes become relevant (Brownstein and Tarr, 1979; Müller-Petke et al., 2015). As a significant consequence, the common interpretation schemes to estimate pore size and hydraulic conductivity are not valid anymore. Past studies dealing with iron mineral coatings reported the occurrence of slow diffusion conditions during their NMR experiments (Keating and Knight, 2010; Grunewald and Knight, 2011). Our objective is to learn how to interpret NMR data also under these conditions and how to estimate hydraulic parameters from it. Therefore, the goals of this study are as follows:

1. to investigate the NMR relaxation behaviour as a function of the content of paramagnetic iron oxide for large pores;

2. to correlate NMR relaxation parameters with hydraulically effective parameters;

3. to assess the model published by Müller-Petke et al. (2015) in the context of iron-coated sediments, which

is the first NMR interpretation approach that considers higher relaxation modes.

We investigate two different sets of iron-oxide-coated samples. The first set consists of commercially available filter sand that was coated with different amounts of synthetic ferrihydrite and goethite. Using this set (Set A), we study the general impact of increasing iron concentration on the NMR relaxation behaviour and investigate how sensitive the measured NMR signature is with regard to the mineral type. The second set consists of filter sand and gravel with natural iron oxide incrustations and material taken from the clogging experiments of Weidner (2016), who investigated the influence of chemical iron-clogging on the hydraulic conductivity of gravel pack material in a sand tank model. The iron oxide content of these samples consists of different amounts of ferric oxide minerals, including ferrihydrite and goethite. Using this set (Set B), we test the general potential of NMR to provide a reliable proxy for hydraulic conductivity even with the content of individual paramagnetic iron oxides varying arbitrarily.

2 Basics of NMR relaxation in porous media

2.1 Principle of NMR relaxometry

The measurement principle is based on the manipulation of hydrogen protons (e.g. in water molecules). They exhibit a magnetic momentum due to their proton spins. When an ensemble of proton spins is exposed to a permanent magnetic field B_0, an additional (nuclear) magnetisation M is formed and aligned with B_0. By electromagnetic stimulation (excitation) using an external field B_1 that alternates the Larmor frequency of proton spins, M can be forced to deflect from its equilibrium position. After shutting off the excitation, the movement of M back to equilibrium is observed. This process is called NMR relaxation and the resulting signal, recorded as induced voltage in a receiver coil, is an exponential decrease (transverse or T_2 relaxation) when measured perpendicular to B_0. When observed parallel to B_0, the signal increases correspondingly (longitudinal or T_1 relaxation). Detailed information on theory and measurement techniques is found in, for example, Coates et al. (1999) and Dunn et al. (2002).

2.2 NMR relaxation in general

Because only the hydrogen proton spins of the pore water molecules contribute to the NMR signal, its amplitude is a measure for the water content of the investigated material, while the relaxation behaviour encodes relevant information on the pore environment. The NMR signal E (V) as a function of the measurement time t is described by

$$E(t) = E_0 \left[1 - \sum_n I^n \exp\left(-\frac{t}{T_1^n}\right) \right] \tag{1}$$

and

$$E(t) = E_0 \sum_n I^n \exp\left(-\frac{t}{T_2^n}\right) \tag{2}$$

for the T_1 and T_2 relaxation, respectively. E_0 is the initial amplitude (V), while I^n and T_i^n ($i = 1, 2$) denote the relative intensity (no units) and relaxation time (s) of the nth relaxation regime.

When considering the T_1 relaxation, the relaxation rate $1/T_1^n$ is given by

$$\frac{1}{T_1^n} = \frac{1}{T_{1,\text{bulk}}} + \frac{1}{T_{1,\text{surf}}^n}, \tag{3}$$

where $1/T_{1,\text{bulk}}$ and $1/T_{1,\text{surf}}^n$ describe the relaxation rates of the pure pore water excluding the influence of the pore walls (bulk relaxation) and the interaction of the proton spins with the pore surface (surface relaxation), respectively. For the general description of the T_2 relaxation, an additional term must be included:

$$\frac{1}{T_2^n} = \frac{1}{T_{2,\text{bulk}}} + \frac{1}{T_{2,\text{surf}}^n} + \frac{1}{T_{2,\text{diff}}}. \tag{4}$$

The rates $1/T_{2,\text{bulk}}$ and $1/T_{2,\text{surf}}^n$ are the same as for the T_1 relaxation, whereas the $1/T_{2,\text{diff}}$ considers the case of an inhomogeneous B_0 field. The diffusion relaxation must be taken into account, if a significant quantity of ferromagnetic minerals is present (Keating and Knight, 2007, 2008) or if the sensitive volume of the measurement includes a significant gradient in B_0 (Blümich et al., 2008; Perlo et al., 2013). However, for the estimation of hydraulic properties from NMR, the surface relaxation is the most interesting phenomenon.

Brownstein and Tarr (1979) derived the NMR relaxation behaviour in restricted environments for simple pore geometries (planar, cylindrical, and spherical). In this study, we consider the corresponding relaxation inside a cylindrical capillary with radius r_c, which exhibits different relaxation modes:

$$T_{i,\text{surf}}^n = \frac{r_c^2}{D \xi_n^2} \text{ with } \xi_n \frac{J_1(\xi_n)}{J_0(\xi_n)} = \frac{\rho_i r_c}{D}. \tag{5}$$

D refers to the self-diffusion coefficient of water ($m^2 s^{-1}$) and ρ_i to the surface relaxivity ($m s^{-1}$) for either the longitudinal ($i = 1$) or the transverse ($i = 2$) relaxation, which is a material constant describing the influence of paramagnetic minerals at the pore surface. J_0 and J_1 are the Bessel functions of the zeroth and first order, respectively. The quantities ξ_n can only be found by calculating the positive roots of

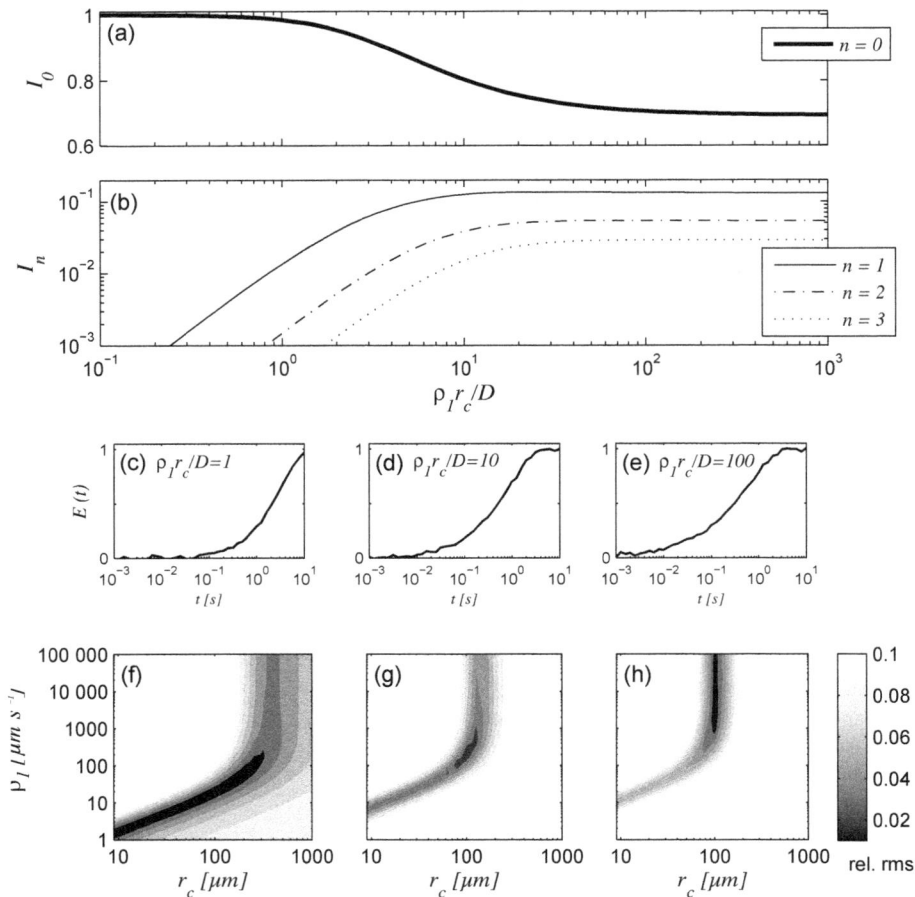

Figure 1. (a, b) Intensities of the zeroth to third relaxation modes as functions of the relationship $\rho_1 r_c/D$ visualising the different diffusion regimes in which NMR relaxation can take place, (c)–(e) simulated T_1 relaxation data for a capillary with (c) $r_c = 100\,\mu m$ and $\rho_1 = 20\,\mu m\,s^{-1}$, (d) $r_c = 100\,\mu m$ and $\rho_1 = 200\,\mu m\,s^{-1}$, and (e) $r_c = 100\,\mu m$ and $\rho_1 = 2000\,\mu m\,s^{-1}$, (f)–(h) corresponding results of a parameter search regarding r_c and ρ_1. The NMR time series was contaminated by Gaussian-distributed random noise with an amplitude of 0.01.

the corresponding equation numerically. The intensities I^n are given by

$$I^n = \frac{4 J_1^2 (\xi_n)}{\xi_n^2 \left[J_0^2 \xi_n + J_1^2 (\xi_n) \right]}. \tag{6}$$

According to Brownstein and Tarr (1979), the term $\rho_i r_c/D$ in Eq. (5) defines a controlling criterion that distinguishes between the fast ($\rho_i r_c/D \ll 1$), intermediate ($1 < \rho_i r_c/D < 10$), and slow ($\rho_i r_c/D > 10$) diffusion regimes. Figure 1a and b demonstrate the relative intensities I^n of the zeroth to third modes as functions of $\rho_i r_c/D$ for all diffusion regimes. Obviously, the zeroth mode I^0 is the only relevant relaxation component taking place in the fast diffusion range, because the intensities of the higher modes can be neglected, i.e. the relaxation is mono-modal inside the considered pore. The phenomenological explanation for this feature is that all proton spins in the pore space diffuse fast enough to sample the entire pore surface dur-

ing the NMR relaxation measurement, which is the case for small pores and low surface relaxivities. The common empirical approaches to provide hydraulic conductivity estimates (e.g. Kenyon, 1997; Coates et al., 1999; Knight et al., 2016) and pore size distributions (e.g. Hinedi et al., 1997; Costabel and Yaramanci, 2013) are only valid if this condition is satisfied: the zeroth mode in Eq. (5) simplifies to $T_{i,\mathrm{surf}}^0 = r_c/2\rho_i$ and, given that ρ_i can be determined by calibration, becomes a unique proxy for a certain pore (capillary) radius.

Outside the fast diffusion regime, the intensities I^n for $n > 1$ increase (Fig. 1a and b), while I^0 decreases asymptotically to about 0.7. In materials with large pores and/or high surface relaxivities, the self-diffusion of the proton spins is slow in regard to the mean distance to the pore surface and thus, the excited protons do not equally get in touch with the pore surface. Protons in the direct vicinity of the surface exchange their spin magnetisation faster than those within the pore body. The consequence is a multi-exponential (i.e. multi-modal) relaxation inside the pore. The theory of Brownstein and Tarr (1979) leads to the simplification of $\xi_n = (n + 1/2)^2\pi^2$ in Eq. (5) describing the asymptotic behaviour in the slow diffusion regime. This is in principle a significant advantage regarding the estimation of pore radii from relaxation times, because a calibration regarding ρ_i is not necessary. However, natural unconsolidated sediments exhibit a large range of pore sizes, which are seldom completely in the slow diffusion regime. Thus, a close description of the problem is desired that considers all diffusion regimes at once.

2.3 Analysis of relaxation modes

The pore space of a well-sorted porous material has a narrow pore size distribution that can be described using a single effective pore radius (r_{eff}). For this case, Müller-Petke et al. (2015) showed that the consideration of relaxation modes as defined in Eqs. (5) and (6) leads to an unambiguous prediction of pore radius and surface relaxivity in the intermediate diffusion regime. In this study, we use this concept to interpret, i.e. to approximate, our NMR relaxation measurements. As demonstrated in the following section, the investigated sample material in this study allows the assumption of a single r_{eff} to describe the pore space. We accept the limitation on a single effective pore radius for the benefit of a closed model that includes the relaxation modes outside the fast diffusion regime on the one hand and that does not demand a priori information on the diffusion regime or calibration of ρ_i on the other.

However, depending on the actual diffusion regime of the sample, the performance of the approximation procedure as well as the general results differ significantly. To demonstrate the corresponding effects, we calculated the synthetic T_1 relaxation response signals according to Eqs. (1), (5), and (6) for a cylindrical pore with a radius $r_c = 100\,\mu\mathrm{m}$ and surface relaxivities $\rho_1 = 20, 200, 2000\,\mu\mathrm{m\,s}^{-1}$. The positions of these three parameter combinations in Fig. 1a and b show that they represent one specimen for each relevant setting of the relaxation modes: the first at $\rho_i r_c/D = 1$, where I^0 is close to 1; the second at $\rho_i r_c/D = 10$, where the corresponding I^0 lays inside the decreasing range; and the third at $\rho_i r_c/D = 100$, where I^0 has reached the asymptote. The initial amplitudes E_0 of the synthetic signals were set to 1 and the resulting synthetic signals are exposed to a Gaussian-distributed noise with an amplitude of 0.01 (Fig. 1c–e).

Figure 1f to h show the results of a parameter search for each of the three cases as surface plots (i.e. their objective functions), where the surface height demonstrates the relative root mean square (rms) value of each combination of ρ_1 and r_c within the search region. The black region in each figure demonstrates the area, where the resulting rms value is 0.01, i.e. where the corresponding parameter combinations lead to a reliable approximation of the original signal within its noise level. According to the findings of Müller-Petke et

al. (2015), a unique solution for both parameters can only be found for the signal at $\rho_i r_c/D = 10$ (Fig. 1g). The fast diffusion regime in Fig. 1f is characterised by an ambiguous region demonstrating the linear relationship of ρ_1 and r_c, while the solution of the third signal at $\rho_i r_c/D = 100$ is independent of ρ_1 (Fig. 1h). Two important facts can be deduced from Fig. 1h: first, by performing a parameter search for NMR relaxation measurements under very slow diffusion conditions, only a minimum of ρ_1 can be determined, and, second, an adequate approximation algorithm based on the mode interpretation of NMR relaxation will always provide a reliable estimate of r_c outside the fast diffusion region, while the corresponding ρ_1 estimate becomes more and more inaccurate when passing through the slow diffusion regime.

In contrast to ferromagnetic impurities that mainly affect the diffusion relaxation by small-scaled disturbances of the magnetic fields involved, the appearance of purely paramagnetic iron mineral coatings is expected to cause an increase in ρ_i and thus a faster relaxation (e.g. Foley et al., 1996; Keating and Knight, 2007). However, iron oxides are known to have large surface areas (e.g. Houben and Kaufhold, 2011) and will consequently affect the NMR relaxation also by an increasing pore surface-to-volume ratio, S/V (Foley et al., 1996; Müller-Petke et al., 2015). It is generally impossible to relate an observed increase in NMR relaxation unambiguously to either an increase in ρ_i or to an increase in S/V without additional information. Along with the general behaviour of relaxation modes, numerical modelling of Müller-Petke et al. (2015) demonstrated that an increasing roughness of the surface inside a capillary with otherwise low and constant ρ_i leads to a similar relaxation as an increasing surface relaxivity, while keeping the radius unchanged. They introduced and defined the apparent surface relaxivity $\rho_{i,\mathrm{app}}$ in combination with an apparent pore radius $r_{\mathrm{app}}^{\mathrm{NMR}}$ to explain NMR relaxation of porous media with narrow pore size distribution. Following their suggestion, we define $\rho_{i,\mathrm{app}}$ to include both the effect of an increasing ρ_i and the corresponding increase in pore surface roughness due to iron oxide coating, while $r_{\mathrm{app}}^{\mathrm{NMR}}$ is considered to be the mean radius of the corresponding capillary. The hypothesis demands the assumption that the coating and the corresponding distribution of $\rho_{i,\mathrm{app}}$ is homogeneously distributed. This is a crucial point, because a perfect homogeneous distribution of iron precipitation on the pore scale due to natural chemical or microbiological processes or even synthetic chemical treatment is questionable. However, regarding the slow NMR relaxation in coarse sediments it is expected that, during the NMR measurement, the diffusing spins statistically sample possible inhomogeneities in the distribution of ρ_i or $\rho_{i,\mathrm{app}}$ inside the pore space uniformly enough to allow the assumption of a mean surface relaxivity (Kenyon, 1997; Grunewald and Knight, 2011; Keating and Knight, 2012). An important objective of this study is the comparison of $r_{\mathrm{app}}^{\mathrm{NMR}}$ with the effective hydraulic pore radius r_{eff}.

3 Material and methods

3.1 Samples with controlled synthetic ferrihydrite and goethite coating

In the first experimental step, the focus was set on a simplified binary system consisting of (a) a relatively uniform carrier phase, quartz in the form of commercially available filter gravel, and (b) synthetically produced iron oxides. For the latter, ferrihydrite and goethite mineral phases were studied separately, both of which are common constituents in soils and aquifers but also in incrustations. Synthetic iron oxides were used because of their controlled crystallite size and composition (Schwertmann and Cornell, 2000). Ferrihydrite is a poorly crystalline mineral that usually precipitates as the first stable oxidation product when dissolved ferrous iron comes into contact with oxygen. Since ferrihydrite is thermodynamically meta-stable, it will convert over time into the more stable goethite (e.g., Houben and Kaufhold, 2011). This process is strongly accelerated at higher temperatures ($> 50\,°C$) and involves a significant reduction of specific surface area and therefore water content, density, and chemical reactivity. Thus, this study does not only encompass two of the most important iron oxides but, at the same time, two different stages of crystallinity, age, and reactivity.

Two series of artificially coated filter sand samples (Set A) were prepared by precipitating the Fe(III)-minerals ferrihydrite and goethite onto quartz following Schwertmann and Cornell (2000). Therefore, iron nitrate nonahydrate ($Fe(NO_3)_3 \cdot 9H_2O$; CAS: 7782-61-8, technical purity, BDH Prolabo) was dissolved in twice de-ionised water to attain a $1\,mol\,L^{-1}$ solution. A $5\,mol\,L^{-1}$ potassium hydroxide solution (KOH, CAS: 1310-58-3, Bernd Kraft) was used to trigger precipitation of ferrihydrite ($Fe_5HO_8 \cdot 4H_2O$). The desired contents of iron in the filter sands were realised by varying the amounts of the two solutions, added to a fixed amount of filter sand. After precipitation the residual solution was carefully exchanged by washing with de-ionised water. For transformation of ferrihydrite to goethite (α-FeOOH), a second batch of ferrihydrite was held in a closed glass bottle at $70\,°C$ for $60\,h$. The applied recipes for ferrihydrite and goethite are based on the collection of standard synthesis procedures compiled in the reference book by Schwertmann and Cornell (2000). They have been successfully applied in numerous studies (e.g. Janney et al., 2000; Houben, 2003b; Houben and Kaufhold, 2011).

After preparation, the sample material was filled into circular petri dishes with a diameter of $50\,mm$ and a height of $15\,mm$ to perform the initial NMR measurements. Most of the iron particles settled to the bottom and formed a gradient in iron concentration inside the dishes, which could visually be observed for most of the samples due to an obvious increase in reddish colour from top to bottom. Initial NMR measurements were performed to qualitatively analyse the vertical distribution of the iron content. Therefore, mea-surements at different heights of the sample holders were conducted. However, for the quantitative analysis of NMR parameters, the samples were homogenised before the final NMR measurements, because it was not possible to determine the amount of iron as a function of height inside the sample holders by chemical analyses. To homogenise the iron content inside the petri dishes, the material was exposed to the atmosphere for 1 day, where it evaporated to a certain state of partial saturation (resulting saturation: 0.2 to 0.5), mixed, and filled into dishes with a diameter of $50\,mm$ and a height of $10\,mm$. Afterwards, samples were dried completely to ensure a proper coating of the pore walls with the iron particles. To maintain a homogeneous iron distribution throughout the sample and a better adhesion to the quartz surface, the material was moistened (de-ionised water) and dried out again. This procedure was repeated 4 times for each sample. Finally, the samples were completely saturated with de-ionised water prior to the NMR measurements.

After the final NMR measurements, the samples were air-dried again to determine their porosity Φ by weight. Afterwards they were subdivided for the controlling analysis. The iron content of each sample was analysed chemically to identify whether and to what extent the precipitation had led to the desired results. This was done by analysing the amount of dithionite-soluble iron, following the method of Mehra and Jackson (1960). The oxidic iron coatings that are expected to affect the NMR results are re-dissolved with dithionite solution and quantified by measuring the iron concentration in the solution. The total iron content was investigated by X-ray fluorescence analysis (XRF, using a PANalytical Axios and a PW2400 spectrometer) for verification. The latter method is expected to yield slightly higher iron contents, because XRF also captures the iron content bound in silicates of the filter sand or gravel grains. The difference for the samples of Set A indicates an amount of siliceous iron in the range of 0.5 to $0.7\,g\,kg^{-1}$. The further analysis is thus based on the actual measured contents of dithionite-soluble iron. The grain size distributions were determined using a CAMSIZER (Retsch GmbH). The specifications of the samples are summarised in Table 1. The comparison of the desired with the actually achieved Fe contents indicates that, during the exchange of the remaining synthesis solutions ($Fe(NO_3)_3$ and KOH) with H_2O_{dest}, some of the fine precipitates have been washed out. A part of each sample was also prepared for the determination of the specific surface area using the BET method (Brunauer et al., 1938). However, the corresponding results fell below the accuracy limit of the device and are not reliable. Obviously, the contents of iron oxide in the investigated samples are too small and the surface area is still dominated by the quartz grains.

3.2 Samples with natural iron coating

A second set of samples with natural iron coatings was also studied (Set B, Table 2). This set consists of gravel samples

Table 1. List of samples with synthetic ferrihydrite (F) and goethite (G) coating (Set A).

Sample	Desired Fe content (g kg^{-1})	Total Fe content (XRF) (g kg^{-1})	Dithionite-soluble Fe content (g kg^{-1})	ϕ (NMR samples) (m^3 m^{-3})	d_{60}/d_{10} (μm μm^{-1})	d_{GSD} (μm)	r_{eff} (μm)
F1	10.00	5.88	5.29	0.36	3.27	508	95
F2	5.00	2.94	2.35	0.38	1.43	838	172
F3	2.00	1.26	0.62	0.45	1.40	944	258
F4	1.00	1.05	0.45	0.43	1.42	892	221
F5	0.50	0.91	0.29	0.42	1.41	909	221
F6	0.20	0.77	0.19	0.40	1.42	906	204
F7	0.10	0.63	0.14	0.39	1.43	901	189
G2	5.00	2.73	2.14	0.39	1.44	835	175
G3	2.00	1.40	0.76	0.35	1.42	927	167
G4	1.00	0.98	0.45	0.45	1.41	920	253
G5	0.50	0.91	0.32	0.36	1.45	902	167
G6	0.20	0.70	0.17	0.35	1.44	909	162
G7	0.10	0.77	0.15	0.37	1.39	936	185
S0*	0.00	0.84	0.11	0.39	1.47	904	196

* S0 refers to the original uncoated filter sand.

from laboratory well clogging experiments (Weidner, 2016), but also encrusted filter sand and gravel samples taken from excavated wells. The analyses were the same as for Set A.

3.3 Estimation of effective pore radius and hydraulic conductivity from grain size distribution

To obtain consistent reference values for comparison with the NMR results, we estimated the effective pore radius from the effective grain diameter d_{GSD} as defined by Carrier (2003), who suggested the use of the equations of Kozeny (1927) and Carman (1939) to estimate the hydraulic conductivity from grain size distribution (GSD) data:

$$d_{GSD} = \left(\sum_i \frac{f_i}{\sqrt{D_{li} D_{ui}}} \right)^{-1}, \tag{7}$$

where f_i refers to the ith weight fraction of grains within the respective sieve size limits D_{li} and D_{ui} with $\sum_i f_i = 1$.

To estimate the effective pore radius r_{eff} from d_{GSD}, we determine the ratio of the wetted surface to the pore volume (= specific surface) for both the capillary geometry of our pore model and the spherical geometry assumed for the effective grain diameter:

$$\frac{\text{pore surface}}{\text{pore volume}} = \frac{2\phi}{r_{eff}} = \frac{6(1-\phi)}{d_{GSD}}, \tag{8}$$

with ϕ being the porosity. The effective pore radius is then given by the following:

$$r_{eff} = \frac{1}{3} \frac{\phi}{1-\phi} d_{GSD}. \tag{9}$$

The Kozeny–Carman equation, when considering a cylindrical capillary with effective radius r_{eff}, is defined as follows (e.g. Pape et al., 2006):

$$K_{KC} = \frac{\varrho g}{\eta} \frac{1}{8\tau} \phi r_{eff}^2. \tag{10}$$

The parameter τ refers to the tortuosity (no units), g to the gravity acceleration (9.81 m s^{-2}), and ϱ and η to the density (1000 kg m^{-3}) and dynamic viscosity (1 g m^{-1} s^{-1}) of the pore water, respectively. The tortuosity is set to 1.5 in this study, which is a reliable estimate for coarse sand and gravel (e.g. Pape et al., 2006; Dlugosch et al., 2013).

An alternative to the semi-empirical Kozeny–Carman equation is the well-known empirical formula of Hazen (1892). The effective measure in this approach is assumed to be the grain diameter corresponding to the 10 wt % percentile of the cumulative GSD (d_{10}). The corresponding estimates of hydraulic conductivity K_{Hz} were used as an additional set of reference values.

3.4 NMR measurements

As described above in Sect. 3.1, the stimulated precipitation yielded an obvious vertical gradient in iron oxide content. To identify the corresponding level of heterogeneity and to control and verify the homogeneity of the iron oxide distribution after the final mixing, an NMR device with vertical sensitivity, i.e. the ability to apply distinct measurements at different heights of the sample holder had to be applied. Using a common NMR core analyser, the entire specimen is measured at once, which can lead to a misinterpretation if different relaxation regimes overlap. Therefore, the experiments in this study were realised using a single-sided NMR

Table 2. List of samples with artificial and natural iron clogging (Set B).

Sample	Total Fe content (XRF) (g kg^{-1})	Dithionite soluble Fe-content (g kg^{-1})	ϕ (NMR samples) (m^3 m^{-3})	d_{60}/d_{10} (μm μm^{-1})	d_{GSD} (μm)	r_{eff} (μm)
HB-Z_0[1]	0.42	0.13	0.39	1.36	1222	261
HB-Z_1[2]	7.20	7.12	0.39	1.44	935	201
HB41_0[3]	0.28	0.10	0.39	1.43	1164	247
HB41_1[2]	2.52	2.39	0.41	1.44	1028	236
HB41_2[2]	7.90	7.85	0.40	1.48	900	197
HB41_3[2]	5.74	5.64	0.40	1.46	823	184
GW3151_0[2,4]	0.28	0.12	0.38	1.69	1037	213
GW3151_1[2]	3.64	3.28	0.38	1.46	1180	244
GW5051_0[1]	1.26	1.08	0.35	1.68	1010	184
GW5051_1[2]	4.06	3.88	0.36	1.70	856	158
GW3120_0[2,4]	0.49	0.25	0.36	1.74	1123	211
GW3120_1[2]	8.18	8.04	0.36	1.85	917	172
GW3120_2[2]	14.76	14.80	0.39	1.70	717	155
DF0[3]	7.48	5.27	0.36	1.51	1719	328
DF11[3]	10.77	8.26	0.36	1.29	2271	420
DF13A[3]	10.77	8.12	0.39	1.29	2269	474
DF13B[3]	11.33	9.05	0.37	1.36	2092	404
FD0[3]	5.87	4.45	0.42	1.36	1954	470
FD12A[3]	10.00	8.51	0.40	1.43	1925	436
FD12B[3]	9.02	7.52	0.38	1.39	1958	404
WS0[3,4]	0.42	0.17	0.42	1.58	1634	391
WS4[3]	8.74	8.40	0.40	1.69	1169	258
WS8[3]	4.69	4.39	0.41	1.57	1586	373

[1] Samples of filter sand and gravel without iron coating taken at dewatering wells excavated in German lignite open pits (HB: Hambach, GW: Garzweiler). [2] Samples of filter sand and gravel with natural iron coating taken at dewatering wells excavated in German lignite open pits. [3] Samples of filter sand and gravel with artificial iron coating generated in well clogging experiments (Weidner, 2016) with original material DF0 and FD0 as used in dewatering wells in German lignite mining from three different gravel pits (DF: Dorsfeld, FD: Frimmersdorf, WS: Weilerswist). [4] Before analysis these samples were treated with dithionite to remove existing surface iron oxides in order to recreate the original state.

apparatus (NMR Mouse, Magritek) with strong sensitivity to vertical changes inside the sample (Fig. 2). Four permanent magnets for the B_0 and the measurement coil for the B_1 field are arranged in a way that the sensitive volume is as a slice with a thickness of 200 μm and a footprint of about 40 by 40 mm (Kolz et al., 2007; Blümich et al., 2008). The operating frequency is 13.05 MHz. The sample is placed on a table, while the sensor is mounted on a platform adjustable in height, i.e. to move the sensitive volume over the sample (along the z axis) with an accuracy of a few micrometres.

Although homogeneous in the plane parallel to the B_1 coil, the B_0 field strength decreases with increasing distance to the magnets, which yields a strong B_0 gradient in the z direction (mean gradient according to user's manual: 273 kHz mm^{-1}) inside the sensitive slide. Consequently, the T_2 measurements (CPMG sequence, for details please see Coates et al., 1999 and Dunn et al., 2002) are dominated by the diffusion relaxation rate. In principle, this effect can be corrected to identify the proportion of surface relaxation in the data (Keating and Knight, 2008). However, testing and discussion of the quality and potential of the additional measurements and calculations necessary for this correction are beyond the scope of this paper. Thus, we use the T_2 measurements only for determining the NMR porosity Φ_{NMR} from the initial amplitude of the corresponding exponential decay. Due to the linearity between NMR signal and water content inside the sensitive volume of the measurement (e.g. Costabel and Yaramanci, 2011; Behroozmand et al., 2014), Φ_{NMR} can simply be determined by the ratio of the initial amplitude of the investigated sample and that of pure water in a sample holder with exactly the same dimensions. The CPMG measurements were conducted with an echo time of 66 μs, while the total number of echoes was varied individually between 3000 and 9000. The corresponding measurement times vary in a range of about 0.2 to 0.6 s.

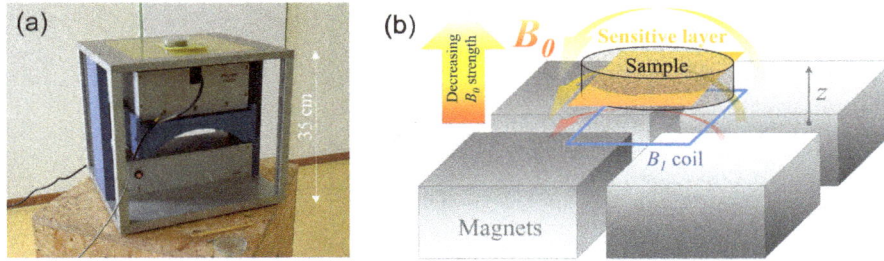

Figure 2. (a) Measurement device and **(b)** schematic showing the configuration of the permanent B_0 magnets, B_1 coil, and the resulting sensitive layer.

For investigating the impact of the iron oxide coating, we use the T_1 relaxation, which is unaffected by gradients in B_0. These measurements are realised as saturation recovery (SR) measurements (details see Coates et al., 1999 and Dunn et al., 2002). Each record consists of 50 single recovery times, which are logarithmically spaced along the measurement time axis. The exact positioning of the recovery times was adjusted for each sample to realise a similar distribution of time samples from zero to equilibrium nuclear magnetisation, which was estimated beforehand by screening SR measurements with a reduced number of time samples (15) and stacks. The maximum observation time for the final SR measurements was set 5 times higher than the prior T_1 estimates. For each sample, SR measurements at different heights were conducted using 1 mm steps in the range of $z = 3$ to 15 mm before and $z = 3$ to 10 mm after homogenisation. In this way, the vertical distribution of iron inside the samples before homogenisation and the natural scattering of the NMR parameters after homogenisation were taken into account. For the latter, mean values and double standard deviations (95 % confidence interval) were calculated from the measurements at different heights. After the T_1 measurements, a small sample of pore water (a few tenths of a millilitre) was extracted from the samples using a pipette in order to measure T_{bulk}. In some cases the extracted amount of pore water was not high enough to achieve a sufficient signal-to-noise ratio for an accurate NMR measurement. However, the T_{bulk} values of the successful measurements did not vary significantly among the samples. Consequently, for the analysis of the relaxation behaviour (Eq. 3) we use a mean T_{bulk} (2.46 ms $\pm$ 0.07 ms) for all samples.

Because the NMR porosity was determined from the T_2 measurements, it was not necessary to take the initial amplitude of the T_1 measurements into account. Thus, each SR time series was normalised to 1 prior to the final signal approximation. Although the main focus of our interpretation is on the approximation using the relaxation modes, we also fitted the data using the commonly used multi-exponential spectral inversion for comparison. As an example, Fig. 3a shows all T_1 measurements of the homogenised sample F4, i.e. all repetitions at different heights, and their approxima-

tions using the spectral approach. The corresponding spectra, depicted in Fig. 3b, demonstrate that the probability functions of all repeated T_1 data are in good agreement. They show a dominating peak with a maximum at about 1.3 s and a smaller peak around 0.1 s.

3.5　Testing for NMR diffusion regimes

The analysis of relaxation modes is useful only outside the fast diffusion regime. Thus, the question arises as to how the diffusion regime can be tested in practice. According to Kenyon (1997), the diffusion condition inside a pore is defined by the ratio of the time for a proton spin to diffuse across the pore ($= r_c^2/D$) and the surface relaxation time:

$$\kappa = \frac{r_c^2/D}{T_{i,\text{surf}}}. \tag{11}$$

Using the logarithmic mean of the measured relaxation spectra $T_{1,\text{lm}}$, the self-diffusion coefficient of water, and accepting r_{eff} as a reliable estimate of r_c, we combine Eq. (11) with Eq. (3) to determine a measure that can be used for practical testing of the diffusion regime:

$$\kappa \approx \frac{r_{\text{eff}}^2/D}{\left(T_{1,\text{bulk}}T_{1,\text{lm}}\right)/\left(T_{1,\text{bulk}} - T_{1,\text{lm}}\right)}. \tag{12}$$

3.6　Inversion of NMR relaxation modes

The uniformity coefficient is defined by the ratio of the grain diameters corresponding to the 60 and 10 wt % percentiles of the cumulative GSD. For all samples investigated in this study it is very low (i.e. < 5, see Tables 1 and 2), which indicates a narrow grain size and consequently narrow pore size distribution (see also Fig. S1 in the Supplement). Thus, the precondition to use the approach of Müller-Petke et al. (2015) (see Sect. 2.3) to fit and interpret the NMR data is fulfilled. The approximation algorithm, i.e. the data inversion yielding the relaxation modes,

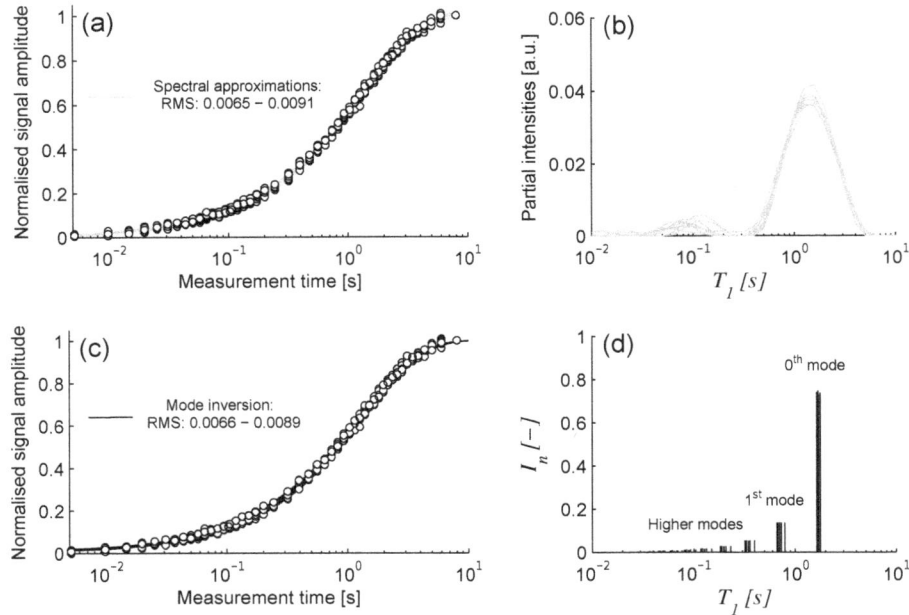

Figure 3. Panels **(a)** and **(c)**: normalised T_1 measurements at different heights of sample F4 after homogenisation and corresponding approximations using **(b)** multi-exponential spectrum and **(d)** relaxation modes.

1. starts using an initial model with given $\rho_{1,\mathrm{app}}$ and $r_{\mathrm{app}}^{\mathrm{NMR}}$;

2. calculates the corresponding multi-exponential NMR response by solving Eqs. (3), (5), and (6);

3. compares the result with the measured NMR signal by means of least squares;

4. modifies the parameters $\rho_{1,\mathrm{app}}$ and $r_{\mathrm{app}}^{\mathrm{NMR}}$ if necessary, that is if the modelled response and the measurement do not coincide; and

5. repeats the procedure until an optimal parameter set $\rho_{1,\mathrm{app}}$ and $r_{\mathrm{app}}^{\mathrm{NMR}}$ is found that explains the data.

We use the nonlinear solver lsqnonlin of the MATLAB® optimisation toolbox (MATLAB®, 2016) for this processing step.

Figure 3c shows the same data as Fig. 3a, but together with the approximations resulting from the relaxation mode inversion that obviously lead to identical fits compared to the spectral inversion. Figure 3d shows the corresponding results in the $I^n - T_1$ domain, that is, the first 10 modes for each measurement as separate spectral lines. The accuracy of the approximations using the relaxation modes represented by the corresponding rms values are similar to the ones of the spectral inversion.

4 Results and discussion

4.1 NMR-based porosity measurements

As mentioned above, to determine Φ_{NMR} of a sample, an additional NMR measurement using pure water is necessary. Figure 4a shows the T_2 data of sample F4 (synthetic ferrihydrite on quartz) and pure water. Due to the diffusion relaxation, the latter exhibits a relaxation time of less than 0.2 s, which is much shorter than that usually measured for water (2–3 s) in a homogeneous B_0. Because the initial signal amplitudes are not affected by the B_0 gradient, Φ_{NMR} can nevertheless be estimated from the T_2 data. Figure 4b shows the NMR-based porosities of all samples after homogenisation compared to those measured by weight. The NMR porosities coincide with the reference values within their uncertainties, which are determined as doubled standard deviations (95 % confidence interval) of the measurement repetitions at different sample heights. However, the uncertainties of the Φ_{NMR} estimates measured using the single-sided NMR device in this study are larger than those of past studies, where conventional laboratory NMR techniques are applied (e.g. Costabel and Yaramanci, 2011; Behroozmand et al., 2014). The reason for this is the relatively thin sensitive slice of 200 µm in combination with the investigated coarse material exhibiting mean r_{eff} values of 95 to 474 µm (see Tables 1 and 2). The inaccuracy of the porosity estimates must be accepted as a natural consequence of the fact that some of the observed pores exceed the z dimension of the probed reference volume (e.g. Costanza-Robinson et al., 2011).

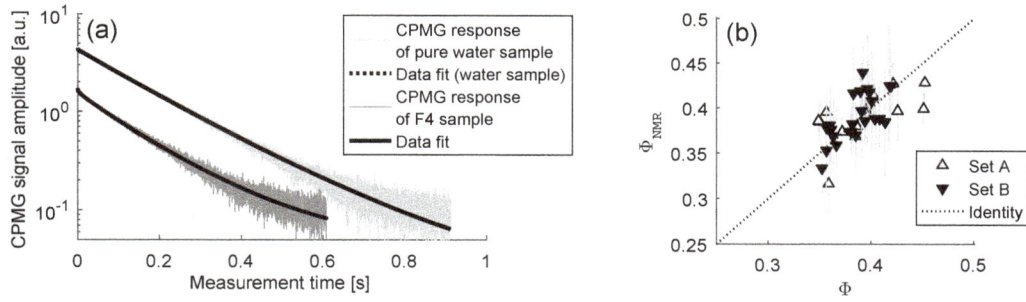

Figure 4. (a) T_2 measurement of sample F4 compared to pure water; **(b)** NMR-based porosity measurements compared to gravimetrical porosity for all samples.

4.2 The logarithmic mean of relaxation as qualitative measure for iron content at the pore walls

A photograph of sample F4 after the ferrihydrite precipitation is shown in Fig. 5a. The reddish section indicates that most ferrihydrite particles settled at the bottom of the petri dish. The same phenomenon was optically observed for almost all samples of Set A. Even though this separation was not visibly apparent in samples F1, F2, and G2 with the highest iron contents, we still expected a gradient in the iron content with z direction for these samples as well. Although not quantifiable to date, it is expected that the mean NMR relaxation time depends on the amount of paramagnetic iron oxides in the pore space (Keating and Knight, 2007). Thus, we performed initial NMR measurements (T_1 and T_2) to qualitatively analyse the level of inhomogeneity in the vertical ferrihydrite and goethite distributions by comparing the NMR parameters at different heights over the sample holders. Figure 5b and c depict the NMR data of sample F4 and those of the pure uncoated filter sand (sample S0), that is, the corresponding porosity determined from the E_0 amplitude of the CPMG data and the distributions of the logarithmic mean relaxation times ($T_{1,lm}$ and $T_{2,lm}$), respectively. Apart from a decrease at the top, the porosity distributions of both samples are homogeneous. It is likely that the decrease at the top is caused by evaporation caused by an imperfect sealing of the sample. The same feature was observed for all samples of Set A to varying extent. Figures S2–S16 show the photographs of all samples compared to the corresponding distributions of porosity and mean relaxation times. Some of the samples also show a significant decrease in porosity at the bottom of the sample holder, which is caused by small iron oxide particles accumulating in the voids between the quartz grains.

Whereas both the $T_{1,lm}$ and $T_{2,lm}$ distributions of the uncoated sample S0 appear to be homogeneous throughout the z axis, the general trend in the distributions of sample F4 is a gradual decrease from top to bottom (Fig. 5c), indicating the increase in surface relaxation with increasing ferrihydrite content. The difference between T_1 and T_2 is about 1 order of

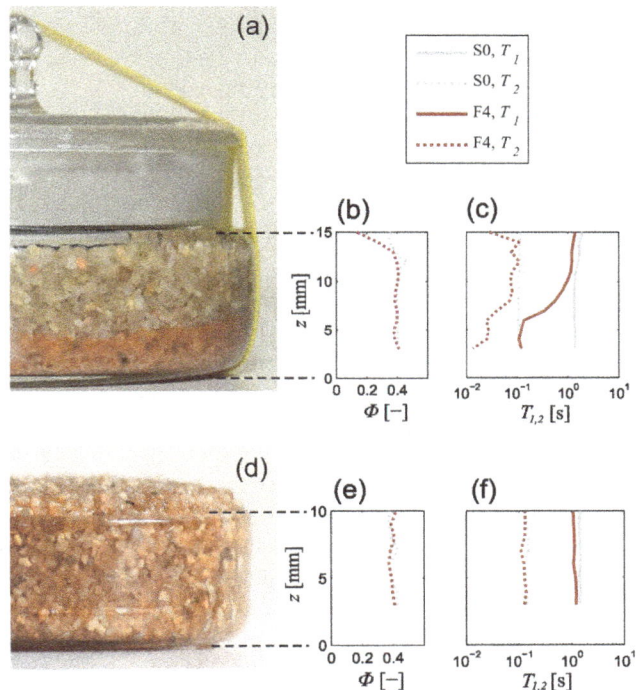

Figure 5. (a) Sample F4 after chemical treatment and precipitation of ferrihydrite particles at the bottom of the sample holder. Panels **(b)** and **(c)**: vertical distributions of corresponding porosities Φ and mean relaxation times T_1 and T_2, compared to those of untreated sand S0. Panels **(d)**–**(f)**: sample F4 after homogenisation and corresponding distributions of Φ and $T_{1,2}$.

magnitude, which is caused by the high diffusion relaxation rate in the inhomogeneous B_0 field of the single-sided NMR apparatus, as expected (see Sect. 3.4). When comparing the $T_{1,lm}$ and $T_{2,lm}$ curves of F4 with S0, it seems that no ferrihydrite remains at the top, because here the curves of both samples are almost in agreement. Although we cannot quantify the ferrihydrite content as a function of z by chemical analyses, we note that the logarithmic means of both T_1 and

Table 3. Estimates of κ according to Eq. (12) for the samples with artificial ferrihydrite and goethite coatings (Set A).

Sample	κ
F1	11.6
F2	16.3
F3	19.0
F4	10.8
F5	8.6
F6	7.0
F7	6.5
G2	16.4
G3	10.0
G4	11.6
G5	5.2
G6	4.5
G7	5.0
S0	5.4

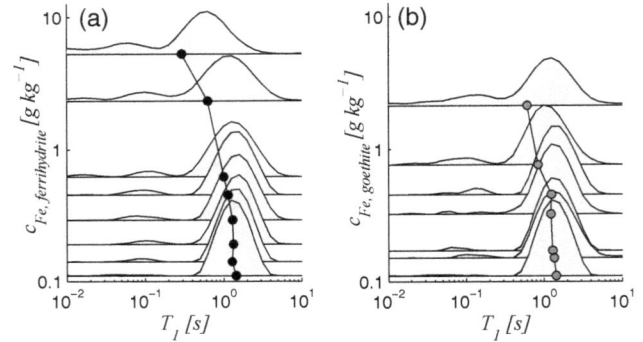

Figure 6. Relaxation time spectra as functions of Fe content for **(a)** ferrihydrite and **(b)** goethite samples (Set A); the circles mark the logarithmic mean for each spectrum.

T_2 are qualified proxies for the corresponding iron content distributions.

To relate the measured NMR parameters with the iron content, the samples had to be homogenised (see Sect. 3.1). Obviously, both the $T_{1,\mathrm{lm}}$ and $T_{2,\mathrm{lm}}$ distribution of the homogenised F4 sample are almost constant with z (Fig. 5d–f). The $T_{1,\mathrm{lm}}$ values of F4 are generally smaller than the ones of S0. In contrast, the $T_{2,\mathrm{lm}}$ distributions of F4 and S0 are almost identical, which is due to the influence of the high diffusion relaxation that masks the impact of the ferrihydrite content on the surface relaxation. As for the inhomogeneous sample, the porosity distributions of F4 and S0 are almost identical, i.e. an obvious impact of the increased content of ferrihydrite on the porosity is not observed. The process of homogenisation was applied and controlled for each sample of Set A. Figures S17–S31 show the corresponding distributions of porosity and mean relaxation times as functions of sample height for all samples. The remaining scattering of the z-dependent NMR parameters is considered as uncertainty intervals depicted by error bars (95 % confidence intervals) in the following analysis.

In Fig. 6, we show the relaxation time spectra of all samples of Set A and their corresponding mean values as a function of iron content. The principle trend is the same for both minerals. For iron contents smaller than approximately 0.7 g kg^{-1}, the main peak (between approximately 0.5 and 4 s) does not change significantly, whereas the logarithmic mean slightly decreases with increasing iron content in the same range. This increase is caused by an increase in the smaller peak (between approximately 0.05 and 0.2 s). If the iron content increases further to values of 1 g kg^{-1} and higher, the main peak shifts towards shorter time periods, while the increase in the smaller peak continues. Considering the classical interpretation of NMR relaxation spectra, it is not clear at this point whether the described changes of the

spectra with increasing iron content are caused by an increasing amount of small pores (possibly within the iron minerals at the pore walls), by enhanced surface relaxivity (due to the increasing amount of paramagnetic coating), or by a combination of both. However, because all samples, including the initial iron-free sand, are outside the fast diffusion regime (see Table 3), we must also consider that the increase in the smaller peak might be due to the increasing occurrence of the higher relaxation modes. Since it is not possible to distinguish between the existence of relaxation modes and different pore sizes when considering the spectral approximation approach, we analyse the relaxation modes in the next section by considering a bundle of capillaries with identical pore radius (= apparent pore radius $r_{\mathrm{app}}^{\mathrm{NMR}}$; see details in Sec. 2.3). This assumption is acceptable because the grain size distribution and consequently also the pore size distribution is narrow for the well-sorted materials studied here, which is proven by their small uniformity coefficient d_{60}/d_{10} (see Tables 1 and 2).

4.3 The relaxation modes as quantitative measure for iron content at the pore walls

The relaxation mode inversion was performed for all T_1 data of Set A and B samples. When considering the relaxation modes (see Sect. 2.3), the underlying model consists of the apparent pore radius $r_{\mathrm{app}}^{\mathrm{NMR}}$ of a virtual capillary with a circular cross section and a rough surface, the NMR sink rate of which is described by the apparent surface relaxivity $\rho_{1,\mathrm{app}}$ (Müller-Petke et al., 2015). The corresponding $r_{\mathrm{app}}^{\mathrm{NMR}}$ and $\rho_{1,\mathrm{app}}$ results for Set A are presented in Fig. 7a and b, respectively. All results of the individual measurements for each sample (= measurement at different heights) are depicted in order to avoid error bars in the logarithmic plot. We note that $r_{\mathrm{app}}^{\mathrm{NMR}}$ generally tends to smaller values for increasing iron content. However, the trend is only obvious for the iron contents higher than 0.5 g kg^{-1}. At least for the ferrihydrite series, the $r_{\mathrm{app}}^{\mathrm{NMR}}$ values even increase slightly for small iron

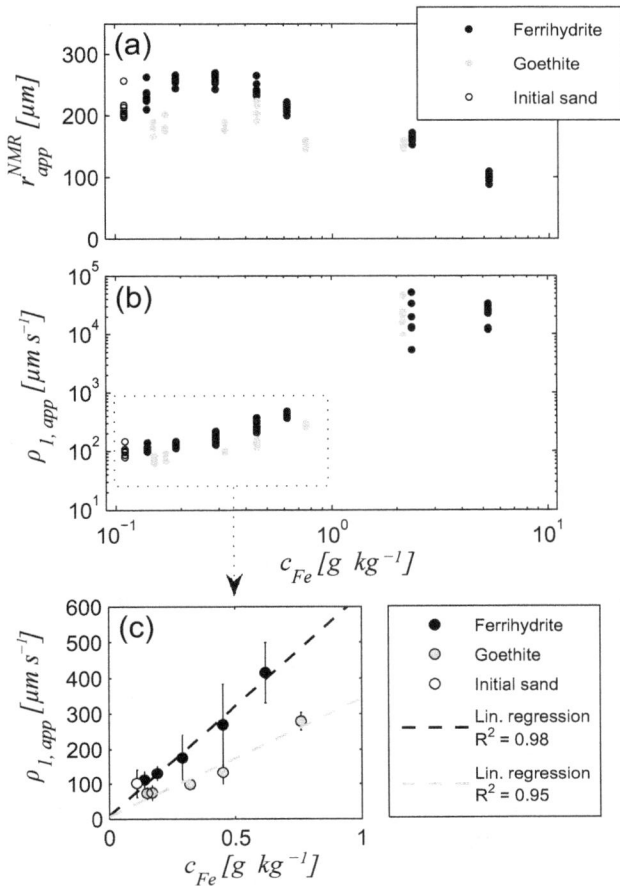

Figure 7. Results of relaxation mode inversion for the ferrihydrite and goethite data sets (Set A): **(a)** apparent pore radius r_{app}^{NMR} and **(b)** apparent surface relaxivity $\rho_{1,app}$ as functions of iron content, **(c)** the mean values and 95 % confidence intervals as error bars for Fe contents smaller than $1\,\mathrm{g\,kg^{-1}}$ and corresponding linear regression lines; regression coefficient for the ferrihydrite series: $646\,\mathrm{\mu m\,s^{-1}\,ppm^{-1}}$ (offset: $8.7\,\mathrm{\mu m\,s^{-1}}$) and for goethite series: $349\,\mathrm{\mu m\,s^{-1}\,ppm^{-1}}$ (offset: $9.5\,\mathrm{\mu m\,s^{-1}}$).

large uncertainties, because these reach the range where correct $\rho_{1,app}$ estimates cannot reliably be provided anymore (see Fig. 1 and corresponding discussion).

It is expected that a linear dependence between the surface relaxivity and the content of paramagnetic impurities at the pore walls exists (Foley et al., 1996). To test this expectation for the apparent surface relaxivity, Fig. 7c provides a focus on the data with accurate $\rho_{1,app}$ estimates, i.e. the data of samples with iron contents $< 1\,\mathrm{g\,kg^{-1}}$. The linear regression can be verified with R^2 values of 0.98 and 0.95 for the ferrihydrite and the goethite series, respectively. We note that the $\rho_{1,app}$ estimates for the goethite series are smaller than those for the ferrihydrite series by a factor of 1.85. We assume that this is an effect of the specific surface area of goethite being about up to 5 times smaller than that of ferrihydrite (goethite $\approx$ 20–80 $\mathrm{m^2\,g^{-1}}$ vs. ferrihydrite $\approx$ 180–300$\mathrm{m^2\,g^{-1}}$; Cornell and Schwertmann, 2003; Houben and Kaufhold, 2011). The larger specific surface of ferrihydrite leads to a higher surface roughness of the pore wall coating. As explained in Sect. 2.3, the apparent surface relaxivity does not distinguish between the increase in the surface roughness and increase in the actual surface relaxivity due to paramagnetic impurities at the pore wall. Because both are naturally linked to each other in an iron mineral by its individual surface area, we also expect an indirect sensitivity of $\rho_{1,app}$ on the type of iron mineral, i.e. on the composition of the iron oxide assemblage, if considering natural samples. However, to verify this assumption more iron oxides and their influence on the NMR relaxation modes must be studied in the future. Moreover, an accurate inspection of Fig. 7c leads to the assumption that a slight systematic discrepancy from linearity exists for both data sets. We hypothesise that this phenomenon is also caused by the influence of the surface roughness. We have found quadratic relationships yielding regression coefficients of 1 for both data sets. However, each of our data sets consists of just five points, which is not sufficient to validate this finding. Further research is necessary to quantify the influence of the surface roughness on the apparent surface relaxivity for natural iron coatings.

4.4 Comparison of NMR-effective pore radius and hydraulic parameters

Whether the NMR-based estimates of r_{app}^{NMR} can be considered to be reliable estimates of the effective hydraulic radius r_{eff} is examined in the cross plot in Fig. 8. The linear correlation between the two is verified with an R^2 of 0.58 when considering a constant offset (regression coefficient: 0.79) and 0.53 when enforcing the point [0, 0] in the fitting algorithm. The regression coefficient of the latter is very close to identity with 1.02.

Figure 9 correlates r_{app}^{NMR} and the corresponding estimates of hydraulic conductivity K_{NMR} with the reference values of hydraulic conductivity K for both Sets A and B. The K_{NMR} values were estimated according to Eq. (10) using the

contents, whereas the r_{app}^{NMR} of the goethite series remains more or less constant. The reason for this variation is likely due to the repacking of the samples after iron oxide precipitation. Considering an initially homogeneous porosity before iron precipitation, one would expect a decrease in porosity with an increasing amount of iron oxide. However, due to the repacking, each sample exhibits an individual porosity. Consequently, the apparent radius, no matter whether it was estimated by NMR or from GSD, also reflects the porosity variations, which covers the dependence on the iron content to some extent. Thus, the expected increase in r_{app}^{NMR} becomes visible only for the higher iron contents. Interestingly, the estimates of $\rho_{1,app}$ seem to be independent from the individual porosities. Figure 7b shows a monotonous increase in $\rho_{1,app}$ with iron content, at least for the samples with iron contents of $< 1\,\mathrm{g\,kg^{-1}}$. For the higher iron contents, $\rho_{1,app}$ exhibits

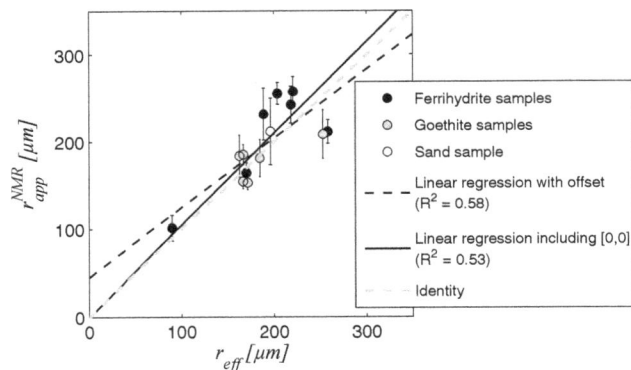

Figure 8. Correlation of effective radius estimates from grain size distribution r_{eff} and the apparent radius estimates from NMR r_{app}^{NMR}, the regression coefficient for fitting with constant offset is 0.79 and for fitting without offset, i.e. including the point [0, 0], is 1.02.

porosities determined from the T_2 measurements discussed with regard to Fig. 4. Because measurements of K are only available for eight samples of Set B, we use the K estimates derived from the GSD (Sect. 3.3) as reference values for all investigated samples, i.e. K_{KC} according to Eq. (10) in Fig. 9a and K_{Hz} according to Hazen (1892) in Fig. 9c. For both approaches, the correlation between r_{app}^{NMR} and K is verified with an R^2 of 0.66 and 0.57, when considering a power law to describe the relation mathematically (Fig. 9a and b). The assumption of a power law is suggested by the Kozeny–Carman equation (Eq. 10), where the exponent of the pore radius should be 2. The actual exponent for our data set reaches slightly higher values of 2.41 (K_{KC}) and 2.20 (K_{Hz}). The linear regression between K_{NMR} with K_{KC} and K_{Hz} (Fig. 9c and d) is verified with an R^2 of 0.47 and 0.38, while the corresponding regression factors are 0.85 and 2.45, respectively.

4.5 Discussion on field applicability

The relaxation analysis in this study is limited to T_1 data, the measurement of which, in boreholes and on the surface, is time-consuming and therefore often inefficient to date. Besides improving the performance of T_1 measurements, future research activities in the given context will also focus on T_2 relaxation measurements, which are often the preferred choice in practical applications. Considering the NMR relaxation theory, the findings of this study regarding the influence of the iron-coated pore surface on T_1 are expected to be valid for T_2 as well. However, the exact analysis of T_2 data regarding higher relaxation modes is crucial if measured in inhomogeneous B_0, because the diffusion relaxation will mask the effect of the modes to some extent. This is expected to be the case for the measurement device used in this study but is also for borehole NMR (e.g. Sucre et al., 2011; Perlo et al., 2013). Moreover, the data quality of field and borehole measurements is lowered compared to laboratory data by

environmental electromagnetic noise. Future research in the framework of iron-coated soils and sediments will therefore focus on potential approaches to correct the influence of the diffusion relaxation rate caused by external field gradients and to identify and characterise the occurrence of relaxation modes in T_2 data under field conditions. However, this study demonstrates that the NMR method is principally applicable to locate and hydraulically characterise zones with iron oxide accumulation in the pore space. In addition, NMR can provide indications for a beginning iron coating by changes in the apparent surface relaxivity, even before the effective hydraulic radius decreases, i.e. before a serious hydraulic clogging takes place.

5 Conclusions

NMR relaxation data of water-saturated sand and gravel are very sensitive to the amount of paramagnetic iron oxides. Here, this is confirmed using samples with synthetic ferrihydrite and goethite coatings as well as filter sand and gravel pack samples with varying contents of different natural iron oxides. We showed that the mean relaxation time can serve as a robust qualitative measure for the inhomogeneous distribution of iron content inside a sample. When focusing on the quantification of NMR parameters as a function of the iron content, the inversion of NMR data considering higher relaxation modes (Brownstein and Tarr, 1979; Müller-Petke et al., 2015) turns out to be a powerful tool, as long as the NMR relaxation takes place outside the fast diffusion regime, which is true for all samples investigated in this study. First, the inherent estimates of apparent surface relaxivity represent a qualified measure that linearly depends on the iron content, at least for values $< 1\,\mathrm{g\,kg}^{-1}$ for our data, above which the surface relaxivity cannot be estimated precisely. However, a further increase in iron content above that limit is nevertheless indicated by a decrease in the NMR-based estimate of apparent pore radius. Second, the corresponding NMR-based apparent pore radius is shown to be a reliable proxy for the effective hydraulic radius, which was verified in this study by comparison with reference estimates from grain size distributions. An important consequence of this finding is that estimates of hydraulic conductivity can be provided from NMR outside the fast diffusion regime without any calibration.

The need for future research must be noted. Besides the limitation on intermediate and slow diffusion regimes, relaxation mode inversion as suggested in this paper is only reliable for well-sorted material with narrow pore size distributions. Otherwise the assumption of a single effective radius might not be true. Future studies will consider the existence of both different characteristic pore sizes and higher relaxation modes. In contrast to the experimental design used here, these studies must combine NMR and direct hydraulic measurements, because broad distributions of grains can systematically bias the results of simple hydraulic models based

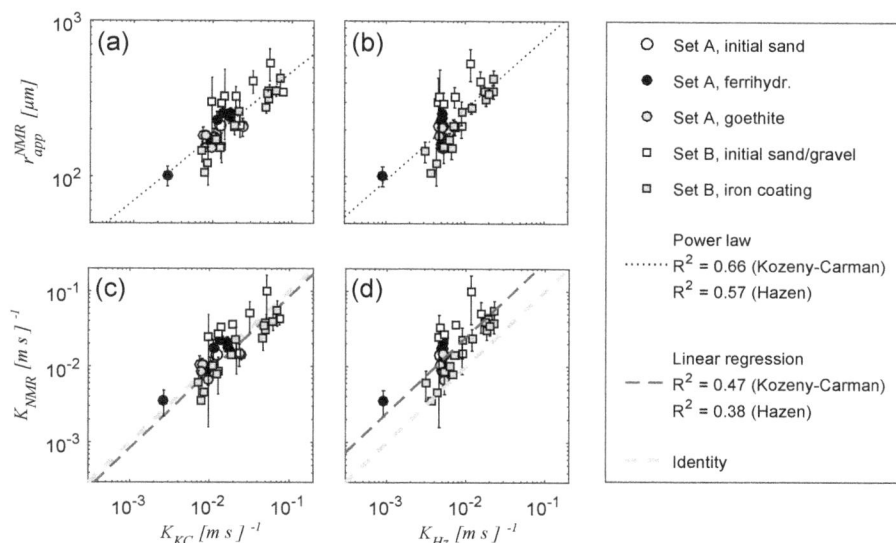

Figure 9. Correlation of NMR-based estimates of apparent radius r_{app}^{NMR} (**a, b**) and hydraulic conductivity K_{NMR} (**b, c**) with reference values for hydraulic conductivity, which are estimated from grain size distribution according to (**a, c**) Kozeny (1927) and Carman (1939) and (**b, d**) to Hazen (1892).

on texture (e.g. Boadu, 2000). Corresponding reference analysis regarding the pore size distribution might consist of imaging analysis or pressure-based water retention measurement.

The findings of this study are promising and interesting within the framework of hydraulic characterisation of aquifers or soils with significant content of paramagnetic iron oxides. The NMR method can complement other geophysical methods in the detection of natural iron oxide accumulations, such as bog iron, laterites, iron-rich palaeo-soils, and hardpan, provided that they are water-saturated. Moreover, a new potential application field for borehole NMR can be established: the identification and localisation of beginning iron incrustation in wells and/or the efficiency control of rehabilitation measures. Our future research activities will focus on the development of a corresponding methodology.

Author contributions. GH initiated and motivated the study and organised the hydrochemical treatment and reference analyses. MMP developed the software for the NMR mode inversion. CW developed and conducted the experiments for the iron oxide precipitation and organised and characterised the sample material. SC developed and performed the NMR experiments and prepared the paper with contributions of all authors.

Competing interests. The authors declare that they have no conflict of interest.

Acknowledgements. We thank the Institute of Hydrogeology and the Institute of Hydraulic Engineering and Water Resources Management of the RWTH Aachen University and the RWE Power AG for providing us with sample material, Stephan Kaufhold and Jens Gröger-Trampe for their advice and support on the geochemical analysis, and Raphael Dlugosch for fruitful discussions on the interpretation of the NMR data.

Edited by: Christine Stumpp

References

Abdel Aal, G., Atekwana, E., Radzikowski, S., and Rossbach, S.: Effect of bacterial adsorption on low frequency electrical properties of clean quartz sands and iron-oxide coated sands, Geophys. Res. Lett. 36, L04403, https://doi.org/10.1029/2008GL036196, 2009.

Atekwana, E. A. and Slater, L. D.: Biogeophysics: A new frontier in Earth science research, Rev. Geophys., 47, RG4004, https://doi.org/10.1029/2009RG000285, 2009.

Behroozmand, A. A., Keating, K., and Auken, E.: A Review of the principles and applications of the NMR technique for near-surface characterization, Surv. Geophys., 36, 27–85, https://doi.org/10.1007/s10712-014-9304-0, 2014.

Blümich, B., Perlo, J., and Casanova, F.: Mobile single-sided NMR, Progr. Nucl. Magnet. Reson. Spectrosc., 52, 197–269, 2008.

Boadu, F. K.: Hydraulic Conductivity of Soils from Grain-Size Distributions: New Models, J. Geotech. Geoenviron. Eng., 126, 739–746, 2000.

Brownstein, K. R. and Tarr, C. E.: Importance of classical diffusion in NMR studies of water in biological cells, Phys. Rev. A, 19, 2446–2453, 1979.

Brunauer, S., Emmett, P. H., and Teller, E.: Adsorption of Gases in Multimolecular Layers, J. Am. Chem. Soc., 60, 309–319, https://doi.org/10.1021/ja01269a023, 1938.

Bryar, T. R. and Knight, R. J.: Sensitivity of nuclear magnetic resonance relaxation measurements to changing soil redox conditions, Geophys. Res. Lett., 29, 2197, https://doi.org/10.1029/2002GL016043, 2002.

Bryar, T. R. and Knight, R. J.: Laboratory studies of the detection of sorbed oil with proton nuclear magnetic resonance, Geophysics, 68, 942–948, 2003.

Bryar, T. R., Daughney, C. J., and Knight, R. J.: Paramagnetic effects of iron(III) species on nuclear magnetic relaxation of fluid protons in porous media, J. Magnet. Reson., 142, 74–85, 2000.

Carman, P. C.: Permeability of saturated sands, soils and clays, J. Agr. Sci., 29, 262–273, 1939.

Carrier, W. D.: Goodbye, Hazen; Hello, Kozeny–Carman, J. Geotech. Geoenviron. Eng., 129, 1054–1056, 2003.

Coates, G., Xiao, L., and Prammer, M.: NMR Logging Principles and Application, Halliburton Energy Services, Houston, 1999.

Colombo, C., Palumbo, G., He, J.-Z., Pinton, R., and Cesco, S.: Review on iron availability in soil: interaction of Fe minerals, plants, and microbes, J. Soils Sediments, 14, 538–548, https://doi.org/10.1007/s11368-013-0814-z, 2014.

Cornell, R. M. and Schwertmann, U.: The iron oxides: structure, properties, reactions, occurences and uses, Wiley-VCH, Weinheim, 703 pp., 2003.

Costabel, S. and Yaramanci, U.: Relative hydraulic conductivity and effective saturation from Earth's field nuclear magnetic resonance – a method for assessing the vadose zone, Near Surf. Geophys., 9, 155–167, https://doi.org/10.3997/1873-0604.2010055, 2011.

Costabel, S. and Yaramanci, U.: Estimation of water retention parameters from nuclear magnetic resonance relaxation time distributions, Water Resour. Res., 49, 2068–2079, https://doi.org/10.1002/wrcr.20207, 2013.

Costabel, S., Siemon, B., Houben, G., and Günther, T.: Geophysical investigation of a freshwater lens on the island of Langeoog, Germany – Insights from combined HEM, TEM and MRS data, J. Appl. Geophys., 136, 231–245, https://doi.org/10.1016/j.jappgeo.2016.11.007, 2017.

Costanza-Robinson, M. S., Estabrook, B. D. and Fouhey, D. F.: Representative elementary volume estimation for porosity, moisture saturation, and air-water interfacial areas in unsaturated porous media: Data quality implications, Water Resour. Res., 47, W07513, https://doi.org/10.1029/2010WR009655, 2011.

Cullimore, D. R.: Microbiology of well biofouling, CRC Press, Boca Raton, FL, 456 pp., 2000.

Cundy, A. B., Hopkinson, L., and Whitby, R. L. D.: Use of iron-based technologies in contaminated land and groundwater remediation: A review, Sci. Total Environ., 400, 42–51, 2014.

Dlugosch, R., Günther, T., Müller-Petke, M., and Yaramanci, U.: Improved prediction of hydraulic conductivity for coarse-grained, unconsolidated material from nuclear magnetic resonance, Geophysics, 78, EN55–EN64, 2013.

Dunn, K., Bergman, D. J., and LaTorraca, G. A.: Nuclear Magnetic Resonance – Petrophysical and Logging Applications, in: Handbook of Geophysical Exploration: Seismic Exploration, Vol. 32, Elsevier Science, Oxford, 312 pp., 2002.

Emerson, D., Fleming, E. J., and McBeth, J. M.: Iron-Oxidizing Bacteria: An Environmental and Genomic Perspective, Annu. Rev. Microbiol., 64, 561–583, 2010.

Foley, I., Farooqui, S. A., and Kleinberg, R. L.: Effect of paramagnetic ions on NMR relaxation of fluids at solid surfaces, J. Magnet. Reson. Ser. A, 123, 95–104, 1996.

Geroni, J. N. and Sapsford, D. J.: Kinetics of iron (II) oxidation determined in the field, Appl. Geochem., 26, 1452–1457, 2011.

Grunewald, E. and Knight, R.: A laboratory study of NMR relaxation times in unconsolidated heterogeneous sediments, Geophysics, 76, G73–G83, 2011.

Hazen, A.: Some physical properties of sands and gravels, with special reference to their use in filtration, 24th Annual Rep., Pub. Doc. No. 34, Massachusetts State Board of Health, Massachusetts, 539–556, 1892.

Hertzog, R. C., White, T. A., and Straley, C.: Using NMR decay-time measurements to monitor and characterize DNAPL and moisture in subsurface porous media, J. Environ. Eng. Geophys., 12, 293–306, 2007.

Hinedi, Z. R., Chang, A. C., and Anderson, M. A.: Quantification of microporosity by nuclear magnetic resonance relaxation of water imbibed in porous media, Water Resour. Res., 33, 2697–2704, 1997.

Houben, G. J.: Iron oxide incrustations in wells. Part 1: Genesis, mineralogy and geochemistry, Appl. Geochem., 18, 927–939, 2003a.

Houben, G. J.: Iron oxide incrustations in wells. Part 2: Chemical dissolution and modeling, Appl. Geochem., 18, 941–954, 2003b.

Houben, G. J. and Kaufhold, S.: Multi-Method characterization of the ferrihydrite to goethite transformation, Clay Minerals, 46, 387–395, 2011.

Houben, G. J. and Weihe, U.: Spatial distribution of incrustations around a water well after 38 years of use, Ground Water, 48, 53–58, 2010.

Janney, D. E., Cowley, J. M., and Buseck, P. R.: Transmission Electron Microscopy of Synthetic 2- and 6-line Ferrihydrite, Clays Clay Miner., 48, 111–119, 2000.

Kappler, A. and Straub, K. L.: Geomicrobiological Cycling of Iron, Rev. Mineral. Geochem., 59, 85–108, 2005.

Keating, K. and Knight, R.: A laboratory study to determine the effects of iron oxides on proton NMR measurements, Geophysics, 72, E27–E32, 2007.

Keating, K. and Knight, R.: A laboratory study of the effect of magnetite on NMR relaxation rates, J. Appl. Geophys., 66, 188–196, 2008.

Keating, K. and Knight, R.: A laboratory study of the effect of Fe(II)-bearing minerals on nuclear magnetic resonance (NMR) relaxation measurements, Geophysics, 75, F71–F82, 2010.

Keating, K. and Knight, R.: The effect of spatial variation in surface relaxivity on nuclear magnetic resonance relaxation rates, Geophysics, 77, E365–E377, 2012.

Kenyon, W. E.: Petrophysical Principles of Applications of NMR Logging, Log. Analyst., 38, 21–43, 1997.

Knight, R., Walsh, D. O., Butler Jr., J. J., Grunewald, E., Liu, G., Parsekian, A. D., Reboulet, E. C., Knobbe, S., and Barrows, M.: NMR logging to estimate hydraulic conductivity in unconsolidated aquifers, Groundwater, 54, 104–114, https://doi.org/10.1111/gwat.12324, 2016.

Kolz, J., Goga, N., Casanova, F., Mang, T., and Blümich, B.: Spatial localization with single-sided NMR sensors, Appl. Magn. Reson., 32, 171–184, 2007.

Kozeny, J.: Uber kapillare Leitung des Wassers im Boden: Sitzungsberichte, in: Mathematisch-Naturwissenschaftliche Klasse Abteilung IIa, Akademie der Wissenschaften, Wien, 136, 271–306, 1927.

Larese-Casanova, P., Kappler, A., and Haderlein, S. B.: Heterogeneous oxidation of Fe(II) on iron oxides in aqueous systems: Identification and controls of Fe(III) product formation, Geochim. Cosmochim. Ac., 91, 171–186, 2012.

Larroque, F. and Franceschi, M.: Impact of chemical clogging on de-watering well productivity: numerical assessment, Environ. Earth Sci., 64, 119–131, 2011.

Legchenko, A., Baltassat, J.-M., Bobachev, A., Martin, C., Henri, R., and Vouillamoz, J.-M.: Magnetic resonance sounding applied to aquifer characterization, Groundwater, 42, 363–373, 2004.

MATLAB®: 9.0.0.341360 (R2016a), Optimization Toolbox TM User's Guide, The MathWorks, Inc., Natick, Massachusetts, 2016.

Medina, D. A. B., van den Berg, G. A., van Breukelen, B. M., Juhasz-Holterman, M., and Stuyfzand, P. J.: Iron-hydroxide clogging of public supply wells receiving artificial recharge: near-well and in-well hydrological and hydrochemical observations, Hydrogeol. J., 21, 1393–1412, 2013.

Mehra, O. P. and Jackson, M. L.: Iron oxide removal from soils and clays by a dithionite-citrate system buffered with sodium bicarbonate, Clays Clay Miner., 5, 317–327, 1960.

Mohnke, O., Jorand, R., Nordlund, C., and Klitzsch, N.: Understanding NMR relaxometry of partially water-saturated rocks, Hydrol. Earth Syst. Sci., 19, 2763–2773, https://doi.org/10.5194/hess-19-2763-2015, 2015.

Müller-Petke, M., Dlugosch, R., Lehmann-Horn, J., and Ronczka, M.: Nuclear magnetic resonance average pore-size estimations outside the fast-diffusion regime, Geophysics, 80, D195–D206, 2015.

Pape, H., Tillich, J. E., and Holz, M.: Pore geometry of sandstone derived from pulsed field gradient NMR, J. Appl. Geophys., 58, 232–252, 2006.

Perlo, J., Danieli, E., Perlo, J., Blümich, B., and Casanova, F.: Optimized slim-line logging NMR tool to measure soil moisture in situ, J. Magnet. Reson., 233, 74–79, 2013.

Pham, A. N. and Waite, T. D.: Oxygenation of Fe(II) in natural waters revisited: Kinetic modeling approaches, rate constant estimation and the importance of various reaction pathways, Geochim. Cosmochim. Ac., 72, 3616–3630, 2008.

Prammer, M. G., Drack, E. D., Bouton, J. C., and Gardner, J. S.: Measurements of clay-bound water and total porosity by magnetic resonance logging, SPE paper 36522, Society of Petroleum Engineers, 6–9 October 1996, Denver, Colorado, 311–320, https://doi.org/10.2118/36522-MS, 1996.

Sapoval, B., Russ, S., Petit, D., and Korb, J. P.: Fractal geometry impact on nuclear relaxation in irregular pores, Magnet. Reson. Imag., 14, 863–867, 1996.

Schwertmann, U. and Cornell, R. M.: Iron Oxides in the Laboratory – Preparation and Characterization, Wiley-VCH, Weinheim, 188 pp., 2000.

Stumm, W. and Lee, G. F.: The chemistry of aqueous iron, Schweiz. Z. Hydrol., 22, 295–319, 1960.

Stumm, W. and Sulzberger, B.: Cycling of iron in natural environments: Considerations based on laboratory studies of heterogeneous redox processes, Geochim. Cosmochim. Ac., 56, 3233–3257, 1991.

Sucre, O., Pohlmeier, A., Miniere, A., and Blümich, B.: Low-field NMR logging sensor for measuring hydraulic parameters of model soils, J. Hydrol., 406, 30–38, https://doi.org/10.1016/j.jhydrol.2011.05.045, 2011.

Tuhela, L., Carlson, L., and Tuovinen, O. H.: Biogeochemical transformations of Fe and Mn in oxic groundwater and well water environments, J. Environ. Sci. Health Pt. A, 32, 407–426, 1997.

Van Dam, R. L., Schlager, W., Dekkers, M. J., and Huisman, J. A.: Iron oxides as a cause of GPR reflections, Geophysics, 67, 536–545, 2002.

Weidner, C.: Experimental Modelling and Prevention of Chemical Fe-Clogging in Deep Vertical Wells for Open-Pit Dewatering, PhD Thesis, RWTH Aachen University, Aachen, 219 pp., 2016.

Weidner, C., Henkel, S., Lorke, S., Rüde, T. R., Schüttrumpf, H., and Klauder, W.: Experimental modelling of chemical clogging processes in dewatering wells, Mine Water Environ., 31, 242–251, 2012.

12

Drought severity–duration–frequency curves: a foundation for risk assessment and planning tool for ecosystem establishment in post-mining landscapes

D. Halwatura[1]**, A. M. Lechner**[2,3]**, and S. Arnold**[1]

[1]Centre for Mined Land Rehabilitation, Sustainable Minerals Institute, the University of Queensland, Brisbane, Australia
[2]Centre for Social Responsibility in Mining, Sustainable Minerals Institute, the University of Queensland, Brisbane, Australia
[3]Centre for Environment, University of Tasmania, Hobart, Australia

Correspondence to: D. Halwatura (d.halwatura@uq.edu.au)

Abstract. Eastern Australia has considerable mineral and energy resources, with areas of high biodiversity value co-occurring over a broad range of agro-climatic environments. Lack of water is the primary abiotic stressor for (agro)ecosystems in many parts of eastern Australia. In the context of mined land rehabilitation quantifying the severity–duration–frequency (SDF) of droughts is crucial for successful ecosystem rehabilitation to overcome challenges of early vegetation establishment and long-term ecosystem resilience.

The objective of this study was to quantify the SDF of short-term and long-term drought events of 11 selected locations across a broad range of agro-climatic environments in eastern Australia by using three drought indices at different timescales: the Standardized Precipitation Index (SPI), the Reconnaissance Drought Index (RDI), and the Standardized Precipitation-Evapotranspiration Index (SPEI). Based on the indices we derived bivariate distribution functions of drought severity and duration, and estimated the recurrence intervals of drought events at different timescales. The correlation between the simple SPI and the more complex SPEI or RDI was stronger for the tropical and temperate locations than for the arid locations, indicating that SPEI or RDI can be replaced by SPI if evaporation plays a minor role for plant available water (tropics). Both short-term and long-term droughts were most severe and prolonged, and recurred most frequently in arid regions, but were relatively rare in tropical and temperate regions.

Our approach is similar to intensity–duration–frequency (IDF) analyses of rainfall, which are crucial for the design of hydraulic infrastructure. In this regard, we propose to apply SDF analyses of droughts to design ecosystem components in post-mining landscapes. Together with design rainfalls, design droughts should be used to assess rehabilitation strategies and ecological management using drought recurrence intervals, thereby minimising the risk of failure of initial ecosystem establishment due to ignorance of fundamental abiotic and site-specific environmental barriers, such as flood and drought events.

1 Introduction

Eastern Australia holds vast mineral and energy resources of economic importance and internationally significant biodiversity (Williams et al., 2002; Myers et al., 2000) occurs over a broad range of agro-climatic environments (Hutchinson et al., 2005; Woodhams et al., 2012). There are also extensive areas of cropping and grazing such as in the Brigalow Belt Bioregion (Arnold et al., 2013) and the wheatbelt regions around Kingaroy and Wagga Wagga (Woodhams et al., 2012) (Table 1, Fig. 1). Lack of water availability is a critical factor for the mining industry, agriculture and biodiversity. For example, water deficit reduces agricultural productivity and increases the risk of failure of ecosystem rehabilitation. Likewise, flooding affects mining as a result of soil erosion in rehabilitation areas or flooded mine sites pre-

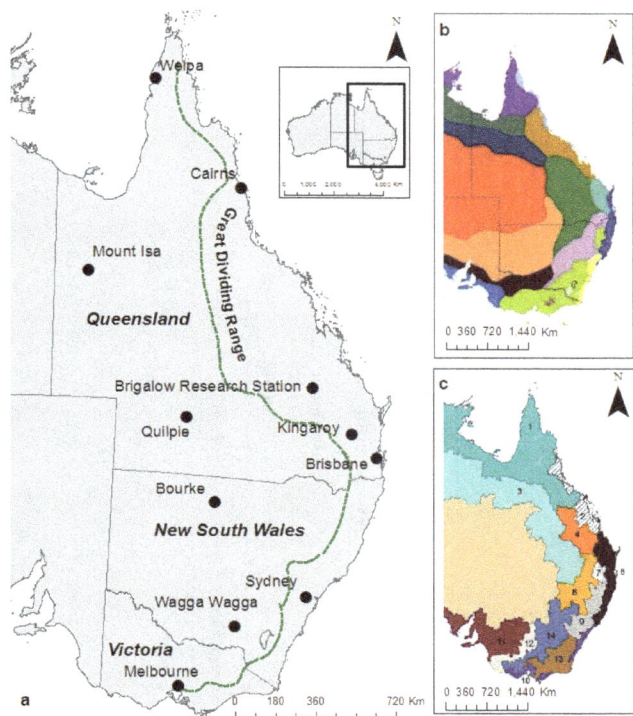

Figure 1. (a) Selected locations of interest with boundaries of (b) agro-climatic classes (Hutchinson et al., 2005) and (c) Australian agricultural environments (Woodhams et al., 2012).

venting production. For some of the agro-climatic regions in eastern Australia, the lack of water is the primary abiotic stressor for (agro)ecosystems throughout the year, whereas for others water availability is at least seasonally limited (Table 1). In the past century, regions across Australia have regularly experienced periods of water deficit (Murphy and Timbal, 2008). Approximately one-third of Australia is arid with rainfall of less than 250 mm per year, and another one-third is semi-arid (250–500 mm yr^{-1}). There are few areas where rainfall exceeds evaporation on an annual basis (Bell, 2001). Drought events are distributed diversely with regard to their duration, severity, and frequency of occurrence over the continent.

Droughts and associated limitations in plant available water determine plant distribution in response to climatic conditions in post-mining landscapes. Ecosystem attributes such as the distribution of native tropical species (Engelbrecht et al., 2007; Kuster et al., 2013), the structure and functioning of forests (Zhang and Jia, 2013; Vargas et al., 2013), biodiversity and ecosystem resilience (Brouwers et al., 2013; Lloret, 2012; Jongen et al., 2013), and primary productivity and respiration of vegetation (Shi et al., 2014) are sensitive to the occurrence of drought events. In the context of mined land rehabilitation, droughts also play a critical role for the early establishment of plants (Nefzaoui and Ben Salem, 2002; Gardner and Bell, 2007) and long-term resilience of novel

(Doley et al., 2012; Doley and Audet, 2013) and/or native ecosystems on post-mining land (Bell, 2001). Across the life span of plants due to their under-developed root system, juvenile vegetation such as seeds, seedlings, and pre-mature rather than climax vegetation is especially vulnerable to lack of water availability (Jahantab et al., 2013; Craven et al., 2013; Arnold et al., 2014a). For climax vegetation, however, medium to long-term drought (greater than 9 months) periods rather than short-term droughts (3 months or less) may critically impact rehabilitation by altering plant communities' species composition (Mariotte et al., 2013; Ruffault et al., 2013).

Droughts are usually characterised through the use of indices, which vary in complexity and data needs. Meteorological or climatological droughts are the simplest and are based on the characterisation of anomalies in rainfall conditions (Anderegg et al., 2013). For meteorological droughts, standardised drought indices such as the Standardized Precipitation Index (SPI), Reconnaissance Drought Index (RDI) and Standardized Precipitation-Evapotranspiration Index (SPEI) provide the means to quantifying the duration and severity, and eventually the frequency or recurrence of drought events (McKee et al., 1993; Tsakiris and Vangelis, 2005; Vicente-Serrano et al., 2010). Although there are numerous comparative studies of drought indices in certain climatic regions such as the Mediterranean, (Paulo et al., 2012; Livada and Assimakopoulos, 2007), the Carpathian region (Spinoni et al., 2013), and other arid locations (Peel et al., 2007; Zarch et al., 2011), none of these indices apply universally to any climate region and it is best for land managers to use a range of drought indices at various temporal scales (Heim, 2002; Spinoni et al., 2013). In many parts of the world evaporation data are unavailable or incomplete and simple rainfall indices such as SPI are most commonly used. In this study, we compare SPI with RDI at the 3-month timescale and SPI and SPEI at the 12-month timescale to determine the difference between using SPI with more complex indices that incorporate evaporation in different climatic regions.

Drought periods can be characterised from a few hours (short-term) to millennia (long-term) depending on the ecological or socio-economic question being addressed. The time lag between the beginning of a period of water scarcity and its impact on socio-economic and/or environmental assets is referred to as the timescale of a drought (Vicente-Serrano et al., 2013). There are three timescales for which drought indices are usually calculated; short-term droughts are 3 months or less; medium-term droughts are between 4 to 9 months; and long-term droughts are 12 months or more (Zargar et al., 2011). Short-term droughts have an impact on water availability in the vadose zone (National Drought Mitigation Center, 2014; Zargar et al., 2011), while long-term droughts also affect surface and ground water resources (National Drought Mitigation Center, 2014; Zargar et al., 2011).

Of key importance for land managers planning for drought events of any timescale is characterising the return period or

Table 1. Climate indices and classification of selected locations across eastern Australia with focus on rainfall.

Location	Length of meteorological data (years)	Climate index		Climate classification system			Potential productive land use[e,d]
		R/PET[a]	R_w/R_s[b]	Köppen–Geiger[c]	Australian agricultural environment[d]	Agro-climatic[e]	
Weipa	1960–1994 (34)	0.99	0.01	Tropical, savannah	Tropics (wet/dry season)	I1 – wet/dry season	Crops, rangeland
Cairns	1965–2013 (48)	0.91	0.10	Tropical, savannah	Tropical coast (wet)	I3 – wet/dry season	Crops, rangeland, sugarcane
Brisbane	1986–2013 (27)	0.55	0.38	Temperate, without dry season	Subtropical coast (wet)	F4 – wet	Horticulture, pasture, sugarcane
Sydney	1970–1994 (24)	0.53	0.51	Temperate, without dry season	Temperate coast east (wet, winter-dominant rainfall)	F3 – wet	Crops, horticulture, pasture
Melbourne	1955–2013 (58)	0.51	0.95	Temperate, without dry season	Temperate coast east (wet, winter-dominant rainfall)	D5 – wet	Crops, forestry, horticulture, pasture
Kingaroy	1967–2001 (34)	0.47	0.34	Temperate, without dry season	Wheatbelt downs (summer-dominant/moderate rainfall)	E4 – water-limited	Cotton, crops, pasture
Brigalow Research Station	1968–2011 (43)	0.32	0.27	Temperate, without dry season	Subtropical plains (summer-dominant/moderate rainfall)	E4 – water-limited	Cotton, crops, pasture
Wagga Wagga	1966–2013 (47)	0.30	1.21	Temperate, without dry season	Wheatbelt east (winter-dominant rainfall)	E3 – water-limited in summer	Crops, horticulture, pasture
Bourke	1967–1996 (29)	0.20	0.61	Arid, steppe	Arid (dry)	E6 – water-limited	Rangeland, wildland
Quilpie	1970–2013 (43)	0.14	0.36	Arid, steppe	Arid (dry)	H – water-limited	Rangeland, wildland
Mount Isa	1975–2013 (38)	0.13	0.05	Arid, steppe	Arid (dry)	G – water-limited	Rangeland, wildland

[a] Unep and Thomas (1992); [b] based on average of 3 months of rainfall during winter (June–August) and summer (December–February); [c] Peel et al. (2007); [d] Woodhams et al. (2012); [e] Hutchinson et al. (2005).

frequency of occurrence of rainfall and drought events. The recurrence interval is defined as the average inter-occurrence time of any geophysical phenomena and is calculated with long-term time series data (Loaiciga and Mariño, 1991). Recurrence intervals of rainfall events greater than the average are commonly used by engineers to derive intensity–duration–frequency (IDF) design estimates for building hydraulic infrastructure such as roofs, culverts, stormwater drains, bridges or water dams (Chebbi et al., 2013; Kuo et al., 2013; Hailegeorgis et al., 2013). IDF design rainfalls are crucial for estimating the risk of hydraulic infrastructure failure and for maximising infrastructure efficiencies (Smithers et al., 2002). Similar to the concept of IDF design rainfall, which aims to quantify the recurrence interval of rainfall events based on their intensity and duration, we apply the same concept to quantify the recurrence intervals of droughts based on their severity and duration, and refer to this as severity–duration–frequency (SDF) design drought. SDF curves have been used to derive drought variables (severity, duration, frequency of occurrence) in different climatic regions (Shiau, 2006; Shiau et al., 2012; Lee and Kim, 2012; Todisco et al., 2013; Mirabbasi et al., 2012) but have rarely been used in ecology, and never been used in relation to rehabilitation and restoration. While IDF design rainfalls are a well-established tool in civil engineering and hydrology, we believe SDF design drought could be used in a similar way to assess the risk of ecosystem rehabilitation failure due to droughts.

This approach contrasts current climate classification methods (Table 1) such as the classification of the Australian agricultural environments (Woodhams et al., 2012) or the Australian agro-climatic classes (Hutchinson et al., 2005) that are used for the management of agricultural land (Audet et al., 2013). These classifications are based on average climatic conditions and may not be adequate for the management of early vegetation re-establishment in post-mining landscapes (Audet et al., 2012, 2013) because of the vulnerability of vegetation to drought events. Although droughts play a critical role in post-mining land restoration in eastern Australia, so far methods for quantifying the frequency of drought events have been rarely applied to assess the risk of failure of ecosystem rehabilitation. In the perspective of mined land rehabilitation, specific metrics of site climate or seasonality are surprisingly rare (Audet et al., 2013).

The objective of our study is to quantify the severity, duration, and frequency (SDF) of short-term and long-term drought events at selected locations across a broad range of agro-climatic environments in eastern Australia (Table 1, Fig. 1). Eastern Australia makes a very good case study for this kind of research as there are a wide range of climates in which data has been gathered using a consistent method by one agency. While other studies assessed the SDF characteristics at locations with the same climate in Iran (Shiau and Modarres, 2009; Shiau et al., 2012), no such investigations

are known for any climatic region in Australia, for the same climate or different climates.

We characterised droughts using the RDI and SPEI for 3- and 12-month timescales respectively, and compared these indices with the SPI at the same timescales. We then linked the univariate distributions of severity and duration calculated with the drought indices to form bivariate distribution functions and estimated the recurrence intervals of droughts. Please note that since the estimated recurrence intervals are based on historic rainfall and evaporation data, our results are descriptive rather than predictive. Nevertheless, our findings are crucial to discuss the potential of design droughts to be applied as a management tool to overcome the challenges of early vegetation establishment and long-term ecosystem resilience in post-mining landscapes. This is because frequency patterns of drought events are ignored in any current rehabilitation guidelines and industry plans, where long-term average rainfall is the only parameter upon which management decisions are based on (Audet et al.,2013).

2　Materials and methods

Estimating SDF curves involves uncertainties associated with the length of the observed rainfall data, the applied drought index, the probability distribution functions used to fit the observed severity and duration, and the estimated copula parameter (Hu et al., 2014). To overcome these uncertainties we tested the applicability of drought indices for locations in different climatic regions by calculating the correlation of three selected drought indices. Likewise, we used the best fitted probability distribution functions and copula for each site. A flow chart of the processing steps is depicted in a schematic diagram (Fig. 2).

We selected 11 sites, for which historical observations of monthly rainfall and evaporation (ranging from 30–60 years) (Table 1) were most comprehensive (more than 97 % coverage) (i.e. longest and most complete – more than 97 % coverage) across eastern Australia (Bureau of Meteorology, 2013). The selected locations covered a broad range of climate classes and environments across eastern Australia (Table 1, Fig. 1).

For each site we compared the simple SPI with the more complex RDI and SPEI drought indices. Amongst the three indices, the SPI is the most widely used and simplest drought index, because it is solely based on long-term rainfall for any period of interest (McKee et al., 1993; Guttman, 1999). However, SPI may not adequately characterise drought events, because it does not incorporate other meteorological data (Vicente-Serrano et al., 2010; Mishra and Singh, 2010). Both the RDI and SPEI integrate potential evaporation and thereby better represent the local water balance (Tsakiris, 2004; Tsakiris and Vangelis, 2005; Tsakiris et al., 2007; Vangelis et al., 2013).

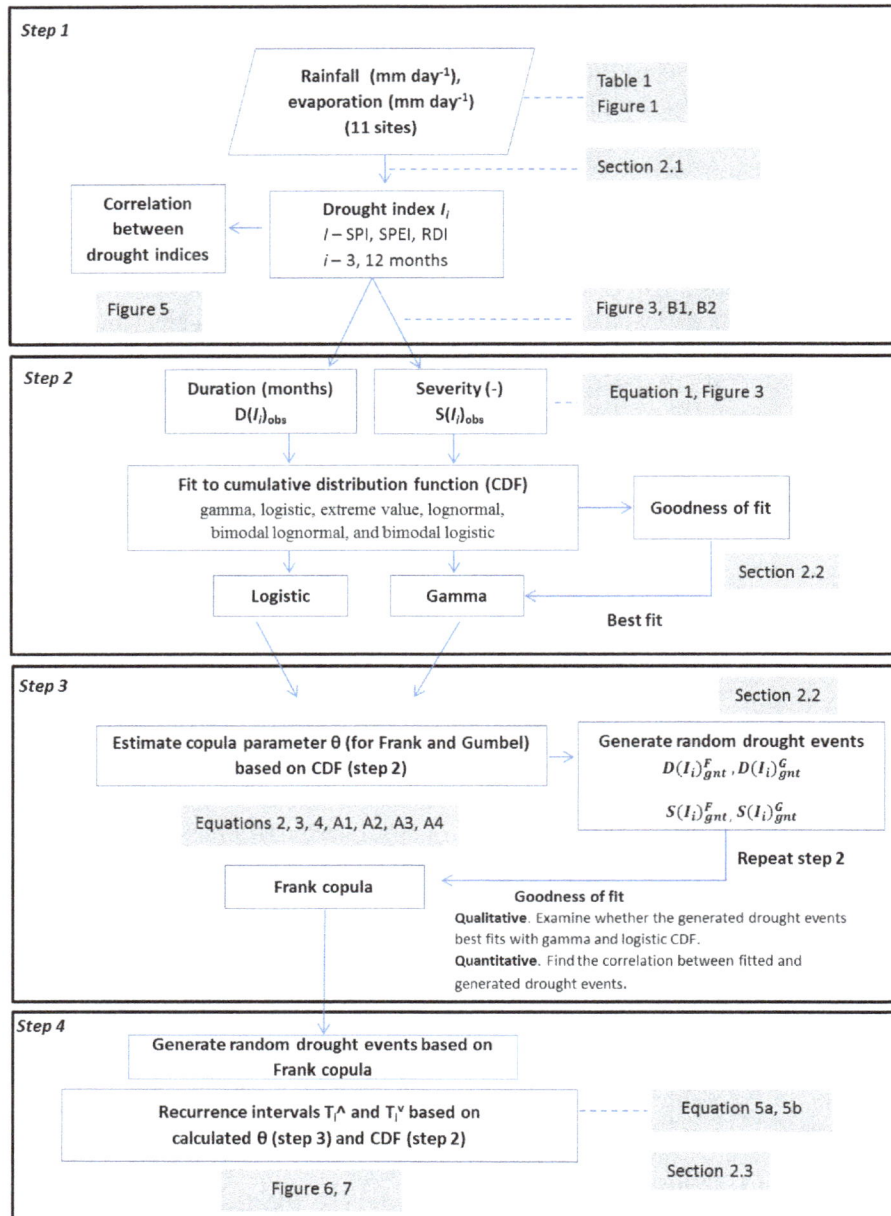

Figure 2. Schematic diagram of steps applied to estimate recurrence intervals of drought events. See Sect. 2 for further details. Step 1: calculate drought index based on monthly rainfall (SPI) and evaporation (RDI, SPEI). Step 2: fit cumulative distribution function (CDF) to estimated drought duration and severity. Step 3: estimate copula parameter based on CDFs. Step 4: calculate recurrence intervals based on CDFs of univariate (severity, duration) distributions and bivariate joint distribution (copula).

The drought indices are calculated using monthly values of rainfall and/or potential evaporation. Amongst the two indices which incorporate potential evaporation, the RDI plays a strong role in detecting maximum drought severities at the medium timescale (3 to 6 months) (Banimahd and Khalili, 2013), while the SPEI plays a strong role in detecting annual drought events by identifying the hydrological summer drought events (Egidijus et al., 2013). There is evidence that SPI overestimates small rainfall scarcity even if excessive rainfall occurs just before the period of interest (Kim et al., 2009). Also for humid climates, there is a good correspondence between the computed SPI_3 and RDI_3 (Khalili et al., 2011). For Mediterranean climate, SPI and SPEI at 9- and 12-month timescales are well correlated (Paulo et al., 2012), and in the Carpathian region SPI, SPEI, and RDI are highly comparable over annual periods (Spinoni et al., 2013). In arid regions, the correlation of SPI and RDI is more considerable at the 3-, 6-, and 9-monthly timescale (Peel et al., 2007; Zarch et al., 2011).

2.1 Step 1: calculate drought indices

The SPI is derived by fitting a probability distribution to the rainfall record and then transforming that to a normal distribution, so that mean and standard deviation of the SPI are zero and one. Positive or negative values of the SPI represent rainfall conditions greater or smaller than average rainfall, respectively (McKee et al., 1993). RDI and SPEI are based on the SPI calculation procedure, except the two indices use the quotient or difference of precipitation and potential evaporation, respectively (Tsakiris et al., 2007; Vicente-Serrano et al., 2010). Equations for the RDI and SPEI are presented in Appendix A. We applied two correlation coefficients to assess the correlations between SPI_3 and RDI_3, and SPI_{12} with $SPEI_{12}$ (step 1 in Fig. 2): Kendall's τ to assess the number of concordances and discordances in paired variables (RDI_3 and SPI_3, $SPEI_{12}$ and SPI_{12}), and Pearson's r to measure linear correlation.

2.2 Step 2: bivariate distribution of drought severity and duration

For each location, we used the estimated drought indices (SPI, RDI, SPEI), hereafter collectively referred to as I, to quantify duration D and severity S (McKee et al., 1993; Vicente-Serrano et al., 2010; Tsakiris and Vangelis, 2005). The duration of any drought was defined as the period of rainfall deficit, i.e. the cumulative time of negative I values preceded and followed by positive I values (Fig. 3). The severity of any drought period starting at the ith month was defined as

$$S = \sum_{i=1}^{D} |-I_i|. \tag{1}$$

We fitted the time series of D and S to a range of cumulative distribution functions (gamma, logistic, extreme value, lognormal, bimodal lognormal, and exponential) and used the function with the best fit for further investigations (step 2 in Fig. 2). The coefficient of determination and 95 % confidence levels were calculated for each distribution in order to select the best distribution.

2.3 Step 3: estimate copula parameter

We used copulas to link the univariate probability distributions of D and S to construct a bivariate joint distribution of D and S (Shiau and Modarres, 2009; Sklar, 1959) (step 3 in Fig. 2). As the choice of copula can be very different from one climate region to another (Khedun et al., 2013) the

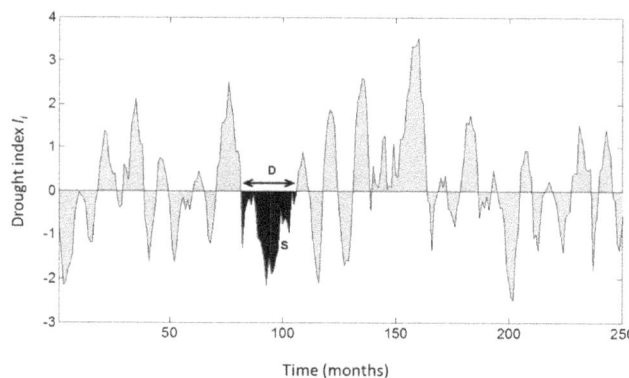

Figure 3. Concept of severity S and duration D of a drought event quantified with drought index I_i, where i refers to any timescale of interest.

present study focused on the Frank and Gumbel copulas (Appendix B), as they perform best when analysing the bivariate drought dependence structure of drought variables such as severity and duration (Ganguli and Reddy, 2012; Reddy and Ganguli, 2012; Shiau, 2006; Lee et al., 2013; Wong et al., 2010; Zhang et al., 2011). The conditional cumulative distribution function $F_{S|D}(s|d)$ which relates to the joint cumulative distribution function (JCDF) of drought severity and duration $F_{S,D}(s, d)$ and the cumulative distribution function (CDF) of drought duration $F_D(d)$ is given by the following relationship (Shiau and Modarres, 2009):

$$F_{S|D}(s|d) = \frac{\partial F_{S,D}(s, d)}{\partial F_{D(d)}}, \tag{2}$$

where $F_D(d)$ is the CDF of drought duration, and $F_{S,D}(s, d)$ is the JCDF of drought severity and drought duration. The JCDF of drought severity and duration in terms of copulas is a function of univariate CDFs of duration and severity:

$$F_{S,D}(s, d) = C(F_S(s), F_D(d)), \tag{3}$$

where $F_S(s)$ and $F_D(d)$ are CDFs for drought severity and duration, respectively, and C is a copula function. The conditional distribution function $F_{S|D}(s|d)$ (Eq. 2) can also be expressed as a function of the copula (Shiau and Modarres, 2009):

$$F_{S|D}(s|d) = \frac{\partial F_{S,D}(s, d)}{\partial F_D(d)} = \frac{\partial C(F_S(s), F_D(d))}{\partial F_D(d)}$$
$$= C_{F_S|F_D}(F_S(s)|F_D(d)). \tag{4}$$

We estimated the copula parameters using the inference function for margins (IFM) (Joe, 1997). The IFM comprises two separate valuation stages. First, the maximum likelihood estimation of each univariate distribution is performed, and then the copula dependence parameter is estimated to derive the joint drought duration and severity distributions (Shiau, 2006; Shiau and Modarres, 2009; Mirabbasi et al., 2012; Shiau et al., 2007).

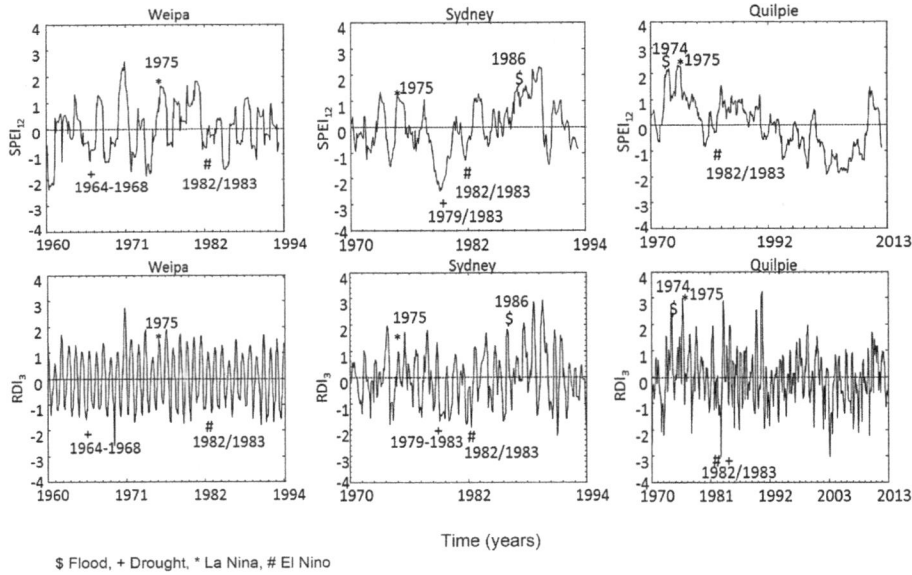

Figure 4. Calculated $SPEI_{12}$ (upper row panels) and RDI_3 (lower row panels) for Weipa, Sydney and Quilpie including major weather events. The same indices are depicted for all other selected locations in Appendix B.

2.4 Step 4: derive recurrence intervals

We used the estimated copula parameters to generate random drought events. Severity and duration of the generated random droughts were then fitted to cumulative distribution functions in the same manner as in step 2 (Fig. 2, step 3) to test which estimated copula parameters result in a distribution that best fit the generated random drought variables. The estimated copula parameters were also assessed quantitatively through calculating the correlation between generated random drought events and the estimated gamma (S) and logistic (D) cumulative distribution functions.

The generated random numbers were then used to calculate the recurrence intervals. Recurrence intervals of bivariate drought events is a standard metric for hydrological frequency analysis (Yoo et al., 2013; Hailegeorgis et al., 2013) and water resources management (Shiau and Modarres, 2009; Mishra and Singh, 2010). For each location, we calculated the recurrence interval of drought events exceeding any severity *or* duration of interest, denoted by the logical operator "∨":

$$T_I^{\vee} = \frac{1}{P(S \geq s \vee D \geq d)} = \frac{1}{1 - C\,[F_S(s), F_D(d)]}, \quad (5a)$$

where I is one of the drought indices of interest, i.e. the 12-monthly $SPEI_{12}$ or SPI_{12}, or the 3-monthly RDI_3 or SPI_3. Alternatively, the recurrence interval of drought events exceeding any severity *and* duration of interest, denoted by the logical operator "∧", was calculated as

$$T_I^{\wedge} = \frac{1}{P(S \geq s \wedge D \geq d)}$$
$$= \frac{1}{1 - F_S(s) - F_D(d) + C\,[F_S(s), F_D(d)]}. \quad (5b)$$

For the sake of simplicity, we only present and discuss $T_I^{\vee}$, whereas $T_I^{\wedge}$ is presented in Appendix D.

3 Results

For both indices, RDI and SPEI, and all selected sites, the gamma and logistic distributions fitted best to the observed drought severity and duration, respectively ($R^2 > 0.98$ for both variables, $p < 0.05$) (Appendix F). Likewise, the same distributions fitted best to the drought severity and duration of the generated drought events based on the Frank rather than the Gumbel copula ($R^2 > 0.90$, $p < 0.05$) (Appendix F).

Based on the drought indices RDI_3 and $SPEI_{12}$ we detected distinct drought patterns across the selected sites at short and long-term scales, respectively. As an example of differences between tropical, temperate and arid rainfall conditions, Fig. 4 depicts calculated time series of RDI_3 and $SPEI_{12}$ for Weipa, Sydney and Quilpie, respectively (see Appendix C for rest of the sites).

Short-term droughts were most severe and prolonged in tropical Weipa and Cairns, and temperate Wagga Wagga (Table 2). However, in contrast to Wagga Wagga, the two tropical locations were characterised by distinct seasonality patterns and very low variation as indicated by the low ratio of winter to summer rainfalls (Table 1) and low coefficients of variation in severity and duration (Table 2). The highest vari-

Table 2. Mean severity μ_S and duration μ_D of selected locations across eastern Australia, and corresponding coefficient of variation CV_S and CV_D for short-term (RDI_3) and long-term ($SPEI_{12}$) droughts.

Location	RDI_3				$SPEI_{12}$			
	μ_S	CV_S	μ_D	CV_D	μ_S	CV_S	μ_D	CV_D
Weipa	5.2	0.2	5.8	0.1	8.4	1.1	10.4	0.8
Cairns	4.7	0.4	6.4	0.3	9.6	1.3	12.5	1.0
Brisbane	3.1	3.3	3.6	0.8	11.2	0.9	13.3	0.8
Sydney	3.4	0.9	4.4	0.6	6.5	1.7	8.9	0.9
Melbourne	4.5	0.7	5.8	0.5	14.5	1.9	18.6	1.6
Kingaroy	2.8	1.2	3.7	0.8	7.0	1.1	8.3	0.8
Brigalow Research Station	3.4	1.0	4.4	0.9	8.0	1.3	10.2	1.0
Wagga Wagga	5.2	0.8	6.2	0.6	8.6	1.8	13.8	1.1
Bourke	2.8	3.9	3.9	1.1	8.2	2.0	9.9	1.5
Quilpie	3.5	1.1	4.6	0.7	18.8	2.1	21.8	1.5
Mount Isa	3.8	0.7	4.9	0.5	11.1	1.2	14.4	0.9

ation in severity was detected in arid Bourke and temperate Brisbane (Table 1).

Long-term droughts were most severe and prolonged in arid Quilpie (Table 2) and rare in temperate Melbourne. Likewise, severity and duration varied most at the two locations, together with arid Bourke. While severity and duration were moderately high in arid Mount Isa and temperate Brisbane, both parameters were low across the other selected temperate and tropical locations (Table 2).

No significant differences were detected ($p < 0.05$ at 95 % confidence level) between RDI_3 and SPI_3, and $SPEI_{12}$ and SPI_{12} (Fig. 5 and Appendix E). Correlation between RDI/SPEI and SPI was greatest for tropical Cairns and Weipa, and lowest for arid Bourke and Quilpie (outliers in Fig. 5). Interestingly, although Mount Isa was the most arid location ($R/PET = 0.13$, Table 1), the correlations between drought indices were relatively strong with values of 0.903 (Pearson's r) and 0.759 (Kendall's τ) for long-term droughts.

For each location, the recurrence intervals of drought events exceeding any severity *or* duration of interest are depicted in Fig. 6 for short-term droughts (based on RDI_3) and Fig. 7 for long-term droughts (based on $SPEI_{12}$). Short-term droughts recurred most frequently in arid Mount Isa and were relatively rare in tropical Weipa and Cairns, and temperate Sydney. For example, in Mount Isa a drought with severity of 14 or duration of 17 months[1] recurred once in 50 years, whereas the same drought recurred only once in 100 000 years in Weipa, 300 years in Cairns, and 100 years in Sydney (Fig. 6). Long-term droughts recurred most frequently in arid Quilpie, where droughts with severity of 18 or duration of 10 months recurred once in 2 years. In Kingaroy and Sydney, the same drought recurred only once in 4 and 5 years, respectively (Fig. 7). Interestingly, although average long-term droughts were very severe and prolonged in Mel-

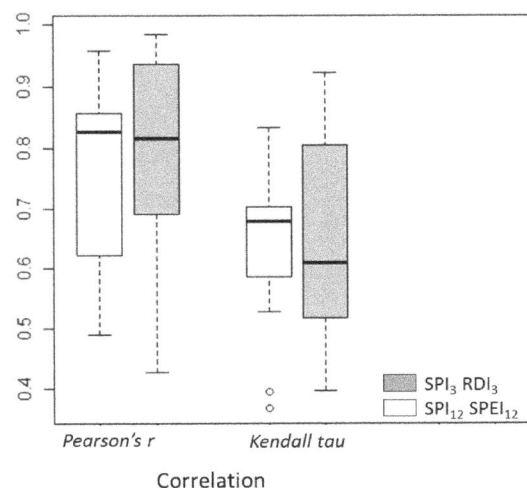

Figure 5. Correlation between SPI_3 and RDI_3, and SPI_{12} and $SPEI_{12}$ based on the correlation coefficient Pearson's r and Kendall's τ. The outliers represent the very dry locations of Bourke and Quilpie.

bourne (Table 2), they only recurred once in 30 to 50 years. We found similar qualitative patterns in all locations for recurrence intervals of droughts exceeding any severity *and* duration of interest (Appendix D).

4 Discussion

In this study we estimated the recurrence intervals of short- and long-term droughts based on meteorological drought indices and copulas (i.e. bivariate probability distributions). For both timescales, the correlation between the simple SPI (rainfall) and the more complex SPEI or RDI (rainfall and evaporation) was much stronger for the tropical and temperate locations (e.g. Cairns, Weipa, Brigalow) than for the arid

---

[1] Drought events are calculated by 3-month (short-term) and 12-month (long-term) running precipitation totals (Guttman, 1999).

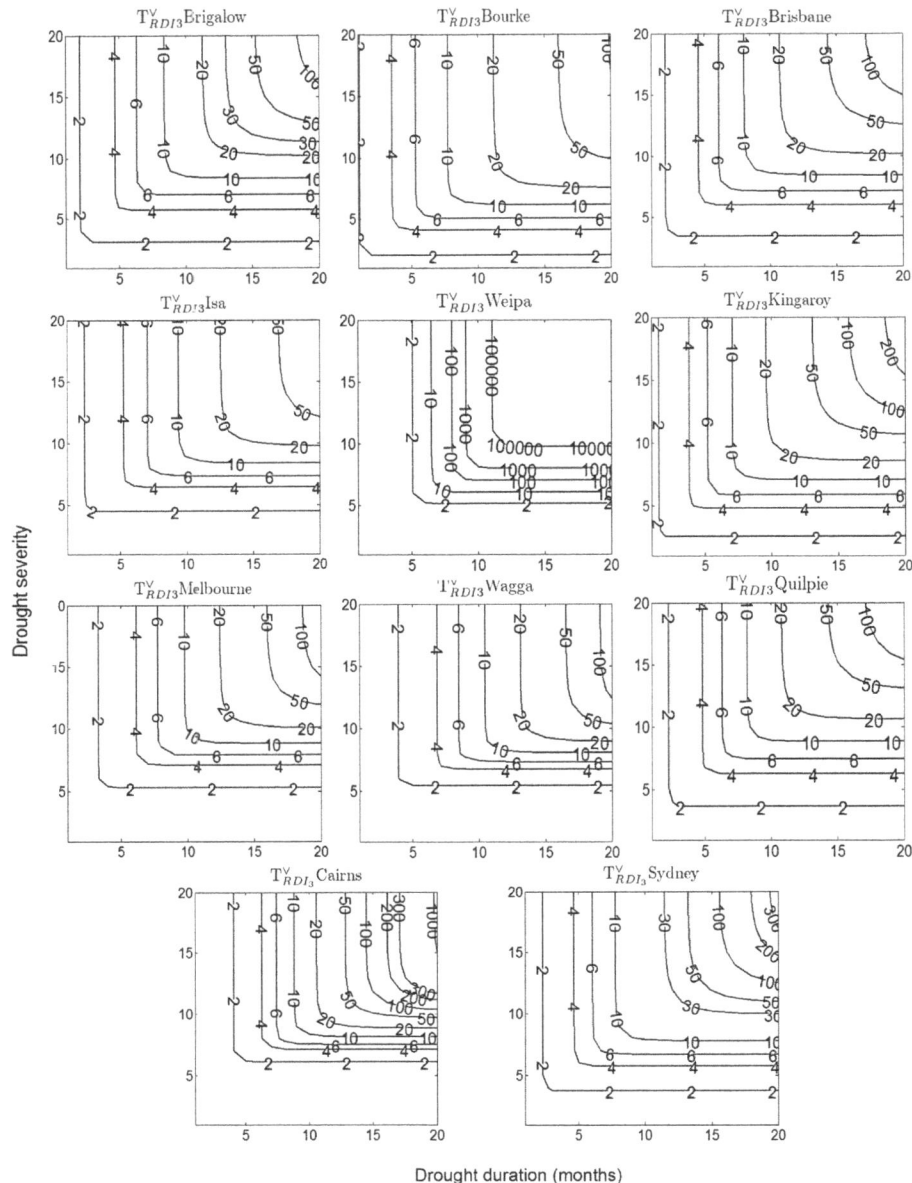

Figure 6. Recurrence interval $T^{\vee}$ (years) of drought events of any severity *or* duration of interest based on the RDI$_3$ (short-term) of historical rainfall.

locations (e.g. Quilpie, Bourke, Wagga Wagga). Extending a former study on abiotic boundaries affecting ecological development of post-mining landscapes (Audet et al., 2013), our findings have critical implications for assessments of rehabilitation success.

4.1 Implications for ecosystem rehabilitation planning

Across eastern Australia, current post-mining land rehabilitation strategies often do not incorporate site-specific rainfall and drought metrics other than the average annual rainfall depth (Audet et al., 2013). However, regionally extreme

rainfall patterns, including both intense rainfall events such as storms or cyclones and prolonged periods of water deficit (droughts), play a critical role in identifying windows of opportunity and/or challenge to the rehabilitation of early-establishment ecosystems (Hinz et al., 2006; Hodgkinson et al., 2010). Furthermore, Audet et al. (2013) suggested that short and long-term ecosystem rehabilitation sensitivity to climate can be effectively determined by the seasonality, regularity, and intensity of weather, combined with both median and standard deviation of periods. In particular, prolonged seasonal drought with high variation and frequently occurring intense rainfall can be used as primary characteristics

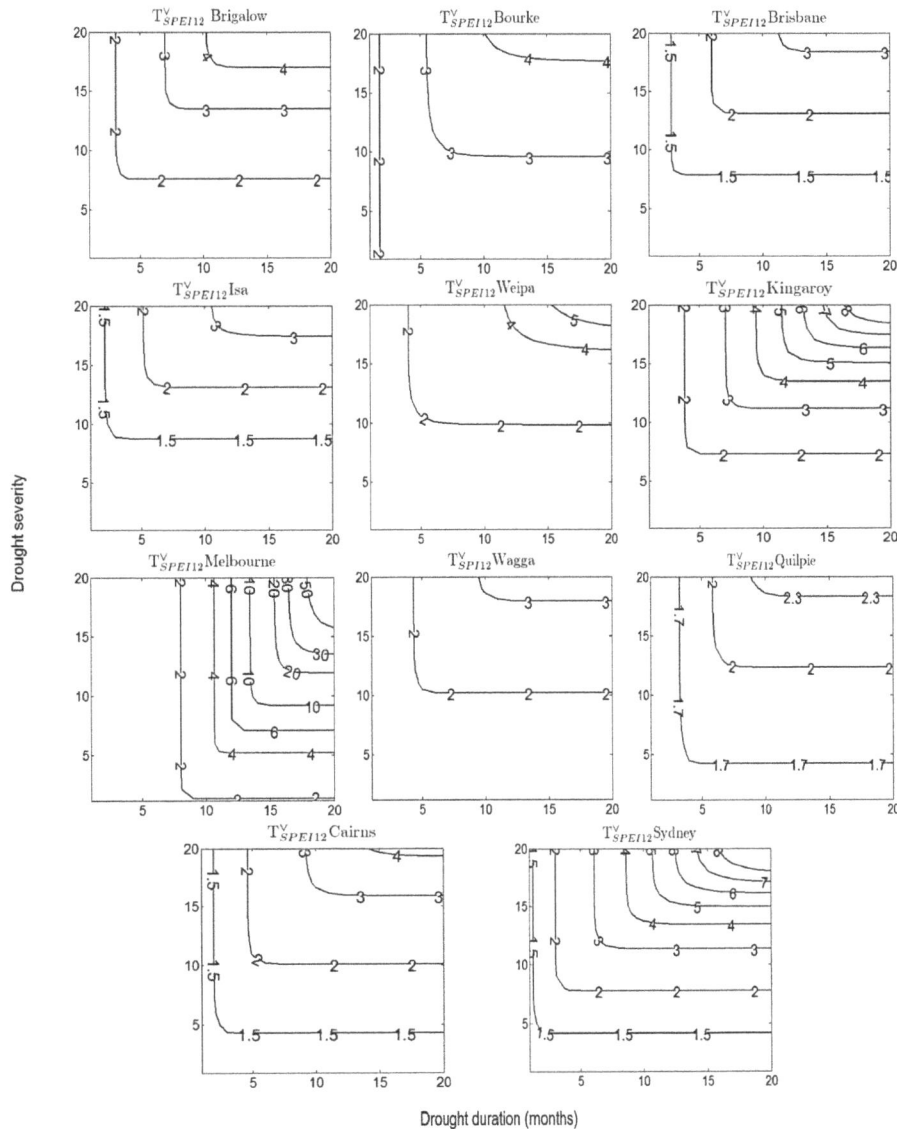

Figure 7. Recurrence interval $T^\vee$ (years) of drought events of any severity *or* duration of interest based on $SPEI_{12}$ (long-term) of historical rainfall.

for determining site sensitivity, while regular rainfall and relatively short periods of water deficit are common characteristics of favourable climate conditions. Based on their findings, Audet et al. (2013) revealed how broad scale rainfall patterns outline climate boundaries that drive rehabilitation sensitivity in arid to temperate locations across eastern Australia. For example, ecosystem rehabilitation in arid regions (Mount Isa, Quilpie, and Bourke) is sensitive to climate as they have highly variable climates (long spell of droughts and high intensity rainfall), which affect the success of rehabilitation.

Commonly, the characterisation of climatic conditions is based on long-term rainfall and does not consider short and long-term drought conditions. Identifying drought and its variables are critical factors in ecosystem rehabilitation because the distribution and health of plant species are vulnerable to droughts and plant available water (Engelbrecht et al., 2007). In our study we presented two hydrological parameters describing the average recurrence intervals of short-term and long-term droughts (Figs. 6 and 7 and Appendix D), which can be used instead of the oversimplified parameters of the median period without rain and standard deviation normally used (Audet et al., 2013).

The design drought tool proposed in this paper is an adaptation of the intensity–duration–frequency (IDF) analysis of rainfall events, a standard tool used by engineers (Hailege-

Table 3. Management actions for addressing specific kinds of drought characteristics identified with SDF curves for the southern hemisphere.

Management domain	Management actions	Type of drought
Plant species selection	Drought tolerant species	LS, LP, SP, SS
	Quickly germinating species	SS
	Species with physical/chemical dormancy	LS, LP
	Shade tolerant species on southern aspects	LS, LP
	Light tolerant species on northern aspects	LS, LP, SP, SS
	Annual grasses	SS, SP
	Perennial grasses	LS, LP, SP, SS
	Trees	LS, LP
Planting/seeding regime	Trees require repeated establishment	LS, LP
	Annual/perennial grasses are successful after rain events	SS, SP
Soil characteristics	Deep topsoil	LS, LP, SP
	Amendments of silt/clay	LS, LP
	Gentle slopes	LS, LP
	Mulching	SS
Irrigation method	Regular irrigation	LS, LP
	Seasonal irrigation	SS, SP
	Critical stage irrigation	LS, LP, SP, SS
	Drainage system	LS, LP

SS – high recurrence of short term (3 months) severe droughts; SP – high recurrence of short term (3 months) prolonged droughts; LS – high recurrence of long term (12 months) severe droughts; LP – high recurrence of long term (12 months) prolonged droughts.

orgis et al., 2013; Chebbi et al., 2013). Our new term "design droughts", characterised by drought severity–duration–frequency (SDF), is based on the severity of droughts (cumulative negative values of a particular drought; see Fig. 3) as opposed to IDF, which is based on the intensity of the rainfall. Design droughts allow for drought severity, duration and frequency to be considered in order to determine the risk of failure of current mining operations (Mason et al., 2013; Burton et al., 2012), and to design robust ecosystem components in the face of the local climate variability (Audet et al., 2013). Unlike degraded land (in the sense of gradual loss of ecosystem productivity) in post-mining landscapes, most ecosystem components are impacted by mining activities; particularly landform, hydrology, and ecosystem structure (Arnold et al., 2014b). Therefore, successful rehabilitation of post-mining land requires the sensible selection of plant species, as well as planting/seeding regime, soil characteristics, irrigation method, and landform characteristics (Table 3). For example, same vegetation types cannot establish if a drought event exceeds values of specific duration or severity (Arnold et al., 2014a). The recurrence intervals can provide the probability of a drought occurring at this duration or severity, and thus the risk of establishment failure can be assessed. This is important for rehabilitation managers who can conduct a cost–benefit analysis to decide whether the costs of constructing mitigation methods (such as irrigation) are comparable with the costs of potential failure of multiple revegetation attempts.

Together, design rainfalls (IDF) and droughts (SDF) should be the primary determinants of rehabilitation strategies and eventually help guide rehabilitation planning, where environmental conditions have an impact on current mining operations. In accordance with IDF parameters of similar locations across eastern Australia (Audet et al., 2013), temperate and tropical environmental conditions (Table 1) are favourable for rehabilitation, i.e. recurrence intervals of droughts are large (Figs. 6 and 7 and Appendix D). By contrast, re-establishment of ecosystems are prone to failure in arid conditions, where droughts recur more frequently (i.e. low recurrence intervals).

At locations with distinct patterns of winter and summer rainfall, such as Weipa, Cairns, Mount Isa, or the Brigalow Belt, seasonality is the primary determinant of drought occurrence (Table 1). The short-term drought index (RDI_3) detects most severe and prolonged droughts in tropical Weipa and Cairns (Table 2), where rainfall is low in winter and high in summer. Annually recurring seasonal patterns also explain the low variability of short-term drought severity and duration. In contrast the long-term drought index ($SPEI_{12}$) detects most severe and prolonged droughts in arid Quilpie and Mount Isa, as well as temperate Melbourne (Table 2). Major weather events such as El Niño and La Niña from recent decades coincided with low and high drought indices, respectively (Fig. 4 and Appendix C).

We compared SPI with SPEI or RDI to determine the potential of using SPI (only based on rainfall data) over SPEI or RDI (both based on rainfall and evaporation data). This might be of interest for many parts of the world, where evaporation data are unavailable or incomplete and therefore simple rainfall indices are most commonly used. Our analysis revealed that Pearson's r and Kendall's τ correlations were strong across selected locations (Fig. 5 and Appendix E), indicating the potential of the simple SPI to serve as a surrogate for the more complex RDI and SPEI. For temperate and tropical environments, such as Cairns, Weipa, or Brisbane, the more complex RDI and SPEI can be replaced by the simple SPI if evaporation data is not available (Fig. 5 and Appendix E). By contrast, in arid Bourke, Quilpie, or Mount Isa, correlations between SPI and the more complex indices were weaker, because evaporation plays a critical role in arid climates rather than in tropics and temperate regions. In these arid and water-limited locations (Table 1) we recommend using SPEI and RDI[2] and also to conduct intensive monitoring of ecosystem development in relation to empirical weather data to measure evaporation directly, e.g. pan evaporation (Lugato et al., 2013; Clark, 2013), or indirectly, e.g. based on radiative and aerodynamic variables (Allen et al., 1998).

4.2   SDF curves as an early risk assessment tool

Risk assessment based on the design rainfall concept is commonly used as a standard tool by engineers to design infrastructure such as storm water drains, flood mitigation levees, or retarding dams (Chebbi et al., 2013; Hailegeorgis et al., 2013). This research paper aims to demonstrate how these concepts can be used for ecosystem rehabilitation, providing a quantitative estimate of ecosystem rehabilitation failure due to water deficit. Traditionally, ecologist and land managers often use the mean annual rainfall as a co-classifier of biogeographic regionalisation. However, annual rainfall alone cannot account for the vulnerability of a site to nondisruptive water supply, the frequency of water limitations, and seasonality (Audet et al., 2013). For example, although mean annual rainfall is lowest in Bourke, the SDF analysis reveals that severe and prolonged droughts occur most frequently in Mount Isa. This is because in Mount Isa on average 23 out of 100 days are with no rainfall, as most of the rainfall occurs in summer as storm events greater than 100 mm (Table 1) (Bureau of Meteorology, 2013). Ecosystem rehabilitation may fail if management actions are based only on the annual rainfall without considering the nature of drought events (i.e. the rate of recurrence of prolonged and severe droughts) (Table 3).

Quantitatively, risk is the product of the probability of an event occurring and the consequences of an event on assets (Athearn, 1971). In the context of post-mining land rehabilitation, the recurrence intervals quantify the probability of occurrence of drought events. If the consequences of drought events for ecosystems are known (Wilhite et al., 2007; Williamson et al., 2000) the risk of ecosystem rehabilitation failure can be quantified. Consequences will typically have to be determined in relation to site specific attributes such as plant species, soil, irrigation, etc. (Table 3). Likewise, the consequences can also be related to the costs of rehabilitation. For example, for frequently recurring droughts of high severity and duration, irrigation may be a cost-efficient alternative to repeatedly replanting at a rehabilitation site due to establishment failure. These consequences in relation to severity and duration may be identified from the literature, field trials or be derived from expert opinion. A key aspect of our study is that SDF curves provide the probability of occurrence of drought events with a specific duration and severity.

4.3   Application of design droughts to rehabilitation planning

One of the major outcomes of this study is to support land managers and/or rehabilitation practitioners to make fundamental decisions on appropriate management actions in the context of drought frequency. For rehabilitation to be successful in the face of severe and prolonged droughts, there are a range of management domains and management actions that need to be considered in response to recurrence intervals, drought severity, and drought duration (Table 3). These management actions can be categorised into four domains: plant species selection; planting/seeding regime; soil characteristics; and irrigation method.

Meteorological droughts indicate deviations of rainfall and/or evaporation relative to the long-term average. Native climax vegetation, which is well adapted to the local climate, is hardly sensitive to these anomalies. However, within the process of post-mining land rehabilitation, establishment of well-adapted climax vegetation is impossible. In fact, post-mining ecosystem rehabilitation is very sensitive to decisions made on the re-established topography and soil characteristics, as well as planting/seeding regimes and irrigation methods (Table 3). In this regard, the frequency of meteorological droughts relative to long-term conditions is the critical driver of these management decisions. For example, seedling establishment might fail under conditions of frequently occurring short-term droughts, even if the absolute rainfall in between droughts is high. Under these conditions, landform and soil need to be restored so that the periods of water limitation can be minimised.

Selection of suitable plant species based on drought type is one of the key management actions for successful rehabilitation. Some management actions can be applied to all drought types (LS, LP, SS, SP in Table 3). These include (i) planting of drought tolerant species (e.g. *Acacia* spp., *Banksia* spp., *Casuarina* spp.) at (ii) northern aspects to address drier conditions that result from higher solar radiation causing increased evaporation (Sternberg and Shoshany, 2001), and

---

[2]Note that the definition and quantification of drought are normative. In this regard, our results indicate under what climatic conditions SPEI and RDI can be replaced by SPI, rather than which index is the best one for each location.

(iii) planting of perennial grasses (*Eragrostis* spp., *Themeda* spp.; Bolger et al., 2005), which may not be affected by long-term water deficits. At locations with frequently recurring long-term (12-monthly timescale) droughts of high severity and durations (LS, LP in Table 3), e.g. in Mount Isa and Quilpie, seeding of species with physical/chemical dormancy may increase the probability of germination during favourable periods (Hilhorst, 1995; Arnold et al., 2014b). Additionally, a southern aspect may require drought tolerant species to increase survival of plant communities (Sternberg and Shoshany, 2001). However, these species need to be shade tolerant as southern aspects get less solar radiation in winter. At locations with frequently recurring short-term (3-monthly timescale) droughts of high severity but short duration, with rainfall throughout the year (SS in Table 3), e.g. in Wagga Wagga, annual grasses and seeds with short germination periods may be suitable.

Soil characteristics play a critical role for plant available water and a number of strategies may need to be employed to make soil more favourable to plant establishment. Except for mulching, all soil management actions can be applied to locations with high recurrence of long-term, severe, and prolonged droughts (LS, LP in Table 3), e.g. in Quilpie and Mount Isa. For locations with high recurrence of short-term, and prolonged droughts (SP in Table 3), such as Melbourne, increasing depth of topsoil can increase water holding capacity (Audet et al., 2013; Bot and Benites, 2005). Similarly, by mixing silt and clay soil in the topsoil and reducing slope gradients may facilitate infiltration and increase soil water retention capacity (Audet et al., 2013). For tropical locations with high recurrence of short-term (3-monthly timescale), severe, and prolonged droughts (SS, SP in Table 3), e.g. in Cairns and Weipa, ground cover such as mulch and fast growing vegetation cover (e.g. Buffel grass) may reduce evaporation and maintain soil moisture to facilitate the establishment of drought sensitive and slow growing species (Blum, 1996).

Utilising irrigation methods for specific site characteristics is a cost-effective strategy for any rehabilitation plan. Regular irrigation with proper drainage systems that distributes water is an effective strategy in locations with high recurrence of long-term, severe, and prolonged droughts (LP, LS in Table 3). For locations with high recurrence of short-term, severe, and prolonged droughts (SS, SP in Table 3), with seasonal rainfall (e.g. Brisbane, Sydney, Kingaroy, Brigalow), seasonal irrigation and irrigation at critical stages of plant growth (Blum, 1996), such as during periods of germination, and root or pod development periods are efficient actions to ensure plant survival throughout drought spells.

4.4   Future research

The method outlined in this study provides a useful tool for land managers to address site-based climatic conditions. Future research needs to build on this tool, as well as address the limitations of our method based on meteorological drought indices inferred from point observations. This research may assess: (i) the relationship between meteorological and agricultural drought indices; (ii) regional scale mapping of drought indices; and (iii) the predictive power of design droughts.

While the applied drought indices are robust indicators of meteorological droughts (Mishra and Singh, 2010; Quiring, 2009), they are limited to detecting anomalies from historic rainfall patterns. Soil plays a critical role for any ecosystem development, particularly with regard to ecosystem rehabilitation in post-mining land (Arnold et al., 2013), as soil properties translate rainfall into plant available water (Zhang et al., 2001; Huang et al., 2013). Future drought analysis would benefit from integrating soil properties such as depth, texture, salinity, or organic matter content into drought indices to describe agricultural droughts (Khare et al., 2013; Baldocchi et al., 2004; Woli et al., 2012). Soil texture and depth are critical factors in highly seasonal climates, where soil water storage overcomes periods of water deficit (Prentice et al., 1992; Bot and Benites, 2005). However, using simple and easily accessible meteorological data is a critical step forward to making it easier for mine rehabilitation managers to adopt the concept of using SDF curves as early risk assessment tools.

Although the selected locations can be considered representative of the agro-climatic environments across eastern Australia (Fig. 1), our analysis is strictly valid for the selected point data and therefore site-specific. Future work should not only integrate the above-mentioned soil component but also extend drought analyses across Australia using gridded weather data from the Bureau of Meteorology (2014). Future investigations could assess possible trends in temporal changes of recurrence intervals by dividing historic time series of rainfall and evaporation into subsets and replicate the analysis for each subset (Li et al., 2014; Darshana et al., 2013; Jacobs et al., 2013; Halwatura et al., 2015).

5   Conclusions

The study revealed site-specific patterns of recurrence intervals of short-term and long-term droughts across eastern Australia. Severe and prolonged short-term droughts recurred most often in tropical climates and temperate Wagga Wagga, while severe and prolonged short-term droughts recurred most often in arid conditions and temperate Melbourne. Design droughts can be applied to quantify the frequency of drought events – characterised by severity and duration – at different timescales. This is a critical step forward to consider drought in risk assessments for rehabilitation of post-mining ecosystems. Together with design rainfalls, design droughts should be used to assess rehabilitation strategies and ecological management based on drought recurrence intervals, thereby minimising the risk of failure of initial ecosystem establishment due to ignorance of fundamental abiotic and site-specific environmental barriers.

Appendix A: RDI and SPEI

A1 RDI

The standardised $\mathrm{RDI_{st}}$ is given as

$$\mathrm{RDI_{st}}(k) = \frac{y_k - \overline{y}_k}{\hat{\sigma}_k}, \qquad \text{(A1)}$$

with

$$y_k = \ln \frac{\sum\limits_{j=1}^{j=k} P_j}{\sum\limits_{j=1}^{j=k} \mathrm{PET}_j} \qquad \text{(A2)}$$

where $\hat{\sigma}$ is the standard deviation, y_k is month k of year y, $\overline{y}_k$ is the arithmetic mean of y_k, $\hat{\sigma}_k$ is the standard deviation of y_k, and P_j and PET_j are precipitation and potential evapotranspiration for the jth month of the hydrological year, respectively (Tsakiris and Vangelis, 2005).

A2 SPEI

The SPEI is calculated as

$$\mathrm{SPEI} = W - \frac{C_0 + C_1 W + C_2 W^2}{1 + d_1 W + d_2 W^2 + d_3 W^3} \qquad \text{(A3)}$$

with

$$W = \sqrt{-2\ln(P)} \text{ for } P \leq 0.5, \qquad \text{(A4)}$$

where P is the probability of exceeding a determined value of the difference between the precipitation and potential evapotranspiration $(P = 1 - F(x))$. If $P > 0.5$, then P is replaced by $1 - P$ and the sign of the resultant SPEI is reversed. The constants are $C_0 = 2.515517$, $C_1 = 0.802853$, $C_2 = 0.010328$, $d_1 = 1.432788$, $d_2 = 0.189269$, and $d_3 = 0.001308$ (Vicente-Serrano et al., 2010).

Appendix B: Mathematical description of Gumbel and Frank copula (Shiau, 2006)

B1 Gumbel copula

$$C(u, v) = \exp\left\{-\left[(-\ln u)^\theta + (-\ln v)^\theta\right]^{\frac{1}{\theta}}\right\}, \quad \theta \geq 1 \qquad \text{(B1)}$$

$$c(u, v) = C(u, v) \frac{\left[(-1\ln u)^\theta (-\ln v)^{\theta-1}\right]}{uv} \left[(-1\ln u)^\theta (-\ln v)^\theta\right]^{\frac{2}{\theta}-2}$$
$$\cdot \left\{(\theta - 1)\left[(-\ln u)^\theta + (-\ln v)^\theta\right]^{-\frac{1}{\theta}} + 1\right\} \qquad \text{(B2)}$$

B2 Frank copula

$$C(u, v) = -\frac{1}{\theta} \ln\left[1 + \frac{\left(e^{-\theta u} - 1\right)\left(e^{-\theta v} - 1\right)}{e^{-\theta} - 1}\right], \quad \theta \neq 0 \qquad \text{(B3)}$$

$$c(u, v) = -\frac{\theta e^{-\theta(u+v)}\left(e^{-\theta} - 1\right)}{\left[e^{-\theta(u+v)} - e^{-\theta u} - e^{-\theta v} + e^{-\theta}\right]^2} \qquad \text{(B4)}$$

Appendix C: Time series of drought indices and major weather event

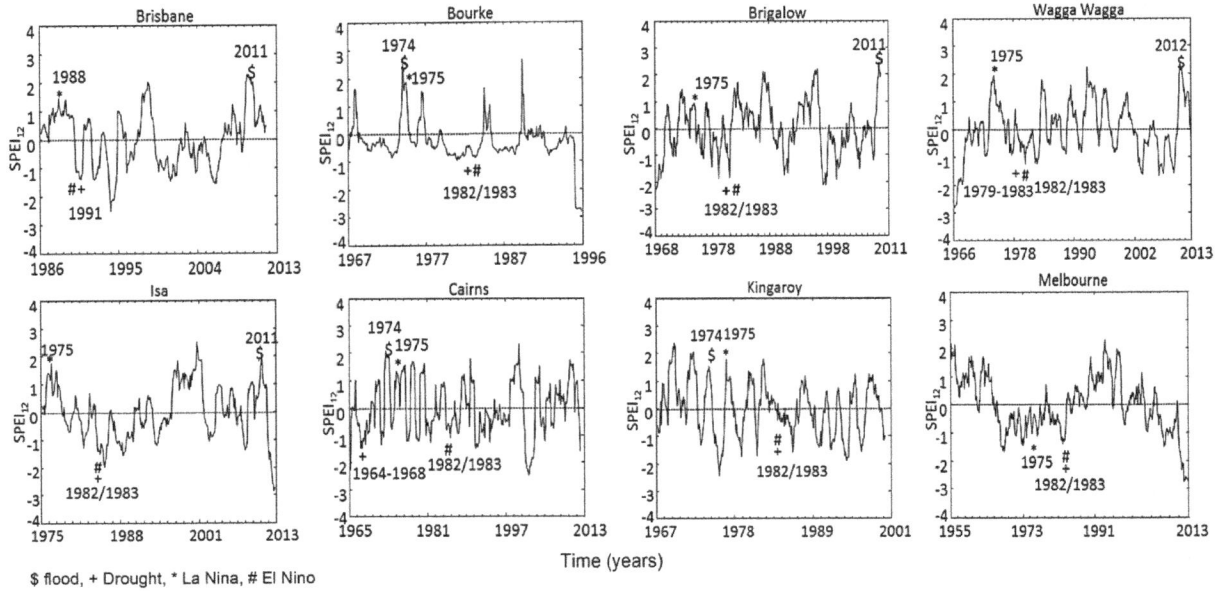

Figure C1. Calculated $SPEI_{12}$ for selected locations across eastern Australia.

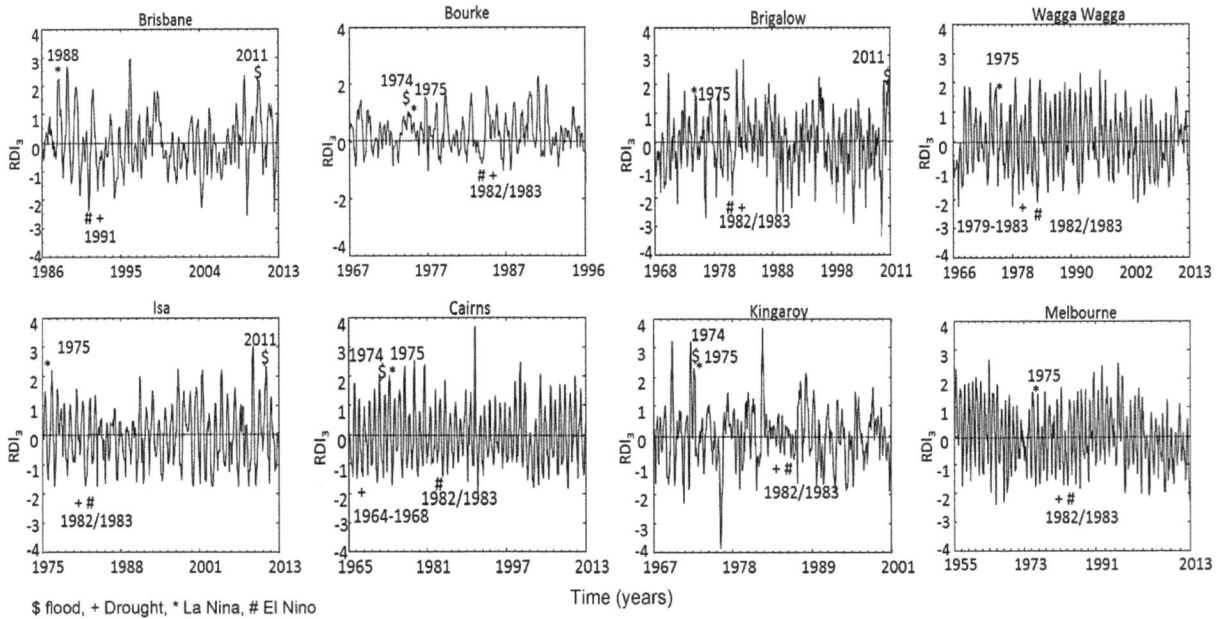

Figure C2. Calculated RDI_3 for selected locations across eastern Australia.

Appendix D: Recurrence intervals of drought events
with any severity and duration of interest

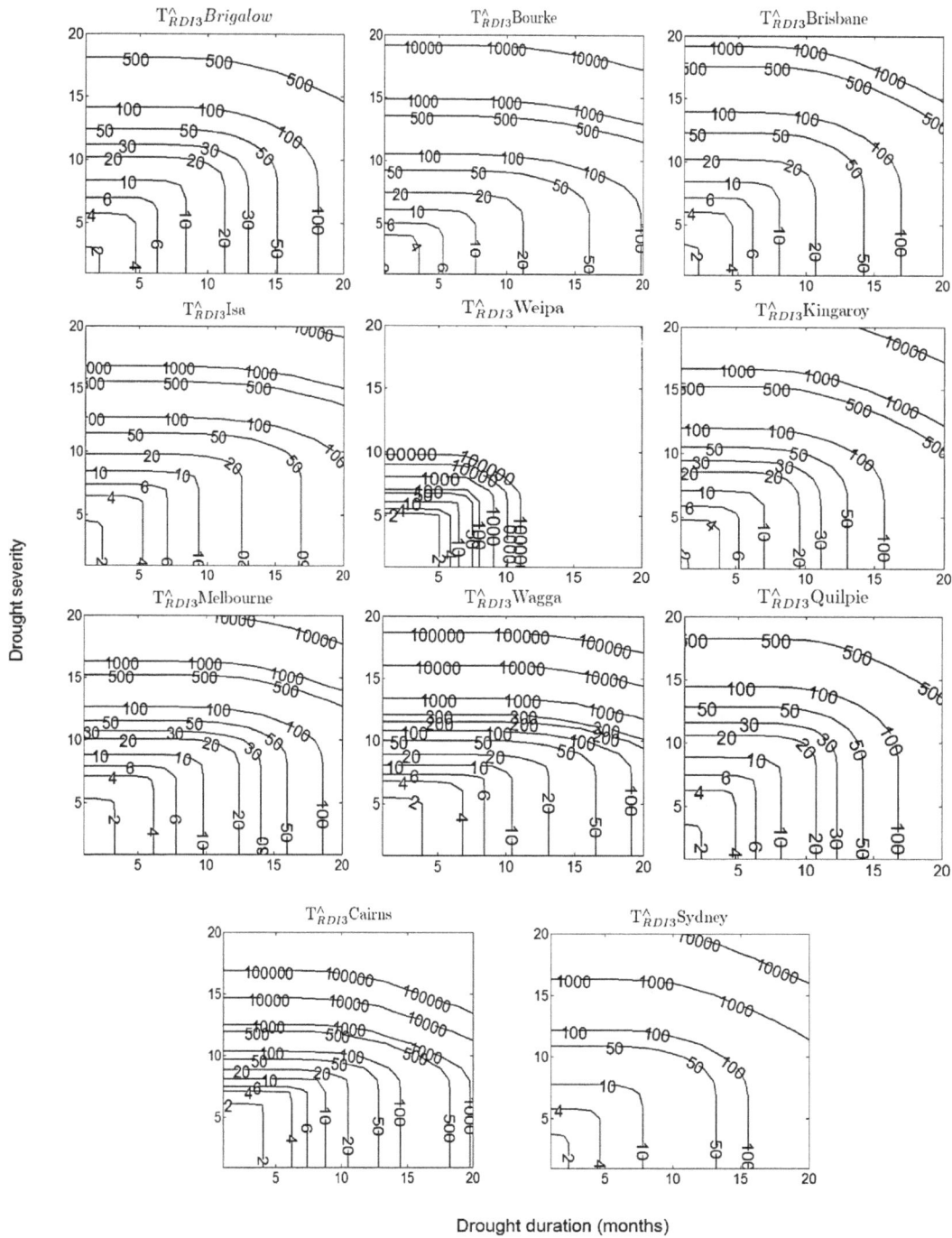

Figure D1. Recurrence intervals $T^{\wedge}$ (years) of drought events with any severity *and* duration of interest based on RDI_3 (short-term) of historical rainfall.

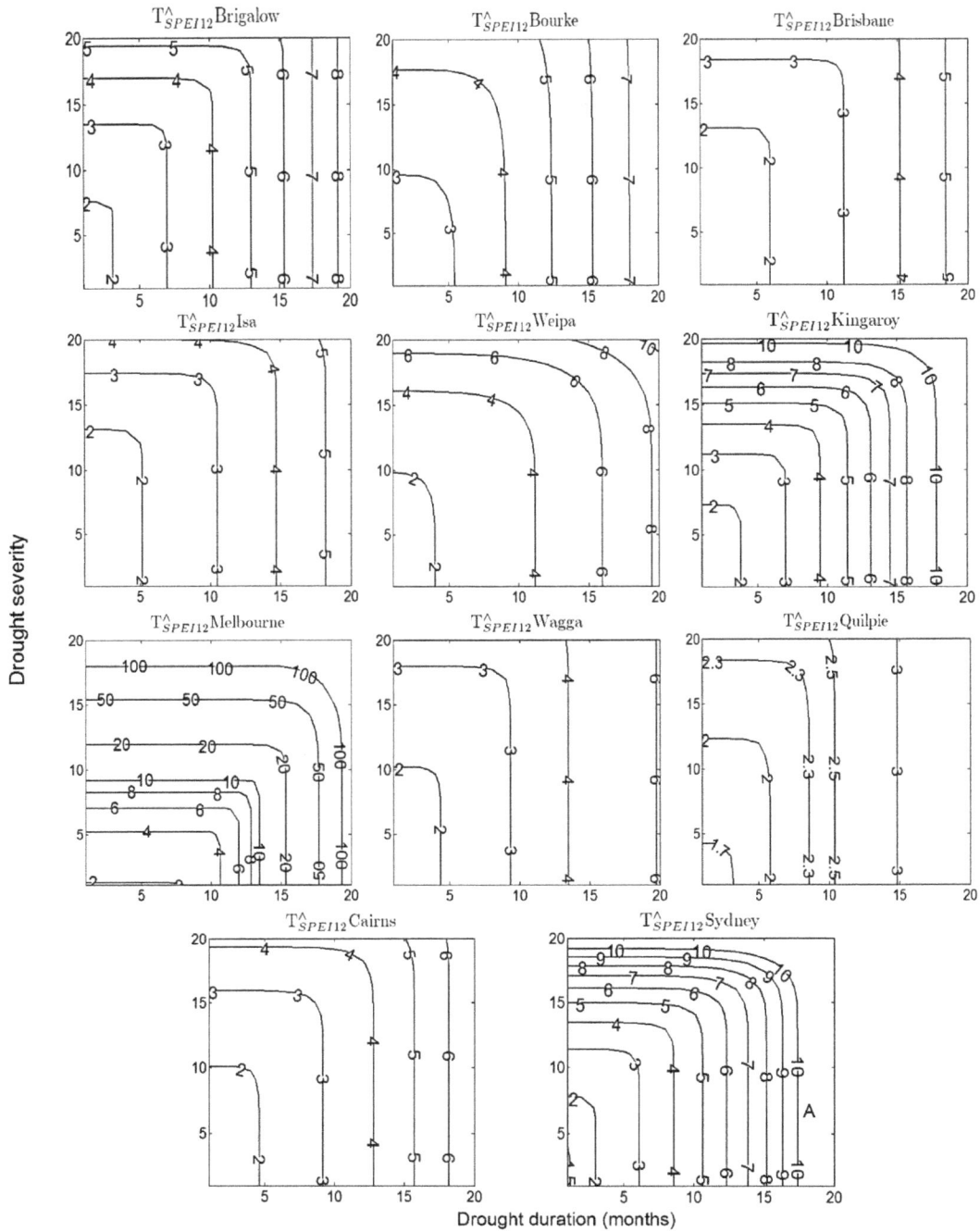

Figure D2. Recurrence intervals $T^{\wedge}$ (years) of drought events with any severity *and* duration of interest based on SPEI_{12} (long-term) of historical rainfall.

Appendix E: Coefficient values of Pearson's r and Kendall's τ for SPI$_3$ vs. RDI$_3$, and SPI$_{12}$ vs. SPEI$_{12}$

Table E1. Coefficient values of Pearson's r and Kendall's τ for SPI$_3$ vs. RDI$_3$, and SPI$_{12}$ vs. SPEI$_{12}$. Correlations were lowest for arid Bourke and Quilpie (bold values).

Location	SPI$_3$ vs. RDI$_3$		SPI$_{12}$ vs. SPEI$_{12}$	
	Pearson's r	Kendall's τ	Pearson's r	Kendall's τ
Weipa	0.98	0.92	0.83	0.68
Cairns	0.98	0.90	0.96	0.83
Brisbane	0.81	0.62	0.68	0.68
Sydney	0.82	0.61	0.90	0.71
Melbourne	0.98	0.90	0.82	0.70
Kingaroy	0.77	0.54	0.87	0.68
Brigalow	0.90	0.71	0.83	0.64
Wagga Wagga	0.69	0.68	0.84	0.71
Bourke	**0.43**	**0.54**	**0.51**	**0.53**
Quilpie	**0.57**	**0.40**	**0.49**	**0.40**
Mount Isa	0.78	0.60	0.72	0.67

Appendix F: R^2 and p values for fitted cumulative distribution functions and Copula parameters

Table F1. R^2 and p values for fitted cumulative distribution functions and Copula parameters for the studied sites.

| Station | Cumulative distribution functions | | | | | | | | | | | | Copula | | | |
| | Exponential | | Logistic | | Lognormal | | Bimodal lognormal | | Gamma | | Extreme value | | Gumbel | | Frank | |
	R^2	p	R^2	p	R^2	p	R^2	p	R^2	p	R^2	p	R^2	p	R^2	p
Weipa	0.24	0.00	0.99	0.00	0.00	0.31	0.00	0.57	0.99	0.00	0.60	0.00	0.97	0.00	1.00	0.00
Cairns	0.00	0.20	1.00	0.00	0.00	0.52	0.00	0.68	1.00	0.00	0.53	0.00	0.98	0.00	0.99	0.00
Brisbane	0.00	0.30	1.00	0.00	0.00	0.61	0.00	0.61	0.98	0.00	0.57	0.00	0.96	0.00	0.98	0.00
Sydney	0.31	0.00	0.99	0.00	0.00	0.64	0.00	0.52	1.00	0.00	0.55	0.00	0.97	0.00	1.00	0.00
Melbourne	0.25	0.00	0.99	0.00	0.00	0.63	0.00	0.64	0.99	0.00	0.42	0.00	0.96	0.00	1.00	0.00
Kingaroy	0.00	0.08	1.00	0.00	0.00	0.42	0.00	0.43	0.99	0.00	0.68	0.00	0.98	0.00	0.99	0.00
Brigalow	0.00	0.06	0.96	0.00	0.00	0.64	0.00	0.26	0.99	0.00	0.62	0.00	0.96	0.00	1.00	0.00
Wagga Wagga	0.00	0.15	0.96	0.00	0.00	0.61	0.00	0.54	0.91	0.00	0.43	0.00	0.97	0.00	1.00	0.00
Bourke	0.00	0.21	0.94	0.00	0.00	0.31	0.00	0.34	0.98	0.00	0.62	0.00	0.95	0.00	1.00	0.00
Quilpie	0.12	0.00	0.98	0.00	0.00	0.15	0.00	0.29	0.99	0.00	0.53	0.00	0.96	0.00	0.99	0.00
Mount Isa	0.20	0.00	0.99	0.00	0.00	0.56	0.00	0.46	0.97	0.00	0.68	0.00	0.95	0.00	1.00	0.00

Acknowledgements. This study was made possible by the University of Queensland Post-doctoral Fellowship scheme and Early Career Research Grant awarded to S. Arnold, as well as the International Postgraduate Research Scholarship awarded to D. Halwatura. We thank Tanja Giebner for her dedicated work on deriving the RDI for selected locations. Further, we thank David Doley and Patrick Audet for critical discussions and seven anonymous reviewers for their constructive comments.

Edited by: N. Ursino

References

Allen, R. G., Pereira, L. S., Raes, D., and Smith, M.: Crop evapotranspiration-Guidelines for computing crop water requirements-FAO Irrigation and drainage paper 56, FAO, Rome, Italy, 300 pp., 1998.

Anderegg, L. D., Anderegg, W. R., and Berry, J. A.: Not all droughts are created equal: translating meteorological drought into woody plant mortality, Tree Physiol., 33, 672–683, 2013.

Arnold, S., Audet, P., Doley, D., and Baumgartl, T.: Hydropedology and Ecohydrology of the Brigalow Belt, Australia: Opportunities for Ecosystem Rehabilitation in Semiarid Environments, Gsvadzone, 12, doi:10.2136/vzj2013.03.0052, 2013.

Arnold, S., Kailichova, Y., and Baumgartl, T.: Germination of Acacia harpophylla (Brigalow) seeds in relation to soil water potential: implications for rehabilitation of a threatened ecosystem, Peer J., 2, e268, doi:10.7717/peerj.268, 2014a.

Arnold, S., Kailichova, Y., Knauer, J., Ruthsatz, A. D., and Baumgartl, T.: Effects of soil water potential on germination of codominant Brigalow species: Implications for rehabilitation of water-limited ecosystems in the Brigalow Belt bioregion, Ecol. Eng., 70, 35–42, doi:10.1016/j.ecoleng.2014.04.015, 2014b.

Athearn, J. L.: What is Risk?, J. Risk Insur., 38, 639–645, doi:10.2307/251578, 1971.

Audet, P., Arnold, S., Lechner, A. M., Mulligan, D. R., and Baumgartl, T.: Climate suitability estimates offer insight into fundamental revegetation challenges among post-mining rehabilitated landscapes in eastern Australia, Biogeosciences Discuss., 9, 18545–18569, doi:10.5194/bgd-9-18545-2012, 2012.

Audet, P., Arnold, S., Lechner, A. M., and Baumgartl, T.: Site-specific climate analysis elucidates revegetation challenges for post-mining landscapes in eastern Australia, Biogeosciences, 10, 6545–6557, doi:10.5194/bg-10-6545-2013, 2013.

Baldocchi, D. D., Xu, L., and Kiang, N.: How plant functional-type, weather, seasonal drought, and soil physical properties alter water and energy fluxes of an oak–grass savanna and an annual grassland, Agr. Forest Meteorol., 123, 13–39, doi:10.1016/j.agrformet.2003.11.006, 2004.

Banimahd, S. and Khalili, D.: Factors influencing markov chains predictability characteristics, utilizing SPI, RDI, EDI and SPEI drought indices in different climatic zones, Water Resour. Manage., 27, 3911–3928, doi:10.1007/s11269-013-0387-z, 2013.

Bell, L. C.: Establishment of native ecosystems after mining – Australian experience across diverse biogeographic zones, Ecol. Eng., 17, 179–186, doi:10.1016/S0925-8574(00)00157-9, 2001.

Blum, A.: Crop responses to drought and the interpretation of adaptation, Plant Growth Regul., 20, 135–148, 1996.

Bolger, T. P., Rivelli, A. R., and Garden, D. L.: Drought resistance of native and introduced perennial grasses of south-eastern Australia, Aust. J. Agr. Res., 56, 1261–1267, doi:10.1071/AR05075, 2005.

Bot, A. and Benites, J.: The importance of soil organic matter: key to drought-resistant soil and sustained food and production, FAO, Rome, Italy, 2005.

Brouwers, N., Matusick, G., Ruthrof, K., Lyons, T., and Hardy, G.: Landscape-scale assessment of tree crown dieback following extreme drought and heat in a Mediterranean eucalypt forest ecosystem, Landscape Ecol., 28, 69–80, doi:10.1007/s10980-012-9815-3, 2013.

Bureau of Meteorology: Climate data, http://www.bom.gov.au/climate/data/, last access: April 2013.

Bureau of Meteorology: Australian water availability project 2014, http://www.bom.gov.au/jsp/awap/, last access: January 2014.

Burton, M., Jasmine Zahedi, S., and White, B.: Public preferences for timeliness and quality of mine site rehabilitation. The case of bauxite mining in Western Australia, Resources Policy, 37, 1–9, 2012.

Chebbi, A., Bargaoui, Z. K., and da Conceição Cunha, M.: Development of a method of robust rain gauge network optimization based on intensity-duration-frequency results, Hydrol. Earth Syst. Sci., 17, 4259–4268, doi:10.5194/hess-17-4259-2013, 2013.

Clark, C.: Measurements of actual and pan evaporation in the upper Brue catchment UK: the first 25 years, Weather, 68, 200–208, doi:10.1002/wea.2090, 2013.

Craven, D., Hall, J. S., Ashton, M. S., and Berlyn, G. P.: Water-use efficiency and whole-plant performance of nine tropical tree species at two sites with contrasting water availability in Panama, Trees-Struct. Funct., 27, 639–653, doi:10.1007/s00468-012-0818-0, 2013.

Darshana, Pandey, A., and Pandey, R. P.: Analysing trends in reference evapotranspiration and weather variables in the Tons River Basin in Central India, Stoch. Environ. Res. Risk Assess., 27, 1407–1421, doi:10.1007/s00477-012-0677-7, 2013.

Doley, D. and Audet, P.: Adopting novel ecosystems as suitable rehabilitation alternatives for former mine sites, Ecol. Process., 2, 1–11, doi:10.1186/2192-1709-2-22, 2013.

Doley, D., Audet, P., and Mulligan, D. R.: Examining the Australian context for post-mined land rehabilitation: Reconciling a paradigm for the development of natural and novel ecosystems among post-disturbance landscapes, Agr. Ecosyst. Environ., 163, 85–93, doi:10.1016/j.agee.2012.04.022, 2012.

Egidijus, R., Edvinas, S., Vladimir, K., Justas, K., Gintaras, V., and Aliaksandr, P.: Dynamics of meteorological and hydrological droughts in the Neman river basin, Environ. Res. Lett., 8, 045014, doi:10.1088/1748-9326/8/4/045014, 2013.

Engelbrecht, B. M., Comita, L. S., Condit, R., Kursar, T. A., Tyree, M. T., Turner, B. L., and Hubbell, S. P.: Drought sensitivity shapes species distribution patterns in tropical forests, Nature, 447, 80–82, 2007.

Ganguli, P. and Reddy, M. J.: Risk assessment of droughts in Gujarat using bivariate Copulas, Water Resour. Manage., 26, 3301–3327, 2012.

Gardner, J. H. and Bell, D. T.: Bauxite mining restoration by Alcoa World Alumina Australia in Western Australia: social, political, historical, and environmental contexts, Restor. Ecol., 15, S3–S10, doi:10.1111/j.1526-100X.2007.00287.x, 2007.

Guttman, N. B.: Accepting the Standardized Precipitation Index: a calculation algorithm, J. Am. Water Resour. Assoc., 35, 311–322, 1999.

Hailegeorgis, T. T., Thorolfsson, S. T., and Alfredsen, K.: Regional frequency analysis of extreme precipitation with consideration of uncertainties to update IDF curves for the city of Trondheim, J. Hydrol., 498, 305–318, doi:10.1016/j.jhydrol.2013.06.019, 2013.

Halwatura, D., Lechner, A. M., and Arnold, S.: Design droughts: A new planning tool for ecosystem rehabilitation, Int. J. Geomate, 8, 1138–1142, 2015.

Heim, R. R.: A review of twentieth-century drought indices used in the United States, B. Am. Meteorol. Soc., 83, 1149–1165, 2002.

Hilhorst, H. W.: A critical update on seed dormancy, I. Primary dormancy, Seed Sci. Res., 5, 61–73, 1995.

Hinz, C., McGrath, G., and Hearman, A.: Towards a climate based risk assessment of land rehabilitation, 1st International Seminar on Mine Closure, Australian Centre for Geomechanics, Perth, 407–416, 2006.

Hodgkinson, J. H., Littleboy, A., Howden, M., Moffat, K., and Loechel, B.: Climate adaptation in the Australian mining and exploration industries, CSIRO Climate Adaptation National Research Flagship, 1921605812 – working paper No. 5, http://www.csiro.au/resources/CAF-working-papers.html (last access: May 2014), 2010.

Hu, Y.-M., Liang, Z.-M., Liu, Y.-W., Wang, J., Yao, L., and Ning, Y.: Uncertainty analysis of SPI calculation and drought assessment based on the application of Bootstrap, Int. J. Climatol., doi:10.1002/joc.4091, in press, 2014.

Huang, M., Barbour, S. L., Elshorbagy, A., Zettl, J., and Si, B. C.: Effects of variably layered coarse textured soils on plant available water and forest productivity, in: Four decades of progress in monitoring and modeling of processes in the soil-plant-atmosphere system: Applications and challenges, edited by: Romano, N., Durso, G., Severino, G., Chirico, G. B., and Palladino, M., Procedia Environmental Sciences, Elsevier Science Bv, Amsterdam, 148–157, 2013.

Hutchinson, M. F., McIntyre, S., Hobbs, R. J., Stein, J. L., Garnett, S., and Kinloch, J.: Integrating a global agro-climatic classification with bioregional boundaries in Australia, Global Ecol. Biogeogr., 14, 197–212, 2005.

Jacobs, S. J., Pezza, A. B., Barras, V., Bye, J., and Vihma, T.: An analysis of the meteorological variables leading to apparent temperature in Australia: Present climate, trends, and global warming simulations, Global Planet. Change, 107, 145–156, doi:10.1016/j.gloplacha.2013.05.009, 2013.

Jahantab, E., Javdani, Z., Bahari, A., Bahrami, S., and Mehrabi, A.: Effect of priming treatments on seed germination percentage and rate in the early stages of triticale plants grown under drought stress conditions, Int. J. Agr. Crop Sci., 5, 1909–1917, 2013.

Joe, H.: Multivariate models and dependence concepts, Chapman & Hall, London, 1997.

Jongen, M., Unger, S., Fangueiro, D., Cerasoli, S., Silva, J. M. N., and Pereira, J. S.: Resilience of montado understorey to experimental precipitation variability fails under severe natural drought, Agr. Ecosyst. Environ., 178, 18–30, doi:10.1016/j.agee.2013.06.014, 2013.

Khalili, D., Farnoud, T., Jamshidi, H., Kamgar-Haghighi, A. A., and Zand-Parsa, S.: Comparability analyses of the SPI and RDI meteorological drought indices in different climatic zones, Water Resour. Manage., 25, 1737–1757, 2011.

Khare, Y. P., Martinez, C. J., and Munoz-Carpena, R.: Parameter variability and drought models: A study using the Agricultural Reference Index for Drought (ARID), Agron. J., 105, 1417–1432, doi:10.2134/agronj2013.0167, 2013.

Khedun, C. P., Chowdhary, H., Mishra, A. K., Giardino, J. R., and Singh, V. P.: Water Deficit Duration and Severity Analysis Based on Runoff Derived from Noah Land Surface Model, J. Hydrol. Eng., 18, 817–833, 2013.

Kim, D.-W., Byun, H.-R., and Choi, K.-S.: Evaluation, modification, and application of the Effective Drought Index to 200-Year drought climatology of Seoul, Korea, J. Hydrol., 378, 1–12, doi:10.1016/j.jhydrol.2009.08.021, 2009.

Kuo, C. C., Gan, T. Y., and Chan, S.: Regional Intensity-Duration-Frequency curves derived from Ensemble Empirical Mode Decomposition and Scaling Property, J. Hydrol. Eng., 18, 66–74, doi:10.1061/(asce)he.1943-5584.0000612, 2013.

Kuster, T., Arend, M., Bleuler, P., Günthardt-Goerg, M., and Schulin, R.: Water regime and growth of young oak stands subjected to air-warming and drought on two different forest soils in a model ecosystem experiment, Plant Biol., 15, 138–147, 2013.

Lee, J. H. and Kim, C. J.: A multimodel assessment of the climate change effect on the drought severity–duration–frequency relationship, Hydrol. Process., 27, 2800–2813, doi:10.1002/hyp.9390, 2012.

Lee, T., Modarres, R., and Ouarda, T. B. M. J.: Data-based analysis of bivariate copula tail dependence for drought duration and severity, Hydrol. Process., 27, 1454–1463, doi:10.1002/Hyp.9233, 2013.

Li, Z., Chen, Y. N., Yang, J., and Wang, Y.: Potential evapotranspiration and its attribution over the past 50 years in the arid region of Northwest China, Hydrol. Process., 28, 1025–1031, 2014.

Livada, I. and Assimakopoulos, V. D.: Spatial and temporal analysis of drought in greece using the Standardized Precipitation Index (SPI), Theor. Appl. Climatol., 89, 143–153, doi:10.1007/s00704-005-0227-z, 2007.

Lloret, F.: Vulnerability and resilience of forest ecosystems to extreme drought episodes, Ecosistemas, 21, 85–90, 2012.

Loaiciga, H. and Mariño, M.: Recurrence interval of geophysical events, J. Water Resour. Plan. Manage., 117, 367–382, doi:10.1061/(ASCE)0733-9496(1991)117:3(367), 1991.

Lugato, E., Alberti, G., Gioli, B., Kaplan, J. O., Peressotti, A., and Miglietta, F.: Long-term pan evaporation observations as a resource to understand the water cycle trend: case studies from Australia, Hydrolog. Sci. J., 58, 1287–1296, doi:10.1080/02626667.2013.813947, 2013.

Mariotte, P., Vandenberghe, C., Kardol, P., Hagedorn, F., and Buttler, A.: Subordinate plant species enhance community resistance against drought in semi-natural grasslands, J. Ecol., 101, 763–773, doi:10.1111/1365-2745.12064, 2013.

Mason, L., Unger, C., Lederwasch, A., Razian, H., Wynne, L., and Giurco, D.: Adapting to climate risks and extreme weather: A guide for mining and minerals industry professionals, National Climate Change Adaptation Research Facility, Gold Coast, 76 pp., 2013.

McKee, T. B., Doesken, N. J., and Kleist, J.: The relationship of drought frequency and duration to time scales, Proceedings of the 8th Conference on Applied Climatology, Anaheim, California, 179–183, 1993.

Mirabbasi, R., Fakheri-Far, A., and Dinpashoh, Y.: Bivariate drought frequency analysis using the copula method, Theor. Appl. Climatol., 108, 191–206, 2012.

Mishra, A. K. and Singh, V. P.: A review of drought concepts, J. Hydrol., 391, 202–216, 2010.

Murphy, B. F. and Timbal, B.: A review of recent climate variability and climate change in southeastern Australia, Int. J. Climatol., 28, 859–879, doi:10.1002/joc.1627, 2008.

Myers, N., Mittermeier, R. A., Mittermeier, C. G., Da Fonseca, G. A., and Kent, J.: Biodiversity hotspots for conservation priorities, Nature, 403, 853–858, 2000.

National Drought Mitigation Cente-Interpretation of Standardized Precipitation Index Maps: http://drought.unl.edu/MonitoringTools/ClimateDivisionSPI/Interpretation.aspx, last access: January 2014.

Nefzaoui, A. and Ben Salem, H.: Cacti: Efficient tool for rangeland rehabilitation, drought mitigation and to combat desertification, in: Proceedings of the Fourth International Congress on Cactus Pear and Cochineal, edited by: Nefzaoui, A. and Inglese, P., Acta Horticulturae, 581, 295–315, 2002.

Paulo, A. A., Rosa, R. D., and Pereira, L. S.: Climate trends and behaviour of drought indices based on precipitation and evapotranspiration in Portugal, Nat. Hazards Earth Syst. Sci., 12, 1481–1491, doi:10.5194/nhess-12-1481-2012, 2012.

Peel, M. C., Finlayson, B. L., and McMahon, T. A.: Updated world map of the Köppen-Geiger climate classification, Hydrol. Earth Syst. Sci., 11, 1633–1644, doi:10.5194/hess-11-1633-2007, 2007.

Prentice, I. C., Cramer, W., Harrison, S. P., Leemans, R., Monserud, R. A., and Solomon, A. M.: Special Paper: A global biome model based on plant physiology and dominance, soil properties and climate, J. Biogeogr., 19, 117–134, doi:10.2307/2845499, 1992.

Quiring, S. M.: Monitoring drought: An evaluation of meteorological drought indices, Geogr. Compass, 3, 64–88, doi:10.1111/j.1749-8198.2008.00207.x, 2009.

Reddy, M. J. and Ganguli, P.: Application of copulas for derivation of drought severity-duration-frequency curves, Hydrol. Process., 26, 1672–1685, doi:10.1002/hyp.8287, 2012.

Ruffault, J., Martin-StPaul, N., Rambal, S., and Mouillot, F.: Differential regional responses in drought length, intensity and timing to recent climate changes in a Mediterranean forested ecosystem, Climatic Change, 117, 103–117, doi:10.1007/s10584-012-0559-5, 2013.

Shi, Z., Thomey, M. L., Mowll, W., Litvak, M., Brunsell, N. A., Collins, S. L., Pockman, W. T., Smith, M. D., Knapp, A. K., and Luo, Y.: Differential effects of extreme drought on production and respiration: synthesis and modeling analysis, Biogeosciences, 11, 621–633, doi:10.5194/bg-11-621-2014, 2014.

Shiau, J. T.: Fitting drought duration and severity with two-dimensional Copulas, Water Resour. Manage., 20, 795–815, 2006.

Shiau, J. T. and Modarres, R.: Copula-based drought severity-duration-frequency analysis in Iran, Meteorol. Appl., 16, 481–489, doi:10.1002/met.145, 2009.

Shiau, J.-T., Feng, S., and Nadarajah, S.: Assessment of hydrological droughts for the Yellow River, China, using copulas, Hydrol. Process., 21, 2157–2163, doi:10.1002/hyp.6400, 2007.

Shiau, J.-T., Modarres, R., and Nadarajah, S.: Assessing multi-site drought connections in Iran using empirical Copula, Environ. Model. Assess., 17, 469–482, doi:10.1007/s10666-012-9318-2, 2012.

Sklar, M.: Fonctions de répartition à n dimensions et leurs marges, Université Paris, Paris, 229–231, 1959.

Smithers, J., Pegram, G., and Schulze, R.: Design rainfall estimation in South Africa using Bartlett–Lewis rectangular pulse rainfall models, J. Hydrol., 258, 83–99, 2002.

Spinoni, J., Antofie, T., Barbosa, P., Bihari, Z., Lakatos, M., Szalai, S., Szentimrey, T., and Vogt, J.: An overview of drought events in the Carpathian Region in 1961–2010, Adv. Sci. Res., 10, 21–32, doi:10.5194/asr-10-21-2013, 2013.

Sternberg, M. and Shoshany, M.: Influence of slope aspect on Mediterranean woody formations: Comparison of a semi-arid and an arid site in Israel, Ecol. Res., 16, 335–345, doi:10.1046/j.1440-1703.2001.00393.x, 2001.

Todisco, F., Mannocchi, F., and Vergni, L.: Severity-duration-frequency curves in the mitigation of drought impact: an agricultural case study, Nat. Hazards, 65, 1863–1881, doi:10.1007/s11069-012-0446-4, 2013.

Tsakiris, G.: Meteorological drought assessment, European Research Program MEDROPLAN, Mediterranean Drought Preparedness and Mitigation Planning, Zaragoza, Spain, 2004,

Tsakiris, G. and Vangelis, H.: Establishing a drought index incorporating evapotranspiration, European Water, 9/10, 3–11, 2005.

Tsakiris, G., Pangalou, D., and Vangelis, H.: Regional drought assessment based on the Reconnaissance Drought Index (RDI), Water Resour. Manage., 21, 821–833, 2007.

Unep, N. M. and Thomas, D.: World Atlas of Desertification, Edward Arnold, London, 15–45, 1992.

Vangelis, H., Tigkas, D., and Tsakiris, G.: The effect of PET method on Reconnaissance Drought Index (RDI) calculation, J. Arid Environ., 88, 130–140, doi:10.1016/j.jaridenv.2012.07.020, 2013.

Vargas, R., Sonnentag, O., Abramowitz, G., Carrara, A., Chen, J. M., Ciais, P., Correia, A., Keenan, T. F., Kobayashi, H., and Ourcival, J.-M.: Drought influences the accuracy of simulated ecosystem fluxes: A model-data meta-analysis for Mediterranean Oak Woodlands, Ecosystems, 16, 749–764, 2013.

Vicente-Serrano, S. M., Beguería, S., and López-Moreno, J. I.: A multiscalar drought index sensitive to global warming: The Standardized Precipitation Evapotranspiration Index, J. Climate, 23, 1696–1718, doi:10.1175/2009JCLI2909.1, 2010.

Vicente-Serrano, S. M., Gouveia, C., Camarero, J. J., Beguería, S., Trigo, R., López-Moreno, J. I., Azorín-Molina, C., Pasho, E., Lorenzo-Lacruz, J., Revuelto, J., Morán-Tejeda, E., and Sanchez-Lorenzo, A.: Response of vegetation to drought time-scales across global land biomes, P. Natl. Acad. Sci., 110, 52–57, doi:10.1073/pnas.1207068110, 2013.

Wilhite, D. A., Svoboda, M. D., and Hayes, M. J.: Understanding the complex impacts of drought: a key to enhancing drought mitigation and preparedness, Water Resour. Manage., 21, 763–774, 2007.

Williams, J., Hook, R., and Hamblin, A.: Agro-ecological regions of Australia methodology for their derivation and key issues in resource management, CSIRO Land & Water, Canberra, ACT, 2002.

Williamson, G. B., Laurance, W. F., Oliveira, A. A., Delamônica, P., Gascon, C., Lovejoy, T. E., and Pohl, L.: Amazonian tree mortality during the 1997 El Nino drought, Conserv. Biol., 14, 1538–1542, 2000.

Woli, P., Jones, J. W., Ingram, K. T., and Fraisse, C. W.: Agricultural reference index for drought (ARID), Agron. J., 104, 287–300, 2012.

Wong, G., Lambert, M. F., Leonard, M., and Metcalfe, A. V.: Drought analysis using trivariate Copulas conditional on climatic states, J. Hydrol. Eng., 15, 129–141, 2010.

Woodhams, F., Southwell, D., Bruce, S., Barnes, B., Appleton, H., Rickards, J., Walcott, J., Hug, B., Whittle, L., and Ahammad, H.: Carbon Farming Initiative: A proposed common practice framework for assessing additionality, Canberra, August 2012.

Yoo, J., Kim, U., and Kim, T.-W.: Bivariate drought frequency curves and confidence intervals: a case study using monthly rainfall generation, Stoch. Environ. Res. Risk Assess., 27, 285–295, 2013.

Zarch, M. A. A., Malekinezhad, H., Mobin, M. H., Dastorani, M. T., and Kousari, M. R.: Drought monitoring by reconnaissance drought index (RDI) in Iran, Water Resour. Manage., 25, 3485–3504, 2011.

Zargar, A., Sadiq, R., Naser, B., and Khan, F. I.: A review of drought indices, Environ. Rev., 19, 333–349, 2011.

Zhang, A. and Jia, G.: Monitoring meteorological drought in semiarid regions using multi-sensor microwave remote sensing data, Remote Sens. Environ., 134, 12–23, doi:10.1016/j.rse.2013.02.023, 2013.

Zhang, L., Dawes, W. R., and Walker, G. R.: Response of mean annual evapotranspiration to vegetation changes at catchment scale, Water Resour. Res., 37, 701–708, doi:10.1029/2000WR900325, 2001.

Zhang, Q., Chen, Y. D., Chen, X., and Li, J.: Copula-based analysis of hydrological extremes and implications of hydrological behaviors in the Pearl River basin, China, J. Hydrol. Eng., 16, 598–607, 2011.

13

Developing a drought-monitoring index for the contiguous US using SMAP

Sara Sadri, Eric F. Wood, and Ming Pan

Department of Civil and Environmental Engineering, Princeton University, 59 Olden St, Princeton, NJ 08540, USA

Correspondence: Sara Sadri (sadri@princeton.edu)

Abstract. Since April 2015, NASA's Soil Moisture Active Passive (SMAP) mission has monitored near-surface soil moisture, mapping the globe (between 85.044° N/S) using an L-band (1.4 GHz) microwave radiometer in 2–3 days depending on location. Of particular interest to SMAP-based agricultural applications is a monitoring product that assesses the SMAP near-surface soil moisture in terms of probability percentiles for dry and wet conditions. However, the short SMAP record length poses a statistical challenge for meaningful assessment of its indices. This study presents initial insights about using SMAP for monitoring drought and pluvial regions with a first application over the contiguous United States (CONUS). SMAP soil moisture data from April 2015 to December 2017 at both near-surface (5 cm) SPL3SMP, or Level 3, at ~ 36 km resolution, and root-zone SPL4SMAU, or Level 4, at ~ 9 km resolution, were fitted to beta distributions and were used to construct probability distributions for warm (May–October) and cold (November–April) seasons. To assess the data adequacy and have confidence in using short-term SMAP for a drought index estimate, we analyzed individual grids by defining two filters and a combination of them, which could separate the 5815 grids covering CONUS into passed and failed grids. The two filters were (1) the Kolmogorov–Smirnov (KS) test for beta-fitted long-term and the short-term variable infiltration capacity (VIC) land surface model (LSM) with 95 % confidence and (2) good correlation (≥ 0.4) between beta-fitted VIC and beta-fitted SPL3SMP. To evaluate which filter is the best, we defined a mean distance (MD) metric, assuming a VIC index at 36 km resolution as the ground truth. For both warm and cold seasons, the union of the filters – which also gives the best coverage of the grids throughout CONUS – was chosen to be the most reliable filter. We visually compared our

SMAP-based drought index maps with metrics such as the U.S. Drought Monitor (from D0–D4), 1-month Standard Precipitation Index (SPI) and near-surface VIC from Princeton University. The root-zone drought index maps were shown to be similar to those produced by the root-zone VIC, 3-month SPI, and the Gravity Recovery and Climate Experiment (GRACE). This study is a step forward towards building a national and international soil moisture monitoring system without which quantitative measures of drought and pluvial conditions will remain difficult to judge.

1 Introduction

Drought is an extreme condition when water in one or a combination of water stores (e.g., river, lake, reservoir, snowpack, soil water or groundwater) or water fluxes (precipitation, evapotranspiration or runoff) drops below a defined condition for a prolonged period of time (Wilhite and Glantz, 1985; Wilhite, 2000; AMS, 2012). Such a water deficit evolves over weeks to months and can last for months and years. Drought's propagation is silent and often without warning until it impacts human lives and environmental activities (Tallaksen and Van Lanen, 2004). Drought conditions are related to water demand, so local water use plays a central role in defining conditions of scarcity and the resulting impacts. Wilhite and Glantz (1985) classified drought into meteorological, agricultural or hydrological, depending on whether the deficit is measured using precipitation, soil moisture or river discharge, respectively.

The reduced supply of precipitation (and subsequently soil moisture) for crops leads to an agricultural drought that impacts crop yield, inflicting enormous economic impacts on developed countries and the suffering of millions of people in less-developed regions of the world. In the US, since 1996, there has been at least one drought event per year except for the years 1997, 2001, 2004 and 2010, and each year drought cost between USD 1 billion and 14 billion in damages (in 2015 – adjusted dollars) (NOAA, 2018b). In California alone, the 2015 drought was estimated to cause USD 2–5 billion in damages to the agricultural sector (Howitt et al., 2015).

Although the impacts of drought are intimately linked to the vulnerability of a population to adverse conditions (UN/ISDR, 2007) and how society responds within the constraints of changing economies, the timely determination of the current level of agricultural drought aids the decision-making process in order to reduce its impacts. Scientifically based drought-monitoring tools and warning systems assist in the mitigation of the losses caused by droughts and the planing and management of water shortages that will accompany future droughts (Martinez-Fernandez et al., 2016). Such drought-monitoring tools are based on long-term observations of the hydrological variables such as precipitation, streamflow, soil moisture and groundwater.

Pluvial conditions are related to an abundance of precipitation and subsequently wet soil conditions that can adversely affect agriculture by waterlogging the fields or exacerbating flooding from additional rainfall. Thus, for monitoring extremes (either agricultural drought or pluvial conditions), realistic estimation of soil moisture at regional to continental scales is required. Soil moisture is the central source of information, since it reflects recent precipitation and antecedent soil conditions (Sheffield and Wood, 2011). In a sense, soil moisture captures the aggregate balance of all hydrological processes and represents available water, being a buffer between incoming precipitation and throughfall and evapotranspiration and drainage processes (Entekhabi et al., 1996). Unfortunately, soil moisture (and evapotranspiration) are among the least-observed components of the hydrological cycle, especially over large spatial and temporal scales (Reichle, 2017; Sheffield and Wood, 2011).

Many statistical measures or indices for extreme conditions have been developed in the US, particularly for drought conditions. This is due to the slow evolution of drought and its economic and social impact. Currently, no single drought index has been able to adequately capture the severity and intensity of drought and its impact on different groups of users (Heim, 2002). Heim (2002) gives an overview of the major 20th US drought indices. The most common ones are the standardized Precipitation Index (SPI), Palmer Drought Severity Index (PDSI), Standardized Runoff Index (SRI) and the U.S. Drought Monitor (DM or USDM).

The SPI is recognized by the World Meteorological Organization (WMO) as the standard index for quantifying and reporting meteorological drought. It is used to characterize drought on a range of timescales from 1 to 36 months. The raw precipitation is fit to an appropriate distribution function and is then transformed into a standardized normal distribution. The SPI index is expressed as the number of standard deviations by which the anomaly deviates from the long-term mean. On short timescales, the SPI is closely related to soil moisture, while at long timescales, it is related to groundwater. The advantages of the SPI include the following: it only relies on precipitation, it can characterize both drought and pluvial conditions, its computation over different timescales can be related to various water resource stores (such as soil moisture and groundwater), and it is more comparable across regions with different climates than the Palmer Severity Drought Index (PDSI). The key limitation of the SPI is the following: it is sensitive to the quantity of the data used. Usually, 30 years of monthly precipitation data are recommended for fitting the data. Additionally, the SPI is a meteorological tool that measures water supply but does not account for evapotranspiration. This limits its ability to capture the effect of increased temperatures (associated with climate change) on moisture demand and availability. Finally, the SPI does not consider the intensity of precipitation and how it impacts on runoff and streamflow. Overall, the SPI can provide information about anomalies in precipitation, so it needs to be used in combination with other information in order to be useful for agricultural drought assessment (NCAR, 2018).

The PDSI uses precipitation and an estimate of evaporation in conjunction with a water balance model to estimate relative soil dryness and potential evapotranspiration. The original formulation used only the temperature to estimate a potential evapotranspiration, but it is now recognized that an energy-based approach, such as the Penman–Monteith approach, is preferred (Sheffield et al., 2012; Mo and Chelliah, 2006). Since PDSI uses potential evapotranspiration and precedent (prior month) conditions, it takes into account the basic effect of global warming and is effective in determining long-term drought, especially over low and midlatitudes. Key limitations of the PDSI include that the PDSI is not as comparable across regions as the SPI and lacks the ability to handle winter-time conditions that include snowmelt and frozen precipitation, which makes its long-term monitoring problematic. Unlike SPI indices, the PDSI lacks multi-timescale features, making it difficult to correlate with specific water resources like runoff, snowpack and reservoir storage.

The SRI is based on the SPI and a model runoff. The strength of the SRI, as a runoff-based index, is that it can be used to forecast future runoff, and its predictability depends not only on climate outlooks, for which seasonal skill is generally low, but also on hydrologic initial conditions (e.g., spring snow state in the western US). The disadvantage of the SRI is similar to the disadvantage of using any modeled runoff; since modeled runoff cannot be verified everywhere, the runoff-based indices of the SRI reflect the customary uncertainties associated with model outputs (Shukla and Wood, 2008).

The USDM integrates several drought indices and professional input from all levels into a weekly operational drought-monitoring map product (Svoboda, 2000). The limitation of the USDM lies in its attempt to show drought at several temporal scales (from short-term drought to long-term drought) on one map product. Hence, the application of the DM is not for replacing any local or state information or subsequently declared drought emergencies or warnings but is rather for providing a general assessment of the current state of drought around the United States, its Pacific possessions, and Puerto Rico (Svoboda, 2000). Since the USDM relies on professional inputs from the field, it is difficult to have historical consistency (since the professionals change) or to provide forecasts.

Long-term and large-scale observations of soil moisture are scarce in the United States and elsewhere, so datasets produced by the North American Land Data Assimilation System (NLDAS) are valuable alternatives. Currently National Centers for Environmental Prediction (NCEP) offer an NLDAS drought monitor (NOAA. 2018a) based on four land surface models (LSMs): variable infiltration capacity (VIC), Noah, Mosaic and Sacramento. Sheffield et al. (2004) used simulations from the NLDAS VIC model forced with observed precipitation and near-surface meteorology to develop a drought index based on soil moisture. The approach Sheffield et al. (2004) took was to fit the VIC-simulated soil moisture to probability distributions, usually beta distributions, where the percentiles are translated to the index values that range from 0 to 1. Recent drought applications such as the VIC-based Princeton University drought and flood monitoring systems for Africa and Latin America (Sheffield et al., 2014) use the simulated soil moisture, which is mostly based on satellite precipitation (Princeton University Hydrology, 2013).

A major limitation of the indices discussed earlier, as well as of the LSM-based approaches, is a reliance on quality meteorological data. While precipitation is one of the best-observed variables, gauge observations are limited in many regions, especially in much of the developing world. Even when they are available, they are often not in near real time, preventing the computation of indices. This reveals one of the weaknesses of the above indices; their estimates rest on the availability and accuracy of the forcings, specifically precipitation (Reichle, 2017). In places such as the US, where the quality of the precipitation data is quite high, VIC quality is also relatively high (Pan et al., 2016). However, in regions with sparse networks or low accessibility, such as Africa, the VIC quality can be relatively low (Reichle, 2017). Additionally, intercomparison of the four NLDAS models showed that soil moisture differs considerably among models (Robock et al., 2000).

Heim (2002) summarizes four characteristics of a useful operational drought-monitoring system. These include the following: (1) the indices need to be available on a near-real-time basis, (2) the indices need to be monitored on a national scale, which will require the establishment of national networks for some variables, (3) complete and reliable historical data are needed over a common reference period to allow the conversion of the observations into a meaningful form (such as a percentile ranking), and (4) the data need to be adjusted to remove non-climatic influences (such as those arising from water management practices; Friedman, 1957; Heim, 2002).

An alternative approach to using model-derived soil moisture for drought detection and prediction is satellite-derived soil moisture. There are currently four major satellite-based systems that provide soil moisture products at various spatial and temporal resolutions: MetOp with the advanced scatterometer (ASCAT; Brocca et al., 2010; Wagner et al., 2013), the Advanced Microwave Scanning Radiometer AMSR2 of the Japan Aerospace Exploration Agency (JAXA; Parinussa et al., 2015; Wu et al., 2015) with the C- and X-band passive radiometers on the GCOM-W1 satellite that is a follow-on to the AMSR-E sensor, which failed on 4 October 2011 and was part of NASA's Earth Observing System, ESA's Soil Moisture Ocean Salinity (SMOS) L-band radiometer (Pan et al., 2010; Kerr et al., 2012, 2016), and NASA's Soil Moisture Active Passive (SMAP) L-band radiometer (Entekhabi et al., 2010). The radar on SMAP failed after 3 months, but soil moisture estimates based on the radiometer continue to be produced.

Of particular interest, especially for applications in parts of the globe with sparse in situ data, is to have an SMAP-based monitoring product that expresses soil moisture in terms of probability percentiles for dry (drought) or wet (pluvial) conditions (Entekhabi et al., 2010). This study presents insights and the potential of using SMAP for monitoring drought and pluvial regions with a first application over the contiguous United States (CONUS). We fit the soil moisture data from SMAP at both the level 3 5 cm passive radiometer retrievals (SPL3SMP) and the level 4 root-zone product that assimilates the surface SPL3SMP into the Catchment LSM (SPL4SMAU) to beta distributions, construct probability distributions for warm and cold seasons, and measure the reliability of our estimates. Producing soil moisture drought indices at two different soil depths allow for the monitoring of agricultural drought in different stages of development (NDMC, 2018a). This is important, firstly because grid analysis showed that values of a full column soil moisture index can be less, similar, or more than near-surface soil moisture index values. Secondly, depending on the plant development stage, the surface soil moisture or root-zone soil moisture drought index can be more useful in agricultural management. For example, surface soil moisture is important in the germination stage but is less important for managing irrigation or in estimating yields. Deficient topsoil moisture at planting may hinder germination, leading to low plant populations per hectare and a reduction of final yield (NDMC, 2018a). At the same time root-zone moisture at this early stage may not affect final yield, but as the growing season progresses, it becomes more important for plant water needs.

The rest of this paper as follows: the SMAP data are discussed in Sect. 2.1, including a determination of whether

1006 days are sufficient for estimating a drought index. Section 2.2 develops the indices by fitting beta distributions, with upper and lower bounds, to the time series and using the percentiles as the index. Section 2.3 develops a numerical analysis of the adequacy of the SMAP data. In Sect. 3, results of adequacy tests are discussed, and comparisons are made to the currently available drought indices. To help relate the percentiles to the U.S. Drought Monitor, which uses levels D0–D4 to indicate severity, the percentiles are mapped. We also extended our indices to pluvial conditions similar to the maps from the Gravity Recovery and Climate Experiment (GRACE) and Princeton University. Conclusions are brought forth in Sect. 4.

2 Data and methods

2.1 SMAP data

Since April 2015, NASA's SMAP mission has been monitoring near-surface soil moisture, mapping the globe (between 85.044° N and S) using an L-band (1.4 GHz) microwave radiometer in 2–3 days depending on location. The SMAP mission provides a set of operational global data products that include the following:

- Level 3 (SPL3SMP) is a composite based on daily passive radiometer estimates of global land surface soil moisture (nominally 5 cm) that are resampled to a global, cylindrical 36 km Equal-Area Scalable Earth Grid, Version 2.0 (EASE-Grid 2.0; O'Neill et al., 2016). For this study, Version 4 of SPL3SMP is used, which is the release version from the very beginning of the launch of SMAP. The release number changes over time. The R16 version is the latest version released in June 2018. However, in all release versions of SMAP, including Version 4, regions with permanent snow and ice, frozen ground, excessive static or transient open water in the cell, excessive radio-frequency interference (RFI) in the sensor data, and heavy vegetation (vegetation water content > 4.5 kg m^{-2}) are masked out using a passive freeze–thaw retrieval based on the normalized polarization ratio (NPR). Given the 1000 km swath and 98.5 min orbit, the SPL3SMP retrievals are spatially and temporally discontinuous, with 2–3 day gaps depending on location.

- Level 4 (SPL4SMAU) provides estimates of global surface and root-zone soil moisture by assimilating the SMAP L-band brightness temperature data (for which SPL3SMP is the gridded version) from descending and ascending half-orbit satellite passes, every 3 h from approximately 06:00 to 18:00 LST (Local Solar Time), into NASA's Catchment LSM (Reichle, 2017; Reichle et al., 2015). The SPL4SMAU data product is gridded using an Earth-fixed, global, cylindrical 9 km EASE-

Grid 2.0 projection. The LSM component of the assimilation system is driven by a forcing data stream from the global atmospheric analysis system at the NASA Global Modeling and Assimilation Office (Rienecker et al., 2008). Additional corrections are applied using gauge- and satellite-based estimates of precipitation that are downscaled to the temporal and 9 km scale of the model forcing using the disaggregation methods described in Liu et al. (2011) and Reichle et al. (2011). The SPL4SMAU product provides global soil estimates for the surface (0–5 cm) and root zone (0–100 cm) and is an effort to provide continuous, daily information without discontinuous data restrictions due to gaps in the SPL3SMP soil moisture retrievals. Nonetheless, the only product that does not use ancillary meteorological data is the SPL3SMP soil moisture retrievals.

In this study, SPL3SMP products from the 06:00 LST retrievals and SPL4SMAU products from 06:00 LST retrievals are used in the analysis of the soil moisture drought index. Our SMAP data records are from 1 April 2015 to 31 December 2017, which is equivalent to 1006 days.

The approach selected here is somewhat similar to that of Sheffield et al. (2004), where the soil moisture time series are fit to a beta distribution (with upper and lower bounds), and the distribution percentiles are the index values. There are, however, differences in our approach to that of Sheffield et al. (2004). Firstly, the basis of the data used in Sheffield et al. (2004) was simulated soil moisture from VIC, while ours is remotely sensed data. Secondly, to calculate the bounds of beta distribution [a, b], Sheffield et al. (2004) used the first (last) 10 % of the sorted soil moisture values linearly related to the empirical cumulative distribution function. In our study, this approach did not yield useful results with the estimated limits for a (b) for SMAP and often did not cover the full range of observed values, preventing interpretation of the historical data. Our methodology for obtaining beta distribution parameters a and b are discussed in this section.

As mentioned in the introduction by Heim (2002), one of the conditions for an index approach is complete and reliable historical data needed over a common reference period to allow the conversion of the observations into a meaningful form. The short SMAP record length of 1006 days, from 1 April 2015 to 31 December 2017, provides a statistical challenge in estimating the drought and pluvial indices, and thus the reliability assessments related to these extreme conditions are necessary. Therefore, to assess the data adequacy, we used a 1979–2017 VIC LSM simulation over CONUS. The VIC runs were carried out at a 4 km spatial resolution, and for the SPL3SMP comparisons, averaged up to 36 km. Here we refer to it as VIC near surface (VIC-ns). The SPL4SMAU is at 9 km spatial resolution, so VIC data were aggregated from 4 km computing grids and averaged over three soil layers with varying total soil thickness. We refer to it as VIC root zone (or VIC-rz). A statistical comparison is made between fitting a beta distribution to the VIC soil moisture values using only days when SPL3SMP soil moisture retrievals are available and fitting it to the complete 1979–2017 VIC data record. The Kolmogorov–Smirnov (KS) statistical test was used to evaluate the consistency of the beta fitted data. We made the assumption that if grids passed the consistency test using VIC data – i.e., the distributions from the SMAP-period record and the complete record were deemed the same statistically – then the SMAP time series over that grid was sufficient for providing an index. More discussion of these results is given in Sect. 3.

Furthermore, we looked at the frequency distribution of soil moisture data at each grid. The data seemed to be dominated by low soil moisture in the summertime and high soil moisture in the wintertime. Therefore, to capture this interseasonal behavior in soil moisture, we divided the record into a warm season (April–September) and a cold season (October–March). Dividing the year into warm and cold seasons enabled us to track the soil moisture dynamics, and thus the probability distribution and index, seasonally. Ideally, we would have divided it into monthly data but there are insufficient observations.

For our study period, each grid has between 144 and 329 SPL3SMP soil moisture retrievals during the warm season and from 16 to 272 retrievals during the cold season. Figure 1 shows that the number of overpasses per grid is related to the latitude, with higher latitudes having a higher number of overpasses, and to the season, with fewer values retrieved during winter, especially in the western US, due to snow cover and frozen ground. For the LSPL4SMAU root zone, there are 457 records for the cold season and 549 records for the warm season for each grid.

2.2 Fitting the beta distribution to the SMAP time series

The beta distribution is a family of continuous distributions with two shape parameters (p and q). It generalizes to a bounded distribution on the interval of $[a, b]$, where a and b usually take on the values of 0 and 1. The beta distribution is flexible enough to model a wide variety of shapes. In our study, we compared the beta distribution to several parametric distributions (including normal and Gumbel), but the beta distribution showed the best goodness of fit. Furthermore, given the bounded nature of the distribution, it is often used as the model of choice for modeling soil moisture time series (Sheffield et al., 2004). The general formula for the beta probability density function (pdf) is:

$$f(x) = \frac{(x-a)^{(p-1)}(b-x)^{(q-1)}}{B(p,q)(b-a)^{p+q-1}},$$
$$a \leq x \leq b, \quad p, q > 0, \tag{1}$$

where p and q are shape parameters, and a and b are lower and upper bounds, respectively, of the distribution. In the

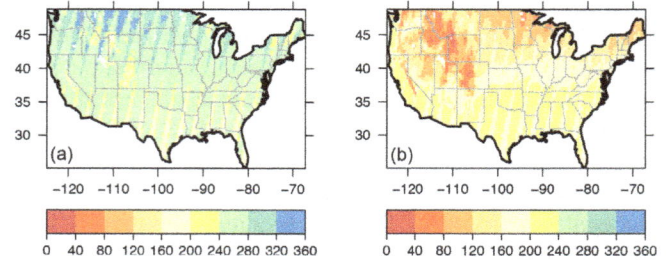

Figure 1. Number of retrievals for each season. **(a)** Warm season (1 April–30 September); **(b)** cold season (1 October–31 March).

case where $a = 0$ and $b = 1$, this becomes a standard beta distribution (NIST, 2013). $B(p, q)$ is a beta constant computed with the formula

$$B(p,q) = \int_0^1 t^{p-1}(1-t)^{q-1}\mathrm{d}t. \tag{2}$$

A main challenge is to fit the four parameters of the beta distribution, given a set of empirical observations. Sheffield et al. (2004) used the method of moments to fit the beta distribution to historical soil moisture simulations from the VIC LSM. They computed the first three moments and minimized the difference between the distribution estimates and sample estimates, since they were over-constrained. We also used the standard method of moments to calculate the parameters p and q. But for each grid location, we fit the beta distribution to six sets of data related to the SPL3SMP product: (1) short warm season VIC, (2) short warm season SMAP (1 April–30 September for 2015, 2016 and 2017; 18 months), (3) long warm season VIC (1 April–30 September, 1979–2017; 129 months), (4) short cold season VIC, (5) short cold season SMAP (1 October–31 March, 2015–2016 and 1 October–31 December 2017; 15 months) and (6) long cold season VIC (1 October–31 March for 1979 and 2016 and 1 October–31 December for 2017; 126 months), using the first and second moments $\mu = p/(p+q)$ and $CV = \mu/\sigma$, where p and q are parameters and its standard deviation is defined as

$$\sigma = \sqrt{\frac{p \times q}{(p+q)^2 \times (p+q+1)}}. \tag{3}$$

For the SPL4SMAU root-zone soil moisture product, the beta distribution was fit to the warm season and cold season using all 457 and 549 records, respectively.

Figure 2 shows the 20th percentile, average and 80th percentile soil moisture data in the warm season and cold season for the SPL3SMP 5 cm soil moisture product, and this is shown similarly in Fig. 3 for the SPL4SMAU root-zone product after data were fit to the beta distribution.

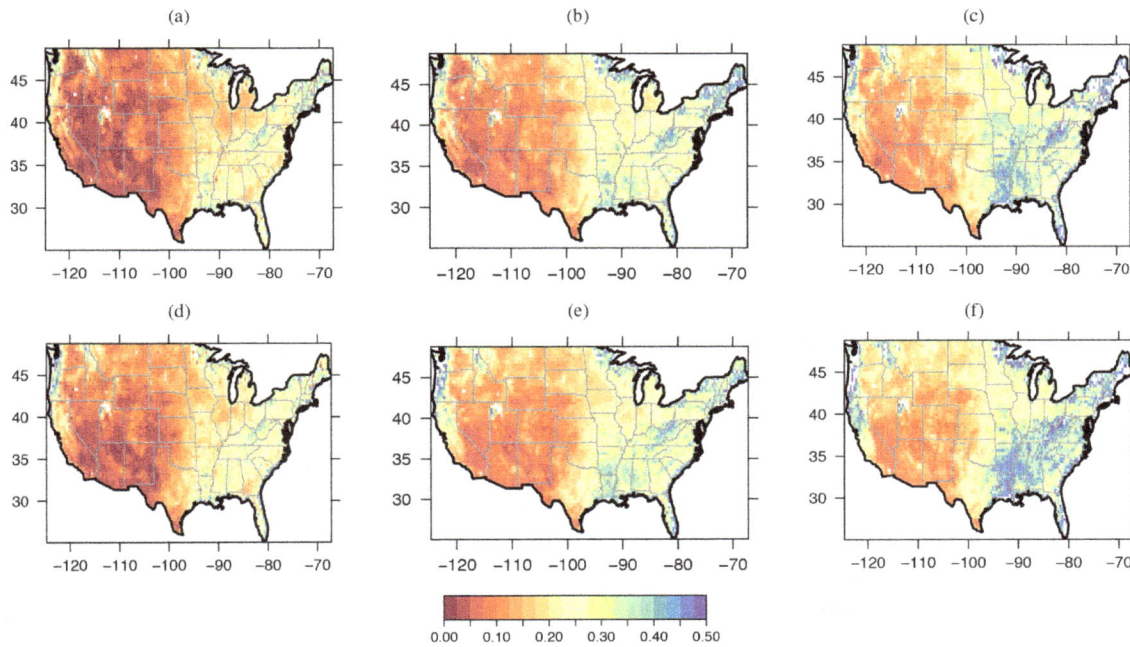

Figure 2. (a–c) SMAP soil moisture values for the warm season during summer for SPL3SMP top 5 cm soil moisture **(a)** at the 20th percentile, **(b)** at the average soil moisture, and **(c)** at the 80th percentile; **(d–f)** same as the top row, but for the cold season. Total period is from 1 April 2015 to 31 December 2017. The soil moisture unit is $m^3 m^{-3}$.

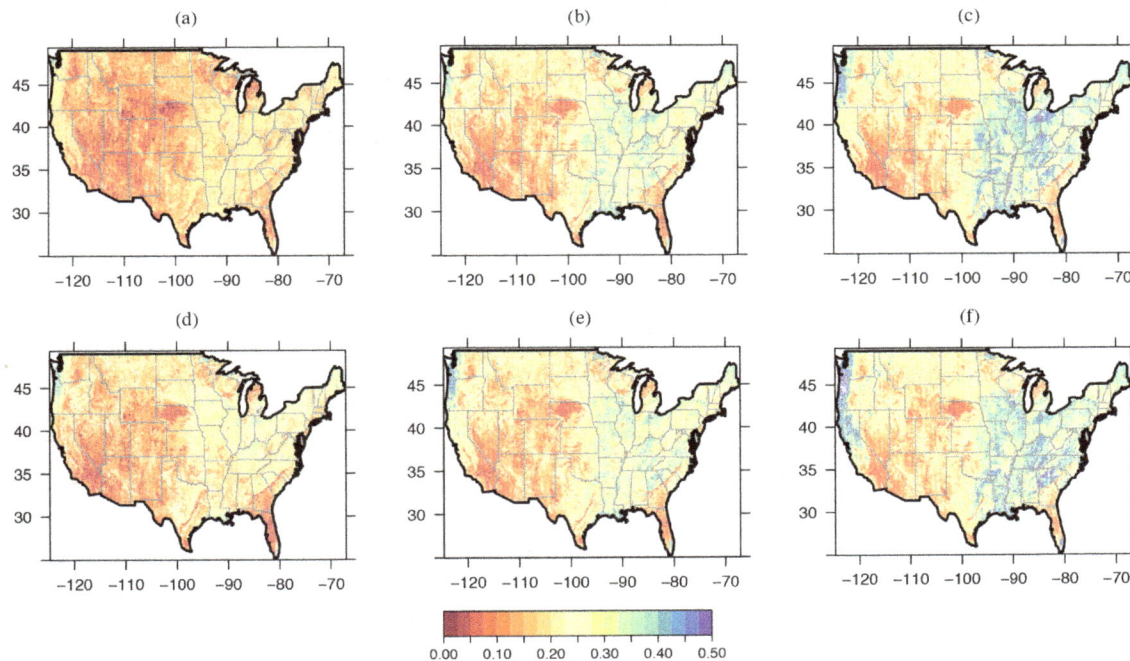

Figure 3. Same as shown in Fig. 2, but for SPL4SMAU (root-zone soil moisture).

2.3   Data adequacy filters

An insufficient SMAP record length may result in unreliable index values. To be meaningful in using short SPL3SMP data for making confident predictions, we will analyze which grids have the highest certainty in our SMAP drought index. That is, we perform adequacy analysis and define filters that separate grids with high reliability in drought monitoring and prediction from ones where we do not expect our predictions to be as accurate. We first define two filters which can separate the 5815 grids covering CONUS into grids that passed and failed quality control. The two filters are as follows:

1. the KS test for beta-fitted long-term and short-term VIC with 95 % confidence;

2. good correlation (≥ 0.4) between beta-fitted VIC and beta-fitted SPL3SMP.

Below we expand upon these two filters and then show how we used them to numerically find the best SPL3SMP filter. We also investigate if combinations of the filters are superior to the individual filters taken alone.

2.3.1 Kolmogorov–Smirnov (KS) filter

The KS test is a well-known nonparametric statistical test that compares whether two samples are coming from the same continuous distribution. We used the KS test for each grid, comparing the modeled beta distribution of the long-term VIC with the modeled beta distribution of the short-term VIC, in both warm and cold seasons. This shows if the long-term and short-term distributions are statistically indistinguishable. If this strong condition is satisfied for a grid, then it is reasonable to assume, for that grid, that the short SMAP time series would be consistent with a hypothetical long SMAP time series. The null hypothesis – that the underlying beta distribution of short-term soil moisture data is the same as the underlying beta distribution of long-term soil moisture data for VIC – is rejected for values of the KS statistic D that exceed a critical value at the 95 % significance level: $D_{\text{critical}} = \frac{1.36}{\sqrt{n}}$, where n is the number of observed variable (Lindgren, 1962).

2.3.2 Correlation filter

As mentioned earlier, one of the key assumptions of this paper is that if the beta distribution fit to the short-term VIC series is statistically consistent with beta fit to the long-term VIC time series, then we assume that the short-term beta-fitted SMAP series is consistent with the hypothetical long-term beta-fitted SMAP time series. This is possible because VIC modeled soil moisture is validated by ground measurements (Pan et al., 2016; Cai et al., 2017), and it is most plausible where the correlation between SPL3SMP and VIC is highest. Correlation maps are shown in Fig. 4 between SPL3SMP and the VIC-ns product for the warm season and cold season periods. This suggests another filter to use: require that the correlation of beta-fitted SPL3SMP and beta-fitted VIC soil moisture be relatively high. We examined the distribution of correlation values across all grids in order to pick the cutoff between high and low correlation. We chose the mean correlation, minus the standard deviation of correlation (across all grids), as a threshold. Thus grids whose correlation is close to average or better than the average pass the filter. For both the warm and cold seasons, this value was

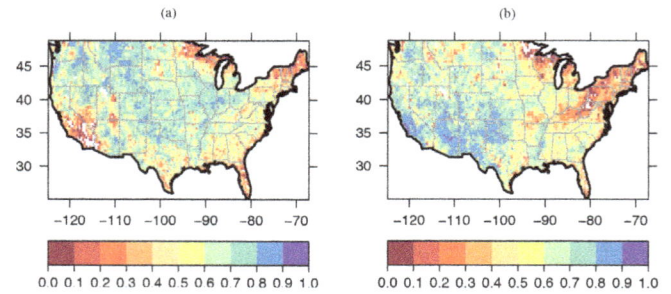

Figure 4. (a) Correlations (R) between VIC and SMAP beta models for the warm season (average $R = 0.57$) and **(b)** cold season (average $R = 0.56$). White regions signify a negative correlation.

very close to 0.4, and as a result, we picked this as the common threshold.

2.3.3 Mean distance (MD)

To evaluate whether the KS-based filter, the correlation filter or a combination of both is best, we define a simple mean distance (MD) metric. Assuming that a VIC index at 36 km resolution is the ground truth, we can calculate a distance between VIC and SMAP. For every day that SMAP provided a retrieval, if SMAP_i is the drought index percentile of grid i that passes the filter, and VIC_i is the VIC drought index percentile of the same grid, and in total n_g grids on day d passed the filter, then the MD_d is defined as the average of absolute distances between the SPL3SMP drought index percentiles and the VIC drought index percentiles. For the candidate date d and for a given filter,

$$\text{MD}_d = \frac{\sum_{i=1}^{n_g} |\text{VIC}_i - \text{SMAP}_i|}{n_g}. \tag{4}$$

In Eq. (4), VIC_i and SMAP_i are VIC and SMAP drought index values for grid i, n_g is the total number of grids that passed the filter, and MD_d is the mean distance for date d.

For each filter, the final pass and fail distance scores are calculated by averaging MD_d values over the number of days, especially for both dry or wet seasons:

$$\text{MD} = \frac{\sum_{i=1}^{n_d} |\text{MD}_d|}{n_d}, \tag{5}$$

where n_d is the total number of days for which the MD_d value is available. While n_g varies every day, since the number of overpasses varies every day, the value of n_d was constant (549 for warm season and 457 for cold season). The MD value obtained from grids that failed a filter is called MD_{fail}, and the MD value from grids that passed a filter is called MD_{pass}. For each filter a difference (Diff) was computed by reducing the MD_{pass} from the MD_{fail}: $\text{Diff} = \text{MD}_{\text{fail}} - \text{MD}_{\text{pass}} > 0$.

2.3.4 Combination filters

In addition to the KS filter and the correlation filter, we investigate two filters defined by the following combination rules:

- *Intersection filter.* A grid cell g passes the intersection filter if it passes both the KS filter *and* the correlation filter. Otherwise, it fails.

- *Union filter.* A grid cell g passes the union filter if it passes *either* filter or both filters. Note that using the union filter gives the best coverage of the grids throughout CONUS, while the intersection filter has the strongest requirements for passing.

3 Results and discussion

3.1 Data adequacy metrics

3.1.1 Correlation filter

Figure 4 shows that the average correlation for both warm and cold seasons is high and is around 0.6. During the warm season, the Central Valley and Southern California, Florida, the northeastern US, the north of Wisconsin, and Minnesota show poor correlation with VIC, at around 0.2. The extent of this poor correlation increases during the cold season for the northeastern US, Wisconsin and Minnesota. Snow season results in poor SMAP coverage during winter time in those areas. In addition, the low number of overpasses (presented in Fig. 1) during winter in the Northeast can play a role in the low amount of data and poor correlation during the cold season. Contrary to the warm season, southern California shows a high a correlation with VIC during the cold season, at around 0.9. We attribute this change from cold season to warm season in southern and southern-central California to the irrigation that SMAP picks up (Lawston et al., 2017) but VIC does not, since the version used here does not have water management effects. A land use and land cover map shows that about one-third of these areas are irrigated vegetation and another third are forests and woodlands (USGS, 2018). There are also as many as 2 million water wells in California that contribute to the irregularity of groundwater and affecting the soil moisture. They range from hand-dug, shallow wells to carefully designed large-production wells drilled to great depths (California Dept. of Water Resources, 2018). More data are needed before we can recognize further attributions to the low correlation between VIC and SMAP in that region. While systematic biases are not revealed in correlations, the temporal consistency among the time series is captured.

3.1.2 KS filter

Figure 5 shows which grids passed the 95 % KS test; there, we have confidence that the SMAP drought (pluvial) indices

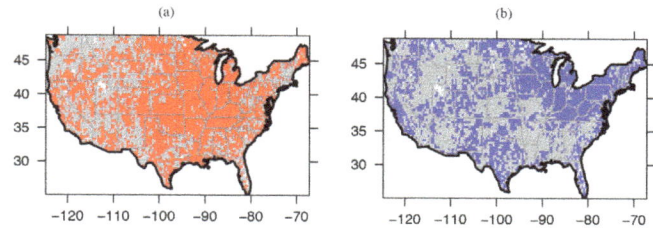

Figure 5. (a) Grids in red show areas whose short-term VIC in warm season data has the same underlying beta distribution as the long-term VIC in warm season data ($n = 3560$ or 68 % of grids are red); **(b)** the same as panel **(a)**, but for cold season period shown in blue ($n = 2927$ or 57 % of grids). Gray areas are grids where the short-term VIC does not have the same beta distribution as their long-term VIC.

provide reliable risk levels given the current period of record. The warm season shows 11 % more grids passing the adequacy test than the cold season. Note that as the record length gets extended, the above analysis needs to be repeated to see if the adequacy changes.

In the warm season, the majority of the grids whose underlying short-term and long-term beta distribution were different were in the western US. The low warm season correspondence in the Pacific Northwest (PNW) region is particularly apparent. The PNW region is covered by dense forests, mountain and heavily regulated agricultural lands by irrigation. This contributes to the fact that most grids in PNW do not pass the KS filter. A pattern of low correspondence over the major mountain areas (e.g., the Rocky Mountains, Sierra and Cascades) is also apparent, given the coarse SMAP brightness temperature (Tb) footprint and dense vegetation.

3.1.3 Combined filters

Figure 6 represents the results of correlation filter and KS filter together for both warm (panel a) and cold (panel b) seasons over all 5815 grids. We use these filters (passed and failed grids) on a daily basis for MD_d measures, though the value changes every day, depending on the number of overpasses for that date. Table 1 summarizes how many grids pass or fail each filter.

3.2 Evaluation of results under different filters

For each filter, the values of MD_d were averaged to calculate MD_{fail} and MD_{pass} for the whole CONUS over the 549 days of the warm season and 457 days of the cold season. The summary result of all four tests is shown in Tables 2 and 3. To test if having a filter is better than having no filter, for each season, we performed a two-sided null hypothesis. The tests used 95 % confidence limits between the MD of all grids – which was 22.7 in the warm season and 22.6 in the cold season – versus the MD of only passed grids. The results

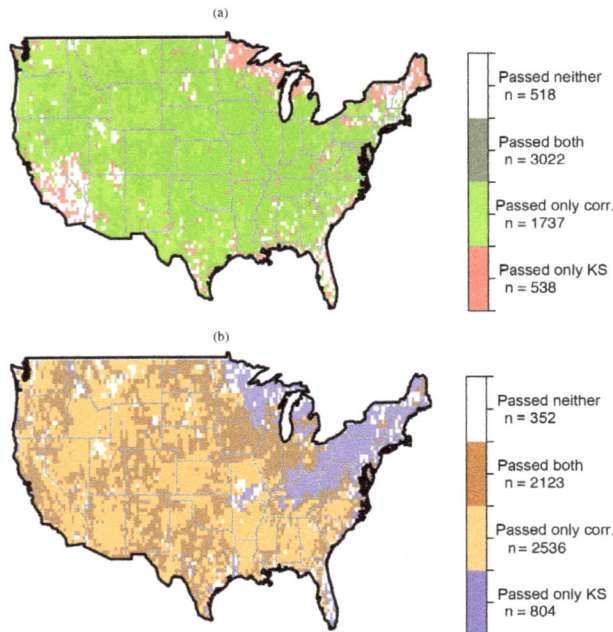

Figure 6. (a) Warm season grids that pass the correlation filter and/or the KS filter. Dark green grids include grids that pass intersection filters. **(b)** Cold season grids that pass the correlation filter and/or the KS filter. Dark orange grids include grids that pass intersection filters. In both figures, white grids show the grids that pass neither filter and will be cross-hatched in index maps.

Table 1. Number of grids, out of total 5815, that fail and pass the quality control for each filter.

n_g	KS filter	Correlation filter	Intersection filter	Union filter
Warm season fail	2255	1056	2793	518
Warm season pass	3560	4759	3022	5297
Cold season fail	2888	1156	3692	352
Cold season pass	2927	4656	2123	5463

Note: per day, the n_g numbers are less because of SMAP overpass missing grids.

showed that all four filters are significantly different than the MD of the whole CONUS. Thus, regardless of the type of the filter, having some sort of filter is better than having no filter.

In the warm season, the KS filter did better (i.e., larger Diff values or better skill in separating high- and low-performance grids) than the correlation filter for only 115 days out of 546 days, mostly in April. For almost half of the dates (260 days out of 546), the union filter did better than the correlation filter. This outperforming of the union filter occurs evenly throughout the warm season.

In the cold season, for only 48 days out of 457 days, the KS filter did better than the correlation filter, and for 198 days, the union filter did better than the correlation filter. These results suggest that for the cold season, the correlation

Table 2. Mean distance (MD) of four tests averaged over 549 days of warm season. Diff is the difference between the first and second row.

	KS filter	Correlation filter	Intersection filter	Union filter
MD_{fail}	24.1	26.5	24.5	26.8
MD_{pass}	21.9	21.9	21.1	22.3
Diff	2.2	4.5	3.4	4.5

Table 3. Mean distance (MD) of four tests averaged over 457 days of cold season. Diff is the difference between the first and second row.

	KS filter	Correlation filter	Intersection filter	Union filter
MD_{fail}	22.8	29.0	24.1	29.2
MD_{pass}	22.4	21.2	20.1	22.1
Diff	0.4	7.8	4.0	7.1

filter is providing the most effective filter. However, if we only accept the grids that pass the correlation filter, we lose 804 grids. This area involved almost all of the northeastern coast and central East Coast as well as northern Wisconsin and northeastern Minnesota. However this is not a concerning problem for drought, since for most of the cold season these areas are covered by snow. We still decided to generate a cold season filter by including the KS filter with the correlation filter, thus we used the union filter for the cold season.

Three considerations for doing so are the following:

1. *The Diff values.* The correlation-filter Diff value and union-filter Diff value during the cold season are similar and close.

2. *The nature of our tests.* It is not that surprising that the correlation filter has a higher Diff than that of the union filter. The MD metric measures how the SMAP index resembles the VIC index. Thus, we find that the most important predictor is that the SMAP values should be correlated with the VIC values.

3. *Optimum coverage.* Although the cold season East Coast drought index is not a matter of concern for this study, cold season soil moisture variability can affect warm season soil moisture and consequently agricultural drought. The goal is to create a filter that does not lose important information while providing the best knowledge of soil moisture data.

During the warm season, most of the grids that failed the test were in southern California and southern Nevada, in the Northeast (New Hampshire, Massachusetts, and Connecticut), and in the Southeast along the eastern coast of Florida. These are attributed to both the lack of correlation between

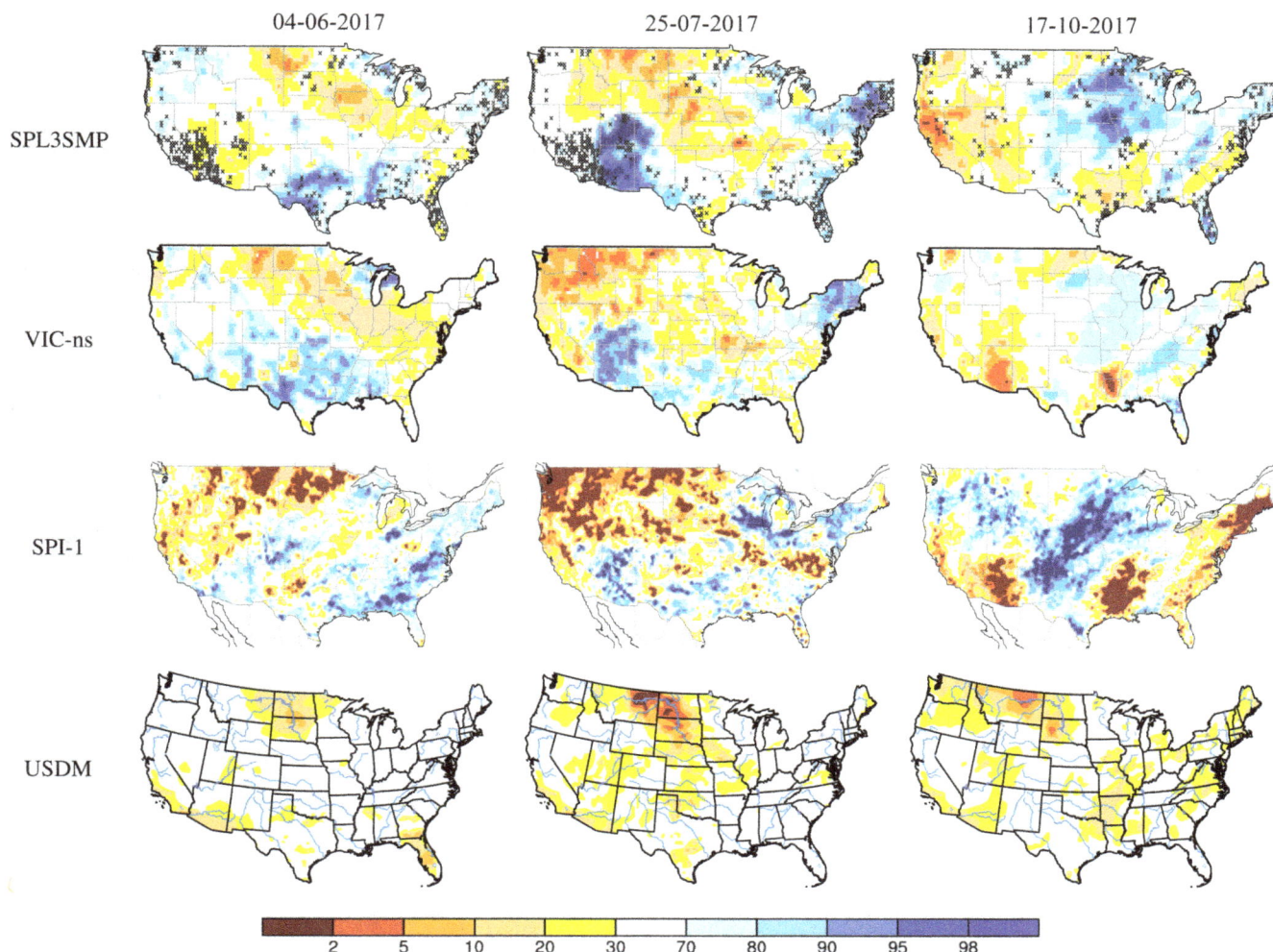

Figure 7. Comparison between SPL3SMP index map and VIC-ns, SPI-1 and USDM in 2017. The black x symbols in the SPL3SMP maps are the grids that passed neither filter and were shown as white grids in Fig. 6. For USDM, drought levels from 30 to 100 are shown in white.

SMAP and VIC and high variability between short-term and long-term soil moisture. These areas show non-stationarity in soil moisture, meaning that soil moisture distribution is subject to change over time, either due to climate or human interventions. During the cold season, most of the areas are covered using the union filter. However, as discussed, we use this filter with caution, knowing that at least according to our numerical analysis, the correlation filter did better than the union filter. The Great Lakes region, Minnesota, and the Mid-Atlantic region do not show a high correlation between VIC and SMAP in the cold season. Snow, heavy canopy and land development cause SMAP retrievals to have errors. In addition, this region does not have a good coverage of soil moisture and has a smaller number of retrievals per grid (Fig. 1). However, the KS filter complements the map by showing that the long-term and short-term VIC during cold season stays pretty stationary over time. This means that the soil mois-

ture in this area has been less subject to change during cold season at least for the past 40 years.

This information can be used to inform an interpretation of SMAP soil moisture percentiles maps based on < 10 years of data, as presented in Figs. 7 and 8 for a selection of soil moisture drought and flood indices. The grids that fail both KS and correlation tests (white grids in Fig. 6) will be flagged and are where we have the highest uncertainty of the quality of the data. This includes about 500 grids in the warm season and about 350 grids in the cold season over the CONUS.

3.3 Comparison of the drought indices

In Figs. 7 to 10, several indices are compared to the SMAP-based drought index. For the surface soil moisture index based on SPL3SMP, we provide a 3-day composite SMAP index to offer more continuous coverage. The union filter is applied to omit the grids that do not have reliable estimates.

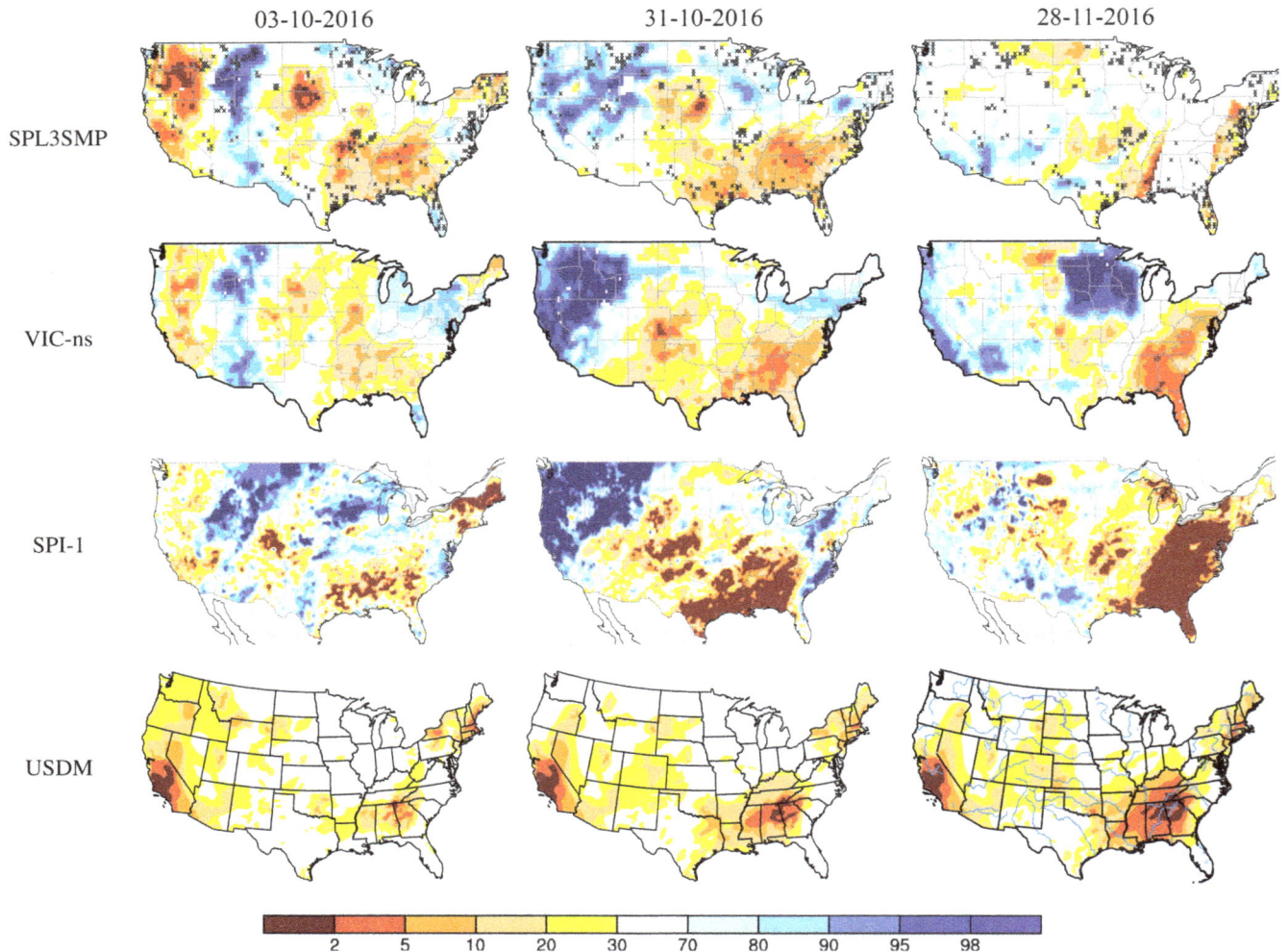

Figure 8. Comparison between SPL3SMP index map and VIC-ns, SPI-1 and USDM in 2016. The black x symbols in the SPL3SMP maps are the grids that passed neither filter and were shown as white grids in Fig. 6. For USDM, drought levels from 30 to 100 are shown in white.

Our index SPL3SMP index maps are compared with the 1-month SPI (SPI-1) index, a VIC-ns index and the USDM. For SMAP soil moisture index based on the SPL4SMAU, comparisons are made with a 3-month SPI (SPI-3) index and a GRACE satellite product. All the products except for GRACE were described in Sect. 1. GRACE is NASA's satellite system that detects small changes in the Earth's gravity field caused by the redistribution of water on and beneath the land surface. Combined with the Catchment LSM using an ensemble Kalman smoother for data assimilation (Zaitchik et al., 2008), GRACE maps root-zone soil moisture and groundwater transformed into percentiles (NDMC, 2018b).

Figures 7 and 9 show drought during the period from 4 June through 17 October 2017, for both the near surface and root zone. In this period, there was one agricultural drought event in Montana and North and South Dakota, with losses exceeding USD 1 billion across the United States

(NOAA, 2018b). The plains of eastern Montana experienced exceptional drought from July to October 2017, and in late October, drought started to end. The peak of the drought was in July 2017 when 20 % of Montana was in severe drought and 23 % of it was in moderate drought. Concurrently, 40 % of North Dakota was in extreme drought, while 70 % of the state was under some level of drought; similarly, 68 % of South Dakota was under severe drought (NOAA, 2018b). Both SPL3SMP and SPL4SMAU index maps seem to catch this drought event, although the event was more pronounced in the root zone than the surface. The maps of these two figures are also in general agreement. It is important to clarify that for 2017 period, the GRACE sensor was failing, and the resulting water storage observations were unreliable. Therefore, the last GRACE gravity field retrieval processing only goes through June 2017. Therefore, GRACE National Drought Mitigation Center (NDMC) results associated with

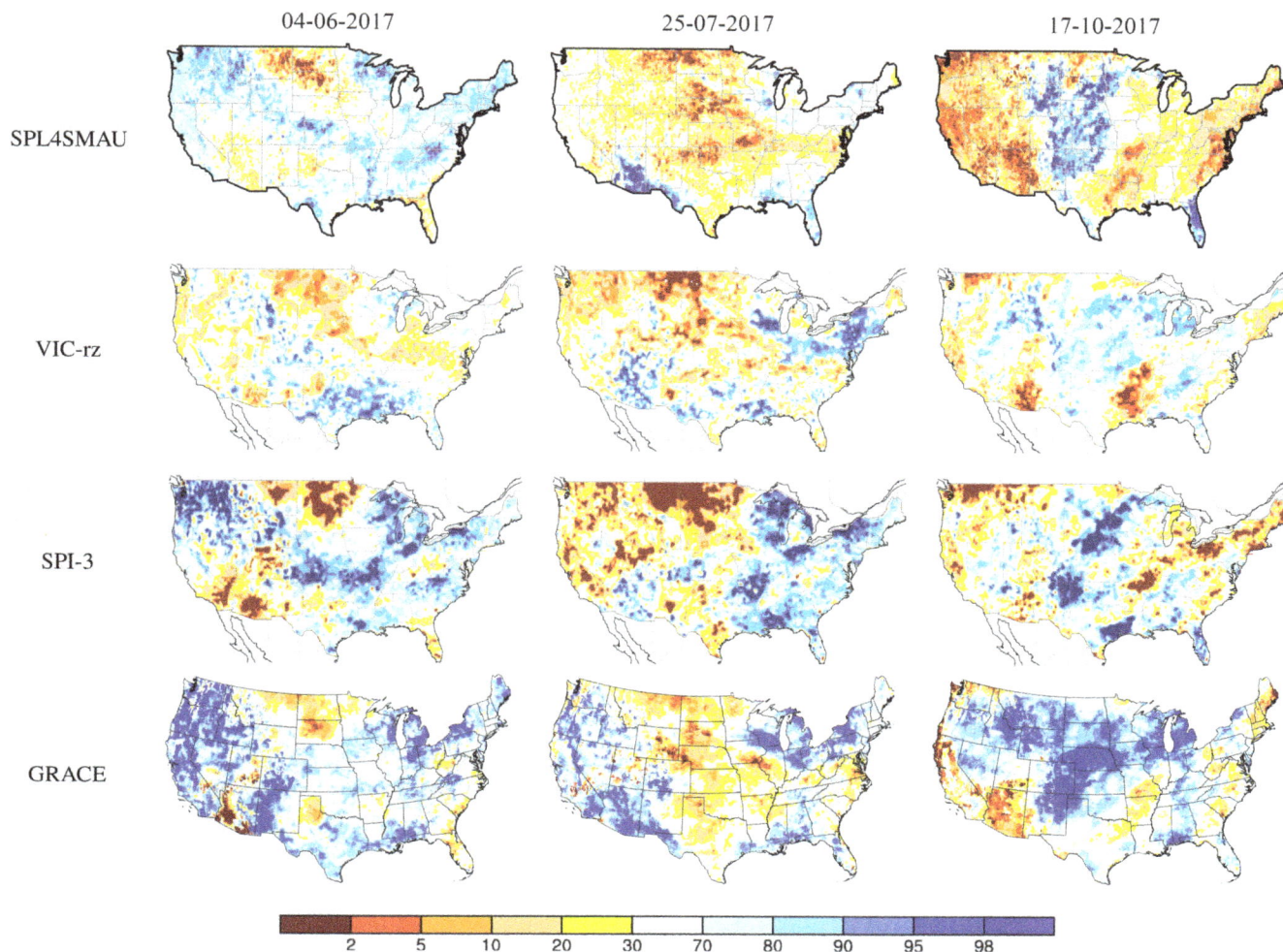

Figure 9. Comparison between SPL4SMAU index map and VIC-rz, SPI-3 and GRACE in 2017.

Fig. 9 are not consistent with other products and likely do not reflect actual GRACE observations for 2017.

In Figs. 8 and 10, drought during the period of 3 October to 8 November 2016 is shown for both near the surface and the root zone. In 2016, there were three drought events in the western, northeastern and southeastern parts of the US, which are captured by both SPL3SMP and SPL4SMAU index maps. The drought had mostly been alleviated in northern California by near-normal precipitation during the 2015–2016 winter and above normal precipitation in fall 2016. The extent that the drought persisted in Southern California after this period it is reflected in total column soil moisture rather than near-surface soil moisture (Fig. 9).

There is a high correspondence among the drought maps, particularly in the development of the drought in the southeastern US during October and November 2016. Due to heavy rainfall along the Mississippi River in November, the drought migrated eastwards. Also, by November 2016 the drought in Southern California was alleviated, which

is picked up by SPL3SMP, SPL4SMAU, VIC-ns and VIC-rz, SP-1 and 3, GRACE, and to a much lesser extent, by the USDM that showed an increasing area under drought on 28 November compared to SPL3SMP, SPL4SMAU, GRACE, or VIC-ns and VIC-rz. Additionally, for the maps that also include wetness (all except USDM), there is a high correspondence of pluvial regions (see Fig. 7).

Most of the grids where we do not have confidence in the accuracy of predictions are in Southern California and Nevada during the warm season (e.g., SPL3SMP index map on 4 June and 25 July 2017; Fig. 7). In fact, there is a visible discrepancy between SPL3SMP and VIC-ns index maps during that period in Southern California. We believe that this is due to a lack of correlation between SPL3SMP and VIC-ns in that area, since VIC does not model regulation. Human interference and the use of groundwater wells during the warm season can play a major part in what VIC models and SMAP see. For that reason, we think SMAP's metrics in the area are more accurate than those from VIC-ns.

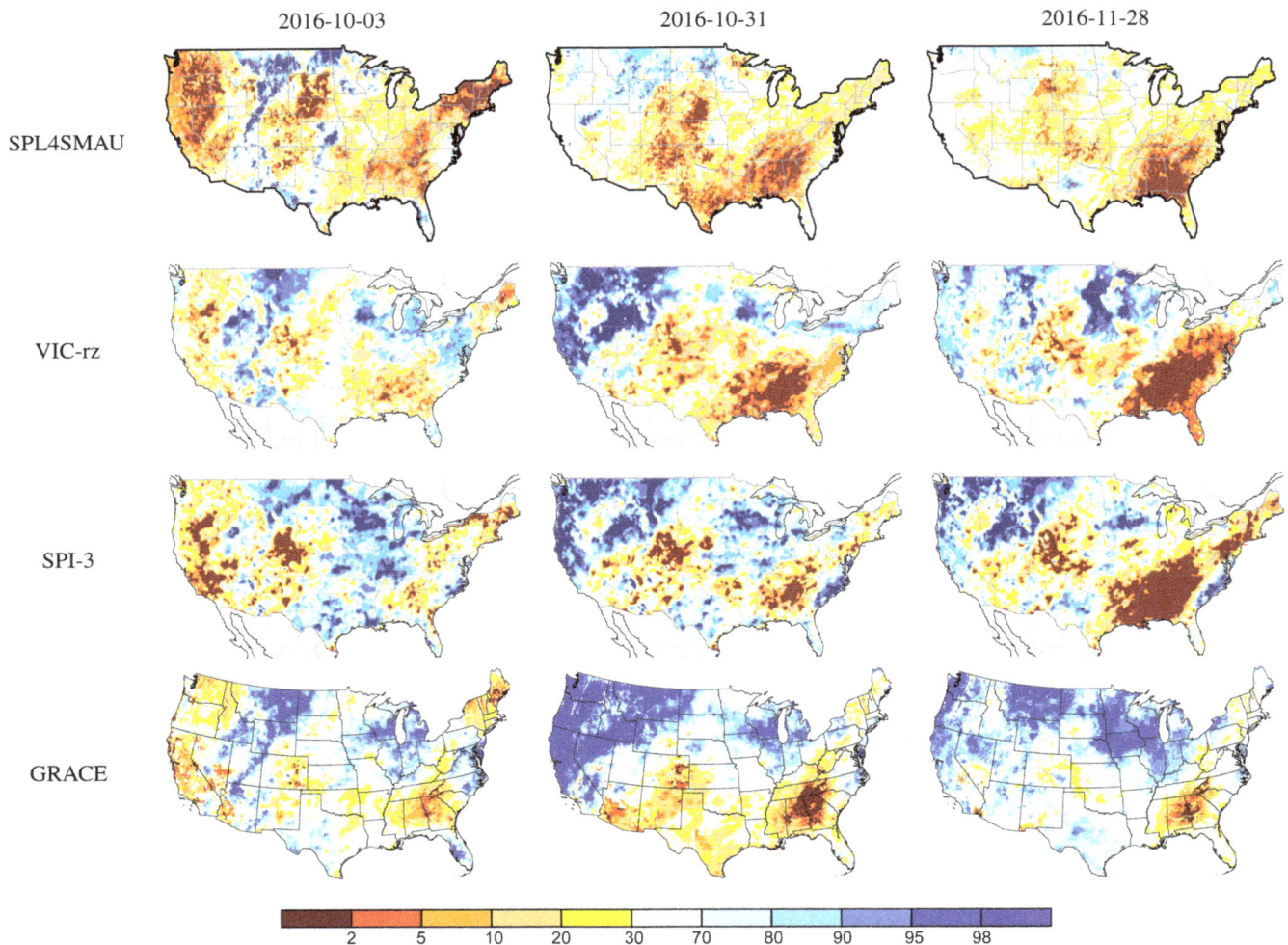

Figure 10. Comparison between SPL4SMAU index map and VIC-rz, SPI-3 and GRACE in 2016.

4   Conclusions

The drought index described in this study provides a reliable estimate of the state of drought on a daily basis for the CONUS, using SMAP. We fitted beta distributions to the SMAP data and used correlation, KS, and a combination of those two filters to numerically assess the adequacy of the short-term SMAP data for each grid cell. The areas that passed neither the KS nor correlation tests were flagged in the final SMAP drought index. These areas are grids where we have less confidence in reliable drought index estimates; they are non-stationary, and thus their soil moisture has been changing over the past 40 years. The flagged grids can be seen as an adjustment to the model to remove non-climatic influences or water management practices, although more in-depth research is needed to confirm such changes. Given the limited scope of the data, the results should be considered a demonstration of the reliability and usefulness of SMAP for

a drought-monitoring product and for implementation into an operational drought-monitoring tool.

Besides drought, SMAP can also identify regions of anomalously wet conditions that can be of great use to water and agricultural managers. Wet indices can indicate potential flood-prone conditions and regions can therefore be put on flood alerts if additional heavy rain occurs. Also, wet conditions can impact farm management, especially in the spring when sowing takes place or during the harvesting period.

Through comparing SMAP-based index maps for drought and wet conditions with other index products, we see a high similarity. Although there can be some errors at different levels, the overall evaluation reveals that SMAP-based drought products can be a viable alternative for drought monitoring in the US. This is advantageous, since SMAP is generated at a daily resolution with almost complete coverage every 3 days. This enables an observation of the effect of fluctuations in other hydrological variables, such as precipitation. In comparison, USDM, GRACE and the SPI have a low temporal resolution, which makes it difficult to study the shorter-term impacts from the other variables on soil moisture.

Both near-surface and root-zone soil moisture drought products can provide important information about the availability of soil moisture at the stage where plants develop in order to cultivate the optimum harvest. Future applications of this study can couple plant growth models with near-surface and root-zone soil moisture drought index products (NDMC, 2018a).

The soil moisture data are a culmination of all hydrological processes and represent available water from incoming precipitation and throughfall to evapotranspiration and drainage processes. The SMAP satellite provides global observations of soil moisture of unprecedented quality. Because SMAP monitors soil moisture directly and provides critical information for drought early warning, it is important that the future developments focus on drought assessment using SMAP in underrepresented parts of the world. Thus the results here provide significant support for a global SMAP drought and pluvial conditions monitoring system. Since SMAP data can be retrieved and maps can be generated in near-real time, it is very promising that a SMAP drought index product can be implemented operationally.

Data availability. The SMAP-based drought index product at the daily resolution for CONUS. The USDM and GRACE maps were provided by the National Drought Mitigation Center at the University of Nebraska–Lincoln and were downloaded from their websites, respectively. The VIC data were provided by Ming Pan from Princeton University's Terrestrial Hydrology Group.

Author contributions. SS carried out the drought index development, developed the confidence analysis and online near-real-time website, and drafted the paper. MP prepared the VIC and SMAP soil moisture data as well as SPI drought index maps. EFW conceived of the study, supervised the project and helped to draft the paper. All authors read and approved the final paper.

Competing interests. The authors declare that they have no conflict of interest.

Acknowledgements. This work was supported by NASA grant CNV1003235. This paper benefited greatly from the reviewers' comments. We thank them for their time and support.

Edited by: Nunzio Romano

References

AMS: Drought, available at: http://glossary.ametsoc.org/wiki/Drought (last access: 30 April 2018), 2012.

Brocca, L., Hasenauer, S., Lacava, T., Melone, F., Moramarco, T., Wagner, W., Dorigo, W., Matgen, P., Martinez-Fernandez, J., Llorens, P., Latron, J., Martin, C., and Bittelli, M.: Soil moisture estimation through ASCAT and AMSR-E sensors: An intercomparison and validation study across Europe, Remote Sens. Environ., 115, 3390–3408, https://doi.org/10.1016/j.rse.2011.08.003, 2010.

Cai, X., Pan, M., Chaney, N. W., Colliander, A., Misra, S., Cosh, M. H., Crow, W. T., Jackson, T. J., and Wood, E. F.: Validation of SMAP soil moisture for the SMAPVEX15 field campaign using a hyper-resolution model, Water Resour. Res., 53, 3013–3028, 2017.

California Dept. of Water Resources: Wells, available at: https://water.ca.gov/Programs/Groundwater-Management/Wells, last access: 25 April 2018.

Entekhabi, D., Rodriguez-Iturbe, I., and Castelli, F.: Mutual interaction of soil moisture state and atmospheric processes, J. Hydrol., 184, 3–17, 1996.

Entekhabi, D., Njoku, E. G., O'Neill, P. E., Kellogg, K. H., Crow, W. T., Edelstein, W. N., Entin, J. K., Goodman, S. D., Jackson, T. J., Johnson, J., Kimball, J., Piepmeier, J. R., Koster, R. D., Martin, N., McDonald, K. C., Moghaddam, M., Moran, S., Reichle, R., Shi, J. C., Spencer, M. W., Thurman, S. W., Tsang, L., and Van Zyl, J.: The Soil Moisture Active Passive (SMAP) Mission Proc., IEEE, 98, 704–716, https://doi.org/10.1109/JPROC.2010.2043918, 2010.

Friedman, D. G.: The prediction of long-continuing drought in south and southwest Texas, Occasional Papers in Meteorology, p. 182, the Travelers Weather Research Center, Hartford, CT, USA, 1957.

Heim Jr., R. R.: A review of twentieth century drought indices used in the United States, B. Am. Meteorol. Soc., 83, 1149–1165, 2002.

Howitt, R. E., Medellin-Azuara, J., MacEwan, D., Lund, J. R., and Daniel, A.: Economic Impact of the 2015 Drought on Farm Revenue and Employment Sumner, Agricultural and Resource Update, Giannini Foundation of Agricultural Economics, University of California, Davis, USA, 2015.

Kerr, Y. H., Waldteufel, P., Richaume, P., Wigneron, J. P., Ferrazzoli, P., Mahmoodi, A., Al Bitar, A., Cabot, F., Gruhier, C., Juglea, S. E., Leroux, D., Mialon, A., and Delwart, S.: The SMOS Soil Moisture Retrieval Algorithm, IEEE T. Geosci. Remote, 50, 1384–1403, https://doi.org/10.1109/TGRS.2012.2184548, 2012.

Kerr, Y. H., Al-Yaari, A., Rodriguez-Fernandez, N., Parrens, M., Molero, B., Leroux, D., Bircher, S., Mahmoodi, A., Mialon, A., Richaume, P., Delwart, S., Pellarin, A. A. B. T., Bindlish, R., Rudiger, T. J. C., Waldteufel, P., Mecklenburg, S., and Wigneron, J.: Overview of SMOS performance in terms of global soil moisture monitoring after six years in operation, Remote Sens. Environ., 180, 40–63, https://doi.org/10.1016/j.rse.2016.02.042, 2016.

Lawston, P. M., Santanello Jr., J. A., and Kumar, S. V.: Irrigation Signals Detected From SMAP Soil Moisture Retrievals, Geophys. Res. Lett., 44, 11860–11867, 2017.

Lindgren, B.: Statistical Theory, Mac-millan, New York, USA, 1962.

Liu, Q., Reichle, R. H., Bindlish, R., Cosh, M. H., Crow, W. T., de Jeu, R., Lannoy, G. J. M. D., Huffman, G. J., and Jackson, T. J.: The contributions of precipitation and soil moisture observations to the skill of soil moisture estimates in a land data assimilation system, J. Hydrometeorol., 12, 750–765, 2011.

Martinez-Fernandez, J., Gonzalez-Zamora, A., Sanchez, N., and A Gumuzzio, .: Satellite soil moisture for agricultural drought monitoring: Assessment of the SMOS derived Soil Water Deficit Index (vol. 177, pg. 277, 2016), Remote Sens. Environ., 183, 368–368, 2016.

Mo, K. C. and Chelliah, M.: The modified Palmer Drought Severity Index based on the NCEP North American Regional Reanalysis, J. Appl. Meteor. Climatol., 45, 1362–1375, 2006.

NCAR: The Climate Data Guide: Standardized Precipitation Index (SPI), available at: https://climatedataguide.ucar.edu/climate-data/standardized-precipitation-index-spi, last access: 2 April 2018.

NDMC: Types of Drought, National Drought Monitoring Center, available at: https://drought.unl.edu/Education/DroughtIn-depth/TypesofDrought.aspx, last access: 17 December 2018a.

NDMC: Groundwater and Soil Moisture Conditions from GRACE Data Assimilation, the National Drought Mitigation Center University of Nebraska-Lincoln, available at: http://nasagrace.unl.edu/Archive.aspx, last access: 17 December 2018b.

NIST: Beta Distribution, available at: http://www.itl.nist.gov/div898/handbook/eda/section3/eda366h.htm (last access: 2 April 2018), 2013.

NOAA: NLDAS Drought Monitor Soil Moisture, available at: http://www.emc.ncep.noaa.gov/mmb/nldas/drought/, last access: 5 February 2018a.

NOAA: U.S. Billion-Dollar Weather and Climate Disasters, national Centers for Environmental Information (NCEI), available at: https://www.ncdc.noaa.gov/billions, last access: 4 October 2018b.

O'Neill, P., Chan, S., Njoku, E. G., Jackson, T., and Bindlish, R.: SMAP L2 Radiometer Half-Orbit 36 km EASE-Grid Soil Moisture, Distributed Active Archive Center Version 4, NASA National Snow and Ice Data Center, Boulder, Colo., USA, https://doi.org/10.5067/XPJTJT812XFY, 2016.

Pan, M., Li, H., and Wood, E.: Assessing the skill of satellite-based precipitation estimates in hydrologic applications, Water Resour. Res., 46, W09535, https://doi.org/10.1029/2009WR008290, 2010.

Pan, M., Cai, X., Chaney, N. W., Entekhabi, D., and Wood, E. F.: An initial assessment of SMAP soil moisture retrievals using high-resolution model simulations and in situ observations, Geophys. Res. Lett., 43, 9662–9668, 2016.

Parinussa, R. M., Holmes, T. R. H., Wanders, N., Dorigo, W. A., and de Jeu, R. A. M.: A Preliminary Study toward Consistent Soil Moisture from AMSR2, J. Hydromet., 16, 932–947, https://doi.org/10.1175/JHM-D-13-0200.1, 2015.

Princeton University Hydrology: http://stream.princeton.edu/CONUSFDM/WEBPAGE/interface.php?locale=en (last access: 15 December 2018), 2013.

Reichle, R. H.: Assessment of the SMAP Level-4 Surface and Root-Zone Soil Moisture Product Using In Situ Measurements, J. Hydrometeorol., 18, 2621–2645, 2017.

Reichle, R. H., Koster, R. D., Lannoy, G. J. M. D., Forman, B. A., Liu, Q., Mahanama, S. P. P., and Toure, A.: Assessment and en-

hancement of MERRA land surface hydrology estimates, J. Climate, 24, 6322–6338, 2011.

Reichle, R. H., Lucchesi, R., Ardizzone, J. V., Kim, G., Smith, E. B., and Weiss, B. H.: Soil Moisture Active Passive (SMAP) Mission Level 4 Surface and Root Zone Soil Moisture (L4SM) Product Specification Document, Tech. Rep. 10 (Version 1.4), NASA Goddard Space Flight Center, Greenbelt, MD, USA, 2015.

Rienecker, M., Suarez, M. J., Todling, R., Bacmeister, J., Takacs, L., Liu, H.-C., Gu, W., Sienkiewicz, M., Koster, R. D., Gelaro, R., Stajner, I., and Nielsen, J. E.: The GEOS-5 Data Assimilation System – Documentation of Versions 5.0.1, 5.1.0, and 5.2.0., NASA Technical Report Series on Global Modeling and Data Assimilation NASA/TM-2008-104606, NASA, vol. 28, 101 pp., 2008.

Robock, A., Vinnikov, K., Srinivasa, G., Entin, J., Hollinger, S., Speranskaya, N., Liu, S., and Namkhai, A.: The Global Soil Moisture Data Bank, B. Am. Meteorol. Soc., 81, 1281–1299, 2000.

Sheffield, J. and Wood, E. F.: Drought: Past Problems and Future Scenarios, 978-1-84971-082-4, EarthScan, London, UK, 2011.

Sheffield, J., Goteti, G., Wen, F., and Wood, E. F.: A simulated soil moisture based drought analysis for the United States, J. Geophys. Res., 109, D24108, https://doi.org/10.1029/2004JD005182, 2004.

Sheffield, J., Livneh, B., and Wood, E. F.: Representation of terrestrial hydrology and large scale drought of the Continental U.S. from the North American Regional Reanalysis, J. Hydrometeor., 13, 856–876, https://doi.org/10.1175/JHM-D-11-065.1, 2012.

Sheffield, J., Wood, E. F., Chaney, N., Guan, K., Sadri, S., Yuan, X., Olang, L., Amani, A., Ali, A., and Demuth, S.: A Drought Monitoring and Forecasting System for Sub-Sahara African Water Resources and Food Security, B. Am. Meteorol. Soc., 95, 861–882, 2014.

Shukla, S. and Wood, A. W.: Use of a standardized runoff index for characterizing hydrologic drought, Geophys. Res. Lett., 35, 1–7, 2008.

Svoboda, M.: An introduction to the Drought Monitor, Drought Network News, 12, 15–20, 2000.

Tallaksen, T. and Van Lanen, H. A.: Hydrological Drought, Processes and Estimation Methods for Streamflow and Groundwater, vol. 48, Elsevier Science, Amsterdam, the Netherlands, 2004.

UN/ISDR: Drought Risk Reduction Framework and Practices: Contributing to the Implementation of the Hyogo Framework for Action. United Nations Secretariat of the International Strategy for Disaster Reduction (UN/ISDR), Tech. Rep. 98+vi pp., United Nations Secretariat of the International Strategy for Disaster Reduction (UN/ISDR), Geneva, Switzerland, 2007.

USGS: National Gap Analysis Program, Land Cover Data Viewer,, available at: https://goo.gl/rntijg, last access: 30 April 2018.

Wagner, W., Hahn, S., Kidd, R., Melzer, T., Bartalis, Z., Hasenauer, S., Figa-Saldaña, J., de Rosnay, P., Jann, A., Schneider, S., Komma, J., Kubu, G., Brugger, K., Aubrecht, C., Züger, J., Gangkofner, U., Kienberger, S., Brocca, L., Wang, Y., Blöschl, G., Eitzinger, J., and Steinnocher, K.: The ASCAT Soil Moisture Product: A Review of its Specifications, Validation Results, and Emerging Applications, Meteorol. Z., 22, 5–33, https://doi.org/10.1127/0941-2948/2013/0399, 2013.

Wilhite, D. A.: Drought as a Natural Hazard: Concepts and Definitions, chap. 1, vol. I, National Drought Mitigation Center at

Digital Commons at University of Nebraska, Lincoln, Routledge, London, UK, 2000.

Wilhite, D. A. and Glantz, M. H.: Understanding the Drought Phenomenon: The Role of Definitions, Water Int., 10, 111–120, 1985.

Wu, Q., Liu, H., Wang, L., and Deng, C.: Evaluation of AMSR2 soil moisture products over the contiguous United States using in situ data from the International Soil Moisture Network, Int. J. Appl. Earth Obs., 45, 187–199, https://doi.org/10.1016/j.jag.2015.10.011, 2015.

Zaitchik, B. F., Rodell, M., and Reichle, R. H.: Assimilation of GRACE Terrestrial Water Storage Data into a Land Surface Model: Results for the Mississippi River Basin, Am. Meteorol. Soc., 9, 535–548, 2008.

14

Estimation of crop water requirements: extending the one-step approach to dual crop coefficients

J. P. Lhomme[1], **N. Boudhina**[1,2], **M. M. Masmoudi**[2], **and A. Chehbouni**[3]

[1]Institut de Recherche pour le Développement (UMR LISAH), 2 Place Viala, 34060 Montpellier, France
[2]Institut National Agronomique de Tunisie (INAT), 43 Avenue Charles Nicolle, 1082 Tunis, Tunisia
[3]Institut de Recherche pour le Développement (UMR CESBIO), 18 Avenue Edouard Belin, 31401 Toulouse, France

Correspondence to: J. P. Lhomme (jean-paul.lhomme@ird.fr)

Abstract. Crop water requirements are commonly estimated with the FAO-56 methodology based upon a two-step approach: first a reference evapotranspiration (ET_0) is calculated from weather variables with the Penman–Monteith equation, then ET_0 is multiplied by a tabulated crop-specific coefficient (K_c) to determine the water requirement (ET_c) of a given crop under standard conditions. This method has been challenged to the benefit of a one-step approach, where crop evapotranspiration is directly calculated from a Penman–Monteith equation, its surface resistance replacing the crop coefficient. Whereas the transformation of the two-step approach into a one-step approach has been well documented when a single crop coefficient (K_c) is used, the case of dual crop coefficients (K_{cb} for the crop and K_e for the soil) has not been treated yet. The present paper examines this specific case. Using a full two-layer model as a reference, it is shown that the FAO-56 dual crop coefficient approach can be translated into a one-step approach based upon a modified combination equation. This equation has the basic form of the Penman–Monteith equation but its surface resistance is calculated as the parallel sum of a foliage resistance (replacing K_{cb}) and a soil surface resistance (replacing K_e). We also show that the foliage resistance, which depends on leaf stomatal resistance and leaf area, can be inferred from the basal crop coefficient (K_{cb}) in a way similar to the Matt–Shuttleworth method.

1 Introduction

The well-known FAO-56 publication on crop evapotranspiration (Allen et al., 1998) is the outcome of a revision project

concerning a previous publication (FAO-24) on the same subject (Doorenbos and Pruitt, 1977). In FAO-56 the current guidelines for computing crop water requirements are presented. Two different ways of calculating crop evapotranspiration are retained and detailed: the single crop coefficient and the dual crop coefficient. In the single crop coefficient approach, crop evapotranspiration under standard conditions is calculated as

$$ET_c = K_c ET_0. \qquad (1)$$

ET_0 is the reference crop evapotranspiration determined from the Penman–Monteith equation and accounts for weather conditions. K_c is the crop coefficient, in which crop characteristics are incorporated and which is supposed to be largely independent of weather characteristics, enabling its transfer from one location to another. In the dual crop coefficient approach, K_c is split into two separate coefficients: one represents crop transpiration K_{cb} (it is called basal crop coefficient) and the other soil evaporation K_e. Thus, crop evapotranspiration under standard conditions is calculated as

$$ET_c = (K_{cb} + K_e) ET_0. \qquad (2)$$

Whereas the values of K_{cb} are tabulated in FAO-56 and easily accessible, those of K_e are the result of a relatively complex and mainly empirical procedure summarized in Appendix A (Allen et al., 1998; Allen, 2000). The basal crop coefficient

K_{cb} is a characteristic value of a given crop, obtained under standard conditions and transferable as such, whereas the value of K_e should be adjusted to the specific conditions under which the crop is grown.

The FAO-56 methodology (single or dual crop coefficients) is commonly called the two-step approach (Shuttleworth, 2007) because ET_0 is first calculated from weather variables and then empirically adjusted using crop-specific coefficients. The empirical character of the FAO methodology has been criticized by many authors for various reasons (Wallace, 1995). Firstly, if crop coefficients mainly depend on crop characteristics, they also vary somewhat with weather variables. This means that transferring their values into locations where weather conditions significantly differ from those under which they were initially determined is risky (Katerji and Rana, 2014). FAO-56 specifies that the tabulated values of crop coefficients are those corresponding to a sub-humid climate and should be modified for more humid or arid conditions according to an empirical formula. Secondly, the origins of K_c–K_{cb} values proposed in FAO-56 are not completely clear: they sometimes appear as a compromise between contradictory data, which makes them subject to caution (Doorenbos and Pruitt, 1977; Shuttleworth and Wallace, 2009; Katerji and Rana, 2014). Thirdly, the relatively complex and mainly empirical procedure to determine the soil evaporation coefficient K_e is another serious issue (Rosa et al., 2012).

Consequently, many authors (e.g. Shuttleworth, 2007) have suggested that a better approach would consist in estimating ET_c as ET_0: i.e. directly by means of the Penman–Monteith equation (Eq. 3), in which the canopy surface resistance (r_s) of a specific crop would play the same role as the crop coefficient K_c.

$$ET_c = \frac{1}{\lambda} \frac{\Delta(R_n - G) + \rho c_p D_a / r_a}{\Delta + \gamma \left(1 + \frac{r_s}{r_a}\right)}. \tag{3}$$

The significance of each variable in Eq. (3) is given in the list of symbols (Table A1). This method is often called the one-step approach, compared to the FAO-56 two-step approach. Shuttleworth (2006) provided a theoretical background, called the Matt–Shuttleworth approach, to transform the currently available crop coefficients (K_c) into effective surface resistances (r_s) to be used with the Penman–Monteith equation. This method, which in principle only applies to the single crop coefficient approach, has been thoroughly examined and discussed by Lhomme et al. (2014) and Shuttleworth (2014).

Given that the familiar Penman–Monteith equation (Eq. 3) is only relevant when soil evaporation is negligible, the problem which arises from a theoretical standpoint is that the dual coefficient of the two-step approach (Eq. 2), which accounts for crop transpiration and soil evaporation, cannot be translated into the one-step approach. A physical model equivalent to the dual coefficient approach would be the one-dimensional two-source model designed for sparse crops by Shuttleworth and Wallace (1985) and revisited by Lhomme et al. (2012). Unfortunately, from an operational standpoint, the practical implementation of this two-source model can be hindered by its mathematical formalism, which is far more complex than the common Penman–Monteith equation. Following the idea of Wallace (1995), who stated that "the key to continued improvement in evaporation modelling is to attempt to simplify these complex schemes while still retaining their essential elements as far as possible", the article aims at showing that the two-source model of evaporation can be transformed into a Penman–Monteith type equation, where foliage transpiration resistance and soil evaporation resistance are included within a bulk surface resistance. Then, it will be shown that the transpiration resistance can be inferred from the basal crop coefficient of the dual approach in a way similar to the Matt–Shuttleworth approach. Numerical simulations will be performed to illustrate the advantages of this new form of the Penman–Monteith equation to estimate crop water requirements with a one-step approach.

2 Theoretical background

2.1 A generalized form of the Penman–Monteith equation

The so-called Penman–Monteith equation (Monteith, 1963, 1965) results from the combination of the convective fluxes emanating from the canopy with the energy balance. Introducing effective resistances within and above the canopy, the convective fluxes of sensible heat (H) and latent heat (λE) can be written in the following way:

$$H = \rho c_p \left(\frac{T_c - T_a}{r_a + r_{a,h}}\right), \tag{4}$$

$$\lambda E = \left(\frac{\rho c_p}{\gamma}\right) \left[\frac{e^*(T_c) - e_a}{r_a + r_{c,v}}\right]. \tag{5}$$

T_a and e_a represent the temperature and the vapour pressure at a reference height (z_r) above the canopy; T_c is the effective temperature of the canopy and $e^*(T_c)$ is the saturated vapour pressure at temperature T_c (the poor definition of T_c is not a key issue since it is eliminated in the final combination equation); $r_{c,v}$ is the effective canopy resistance for water vapour (which includes air and surface resistances within the canopy) and $r_{a,h}$ is that for sensible heat (which includes only air resistances). Both resistances should be logically added to the aerodynamic resistance above the canopy (r_a) calculated between the mean source height (z_m) and the reference height (z_r). In the common Penman–Monteith equation, the air resistances within the canopy ($r_{a,h}$ or the air component of $r_{c,v}$) are neglected or assumed to be incorporated into the aerodynamic resistance r_a. The combination of Eqs. (4) and (5) with the energy balance equation

$(R_n - G = H + \lambda E)$ results in the following equation:

$$\lambda E = \frac{\Delta(R_n - G) + \rho c_p D_a/(r_a + r_{a,h})}{\Delta + \gamma \left(\frac{r_a + r_{c,v}}{r_a + r_{a,h}} \right)}, \tag{6}$$

where D_a is the vapour pressure deficit at reference height and Δ is the slope of the saturated vapour pressure curve at air temperature.

As thoroughly explained in Lhomme et al. (2012, Sect. 4), the within-canopy resistances ($r_{a,h}$ and $r_{c,v}$) can be interpreted using a two-layer representation of canopy evaporation, which takes into account foliage and soil contributions, as visualized in Fig. 1. From a theoretical standpoint, these effective resistances should be calculated as the parallel sum of the component resistances expressed per unit area of land surface: $r_{a,h}$ is the parallel sum of $r_{a,f,h}$ (bulk boundary-layer resistance of the foliage for sensible heat) and $r_{a,s}$ (air resistance between the substrate and the canopy source height); $r_{c,v}$ is the parallel sum of $r_{s,f} + r_{a,f,v}$ and $r_{s,s} + r_{a,s}$ with $r_{s,f}$ the bulk stomatal resistance of the foliage, $r_{s,s}$ the substrate resistance to evaporation and $r_{a,f,v}$ the bulk boundary-layer resistance of the foliage for water vapour. Applying these formulations, however, does not allow the bulk canopy resistance for water vapour ($r_{c,v}$) to be separated into two resistances in series: one for the air and the other for the surface. Consequently, the simple ratio of a surface resistance to an air resistance cannot appear in the denominator of Eq. (6), as in the common formalism of the Penman–Monteith equation (Eq. 3). Yet, this simple ratio is very convenient and useful from an operational standpoint because it allows separating the biological component of the canopy (r_s) from the aerodynamic one (r_a). Nevertheless, this simple ratio and the common form of the Penman–Monteith equation can be retrieved from its generalized form (Eq. 6) by means of a simple assumption, which consists in splitting the effective canopy resistance for water vapour ($r_{c,v}$) into two bulk resistances put in series: one representing the transfer through the surface components ($r_{s,v}$) and the other the transfer in the air within the canopy ($r_{a,v}$):

$$r_{c,v} = r_{s,v} + r_{a,v}. \tag{7}$$

This procedure is not sound from a strict physical standpoint, but the numerical simulations performed below will show that it constitutes a fairly good approximation. Assuming the component resistances within the canopy that act as parallel resistors and the bulk boundary-layer resistances of the foliage for sensible heat and water vapour to be equal ($r_{a,f,h} = r_{a,f,v} = r_{a,f}$), the bulk air and surface resistances can be expressed as the parallel sum of two component resis-

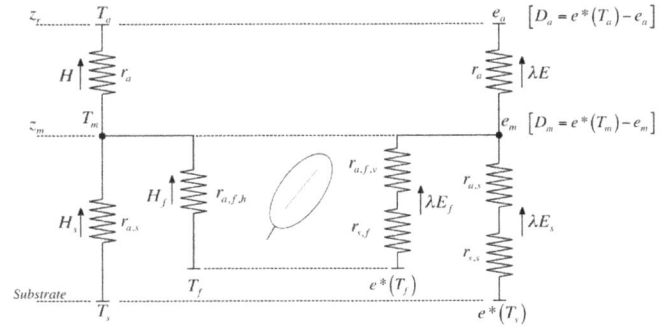

Figure 1. Resistance networks and potentials for a two-layer representation of the convective fluxes (sensible heat and latent heat) within the canopy. The nomenclature used is given in the list of symbols.

tances (see Fig. 1):

$$\frac{1}{r_{a,v}} = \frac{1}{r_{a,h}} = \frac{1}{r_{a,f}} + \frac{1}{r_{a,s}}, \tag{8}$$

$$\frac{1}{r_{s,v}} = \frac{1}{r_{s,f}} + \frac{1}{r_{s,s}}. \tag{9}$$

Consequently, Eq. (6) can be rewritten in a simpler way as

$$\lambda E = \frac{\Delta(R_n - G) + \rho c_p D_a/(r_a + r_{a,h})}{\Delta + \gamma \left(1 + \frac{r_{s,v}}{r_a + r_{a,h}} \right)}. \tag{10}$$

This expression is similar to the traditional Penman–Monteith equation and its surface resistance expressed by Eq. (9) takes into account both foliage transpiration ($r_{s,f}$) and soil surface evaporation ($r_{s,s}$). Equation (10), therefore, can be considered in the one-step approach as a realistic substitute of Eq. (2) in the two-step approach. When all the air resistances within the canopy are neglected (they are generally much smaller than the surface resistances), $r_{a,h} = 0$ and Eq. (10) adopts strictly the same form as the original Penman–Monteith equation.

2.2 Expressing the component resistances

The soil surface resistance ($r_{s,s}$) has a clear mathematical definition based on the inversion of the equation representing the latent heat flux (λE_s) emanating from the soil surface (see Fig. 1):

$$r_{s,s} = \left(\frac{\rho c_p}{\gamma} \right) \frac{\left[e^*(T_s) - e_s \right]}{\lambda E_s}, \tag{11}$$

where e_s is the vapour pressure at the soil surface, the other quantities are defined in the list of symbols. Its calculation, however, is rather challenging. Many parameterizations have been proposed in the literature in the form of empirical functions of near-surface soil moisture (e.g. Mahfouf and Noilhan, 1991; Sellers at al., 1992). But this issue is considered to be out of the scope of the present paper. Because of the stomatal characteristics of the leaves (amphi- vs. hypostomatous), the formulation of foliage resistance can be a little bit tricky and this point has been thoroughly examined by Lhomme et al. (2012). For the sake of convenience, denoting by $r_{s,l}$ the mean two-sided stomatal resistance of the leaves (per unit area of leaf), the bulk surface resistance of the foliage can be simply expressed as

$$\frac{1}{r_{s,f}} = \frac{LAI}{r_{s,l}}, \tag{12}$$

and the bulk boundary-layer resistance of the foliage (for sensible heat and water vapour) is expressed similarly

$$\frac{1}{r_{a,f}} = \frac{LAI}{r_{a,l}}, \tag{13}$$

where $r_{a,l}$ is the leaf boundary layer per unit area of two-sided leaf, calculated by Eq. (B2) in Appendix B. The air resistance between the substrate and the canopy source height ($r_{a,s}$) is given by Eq. (B1) in the same appendix.

According to FAO-56, the aerodynamic resistance above the canopy (r_a) is generally calculated in neutral conditions, without stability correction functions, which is justified by the fact that the sensible heat flux is generally low under standard conditions (no water stress). It is expressed as a simple function of wind speed u_a at reference height z_r:

$$r_a = \left(\frac{1}{k^2 u_a}\right) \ln\left(\frac{z_r - d}{z_{0,m}}\right) \ln\left(\frac{z_r - d}{z_{0,h}}\right), \tag{14}$$

where $d = 0.66 z_h$, $z_{0,m} = 0.12 z_h$, $z_{0,h} = z_{0,m}/10$ (z_h: canopy height) and k is von Karman's constant (Allen et al., 1998). However, given that the canopy roughness length for scalar $z_{0,h}$ is supposed to play the same role as the additional air resistance $r_{a,h}$ appearing in Eq. (10), i.e. accounting for the transfer of sensible and latent heat in the air within the canopy, it would certainly be more judicious to replace $z_{0,h}$ by $z_{0,m}$ in Eq. (14), at least when the Penman–Monteith equation is interpreted in the framework of a two-layer model. It is interesting to note also that the resistance $r_{a,h}$ can be translated into a modified roughness length for scalar $z'_{0,h}$ by writing the air resistance ($r_a + r_{a,h}$) in Eq. (10) in two different forms: one containing the modified roughness length and the other the additional air resistance:

$$\left(\frac{1}{k^2 u_a}\right) \ln\left(\frac{z_r - d}{z_{0,m}}\right) \ln\left(\frac{z_r - d}{z'_{0,h}}\right)$$
$$= \left(\frac{1}{k^2 u_a}\right) \ln^2\left(\frac{z_r - d}{z_{0,m}}\right) + r_{a,h}. \tag{15}$$

Extracting $z'_{0,h}$ from this equation leads to

$$z'_{0,h} = z_{0,m} \exp\left[-\frac{k^2 u_a r_{a,h}}{\ln\left(\frac{z_r - d}{z_{0,m}}\right)}\right]. \tag{16}$$

Consequently, Eq. (10) with $r_{a,h}$ added to r_a can be replaced by the same equation where $r_{a,h} = 0$ but where r_a is calculated by Eq. (14), $z'_{0,h}$ replacing $z_{0,h}$. This parameter will be numerically explored below.

3 The Matt–Shuttleworth approach extended to dual crop coefficients

Similarly to the Matt–Shuttleworth method developed for a single crop coefficient (Shuttleworth, 2006), the problem to tackle now is to infer the values of both surface resistances ($r_{s,f}$ and $r_{s,s}$), which govern respectively foliage and substrate evaporation, from those of crop coefficients (K_{cb} and K_e). As already stated, K_{cb} is a characteristic value of a given crop, tabulated and transferable, whereas K_e is a soil parameter adjustable to the specific conditions under which the crop is grown. Therefore, it is not really relevant to retrieve the soil surface resistance ($r_{s,s}$) from K_e. Nevertheless, the mathematical development being similar, it will be made for both resistances. But first, the issue of the reference height will be recalled.

3.1 Inferring weather variables at a higher level

Given that many crops have a crop height close to (or greater than) the reference height of 2 m, the weather variables involved in the Penman–Monteith equation should be taken at a higher level than the reference height. This point is thoroughly developed in the Matt–Shuttleworth method, where it is suggested that air characteristics be taken at a blending height arbitrarily set at $z_b = 50$ m (Shuttleworth, 2006). Wind speed (u_b) at this height can be inferred from the one (u_a) at reference height (z_r) by means of the following equation based on the log-profile relationship:

$$u_b = u_a \frac{\ln\left(\frac{z_b - d_0}{z_{0m,0}}\right)}{\ln\left(\frac{z_r - d_0}{z_{0m,0}}\right)}, \tag{17}$$

where d_0 is the zero plane displacement height of the reference crop and $z_{0m,0}$ its roughness length for momentum. Similarly, the water vapour pressure deficit at blending height (D_b) can be expressed as a function of the one at reference height (D_a) by

$$D_b = \left(D_a + \frac{\Delta A_0 r_{a,0}}{\rho c_p}\right) \left[\frac{(\Delta + \gamma) r_{a,0,b} + \gamma r_{s,0}}{(\Delta + \gamma) r_{a,0} + \gamma r_{s,0}}\right]$$
$$- \frac{\Delta A_0 r_{a,0,b}}{\rho c_p}, \tag{18}$$

where $A_0 = R_{n,0} - G_0$ is the available energy of the reference crop, $r_{s,0}$ its surface resistance, $r_{a,0}$ the aerodynamic resistance between the reference crop and the reference height, $r_{a,0,b}$ the aerodynamic resistance between the reference crop and the blending height, and Δ calculated at the reference temperature T_a (Lhomme et al., 2014, Eq. 5).

3.2 Retrieving the component surface resistances from crop coefficients

Canopy evapotranspiration is the sum of foliage evaporation (ET_f) and soil surface evaporation (ET_s):

$$ET_c = (K_{cb} + K_e) ET_0 = ET_f + ET_s. \tag{19}$$

The retrieval of surface resistances is obtained by expressing the two component evaporations as a function of their respective surface resistance. In the two-layer representation (Fig. 1), the component evaporations are expressed as a function of the saturation deficit (D_m) at canopy source height ($z_m = d + z_{0,m}$) and the radiation load of each component ($R_{n,f}$ for the foliage and $R_{n,s}$ for the soil surface):

$$ET_f = \frac{1}{\lambda} \cdot \frac{\Delta R_{n,f} + \rho c_p D_m / r_{a,f}}{\Delta + \gamma \left(1 + \frac{r_{s,f}}{r_{a,f}}\right)}, \tag{20}$$

$$ET_s = \frac{1}{\lambda} \cdot \frac{\Delta (R_{n,s} - G) + \rho c_p D_m / r_{a,s}}{\Delta + \gamma \left(1 + \frac{r_{s,s}}{r_{a,s}}\right)}. \tag{21}$$

The saturation deficit at canopy source height can be inferred from the one at reference height (D_a) by means of the following relationship (Shuttleworth and Wallace, 1985, Eq. 8; Lhomme et al., 2012, Eq. 7):

$$D_m = D_a + \frac{[\Delta (R_n - G) - \lambda ET_c (\Delta + \gamma)] r_a}{\rho c_p}. \tag{22}$$

In fact D_a and the corresponding aerodynamic resistance r_a should be preferably replaced by those calculated at the blending height, as discussed above. Following Shuttleworth (2006), the parameter $f = R_n / R_{n,0}$ is introduced to allow for differences in net radiation between the considered crop and the reference crop. Beer's law is used to distribute the net radiation within the canopy as a function of the leaf area index (Eqs. C5 and C6 in Appendix C).

The two surface resistances ($r_{s,f}$ and $r_{s,s}$) can be retrieved from the coefficients K_{cb} and K_e by simply equating Eq. (20) with $K_{cb} ET_0$ and Eq. (21) with $K_e ET_0$, in a way similar to the Matt–Shuttleworth approach (Shuttleworth, 2006). This leads to

$$r_{s,f} = r_{a,f} \left(\frac{\Delta}{\gamma} + 1\right) \left[\frac{(\Delta/\gamma) R_{n,f} + \frac{\rho c_p D_m}{\gamma r_{a,f}}}{(\Delta/\gamma + 1) K_{cb} \lambda ET_0} - 1\right], \tag{23}$$

$$r_{s,s} = r_{a,s} \left(\frac{\Delta}{\gamma} + 1\right) \left[\frac{(\Delta/\gamma)(R_{n,s} - G) + \frac{\rho c_p D_m}{\gamma r_{a,s}}}{(\Delta/\gamma + 1) K_e \lambda ET_0} - 1\right]. \tag{24}$$

Reference crop evapotranspiration ET_0 is calculated as usual (Eq. 3): the available energy and the aerodynamic resistance are those of the reference crop and the surface resistance $r_{s,0}$ has a fixed value of $70 \, s\,m^{-1}$, soil heat flux (G) being generally neglected on a 24 h time step. If the air resistances

within the canopy $r_{a,f}$ and $r_{a,s}$ are supposed to be negligible, Eqs. (23) and (24) transform into much simpler equations:

$$r_{s,f} = \frac{\rho c_p}{\gamma} \frac{D_m}{K_{cb} \lambda ET_0}, \tag{25}$$

$$r_{s,s} = \frac{\rho c_p}{\gamma} \frac{D_m}{K_e \lambda ET_0}. \tag{26}$$

These resistances should be introduced into Eq. (9) and then into the evapotranspiration formula (Eq. 10). It is important to stress that $r_{s,f}$ should be calculated with the standard climatic conditions under which the crop coefficients were obtained, whereas $r_{s,s}$ should be calculated with the actual conditions under which the crop is grown, which is a major difference. When there is no soil evaporation, $K_e = 0$ and $r_{s,s}$ logically tends to infinite.

The fact that surface resistances are necessarily positive imposes a physical constraint on the values of K_{cb} and K_e. These coefficients are necessarily bounded above and should verify the following inequality inferred from Eq. (22), where the saturation deficit D_m is maintained strictly positive with $ET_c = (K_{cb} + K_e) ET_0$:

$$K_{cb} + K_e < \frac{\lambda E_p}{\lambda E_0} \quad \text{with} \quad \lambda E_p = \frac{\Delta f R_{n,0} + \rho c_p D_a / r_a}{\Delta + \gamma}. \tag{27}$$

λE_p represents the "potential" evaporation of the crop, this inequality means that, under given environmental conditions, actual crop evapotranspiration cannot be greater than its potential evaporation, which is logical.

4 Numerical simulations and discussion

4.1 Preliminary considerations

In the numerical simulations carried out below, the daily net radiation of the reference crop ($R_{n,0}$) is estimated following Allen et al. (1998, Eqs. 37, 38 and 39) from the solar radiation taken at sea level and assumed to be at its maximum value, i.e. 75 % of the extraterrestrial solar radiation R_a. Leaf area index (LAI) being a parameter of the two-layer model with an evident link with the basal crop coefficient (K_{cb}), the empirical relationship between them proposed by Allen et al. (1998, Eq. 97), is used in the simulations:

$$K_{cb} = K_{cb,\text{full}} [1 - \exp(-0.7 \, \text{LAI})]. \tag{28}$$

It starts from zero for LAI $= 0$ with an asymptotic trend towards $K_{cb,\text{full}}$ for LAI greater than 3 (for most of cereals $K_{cb,\text{full}} = 1.10$ according to FAO-56). This relationship is close to the one established by Duchemin et al. (2006) on wheat in Morocco. The adjustment of crop coefficient to differing climate conditions is systematically applied in the simulations using the empirical equation given in Allen et al. (1998, Eq. 62).

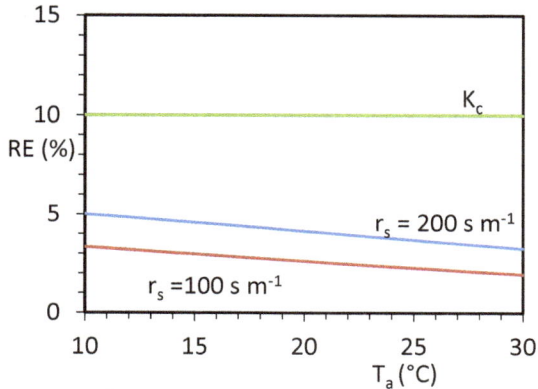

Figure 2. Relative error on crop evapotranspiration ET_c ($RE = 100\delta ET_c/ET_c$) as a function of air temperature (T_a) for a 10 % error on crop coefficient K_c (two-step approach) or on surface resistance r_s (one-step approach) with $z_h = 1$ m and $u_a = 2$ m s^{-1}.

The sensitivity of crop evapotranspiration ET_c to its crop parameter has been previously assessed. In the two-step approach the crop parameter is represented by the crop coefficient K_c and in the one-step approach by the surface resistance r_s. The sensitivity is calculated by differentiating Eqs. (1) and (3), assuming all other variables to be accurately known. This leads respectively to

$$\frac{\delta ET_c}{ET_c} = \frac{1}{K_c}\delta K_c, \tag{29}$$

$$\frac{\delta ET_c}{ET_c} = \frac{-1}{(\Delta/\gamma + 1)r_a + r_s}\delta r_s. \tag{30}$$

ET_c is less sensitive to an uncertainty on r_s than on K_c as shown in Fig. 2. For a 10 % error on K_c, the error on ET_c is 10 %, whereas for the same error on r_s (10 %), the error on ET_c is less than 5 %. This result is an additional argument in favour of the one-step approach.

4.2 Validation of the comprehensive combination equation

Simulations were undertaken to compare the proposed comprehensive Penman–Monteith equation (Eq. 10) with the reference model represented by the full two-layer model detailed in Appendix C. Working on a daily basis, soil heat flux is neglected and the ratio $f = R_n/R_{n,0}$ is taken to be equal to 1 for the sake of convenience. Figure 3 shows the relative error made on crop evapotranspiration as a function of air temperature for different values of leaf area index and a fixed crop height. The relative error is less than 1 % for a large range of air temperature and LAI. So, it is clear that Eq. (10) constitutes an accurate approximation of the two-layer model of evaporation, which justifies a posteriori the theoretical assumption (Eq. 7) made in deriving the formula.

As explained in Sect. 2.2, the modified roughness length $z_{0,h}'$ (Eq. 16) can be used to calculate the aerodynamic re-

Table 1. Typical values at reference height of daily minimum relative humidity ($RH_{n,r}$) and of its daily mean value ($RH_{m,r}$) for three types of climate (from Table 16 in FAO-56).

Climatic classification	$RH_{n,r}$ (%)	$RH_{m,r}$ (%)
Semi-arid (SA)	30	55
Sub-humid (SH)	45	70
Humid (H)	70	85

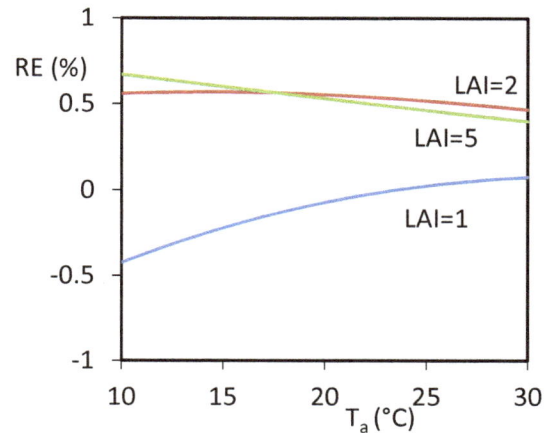

Figure 3. For different LAI, RE on crop evapotranspiration ET_c when it is calculated with the modified Penman–Monteith equation (Eq. 10) compared to the two-layer model used as a reference: $z_h = 1.5$ m, $r_{s,s} = r_{s,l} = 100$ m s^{-1}, under sub-humid conditions with $u_a = 2$ m s^{-1} and $R_a = 40$ MJ m^{-2} d^{-1}.

sistance r_a in Eq. (10), replacing the additional resistance $r_{a,h}$; it is essentially a function of wind speed and crop structural characteristics (LAI and height). Figure 4 shows how the ratio $z_{0,h}'/z_{0,m}$ varies as a function of crop height and wind speed for a fixed LAI (3): it decreases slightly with crop height and more strongly with wind speed, ranging approximately between 0.1 and 0.4. These values are slightly higher than the value of 0.1 commonly used in the FAO-56 calculation of the aerodynamic resistance (Eq. 14). In future, simple statistical parameterizations of this ratio could be developed to facilitate its use in the calculation of the aerodynamic resistance.

4.3 Inferring surface resistance from crop coefficient

Foliage surface resistance $r_{s,f}$ can be inferred from the tabulated value of the basal crop coefficient K_{cb} by means of Eq. (23) or (25). The tabulated value is supposed to be valid under sub-humid conditions and should be corrected under other conditions, as previously mentioned. Inferring soil surface resistance $r_{s,s}$ from soil evaporation coefficient K_e by means of Eq. (24) or (26) is not really relevant since K_e is not a tabulated value. Numerical explorations are carried out under different conditions of air temperature and humidity

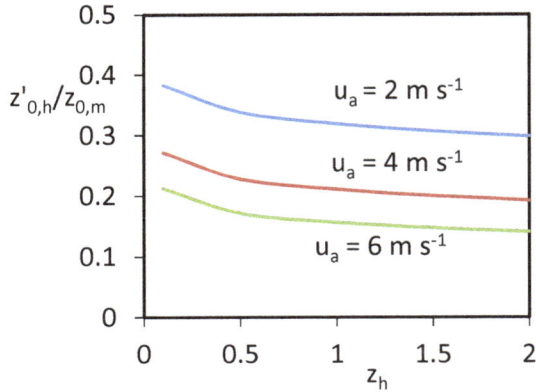

Figure 4. Variation of the ratio between the modified roughness length ($z'_{0,h}$) and the roughness length for momentum ($z_{0,m}$) as a function of crop height (z_h) for different wind speeds at the reference height (u_a) and LAI = 3.

Table 2. For three types of climate (SA, SH, H) and three different temperatures, relative error made on the value of foliage surface resistance ($r_{s,f}$), as inferred from the basal crop coefficient (K_{cb}), when calculated with the simplified formula (Eq. 25) compared to the comprehensive formula (Eq. 23). $K_{cb} = 0.9$, $K_e = 0.1$, $z_h = 1$ m, $u_a = 2$ m s^{-1}, $R_a = 35$ W m^{-2}.

	Air temperature		
	10°C	20°C	30°C
SA	3 %	4 %	6 %
SH	0 %	1 %	2 %
H	−7 %	−5 %	5 %

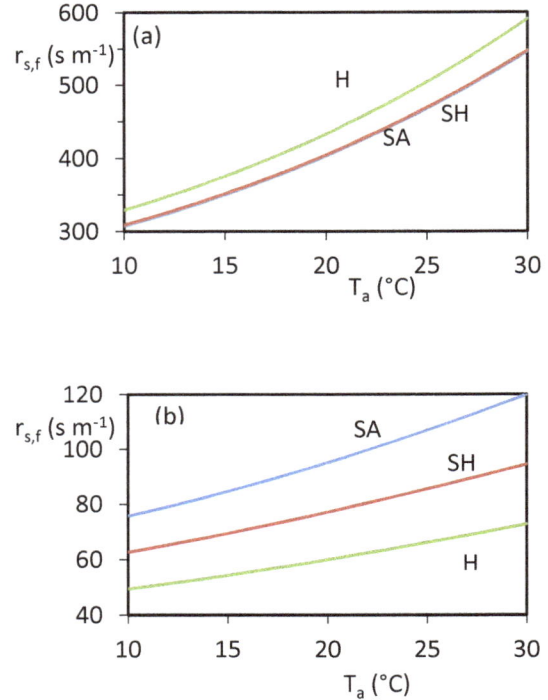

Figure 5. Variation of foliage surface resistance ($r_{s,f}$) inferred from the basal crop coefficient (K_{cb}) as a function of air temperature (T_a) for the three climatic environments (SA: semi-arid; SH: sub-humid; H: humid) described in Table 1 with $u_a = 2$ m s^{-1}, $R_a = 35$ MJ m^{-2} d^{-1} and $K_e = 0.1$: (**a**) initial stage, $z_h = 0.5$ m, $K_{cb} = 0.5$; (**b**) mid-season stage, $z_h = 1.5$ m, $K_{cb} = 1$.

following FAO-56 (Table 16 and Fig. 32), where three types of climate are defined as a function of their relative humidity (Table 1). Figure 5 shows, for these three climatic environments, how the foliage surface resistance ($r_{s,f}$), inferred from the basal crop coefficient (K_{cb}), varies as a function of air temperature. Two contrasting cases are considered with the assumption $f = 1$: one representing the initial stage of an annual crop with $z_h = 0.5$ m and $K_{cb} = 0.5$ (Fig. 5a) and the other case, with $z_h = 1.5$ m and $K_{cb} = 1.0$, representing the mid-season stage (Fig. 5b). These figures clearly show that crop coefficients cannot be easily translated into surface resistances because of the interference of climate characteristics such as air temperature and humidity (as shown here), but also wind speed and solar radiation (not shown) and other factors such as the soil evaporation coefficient (K_e). Table 2 exemplifies for a typical crop and different climatic conditions the relative error made on the value of $r_{s,f}$ when the simplified formulation (Eq. 25) is used instead of the comprehensive one (Eq. 23). The relative error is generally lower than 10 % and much less under sub-humid conditions (around 1 %), which justifies the use of the simplified formula as an accurate approximation.

5 Conclusion and perspectives

We have shown that the FAO-56 dual crop coefficient approach, where the crop coefficient K_c is split into two separate coefficients (one for crop transpiration and another for soil evaporation), can be easily translated into a one-step approach based upon a Penman–Monteith type equation (Eq. 10), its surface resistance being the parallel sum of a soil and foliage resistance. This new form of the Penman–Monteith equation estimates fairly accurately crop evapotranspiration when compared to a full two-layer model. It is also much less sensitive to an error on the crop parameter (represented by the surface resistance) than the FAO-56 methodology based on the crop coefficient. We have also shown that the foliage resistance of the one-step approach can be inferred from the crop coefficients (K_{cb} and K_e) in a way similar to the Matt–Shuttleworth method. The interference of environmental factors, however, makes the calculation somewhat hazardous.

As a consequence of the above development, and following the suggestion already made by Shuttleworth (2014) for computing crop water requirements, we think that the United Nations FAO could find some interest in recommending the use of the one-step approach in replacement of the FAO-56 two-step approach. In the one-step approach, four parameters should be adjusted to a specific crop: its albedo to estimate the net radiation, its aerodynamic resistance and the two components of the surface resistance (soil and vegetation). Albedo varies as a function of green canopy cover (or LAI). The aerodynamic resistance is calculated as a function of crop height (Eq. 14), provided the roughness length is correctly determined (Eq. 16). The soil component of the surface resistance requires a specific parameterization as a function of top soil layer water content. Some empirical parameterizations already exist and should be thoroughly examined and tested. With regard to foliage resistance, although it can be inferred in principle from the basal crop coefficient, it is certainly more recommendable to undertake experimental and bibliographical works in order to determine appropriate values under standard conditions (i.e. non-stressed and well-managed crop). Given that foliage resistance is expressed as the simple ratio of leaf stomatal resistance to leaf area (see Eq. 12) and that LAI is an adjustable and experimentally accessible parameter, one can imagine that the mean leaf stomatal resistance could play the same role in the one-step approach as (and replace) the basal crop coefficient of the two-step approach. Tabulated values for different crops could be supplied and organized by group type in the same way as the crop coefficients in FAO-56. Only one value per crop could be needed, instead of the three values generally provided for crop coefficients, given that LAI values should be able to account for the necessary adjustment to crop cycle characteristics. It is worthwhile stressing, nevertheless, that the leaf stomatal resistance of a given crop under standard conditions (which represents a minimum value) is subject to the influence of other climatic environment parameters than water stress (i.e. temperature, humidity, radiation, CO_2; Jarvis, 1976): its value should be specific to a particular environment and adjustable to other conditions by means of appropriate formulae.

Appendix A: Calculation of the coefficient for soil evaporation (K_e)

According to FAO-56, the daily calculation of K_e is the result of a relatively complex procedure based on Eq. (A1):

$$K_e = \min \left[K_r \left(K_{c,\,max} - K_{cb} \right), f_{ew} K_{c,\,max} \right], \qquad (A1)$$

where K_{cb} is the basal crop coefficient, $K_{c,\,max}$ is the maximum value of $K_c = K_{cb} + K_e$ following rain or irrigation, and K_r is a dimensionless coefficient for the reduction of evaporation due to the depletion of water from the top soil. Its practical calculation relies on a daily water balance computation for the surface soil layer detailed in FAO-56. f_{ew} is the fraction of soil surface from which most evaporation occurs. Its calculation is also detailed in FAO-56. $K_{c,\,max}$ is obtained from the following empirical equation:

$$
\begin{aligned}
K_{c,\,max} \\
= \max \left[\left\{ 1.2 + \left[0.04 \left(u_2 - 2 \right) - 0.004 (RH_{min} - 45) \right] \left(\frac{z_h}{3} \right)^{0.3} \right\}, \right. \\
\left. \{ K_{cb} + 0.05 \} \right],
\end{aligned}
\qquad (A2)
$$

where u_2 is the mean wind speed at 2 m height over grass and RH_{min} is the mean minimum relative humidity.

Table A1. List of symbols.

D_a	Vapour pressure deficit at reference height (Pa)
D_b	Vapour pressure deficit at blending height (Pa)
D_m	Vapour pressure deficit at canopy source height (Pa)
d	Canopy displacement height (m)
ET_0	Reference crop evapotranspiration (mm d^{-1})
ET_c	Crop evapotranspiration under standard conditions (mm d^{-1})
e_a	Vapour pressure at reference height (Pa)
e_m	Vapour pressure at canopy source height (Pa)
$e*(T)$	Saturated vapour pressure at temperature T (Pa)
$f =$	$R_n/R_{n,0}$ (dimensionless)
G	Soil heat flux of a given crop (W m^{-2})
G_0	Soil heat flux of the reference crop (W m^{-2})
K_c	Crop coefficient (dimensionless)
K_{cb}	Basal crop coefficient (dimensionless)
K_e	Coefficient for soil evaporation (dimensionless)
LAI	Leaf area index (m^2 m^{-2})
R_a	Extraterrestrial solar radiation (MJ m^{-2} d^{-1})
R_n	Net radiation of a given crop (W m^{-2})
$R_{n,0}$	Net radiation of the reference crop (W m^{-2})
$R_{n,f}$	Net radiation of the foliage (W m^{-2})
$R_{n,s}$	Net radiation of the soil surface (W m^{-2})
r_a	Aerodynamic resistance between canopy source height and reference height (s m^{-1})
$r_{a,0}$	Aerodynamic resistance of the reference crop (s m^{-1})
$r_{s,0}$	Surface resistance of the reference crop (s m^{-1})
$r_{a,h}$	Bulk air resistance of the canopy defined by Eq. (8) (s m^{-1})
$r_{a,v}$	Defined by Eq. (8) and equal to $r_{a,h}$ if $r_{a,f,v} = r_{a,f,h}$ (s m^{-1})
$r_{s,v}$	Bulk surface resistance of the canopy defined by Eq. (9) (s m^{-1})
$r_{a,f,h}$	Bulk boundary-layer resistance of the foliage for sensible heat (s m^{-1})
$r_{a,f,v}$	Bulk boundary-layer resistance of the foliage for water vapour (s m^{-1})
$r_{a,f} =$	$r_{a,f,h} = r_{a,f,v}$
$r_{a,s}$	Aerodynamic resistance between the soil surface and the source height (s m^{-1})
$r_{s,f}$	Bulk stomatal resistance of the foliage (s m^{-1})
$r_{s,l}$	Mean stomatal resistance of the leaves per unit area of leaf (s m^{-1})
$r_{s,s}$	Soil surface resistance to evaporation (s m^{-1})
T_a	Air temperature at reference height (°C)
T_m	Air temperature at canopy source height (°C)
T_f	Foliage temperature (°C)
T_s	Soil surface temperature (°C)
u_a	Wind speed at reference height (2 m; m s^{-1})
u_b	Wind speed at blending height (50 m; m s^{-1})
z_r	Reference height (m)
z_h	Mean canopy height (m)
z_m	Mean canopy source height (i.e. $d + z_{0,m}$; m)
$z_{0,m}$	Canopy roughness length for momentum (m)
$z_{0,h}$	Canopy roughness length for scalar (m)
c_p	Specific heat of air at constant pressure (J kg^{-1} °C^{-1})
ρ	Air density (kg m^{-3})
γ	Psychrometric constant (Pa °C^{-1})
Δ	Slope of the saturated vapour pressure curve at air temperature (Pa °C^{-1})

Appendix B: Parameterization of air resistances within the canopy

The parameterization commonly used to simulate the component air resistances taken and adapted from Shuttleworth and Wallace (1985), Choudhury and Monteith (1988), Shuttleworth and Gurney (1990), Lhomme et al. (2012). The aerodynamic resistance between the substrate (with a roughness length $z_{0,s}$ of 0.01 m) and the canopy source height $(d + z_{0,m})$ is calculated as the integral of the reciprocal of eddy diffusivity over the height range $[z_{0,s}, d + z_{0,m}]$:

$$
r_{a,s} = \frac{z_h \exp(\alpha_w)}{\alpha_w K(z_h)} \left\{ \exp\left[-\alpha_w z_{0,s}/z_h \right] \right.
$$
$$
\left. - \exp\left[-\alpha_w (d + z_{0,m})/z_h \right] \right\}, \tag{B1}
$$

where z_h is the canopy height, $\alpha_w = 2.5$ (dimensionless) and $K(z_h)$ is the value of eddy diffusivity at canopy height. With the assumption that leaf area is uniformly distributed with height, the leaf boundary-layer resistance (two sides) per unit area of leaf is expressed as a function of wind speed at canopy height $u(z_h)$ as

$$
r_{a,l} = \frac{\alpha_w \left[w/u(z_h) \right]^{1/2}}{4\alpha_0 \left[1 - \exp\left(-\frac{\alpha_w}{2} \right) \right]}, \tag{B2}
$$

w is leaf width (0.03 m) and α_0 is a constant equal to 0.005 (in m s$^{-1/2}$). The eddy diffusivity at canopy height is expressed as $K(z_h) = k^2 u_a(z_h\text{-}d)/\ln[(z_r - d)/z_0]$ and the corresponding wind speed $u(z_h)$ is obtained from an equation similar to Eq. (17).

Appendix C: Formulations of the two-layer model

Following the reformulated expression of the two-layer model proposed by Lhomme et al. (2012), crop evaporation is given by

$$
\lambda E = \left(1 + \frac{\Delta}{\gamma} \right) (P_f + P_s) \lambda E_p
$$
$$
+ \frac{\left(\frac{\Delta}{\gamma} \right) \left(P_f R_{n,f} r_{a,f} + P_s (R_{n,s} - G) r_{a,s} \right)}{r_a}, \tag{C1}
$$

where λE_p represents the potential evaporation expressed as

$$
\lambda E_p = \frac{\Delta(R_n - G) + \frac{\rho c_p D_a}{r_a}}{\Delta + \gamma}. \tag{C2}
$$

The resistive terms are defined as follows:

$$
P_f = \frac{r_a R_s}{R_f R_s + R_a R_f + R_a R_s},
$$
$$
P_s = \frac{r_a R_f}{R_f R_s + R_a R_f + R_a R_s}, \tag{C3}
$$

with

$$
R_a = \left(1 + \frac{\Delta}{\gamma} \right) r_a, \quad R_f = r_{s,f} + \left(1 + \frac{\Delta}{\gamma} \right) r_{a,f},
$$
$$
R_s = r_{s,s} + \left(1 + \frac{\Delta}{\gamma} \right) r_{a,s}. \tag{C4}
$$

Net radiation R_n is partitioned between the foliage and the soil surface as a function of the LAI following Beer's law:

$$
R_{n,s} = R_n \exp(-\alpha \text{LAI}), \tag{C5}
$$
$$
R_{n,f} = R_n \left[1 - \exp(-\alpha \text{LAI}) \right]. \tag{C6}
$$

A typical value of the attenuation coefficient is $\alpha = 0.6$. Soil heat fluxes (G) are generally neglected on a 24 h time step.

Edited by: N. Romano

References

Allen, R. G.: Using the FAO-56 dual crop coefficient method over an irrigated region as part of an evapotranspiration intercomparison study, J. Hydrol., 229, 27–41, 2000.

Allen, R. G., Pereira, L. S., Raes, D., and Smith, M.: Crop evapotranspiration, Irrig. Drainage Paper No 56, United Nations FAO, Rome, 300 pp., 1998.

Choudhury, B. J. and Monteith, J. L.: A four-layer model for the heat budget of homogeneous land surfaces, Q. J. Roy. Meteor. Soc., 114, 373–398, 1988.

Doorenbos, J. and Pruitt, W. O.: Guidelines for predicting crop water requirements, FAO Irrigation and Drainage Paper no. 24. FAO, Rome, 144 pp., 1977.

Duchemin, B., Hadria, R., Erraki, S., Boulet, G., Maisongrande, P., Chehbouni, A., Escadafal, R., Ezzahar, J., Hoedjes, J. C. B., Kharrou, M. H., Khabba, S., Mougenot, B., Olioso, A., Rodriguez, J. C., and Simonneaux, V.: Monitoring wheat phenology and irrigation in Central Morocco: on the use of relationships between evapotranspiration, crop coefficients, leaf area index and remotely-sensed vegetation indices, Agric. Water. Manage., 79, 1–27, 2006.

Jarvis, P. G.: The interpretation of leaf water potential and stomatal conductance found in canopies in the field, Phil. Trans. R. Soc. London, Ser. B, 273, 593–610, 1976.

Katerji, N. and Rana, G.: FAO-56 methodology for determining water requirements of irrigated crops: critical examination of the concepts, alternative proposals and validation in Mediterranean region, Theor. Appl. Climatol., 116, 515–536, 2014.

Lhomme, J. P., Montes, C., Jacob, F., and Prévot, L.: Evaporation from heterogeneous and sparse canopies: on the formulations related to multi-source representations, Bound.-Lay. Meteorol., 144, 243–262, 2012.

Lhomme, J. P., Boudhina, N., and Masmoudi, M. M.: Technical Note: On the Matt-Shuttleworth approach to estimate crop water requirements, Hydrol. Earth Syst. Sci., 18, 4341–4348, doi:10.5194/hess-18-4341-2014, 2014.

Mahfouf, J. F. and Noilhan, J.: Comparative study of various formulations of evaporation from bare soil using in-situ data, J. Appl. Meteorol., 30, 1354–1365, 1991.

Monteith, J. L.: Gas exchange in plant communities, in: Environmental Control of Plant Growth, edited by: Evans, L. T., Academic Press, New York, 95–112, 1963.

Monteith, J. L.: Evaporation and the environment, Symp. Soc. Experimental Biology, 19, 205–234, 1965.

Rosa, R. D., Paredes, P., Rodrigues, G. C., Alves, I., Fernando, R. M., Pereira, L. S., and Allen, R. G.: Implementing the dual crop coefficient approach in interactive software, 1. Background and computational strategy, Agric. Water Manage., 103, 8–24, 2012.

Sellers, P. J., Heiser, M. D., and Hall, F. G.: Relations between surface conductance and spectral vegetation indices at intermediate ($100 \, m^2$ to $15 \, km^2$) length scales, J. Geophys. Res., 97, 19033–19059, 1992.

Shuttleworth, W. J.: Towards one-step estimation of crop water requirements, T. ASABE, 49, 925–935, 2006.

Shuttleworth, W. J. and Wallace, J. S.: Calculating the water requirements of irrigated crops in Australia using the Matt-Shuttleworth approach, T. ASABE, 52, 1895–1906, 2009.

Shuttleworth, W. J.: Putting the "vap" into evaporation, Hydrol. Earth Syst. Sci., 11, 210–244, doi:10.5194/hess-11-210-2007, 2007.

Shuttleworth, W. J.: Comment on "Technical Note: On the Matt-Shuttleworth approach to estimate crop water requirements" by Lhomme et al. (2014), Hydrol. Earth Syst. Sci., 18, 4403–4406, doi:10.5194/hess-18-4403-2014, 2014.

Shuttleworth, W. J. and Gurney, R. J.: The theoretical relationship between foliage temperature and canopy resistance in sparse crops, Q. J. Roy. Meteorol. Soc., 116, 497–519, 1990.

Shuttleworth, W. J. and Wallace, J. S.: Evaporation from sparse crops-an energy combination theory, Q. J. Roy. Meteorol. Soc., 111, 839–855, 1985.

Wallace, J. S.: Calculating evaporation: resistance to factors, Agric. Forest Meteorol., 73, 353–366, 1995.

Permissions

The contributors of this book come from diverse backgrounds, making this book a truly international effort. This book will bring forth new frontiers with its revolutionizing research information and detailed analysis of the nascent developments around the world.

We would like to thank all the contributing authors for lending their expertise to make the book truly unique. They have played a crucial role in the development of this book. Without their invaluable contributions this book wouldn't have been possible. They have made vital efforts to compile up to date information on the varied aspects of this subject to make this book a valuable addition to the collection of many professionals and students.

This book was conceptualized with the vision of imparting up-to-date information and advanced data in this field. To ensure the same, a matchless editorial board was set up. Every individual on the board went through rigorous rounds of assessment to prove their worth. After which they invested a large part of their time researching and compiling the most relevant data for our readers.

The editorial board has been involved in producing this book since its inception. They have spent rigorous hours researching and exploring the diverse topics which have resulted in the successful publishing of this book. They have passed on their knowledge of decades through this book. To expedite this challenging task, the publisher supported the team at every step. A small team of assistant editors was also appointed to further simplify the editing procedure and attain best results for the readers.

Apart from the editorial board, the designing team has also invested a significant amount of their time in understanding the subject and creating the most relevant covers. They scrutinized every image to scout for the most suitable representation of the subject and create an appropriate cover for the book.

The publishing team has been an ardent support to the editorial, designing and production team. Their endless efforts to recruit the best for this project, has resulted in the accomplishment of this book. They are a veteran in the field of academics and their pool of knowledge is as vast as their experience in printing. Their expertise and guidance has proved useful at every step. Their uncompromising quality standards have made this book an exceptional effort. Their encouragement from time to time has been an inspiration for everyone.

The publisher and the editorial board hope that this book will prove to be a valuable piece of knowledge for researchers, students, practitioners and scholars across the globe.

List of Contributors

J. P. Lhomme and C. Montes
IRD (UMR LISAH), 2 Place Viala, 34060 Montpellier, France

Edouard Goudenhoofdt and Laurent Delobbe
The Royal Meteorological Institute of Belgium, Brussels, Belgium

Patrick Willems
Department of Civil Engineering – Hydraulics Division, University of Leuven, Leuven, Belgium

J. H. Sung
Ministry of Land, Infrastructure and Transport, Yeongsan River Flood Control Office, Gwangju, Republic of Korea

E.-S. Chung
Department of Civil Engineering, Seoul National University of Science & Technology, Seoul, 139-743, Republic of Korea

Maryam Barati Moghaddam, Mehdi Mazaheri and Jamal Mohammad Vali Samani
Department of Water Structures, Tarbiat Modares University, Tehran, Iran

Ashley Wright, Jeffrey P. Walker and Valentijn R. N. Pauwels
Department of Civil Engineering, Monash University, Clayton, Victoria, Australia

David E. Robertson
CSIRO, Land and Water, Clayton, Victoria, Australia

H. Medina
Department of Basic Sciences, Agrarian University of Havana, Havana, Cuba

N. Romano and G. B. Chirico
Department of Agricultural Engineering, University of Naples Federico II, Naples, Italy

Bin Xiong, Lihua Xiong, Jie Chen and Lingqi Li
State Key Laboratory of Water Resources and Hydropower Engineering Science, Wuhan University, Wuhan 430072, P.R. China

Chong-Yu Xu
State Key Laboratory of Water Resources and Hydropower Engineering Science, Wuhan University, Wuhan 430072, P.R. China
Department of Geosciences, University of Oslo, 0315 Oslo, Norway

Mie Andreasen, Karsten H. Jensen and Majken C. Looms
Department of Geosciences and Natural Resource Management, University of Copenhagen, Copenhagen, Denmark

Darin Desilets
Hydroinnova LLC, Albuquerque, New Mexico, USA

Marek Zreda
Department of Hydrology and Water Resources, University of Arizona, Arizona, USA

Heye R. Bogena
Agrosphere IBG-3, Forschungszentrum Jülich GmbH, Jülich, Germany

Stephan Costabel
Federal Institute for Geosciences and Natural Resources, Wilhelmstraße 25–30, 13593 Berlin, Germany

Christoph Weidner
Federal Institute for Geosciences and Natural Resources, Stilleweg 2, 30655 Hannover, Germany
North Rhine Westphalian State Agency for Nature, Environment and Consumer Protection, Leibnizstr. 10, 45659 Recklinghausen, Germany

Georg Houben
Federal Institute for Geosciences and Natural Resources, Stilleweg 2, 30655 Hannover, Germany

Mike Müller-Petke
Leibniz Institute for Applied Geophysics, Stilleweg 2, 30655 Hannover, Germany

D. Halwatura and S. Arnold
Centre for Mined Land Rehabilitation, Sustainable Minerals Institute, the University of Queensland, Brisbane, Australia

A. M. Lechner
Centre for Social Responsibility in Mining, Sustainable Minerals Institute, the University of Queensland, Brisbane, Australia
Centre for Environment, University of Tasmania, Hobart, Australia

Sara Sadri, Eric F. Wood and Ming Pan
Department of Civil and Environmental Engineering, Princeton University, 59 Olden St, Princeton, NJ 08540, USA